GILBERT WHITE'S DISCIPLE

To my wife Joanna, my children and grandchildren, with all my love.

*Also
in memory of my father, whose knowledge of natural history
far exceeded my own.*

Portrait of John Leonard Knapp (1767 – 1845), photographed in 1911 but since lost. This is probably the painting of which Knapp wrote 'I have this morning given Mr. Smith a long sitting, and a most frightful Phiz he has produced – to be amended it is to be hoped – he talks of five more sittings…'

GILBERT WHITE'S DISCIPLE

JOHN LEONARD KNAPP
(1767 – 1845)

*Pioneer naturalist, ecologist
and conservationist*

The 'plain countryman' whose JOURNAL OF A NATURALIST *popularised
nature study on both sides of the Atlantic.*

George S.J. White

HOBNOB PRESS
MMXXV

First published in the United Kingdom in 2025

by The Hobnob Press,
8 Lock Warehouse, Severn Road, Gloucester GL1 2GA
www.hobnobpress.co.uk

© George S. J. White 2025

The Author hereby asserts his moral rights to be identified as the Author of the Work.

All rights reserved. No part of this publication may be reproduced, stored in a retrieval system, or transmitted in any form or by any means, electronic, mechanical, photocopying, recording or otherwise, without the prior permission of the publisher and copyright holder.

British Library Cataloguing in Publication Data
A catalogue record for this book is available from the British Library

ISBN 978-1-914407-89-5

Typeset in Adobe Garamond Pro, 11/14 pt
Typesetting and origination by John Chandler

Front cover: Aquatint after a drawing attributed to John Knapp, showing the Shellard's Lane Oak, which stood about half a mile from his house at Alveston in southern Gloucestershire.

Back cover: A young pied flycatcher, adapted from an illustration in John Gould's Birds of Great Britain *(Vol. II, London 1873).*

PREFACE

Susan Augusta Fenimore Cooper (1813–1894), the remarkable American naturalist, writer and philanthropist, republished John Leonard Knapp's last and most successful book *The Journal of a Naturalist* in the United States in 1853.

In her respectful introduction to the work, she wrote

> It will probably be unnecessary to observe that the writer of these remarks must be understood as laying no claim to the honourable position of a teacher, on either of the many branches connected with Natural History; a mere learner herself, she can offer the reader no other guidance than that of companionship, in looking after the birds, or plants, or insects mentioned by Mr. Knapp.

It will be equally clear to the reader of this book, that I too am unskilled in matters of natural history, hoping only to shine a light on the life of a modest, deeply religious man who, during a lifetime of extraordinary commitment to his subject, suffered unbelievable misfortune with his first two publications, but was finally rewarded with a run-away best seller.

No writer had followed the successful format of Rev. Gilbert White's pioneering book *The Natural History of Selborne*, which was first published in 1789. In 1829, John Knapp, a great admirer of White's work, became the first to do so. At a stroke, he reignited popular interest in wildlife and ecology and precipitated the flood of look-alike books on the subject that followed. If television had existed and John Knapp could have overcome his innate reserve, he would have made the perfect wildlife presenter. His exceptional powers of observation and his ability to explain lucidly and simply any aspect of his subject, would have ensured him a job for life. His *Journal of a Naturalist* went into four editions. He hoped for a fifth edition and wrote copious additional notes for it, but for many reasons, it was never published.

Death finally overtook Knapp in 1845 and the notes for his 5th edition were put away, but not destroyed. I have recently transcribed them and encouraged by John Chandler and Louise Ryland-Epton, my publishers, have followed Knapp's written instructions and recreated the final edition that he had longed for. It forms the second part of this book. In addition to being a lifelong naturalist, gardener and talented botanical illustrator, Knapp was also an enthusiastic, thoughtful and sometimes moving poet.

My interest in John Knapp arose initially, when I found a handwritten statutory declaration tucked into the deeds of my house, which is no great distance from Knapp's last home. The declaration is dated 1850 and was drawn up by Knapp's eldest son Arthur, a Bristol solicitor. It explains that in or about the year 1817, a retired naval commander, William Norris Tonge came to live in my house and the Knapp family and the Tonge family became close friends. It seems that sometime after Commander Tonge's death, there was doubt as to the number of his children who were still living, a question that Arthur Knapp was able to answer with certainty.

The room in which I write today was the Tonge family's dining room. It must surely have been witness to the joys and sorrows of both families. Perhaps some of that emotion remains in its walls. Close by are the ruins of the ancient village church, where John Knapp served as churchwarden and where he and his wife, together with their three young daughters, lie buried. It is difficult to live where I do, and to be unaware of his shadow.

When I began my research, I thought that I would be recording the life of an enthusiastic amateur naturalist whose work was read locally, but not much further afield. It never occurred to me that before I had finished, I would discover that his *Journal of a Naturalist*, not only became the national and international best-seller that I have described, but at the time found as diverse a readership as Charles Darwin, William Howley, Archbishop of Canterbury, Davies Gilbert, President of the Royal Society, Professor S.F. Baird, Assistant Secretary of the Smithsonian Institute and David Thoreau, American essayist, poet and philosopher. John Leonard Knapp was nothing if not retiring and so not easy to get to know as an individual, but however unwillingly, he has stepped into the limelight as a man of remarkable talent and determination.

George S.J. White Bt., FSA., 2025.

ACKNOWLEDGEMENTS

THE AUTHOR IS deeply indebted to Dr. Samantha Cavell, Assistant Professor of Military History, Southeastern Louisiana University, Hammond LA, who advised him on John Knapp's brief career in the Royal Navy, to Dr Kirsty McHugh, Curator, John Murray Archive & Publishers' Collections, Archives & Manuscript Collection, National Library of Scotland and to Yvonne Shand, Assistant Curator of the same institution, who went to endless trouble to supply him with copies of Knapp's letters to John Murray and Murray's replies, to Kat Peterson, Archivist, The Society of Antiquaries of London, who directed him to the record of Knapp's election to the Antiquaries' online, The Linnean Society of London, whose online resources have been invaluable, to Lynn Parker, Curator, Illustrations and Artefacts Collections at the Royal Botanic Gardens, Kew, to Pei Chu, Publishing Assistant and Image Licensing Assistant and to Emma Harry, Collection Assistant Illustrations both also of Kew, for photographing the wax bust of John Knapp in their care, to Matt McKenzie of the Paekakiriki Press, for his kindness in sourcing an impossible-to-find illustration, to Rhian Rowson, Natural History Curator, Bristol Museum & Art Gallery, who added greatly to his knowledge of John Knapp's relationship with the Bristol Philosophical Institution and sending him images of Knapp's herbarium, to Elspeth and Philip Grace for generously allowing him to photograph Knapp's birthplace, to Mrs. Ben Jones, for generously allowing him to photograph Knapp's home at Llanfoist, to James Knapp for his generous advice in relation to Knapp family portraits, to Christopher and Virginia Walker, for allowing him to photograph their home, to Dan Geoghegan for his insight into Monmouth society, to Chris Trim for allowing him to photograph the ruins of Old St Helen's Church at Rudgeway and search for the Knapp gravestones, to Jamie Fairchild, for kindly loaning me his rare extra-large 2nd edition of Knapp's Journal, to Jekka McVicar for introducing me to her neighbours, to Val Hicks, churchwarden of St. Mary's Shenley for her help and to Paul Randall for his exceptional generosity in sharing his research, to Col. A.S. Taylor OBE DL Hon Curator, the Herefordshire Light Infantry Museum, Suvla Barracks, Hereford, who advised on Knapp's supposed sojourn in the Herefordshire Militia and finally to Buckinghamshire Archives, who granted him a licence to transcribe Knapp's notes and poetry and reproduce them here. The book of course would never have seen the light of day, first, if C.H. Shaw had not instilled in the author a love of history, when at school in the 1960s, second, if it had not been for the enthusiasm and encouragement of John Chandler and Louise Ryland-Epton of the Hobnob Press, whose idea it was to recreate the previously unpublished fifth edition of John Knapp's *Journal of a Naturalist* from Knapp's original notes, and finally if it was not for the endless patience and love of the author's wife Joanna. She is a true star.

PHOTOGRAPHIC ACKNOWLEDGEMENTS

The author is grateful to the following for permission to reproduce their images:

Bath and North East Somerset Council: 5/1, 6/1; Bedfordshire Archives: 2/3; Buckinghamshire Archives (the author's copyright, but taken by permission of the Archive): A2/1; Bristol Museums: 4/3; Fenimore Art Museum, Cooperstown N.Y: 14/1; The Hunt Institute for Botanical Documentation, Pittsburgh PA: 13/1; The Linnean Society of London: 4/1; Museum of the Société Jersiaise: 2/4; The Morgan Library & Museum. 2006.42:104. Morgan Family Collection: 5/5; North East Wales Archives: 4/2; Paekakariki Press: 5/4; Portsmouth Museums and Records Service: 2/5; Private collections: 2/1, 3/1, 5/2, 5/3, 6/2, 7/2, 7/3, 9/1, 11/1, 13/2, 13/3, 13/4, 14/2, 14/3; Paul Randall, (scanned from O.G. Knapp's *A History of the Chief Families of the Name of Knapp* privately printed 1911): the title page, 1/1, 1/2, 1/3, 1/5, 2/2, 13/4; Wikimedia Commons: Yale Center for British Art, Paul Mellon Collection: 2/6, 8/2, 12/1

The following images are the copyright of the author: 1/4, 1/6, 6/3, 6/4, 6/5, 7/1, 7/4, 8/1, 10/1, 10/2, 11/2, 11/3, 12/2, A2/1.

I

John Leonard Knapp (1767-1845) was descended from a wealthy Buckinghamshire landowning family. His great grandfather John Knapp (1646-1710), though born and brought up at Chilton in Berkshire, made his way to London and there became a drysalter by profession, though a freeman of the Worshipful Company of Wax Chandlers by guild.[1]

Drysalters were dealers in rudimentary chemicals, and while John Knapp lived and worked in Basinghall in the very heart of the City, he owned all or part of a copperas works in Deptford. Here iron (II) sulphide was converted to iron (II) heptahydrate or 'green vitriol' as it was then known, a chemical largely used for fixing dyes in cloth. The fact that Dighton Avery, apprenticed to John Knapp in 1668, was the son of a freeman of the Haberdashers' Company, and that James Dobson, apprenticed to him in 1679 was the son of a free Draper, suggests that the chemical treatment of cloth was the basis of the very considerable fortune he built for himself.[2] Many of John Knapp's numerous Berkshire landowning relatives also had close connections with those two London companies.[3]

1/1 John Knapp (1646-1710), drysalter of London and founder of the Knapp family fortune

With his fortune, John Knapp invested heavily in farms and land in the Buckinghamshire parishes of Great Linford, Little Linford, Loughton, Whaddon and Shenley and bought pearls and precious stones for his wife Katherine. He purchased the living of Shenley, or the right to appoint the rector there, and in his will decreed that his second son Matthew (1684-1752), the naturalist's grandfather, should be 'bred a Divine according to the discipline and doctrine of the Church of England' and should, when qualified, replace the Rector of Shenley then in office.[4] So it was that Rev. Matthew Knapp was

1/2 Little Linford Hall, Buckinghamshire, c. 1911. The senior branch of the Knapp family's principal residence from 1684 until its demolition in 1959.

installed in the Rectory at Shenley in 1709, after attending Trinity College Oxford. The naturalist's father, the Rev Primatt Knapp (1730-1793), third son of Rev. Matthew, succeeded him at Shenley Rectory in 1755, after attending Christ Church, Oxford. It was at Shenley therefore that John Leonard Knapp, third son of Rev. Primatt and the subject of this biography, was privately baptised on 9th May 1767.[5]

Very little is known of Knapp's early childhood or of his parents, although his youngest sibling Leonora (1774 – 1851) may have given some insight when she wrote in 1820, that 'my family were of considerable consequence in the county, and we were all educated with much aristocratic pride. My father was a very pious and very learned man. He had an extensive knowledge of Hebrew and of the oriental languages; and he devoted his leisure time to religious studies, and particularly occupied himself in writing illustrations of the Scriptures from oriental customs.' While her account continued that 'his devotedness to this work rendered him unfit for the office of teacher in his family' and finished with a description of the neglect that

1/3 Rev. Matthew Knapp (1684 – 1752), second son of John Knapp of London and Little Linford and grandfather of the naturalist, who served as Rector of Shenley Church End from 1709 until his death.

1/4 St. Mary's Shenley Church End, Buckinghamshire.

she believed she encountered as the youngest child, her words contrast dramatically with those of her brother, who delighted in his early years, describing them in two lengthy poems that he later published.[6] The first, in blank verse, appeared in print in 1818.

1/5 Rev. Primatt Knapp (1730-1793), father of the naturalist, who served as Rector of Shenley Church End from 1755 until 1793.

How wonderful are man's attachments
To a spot! Without perhaps one tie, but
That illusive thread which holds him down, and
Binds him to the place where first impressions
Rose: and, be our early years to joy or
Sorrow giv'n, yet still thro' life, thro' ev'ry
Vary'd scene some momentary pleasure in
Remembrance lives. There is a sympathy
In man that links him to a fancy'd bliss,
A whispering something that enchants; a
Still small voice that woos and wins to sweetest
Harmony, a willing something in the mind.

The mossy-bank where oft a boy I've sat
And shell'd my brown ripe hazel store, and watch'd
The twitt'ring linnet, dabbling in the rill
Beside, has charms no eye can see but mine,
Can feel, but me; its violets shed a
Sweeter scent, and sweeter woodbines blossom there.[7]

While his 1818 work was a narrative poem and could conceivably be fictional, the second poem, written in rhyming couplets, entitled 'The Progress of a Naturalist' and dedicated to his muse, the goddess Flora, was clearly autobiographical. It was first published in 1830 and begins:

1/6 *The Rectory at Shenley Church End. It was here, next door to St. Mary's church, that John Leonard Knapp was born and brought up*

Rambling alone the woodland lea,
A musing, slender, happy child,
Snatching in haste his flowers wild:
Twined in a wreath, with rushes green,
Cowslips and violets are seen;
Now both his hands he stoops to fill,
Then shouts, and smiles, and gathers still:
He seems the very king of joy –
This young, this gleeful, slender boy.

The wild bee sees, and murmurs round,
To scare him from her fav'rite ground,
In fear, lest he should gather more,
And rob her of her honied store.

In little garden see him toil,
Planting these natives of the soil:
Though tended by his daily care,
The sickly nurslings languish there.

A dusky cloud is come between –
This musing child's no longer seen.[8]

It is obvious from both these works that in his middle years, Knapp remembered his boyhood as an idyl.[9] He had taken a delight in wildflowers at a very early age, and in his 'little garden' had done his childish best to propagate them. He was a born naturalist and a born gardener. He pursued both those occupations relentlessly until his last day.[10]

The early idyl must have drawn to a close ('a dusky cloud is come between…') when he was dispatched to boarding school. While his elder brother Primatt (1764-1838), who was his parents' second son, was sent to study for ordination at Oxford, it was decided that John, the third son, should join the Navy. His early schooling was intended to prepare him for life at sea.[11]

2

ALL ACCOUNTS OF Knapp's early years published to date, have relied on two closely related sources for information. The first was an obituary of Knapp in the *Proceedings of the Linnean Society* of 1845 and the second was a longer version of the same, in *The Gentleman's Magazine*. Current research and especially the discovery of two letters written by Knapp's younger sister Mary Pierce (1770–1860) to his eldest son Arthur (1808-1883) in July and November 1845, reveal that the published obituaries contained errors.[12]

Both obituaries stated that Knapp 'was educated at the grammar-school of Thame in Oxfordshire, but being destined for the navy, left school at an early age. The sea however, disagreeing with his health, he left the navy and afterwards served both in the Hereford and Northampton Militia, in the latter of which, he commanded a troop.'[13] *The Gentleman's Magazine* of 1845 enlarged on this, suggesting that while in the navy, Knapp sailed under Captain

2/1 *Thame Grammar School, where Knapp is reported to have received his early education.*

2/2 *Mary Pierce (1770 – 1860), née Knapp, John Leonard Knapp's younger sister and a woman worthy of a biography of her own.*

Philip Carteret (d. 1796), famous for his two voyages of circumnavigation of the globe, also that he was present at a naval engagement with 'the pirate' John Paul Jones (1747-1792).[14] *The Gentleman's Magazine* added that later, while a Lieutenant in the Herefordshire Militia, Knapp was present with the regiment during the notorious suppression of the riots at Bristol Bridge in September 1793.[15]

It may well be that Arthur Knapp had been asked to supply information about his father's early life immediately after his father's death, and simply did the best he could in the time available. His aunt Mary, the eldest of John Knapp's surviving siblings, was then a widow living abroad on the German / Polish border and so was not easily contacted for advice.[16] When he wrote informing her of her brother's death, he asked for any recollections she may have had, but she replied 'I must be more tranquil in my head before I can give any account of my dear John's youth. I will write some day on that account. I have all my thoughts in confusion and as blood [is] overswimming in my eye – it is painful to write. Adieu my dear Nephew the Love I bore your Father concentres on you, Mary Pierce, Wolstein, 23rd July 1845.'

It was not until four months later that Mary Pierce felt able to provide what information she could, but even then, believed the task to be beyond her. 'As to anecdotes of my brother &c. I am a poor help for I know none. I can only say that he was beloved by everyone … a generous amiable child…' Despite her doubts though, she then went on to provide key facts. The first was that when older, he had 'made great progress at Aspley.' This suggests that rather than attending Thame Grammar School, or perhaps after initial schooling at Thame, he had been sent to Aspley Classical Academy, then a distinguished public school based at Guise House, Aspley Guise. The Academy was only twelve and a half miles northwest of Shenley Rectory. It was known for preparing the sons of the gentry for careers in the services, church and commerce and so it is entirely logical that a boy destined for the navy should have been sent there. Attendance at Aspley may also account for the wide knowledge of the classics and foreign languages that Knapp displayed in later life.

Mary confirmed that her brother did then sail with 'Captain Carteret who has sailed with Captain Cook,' noting that he 'was particularly distinguish'd

2/3 Aspley Guise Classical Academy, which Knapp almost certainly attended in preparation for his entry into the Navy.

by him & received from [him] presents of Voyages.' She added 'he never passed the rank of Midshipman because the seasickness attacked him so violently that vomiting blood, it was necessary to land him, I think on the Irish coast, it was his unlucky Luck, for in a short time after, Carteret's ship was Taken by Paul Jones the great Pirate.' He was therefore not present at the naval engagement that the obituaries described.

2/4 The only known image of Captain Philip Carteret, R.N. (1733-1796) circumnavigator of the Globe. Knapp served under Carteret aged perhaps twelve or fourteen, until forced to give up the sea through extreme sickness.

'At home' she wrote 'he studied Chemistry as an amusement and began to draw animals and Plants with great accuracy – He never had a publick Education i.e. at Eton or Winches: or Ox: but he was sent to Kensington in private instruction & so on until he went into the Worcester Militia – but be he where he would, he was beloved & sought for. Lord Northampton particularly loved him & wished him to quit the Worcester & offer'd him rank in his own Regiment which however he did not accept – you see I am very poor in Biography – & indeed his simple studious life furnished little to be recited – but much to be felt, and infinitely more to be regretted…'

The *London Gazette* confirms that Mary Pierce was correct in her recollection that Knapp had been appointed to the

Worcester Militia as a lieutenant, giving the date as December 1792. No doubt he was recruited to the Worcester because he was living at the time with his father, who had retired from Shenley in 1786, and was now resident at Powick in that county. No evidence can be found that Knapp ever transferred to the Herefordshire Militia, as suggested by the *Linnean* and *Gentleman's Magazine* obituaries. As the Worcester Militia was not present at the suppression of the Bristol Bridge riots in 1793, it is clear that Knapp played no part in that unpleasant event either.

2/5 A Review of the Worcestershire Militia on Southsea Common, painted by Richard Livesay in 1823.

Mary Pearce was however mistaken in her belief that her brother had refused the offer of preferment by Lord Northampton. *The London Gazette* confirms that on April 30th, 1798, he transferred to the Northampton Militia as a lieutenant and was promoted to Captain in the Northampton Supplementary Militia only fourteen days later.[17] There may have been more to his move than simply Lord Northampton's influence. His father had died at Powick in 1793 and shortly after, his eldest brother Nathaniel Matthew died also. In Nathaniel's will Knapp was bequeathed two cottages at Brook End, Shenley, as well as a rent of eight pounds a year from a farm at Caldecot (Buckinghamshire) and small piece of land in the open field at Whaddon.[18] It was perhaps because he now had property interests situated only two or three miles from the Buckinghamshire and Northamptonshire border, that he felt it desirable to take up Lord Northampton's offer.

2/6 Thomas Rowlandson's preliminary sketch for A Review of the Northampton Militia at Brackley.

The numerous county militias, raised to address the ever-present threat of French invasion, were reserve forces, in which men from each parish were obliged to serve for several weeks a year, over a period of between three and five years. Those chosen could opt out if they wished, by providing a substitute, or paying a fine, but military service must have given Knapp a modest income. More importantly, it provided a much-needed purpose after his departure from the Navy, following what appears to have been a serious lower oesophageal tear, brought on by extreme seasickness.

Whatever the reason, it was during this period of part-time military employment, that Knapp began to pursue his private interests in earnest. He described this himself in later years. 'As a junior in my family, no destination in life was found, this would not do – for that I was not fitted – no favours brightened into hopes – no interest led my hand – finding as boyish days advanced to conscious youth, that my race was to be only that of a poor younger brother, one path alone remained – my dog, and my gun, did very well for bodily recreation, but there was a mind that would strive for equality with its fellows, a wish to preserve a reputation guiltless of folly or of vice – so scrambling thro' all the sciences and arts that I could accomplish, stuffed my head with all the grain & husks that fell in my way – Botany – mineralogy – antiquity were my favourite diets. Contriving with my pencil in the evening to scratch out some semblance to the remains of distant ages, which by aid of strong lights, and deep shades, sometimes was affected. I got a Pony, and rambled over much of the northern & midland parts of the Kingdom, searching for plants, minerals or objects that engaged my fancy, whiling away many a summer hour, and as my means became low returned to my home.' He remembered his adventures there with obvious delight 'I think it was in my last excursion northwards, that I passed over that curious line of country extending from Wakefield to Kirby Lonsdale, for the especial object of botanizing in that district, & the forest and mountain of Ingleboro' – and having chosen my haven at what was then a most delightful quiet, unobtrusive, pleasant, cleanly little Inn, at Ingleton, passed many an hour amidst the romantic scenery in the vicinity of that station, rambling after my hurried breakfast up the steepy crags, along the levels of the mountain, startling the guggling moor-fowl from his haunt at the little moory pools, and rills of this famous hill, and scuttling the timid Rabbit from his lair – collecting my plants, delighting in their variety and number – returning when appetite called, to feast on the nice trouts of its rocky streams and Rapids – the baked wild berries from the bogs, & fells – then not again roaming till nightfall in the forest – by aid of my pony visiting all the strange scenery of the Craven, its springs, its coves, its tarns, its ivied fanes, its fortress, & its crag – then sleeping as healthy youth should rest, refreshed as a giant to begin again.'[19]

3

Even today, many are brought to a passion for natural history, conservation and ecology, through early enjoyment of field sports and it seems that Knapp too was able to focus his broader interests in botany, minerology and history, through time spent with friends, his dog and his gun. Shooting gave him the opportunity to increase his botanical knowledge. While out in the countryside he set about collecting and pressing specimens of ferns and over many years used his exceptional talent as a botanical illustrator, to fill five volumes with exquisitely detailed drawings of fungi.[20]

Then, perhaps with the intention of making his name in the subject he so much enjoyed, he decided to use his summer adventures specifically to seek out and collect specimens of every native British grass, to draw each example while it was still fresh, then later to study his preserved collection in microscopic detail and so categorise it with a view to eventual publication. It was not that the subject had been overlooked by others. Many of the great naturalists of the day had included detailed observation of grasses in their general works, but there remained much uncertainty as to the number of native grasses growing in Great Britain, their value as animal fodder, their inter-relationship and indeed their correct botanical classification. Most recently Rev. George Swayne (c.1746-1827), vicar of Pucklechurch and Dyrham in Gloucestershire, and of East Harptree in Somerset, who delighted in the practical cultivation of grasses, had attempted to draw together recorded knowledge of all pasture grasses in his elegant *Gramina Pascula* published in 1790. He illustrated his work however with actual pressed specimens of nineteen grasses, picked and dried by his daughters, with the result that his book was so expensive that it is thought that only twenty copies sold.[21] Knapp now set out to do something very similar, but to encompass all British grasses and crucially, to illustrate his work with printed versions of his own detailed drawings.

Since the paper on which he made his original illustrations is known to have been watermarked 1797, it must be assumed that he began this massive undertaking then.[22] 'It has long been a desideratum that that part of the Triandrious class denominated Grasses, might be assembled,' he wrote.[23]

3/1 Snipe shooting, a sport much enjoyed by Knapp in his youth, which paradoxically greatly increased his passion for nature. Engraving by R. Reeve, from a painting by Dean Wolstenholm, 1806.

In his preface to *Gramina Britannica or Representation of British Grasses*, the magnificent work that he published in royal quarto format in 1804, he mused on the value of this species to mankind.[24] He observed that while many British herbs, once valued for their medicinal properties, were now superseded in importance by mineral preparations and more powerful imported drugs, native grasses remained of the utmost importance 'for the maintenance of quadrupeds, and ultimately for that of man, which were noted for that purpose from the earliest creation, and will continue thus applicable till the final consummation of all things.'

While no detailed records of his excursions have survived, his 1830 poem again provides clues:

I trace him still on Scotia's hills,
By craggy steeps, by mossy rills,
In heathy vale, o'er ferny lands,
'Long dusky shelves, on granite bands
Each nook, each cranny, close explore,
And half the island's sandy shore;

The mead, the woodland, and the plain:
And now the wand'rer's home again.

Clues to many of the places he visited can be found in *Gramina Britannica*. These show that he combed twenty-eight counties in England, Wales and Scotland for his specimens. Some sixty-three descriptions specify the places where his grasses were found. These vary from the crags and mountain tops of Ben Lawers, Ben Ledi, Ben Nevis and other Scottish mountains, which he climbed in June 1802 in the company of the celebrated Scottish naturalist George Don (1764-1814),[25] to the sands of Marazion in Cornwall, the Dunes of Yarmouth, Seaton Snook (at the mouth of the Tees) and the 'ballast hills' at Sunderland, where incoming ships dumped their ballast in readiness to take on outgoing cargo. His most evocative descriptions are those that he provided with the greatest detail. *Panicum Crusgalli* for example, could be found in Battersea Fields, 'opposite the Linseed-mill, in the open field, near a footpath, under the hedge nearest the mill.' *Alopecurus Bulbosa* might be found 'at Yarmouth in Norfolk, in a little meadow behind what is called the 'Cinder Ovens.' *Phleum Bochmeri* flourished 'by the side of the road from Narborough Hall to Marram, near Swaffham in Norfolk.' As for *Polypogon Littoralis*, which he found at Erith, in Kent, he wrote 'we spent three days in search of it. In a grassy field in which the gunpowder magazine stands, will be found a morassy place, filled with *Aster tripollum;* at the edge of this morass, on the side next the high bank of the ditch, which divides it from another fenny meadow, we found, rather plentifully, this Marine Bent-grass.'[26] A pressed specimen of a lichen, *Physcia indet*, which Knapp sent to the Linnean Society, and which still survives in its collection, is labelled 'Pendant from the granite rocks of Lamorna Cove, near the Land's end.'[27] Only on one occasion it seems, did Knapp deliberately withhold the exact details of where a grass might be found, for fear that collectors might be drawn to it. 'We have a new and plentiful habitat for *Cyprinedium*' he wrote, 'but must not divulge it, lest avarice destroy the plant.' A list of all known places visited by Knapp in his search for grasses, is given in Appendix I.

His adventures were not without risk. Although Knapp used the pronoun 'we' when referring to his journeys, he made no mention of any physical companions, except George Don who acted as a guide in Scotland. At that date, the only modes of transport were by foot, horse, coach or sea, all of which were slow and none of which were free of danger, especially for unaccompanied travellers in the remote areas that Knapp chose to visit. As for spiritual companionship, Knapp relied upon his strong Christian faith.

On the final page of *Gramina,* is a paragraph expressing his thanks to those who had helped him. It ends 'and finally, to that Supreme Being, whose mercy hath led his unseeing creature through all his various wanderings, be praise and adoration! Whether I have trembled on the faithless bog – hung over the slippery margin of the unfathomed lake – or climbed the dangerous precipice of alpine steeps, in all hath his providence upheld me; under the shadow of his wings I have found protection, and hallowed be his name!'

4

I see, upon a wide-spread board,
This rambler's rare and cherish'd hoard:
Mysterious grasses scatter'd o'er,
A glass, a press, and books of lore;
With grave, consid'rate care and thought,
Compare the species he has brought;
Then, ranging out each sep'rate race,
He bids their form the pencil trace.
Ah, Vulcan! That thy hateful rage
Should moulder half his studious page.[28]

TAXONOMY, OR THE classification of plants, was a relatively new subject in Knapp's day. Its seeds had been sown in the 17th century, but as the 18th century wore on, and especially after Carl Linnaeus (1707-1778) the Swedish biologist and physician had published his new system of binomial nomenclature, taxonomy became a focus of study not only amongst academics, but the clergy and gentlemen amateurs too.[29] As he continued his travels, Knapp was only too happy to present himself as an amateur and sought advice from experts wherever he could. Amongst those with whom he corresponded were Dr. James Edward Smith (1759-1828), President and founder of the Linnean Society of London, Rev. Hugh Davies, (1739-1821), Rector of Llandegfan with Baumaris in North Wales, a distinguished botanist, and George Don, then Principal Gardener at the Royal Botanic Garden of Edinburgh. It was through these connections that Knapp was elected a Fellow of the Linnean Society in 1796. This was something of a feather in his cap, as Sir Joseph Banks (1743-1820), President of the Royal Society and Honorary Member of the Linnean Society had written that the Society 'will flourish as great care is taken … to keep out improper people.'[30] Fellowship gave Knapp the opportunity to meet and correspond with almost any of the most celebrated botanists of the day. It was through his association with the Society and especially his friendship with Dr Smith, that in due course a grass, *Knappia agrostidea*, was named in his honour.[31]

4/1 Portrait of Dr James Edward Smith (1759-1828), founder of the Linnean Society. Smith (later Sir James) not only encouraged Knapp in his botanical work but named a species of grass Knappia Agrostidea *in his honour.*

Knapp read the works of the best-known botanists and taxonomists voraciously, mentioning many of these in his own published work and in his correspondence. One of his favourite authors was John Ray (1627-1705), whom he described as 'our own immortal Ray.'[32] Ray's first work *Catalogus Plantarum circa Cantabrigiam Nascentium* was published in 1660. *Historia Plantarum* (1685-1703) followed, in which Ray laid out his own pioneering system of classification. Ray's *The Wisdom of God Manifested in the Works of the Creation* of 1691, may have had a particular influence on Knapp, whose underlying theme in all that he wrote himself, was that that every animal, plant and flower was the creation of the Almighty, and that even those that appeared pointless in terms of God's gift to mankind, must have had a purpose too subtle for human understanding.

Knapp also read the works of the man he called 'the amiable' Benjamin Stillingfleet FRS (1702-1771),[33] who had published *Calendar of Flora* in 1755 and *Miscellaneous Tracts* in 1759. He read Antoine Laurant de Jussieu (1748-1836), Professor of Botany at the Muséum National d'Histoire Naturelle in Paris, who published *Genera Plantarum* in 1789. He read William Withering FRS (1741-1799), who published *An Account of the Foxglove and some of its Medical Uses* in 1785 and he read the work of Albrecht von Haller FRS (1708-1777) the Swiss physiologist, naturalist, encyclopaedist and poet. He based much of his early work on his friend Dr J. E. Smith's *Flora Britannica*, published in three volumes in 1800.

Perhaps the reference books that he valued most were those that he mentioned specifically in his will, which was proved on 12th June 1845.[34] They were *The Botanical Magazine*, published from 1787 onwards by William Curtis (1747-1799), *The Natural History of British Insects*, published in ten

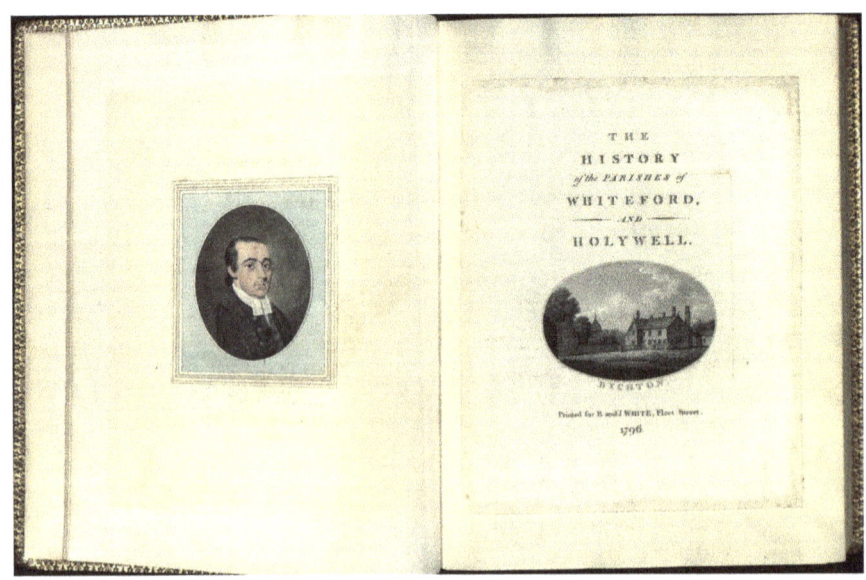

4/2 Rev. Hugh Davies (1739-1821), Rector of Llandegfan with Beaumaris and later of Aber, a much-respected Welsh botanist and author, who corresponded with Knapp, admired his ability as an artist and soon came to regard him as a friend.

volumes from 1792 by Edward Donovan F.L.S. (1768-1837), *English Botany* published in thirty-six volumes from 1790 by James Sowerby (1757-1822), *British Zoology* published in 1776 by Thomas Pennant F.R.S (1726-1798), *The History of British Birds,* published from 1797 by Thomas Bewick (1753-1828), *The Ornithological Dictionary; or Alphabetical Synopsis of British Birds,* published in 1802 by George Montagu F.L.S (1753-1815) and finally the *Encyclopaedia Britannica*. All these must have constituted the magical-sounding 'books of lore' that Knapp mentioned in his poem. A final work, which clearly changed Knapp's entire approach to life was the Rev. Gilbert White's *The Natural History of Selborne,* first published in October 1789, but more of that will be found in later chapters.

In addition to reading the works of others, Knapp consulted their herbaria or collections of preserved plant specimens, where he could. He visited that of William Sherard FRS (1659-1728) who left the unfinished manuscript for his *Pinax* (a list of all known plant species) to Oxford University, together with his herbarium. Knapp was almost overwhelmed by the botanical collection of Thomas Velley of Bath (1748-1806) a retired soldier and gentleman botanist. He wrote after Velley's death 'it may I think be pronounced as highly finished, and the most beautiful of any existing! – The British Phanagamous plants were all preserved by that tedious but most effectual process of Ironing; and are very neatly arranged in folio fasciculi – containing in the whole about 740 specimens – but the ne plus ultra of perfection are his Marine Plants – There are about 590 specimens (mostly folio) placed down in the most exquisite manner, with much manuscript observation – the minuter parts of fructification in several drawn out, and coloured, by the side of the specimens – done by a Bath artist under the inspection of Mr. Velley – the whole bound up in vols. Russia – there are besides about 300 specimens of these marine plants, Fuci, conferva, &c – equally excellent, but not arranged, or bound up – The trouble that these have occasioned him may be estimated by knowing that many of the individuals, were the labour of six, or ten days, placing & preserving on paper. In the cabinet are besides much miscellaneous matter such as botanical collections must of necessity consist.' As if this was not enough, Velley's collection further included 'a prodigious herbarium (perhaps 2 ½ feet thick) containing specimens of foreign plants tolerably preserved, that were formerly the property of Sir Noah Thomas[35] but wether collected by, or for him, I do not know – unarranged marine plants from Botany Bay – from French Guiana – Boxes of Lichens &c &c.'[36] In December 1801 he wrote of his hope that James Edward Smith might allow him sight of his herbarium if he travelled to Norwich to see it. It was perhaps the ultimate collection in

England, as it included the herbarium of Carl Linnaeus himself, that Smith had bought and imported from Sweden in 1784.

It was inevitable perhaps, that Knapp's all-consuming enthusiasm for his subject would occasionally lead him into trouble. Botany at this period was being studied with similar intensity by others, many of whom held very strong views. Some were extremely generous with their knowledge. Others were less inclined to share their scholarship, especially if it led to any questioning of their deeply-held opinions. Jealousy, as in any other area of impassioned research, undoubtedly played its part.

Letters from Knapp now in the archive of the Linnean Society, show that he blundered into at least two such delicate situations. Knapp confessed in 1801 to Dr J. E. Smith, President of the Society, that the Rev. Hugh Davies was 'a little irritated at what originated from a mistake; but I shall write to him in a few days, and trust to remove any unfavourable impressions he may have conceived.'[37] The misunderstanding, whatever it was, seems to have run on for a month, but was finally cleared up by reassurances given to Davies by Smith. Davies was grateful for this, writing to Smith 'Many thanks for your ready answer to my inquiry concerning (now I shall call him friend) Mr. Knapp; I have exchanged a letter or two with him; he seems very conversant with grasses, and (by a specimen he has favoured me with) draws admirably.'[38]

The second situation seems to have been a more highly charged and long-drawn-out drama, which strictly was not Knapp's business. Indeed, he could have kept well clear of it, had he taken care. But it seems, that he did not. The matter involved controversial comments made by William Sole (1741-1802), an apothecary of Bath, about the work of Dr. J.E. Smith. Sole had published *Menthae Brittanicae*, a botanical arrangement of all British mints in 1798. Two years later, Smith published the first volume of his *Flora Britannica*, which did not wholly agree with Sole's earlier work. 'In consequence,' wrote Knapp some years later, 'this irritable man in the heat of his anger, wrote upon the margins of a copy of his *Mentha*, some observations and censures upon the *Flo: Brit:*'[39] These he intended to present to a meeting of the Linnean Society, but sent them first to Thomas Velley, to ask his opinion. Velley did not disagree with the substance of what Sole had written, but heartily disliked its tone. He therefore delayed returning the copy, in the hope that Sole might calm down. Sole however demanded its return but perhaps fortuitously died before he was able to present his apparently explosive criticism to the Society.

At an executors' sale of Sole's library, the book, complete with its incendiary annotations, was sold, being bought by Mr. West, Sole's first employer. Velley at once extracted a promise from the elderly Mr. West that no

4/3 A specimen of Knappia Agrostidea, from Knapp's own Herbarium, now held by Bristol Museum.

one would be permitted to see or copy its annotations and that ultimately West would bequeath the book to Smith. Unexpectedly though, a rough draft of the annotations appeared in another auction. Velley bought the draft, in order to suppress that too, but after the sale was approached by a sinister-sounding clergyman, who, wrote Velley, 'desired me to give him the book, in order that he might transcribe the notes. I told him that I had secured the book expressly for the purpose of preventing their being seen. He replied with some warmth – 'I know where to get it and I will have it."[40]

It seems that Knapp had unwisely relayed tales of these unpleasant events to Dr. Smith, a man who was known to be easily offended.[41] He soon found himself in deep water, obliged to explain and to apologise. 'The observations which the late Mr. Sole left behind him have been much misrepresented,' he wrote to Smith, 'and probably I have been, a little instrumental in giving rise to the opinion that you have formed of them, but I can assure you, there is nothing in his M.S.S. that can be construed as abusive, but were solely the dictates of a mind embittered by disappointment… I can assure you that his notes are in the possession of a most incomparable botanist, and a friend of yours, who would be as delicate of any injurious aspersions upon your judgement, as upon his own; neither would he wrap up any observations in obscurity that might tend to the illustration of his favourite study – an unknown clergyman, did apply to the possessor, for permission to compile the contents of one volume, but failing there, he made application to the holder of the other copy, and was equally unsuccessful; as the object of the purchasers were to bury in oblivion the weakness of declining genius – I do not know if I have been successful in clearing up the subject to your satisfaction, but I trust you will exclude the idea that the subjects of Mr. S latter writings were abuse, consider them only as the harmless remarks of a mind soured by disappointment, and directed by a judgement blinded by bigotry…'[42]

5

Knapp's Linnean Society correspondence shows that in December 1801 and November 1802 he was living or lodging in King Street, Bath, first at No. 13, later at No. 2. The minutes of the Society of Antiquaries of London show that he was still living in or visiting Bath in June 1803, when his name was put forward for election as a Fellow.[43] He was still there in December 1803, when he was actually elected.[44] There are many reasons why he might have been in that fashionable city, but it is likely that his principal purpose was to visit John Hibbert, who lived at No. 8 Chapel Row, Queen Square, no more than two minutes' walk away from both those lodgings. Hibbert was a teacher of drawing, but most significantly was an extremely skilled copper-plate engraver and printer. It was to Hibbert that Knapp took some fifty of his finished drawings of grasses to be hand-engraved on copper, prior to printing, colouring and binding in London, as *Gramina Britannica*, the culmination of his years of research. It must have been an expensive commission, given the size of each plate (9¼ in by 7¼ in), the time, the skill and the care involved in cutting it and the overall number of plates required. He took the remaining drawings, some sixty-nine in number, to Richard Polydore Nodder (1774-1829), a distinguished London engraver and botanical artist to King George III.

5/1 An advertisement placed in Gye's Bath Directory of 1819 by John Hibbert, artist, engraver and copper-plate maker.

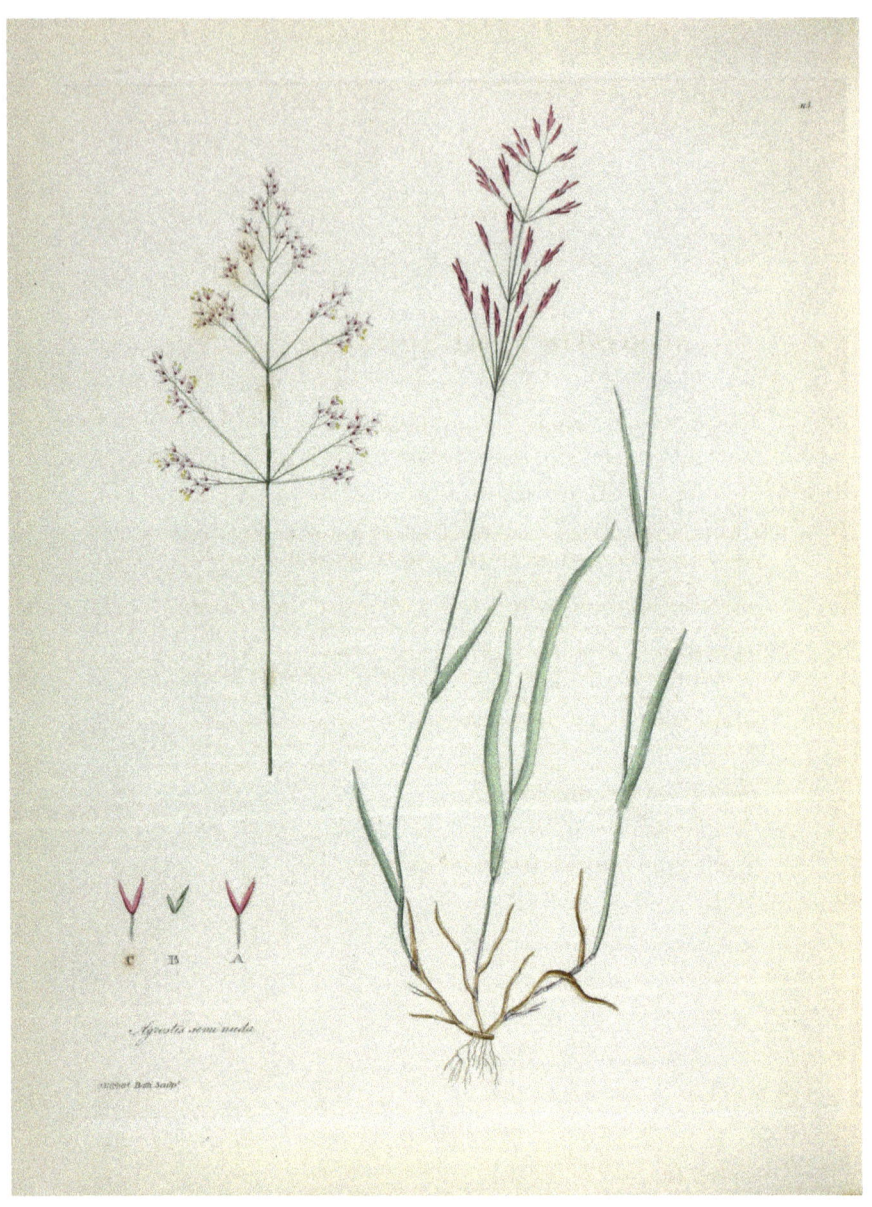

5/2 Agrostis semi-nuda, *an example of a plate engraved for Knapp by John Hibbert of Bath.*

Knapp found a bookseller prepared to publish his work in John White (fl.1785-1816) at the sign of Horace's Head in Fleet Street, London. This was a significant moment in terms of his recognition as an accepted botanical author. John White's father, Benjamin White, who had retired from the business only in 1792, was the younger brother of Rev. Gilbert White (1720-1793), author of *The Natural History of Selborne,* a book so celebrated and pioneering in

5/3 Festuca repens, *an example of a plate engraved for Knapp by Richard Polydore Nodder of London, botanical artist to King George III.*

terms of natural history writing, that it has not been out of print to this day. Benjamin had become a partner of the publisher and bookseller John Whiston in 1756, taking over his firm in 1765. Thereafter he transformed the business into the first specialist publisher of natural history books in Great Britain. His shop became a meeting place for all the great naturalists of the day. It was in his shop that his elder brother Gilbert first met Thomas Pennant (1726-1798) author of *British Zoology* and the Hon. Daines Barrington (1727/8-1800), lawyer, antiquary and naturalist. It was Gilbert White's subsequent letters to both these men, describing his observations at Selborne, that were put together to form the famous book.

John White commissioned Thomas Bensley of Bolt Court, Fleet Street (d.1835) to print *Gramina Britannica*. Bensley was one of the finest printers of his day and at the time that White approached him, had just completed a much-celebrated edition of James Thomson's *The Seasons*, with illustrations by Bartolozzi and Tomkins, 'in royal folio... a volume quite worthy of the warmest eulogies.' From his press too came Thomas Macklin's superb folio Bible, the largest printed Bible of its time, in seven volumes, 'embellished with engravings, from pictures and designs by the most eminent English artists...' These, together with eight volumes of *The Plays of William Shakespeare*, were hailed by contemporaries as the best examples of the printers' art.[45]

5/4 *Thomas Bensley of Bolt Court, printer of Knapp's* Gramina Britannica

Two hundred complete sets of text and illustrations for Knapp's book had been printed by the end of 1804. The title page made the author's intentions clear. The book was to provide detailed illustrations of all British grasses. The 'remarks and occasional descriptions' attached to them were secondary. He made no claim that his work was a botanical or taxonomical textbook, but by quoting the Roman poet and philosopher Lucretius (99BC – c.55BC) indicated what he believed to be its importance. 'Hinc fessae pecudes pingues per pabula laeta / Corpora deponent, et cadens lacteus humor / Uberibus manat distentis' which Knapp translated inelegantly as 'By these both man and beast are fed: by these / The herds fatten, and with encumbered / Bodies ruminate supine amid / The rich luxur'ant herbage: hence flow from / Wide distended udders, nutritious / And milky streams.'

Gramina Britannica, though beautifully presented with hand-coloured illustrations throughout, was not in any way a 'coffee table' book, as it might be described if it was republished as expensively today. It was the first time that painstakingly accurate illustrations of all known British grasses had been made available in one specialist book, allowing landowners to identify the types of grass growing on their land and to consider whether the introduction of different types might benefit their herds.[46] As for the expensive method used to illustrate the book, there was no other means of producing so detailed a pictorial work at that date. Rev. George Swayne had made his book of grasses uneconomic by including real pressed specimens in 1790. Attempts had been made from 1802 to reproduce images of plants and leaves by laying specimens on chemically treated paper and exposing them to light, but these 'photograms' were one-off and quickly faded. It would be thirty years before William Henry Fox Talbot (1800–1877) would perfect his 'calotypes', or black and white photographs capable of being reproduced in small numbers from a paper negative. It would be thirty-nine years before Fox Talbot would succeed in printing conventional photographs in sufficient numbers for publication. It would be well into the 20th century before printed books containing lifelike colour photographs would become a possibility.

John White placed the first advertisement for *Gramina Britannica* in the London newspaper, *The Sun* on Monday 8th April 1805. Two more advertisements followed on the 12th and 13th, one in *The Sun* again, and the other in *The London Courier* and *Evening Gazette*. All three read 'Lately was published in One Volume, Royal Quarto, elegantly printed by Bensley, and illustrated with One Hundred and Nineteen coloured Plates, price Eight Guineas, in Boards, GRAMINA BRITANNICA; or Representations of the British Grasses. With Remarks and occasional Descriptions. By J.L. Knapp, Esq. F.L.S. and A.S. Printed for J. White, Fleet-street.'

Soon after, an anonymous assessment running to six pages, appeared in *The Annual Review and History of Literature*. Its author was someone with a deep knowledge of the subject and an enthusiasm for it that equalled Knapp's. Perhaps it was this that caused him to be less than fair. Despite observing that 'as the title page promises only representations of British Grasses, with remarks and occasional descriptions, we certainly have no right to expect more…' he then devoted the best part of five pages to minutely examining and finding nit-picking fault with precisely those 'remarks and occasional descriptions'. When he did focus on the 'representations,' the core of the book, he was complimentary. 'We have here the satisfaction to meet with a man of fortune, who devotes his leisure arising from his rank in life to the study of nature. Mr. Knapp has the felicity to possess the skill of the designer, as well as the knowledge of the botanist; and having directed his attention particularly to the investigation of grasses, he has traversed almost every part of our island, that he might contemplate the indigenous species of this curious and numerous tribe in their several places of growth and, catch their living forms and manners. Botanical figures are often unavoidably taken from specimens sent in tin boxes to a great distance from their native soil; and, though sufficiently recent to enable the artist to draw them with scientific exactness, cannot be represented by him as they appear in the fields, or woods: for no one can copy what he never saw : and yet each of them has a specific air and character, which an experienced eye discerns at a single glance, before it has time, or is come near enough to them to perceive their peculiar structure. We are informed by the author, that most of the figures in the present work were taken from drawings made by himself 'from plants of his own gathering in their native stations, and to the best of his opinion, judiciously selected;' and, as far as we are acquainted with them, he has happily succeeded in his design to exhibit their living images, with respect to their general outline, and mode of growth. The inflorescence also is delicately and accurately finished…'

Although the reviewer then complained that the plates should have been bigger to accommodate the larger grasses at a greater size, he conceded that the 'figures in general are faithful and elegant, and the colouring in particular is remarkably chaste and natural.' He praised Knapp for describing how the same plant, growing in different situations, can take on a different appearance, saying that he has 'thereby done an acceptable service to the investigating botanist in a branch of his inquiries which has always been the source of much perplexity and confusion.' All this must have pleased Knapp, although whether he would have been happy to be described as 'another man of fortune' is questionable. There is no evidence that at this stage of his life he had come into any great wealth.[47]

The final paragraph of the review would undoubtedly have delighted Knapp, given that he was a man brought up with an unshakable religious faith. 'We cannot take leave of this very elegant and pleasing work' wrote the reviewer, 'without expressing our warmest approbation of the religious spirit which it uniformly breathes. Unlike those naturalists who have studied at the school of Buffon and other pseudo-philosophers,[48] its author does not arrogantly censure as useless everything of which he cannot immediately discover the use, or as a blemish whatever does not exactly correspond with his own limited ideas of excellence; but where he finds so much to admire, justly concludes that all is wise and good; and, where he perceives so many parts of nature directly conducive to the sustenance, convenience, and gratification of man, piously believes that all the rest have a real utility, though their beneficial operation is often conveyed to the appointed lord of this lower world through a long chain of intermediate instruments. This alone is true philosophy. It is equally satisfactory to the understanding, and delightful to the heart.' Perhaps the anonymous reviewer was Samuel Goodenough (1743-1827), keen amateur botanist, who in 1808 became Lord Bishop of Carlisle, or the Hon. Brownlow North (1741-1820), Lord Bishop of Winchester, Prelate of the Order of the Garter and Honorary Member of the Linnean Society.

There is no record of how many copies of *Gramina Britannica* were bound in the first instance or were indeed sold. It is known however that by November 1807, one hundred copies were in the hands of the book binders and the remainder were in the 'ware-room' of the printer. Then disaster struck. At half past eleven on the night of Friday 6th November, a fire broke out in Thomas Bensley's store. 'From the confined situation in which the premises stood, the engines had not that ready access to the fire which they might have had if it were to have occurred in a more open place; but, as it was, they played with a good supply of water to such effect that in less than an hour the fury of the flames was considerably abated, and before two o'clock it was entirely subdued, with the exception of that sort of smothering fire which might naturally be expected where an immense body of closely packed paper had been on fire. By that time the inside of the warehouse was almost completely destroyed... But though some thousand valuable copies were saved from the ravages of the destructive element, the greater part of them were, however paradoxical it may appear, destroyed by the very act of rescuing them from that species of destruction. The water and dirt which was washed along with it from the upper part of the house down on the books, have stained, disfigured and torn some to such a degree as to render them unfit for sale...' Among the losses were 'a beautiful edition of *The Life of the immortal Nelson*... a splendid quarto edition of Thomson's *Seasons*... Smith's beautiful work, *The Antiquities of Westminster*... [and] a fine edition of *Juvenal* is also supposed to be very nearly destroyed or rendered unsaleable.'[49] The remaining copies of John Knapp's *Gramina Britannica* bound and unbound, suffered the same fiery then watery fate.

At first the fire was blamed on boys playing with fireworks, but *The St James's Chronicle* reported that 'from an examination of the premises, we are convinced that the fire did not originate in Mr. Bensley's premises, but those of Messrs. Downer and Sanders, the Ironmongers, whose forge was situated under the warehouse of Mr. Bensley; and besides, that warehouse of Mr. Bensley which has been consumed was no part of his Manufactory, but merely a place of deposit for the finished books, and therefore less liable to accident. The damage is considerable; but all that belongs to Mr. Bensley is insured, also much that was belonging to various Booksellers.'[50] It may

5/5 A watercolour of Bolt Court, painted by J.S. Ogilvy in 1902, showing the restricted entrance from Fleet Street that made fighting fire there almost impossible.

be that John White was not one of these, because in a letter written to the publisher John Murray in September 1827, Knapp recalled not only that he had 'lost the whole impression except 100, [which] were soon disposed of, but my bookseller (White) failing, I received only my dividend with the other creditors.' It is remarkable that in his autobiographical poem of 1838, he limited his description of this total disaster, amounting to the loss of at least seven years' work, to two extraordinarily restrained lines: 'Ah, Vulcan! That thy hateful rage / Should moulder half his studious page.'[51]

6

ANOTHER REASON THAT John Knapp might have been staying in Bath in 1801 and 1802, apart from visiting John Hibbert the engraver in Chapel Row, was that Lydia Frances Freeman (1772-1838), daughter of Arthur Freeman of Ratton, in Sussex, may have been resident there.[52] She was certainly resident in Bath on 16th October 1804, when she and John Knapp married in St. Mary's Chapel, at the corner of Queen Square, directly opposite John Hibbert's shop.[53]

Both *The Gentleman's Magazine* and Knapp's Linnean Society obituary,[54] state that shortly after their marriage, John and Lydia Knapp took up residence at Llanfoist, near Abergavenny. Quite why the membership lists of the Linnean Society for 1805 and 1806 give Knapp's address as Brockford, in Suffolk, must

6/1 St Mary's Chapel, Queen Square, Bath, where John and Lydia Knapp married in October 1804. The chapel was demolished 1875, to widen access to Green Park Station.

6/2 A view over the River Usk towards Abergavenny and the Skirrid, from a drawing by Sir Richard Hoare, published in 1800. The view was taken from a position very close to the Knapp's home at Llanfoist.

remain a mystery, for the Monmouthshire church records show that they were certainly living at Llanfoist on 4th May 1806, when their first child Lydia Margaret was baptised there. In May 1807, a second daughter Frances was baptised at Llanfoist, but lived for only twelve days and was buried in St Faith's churchyard. Their first son, Arthur John was baptised at St Faith's on June 9th, 1808, two daughters, Frances Anne and Catherine Mary were baptised there on August 28th, 1809, and March 31st, 1811, and finally a second son, William Thomas Knapp was baptised there on January 17th, 1813. While each of the baptismal records gives the family's address simply as 'Llanfoist', the last entry is in a differently formatted register and gives their precise address as 'Middle Llanfoist'.

While 'Middle Llanfoist' may sound like a hamlet or part of a larger settlement, it was in fact a historic farmstead of the most romantic kind, remote from the main settlement of Llanfoist Fawr and some half a mile from the church. Some of its land is now developed, as Llanfoist itself expands to the south-east. But in Knapp's day, it nestled in wholly undisturbed countryside, standing amid a myriad of small fields, in the verdant Usk valley. 'The Blorenge', a vast steep-sided hill clothed at its base by Llanfoist Wood, rises to the south-west. Beyond the meandering river Usk to the north, is the

6/3 Middle Llanfoist, where the Knapp family lived from 1806 until May 1813.

ancient, fortified town of Abergavenny and the forbidding mountains behind it. The house itself, though gently modernised in the 18th and 19th centuries, is predominantly mediaeval in origin. Anciently named 'Beili-du', it is said to have been the seat of Hywel y Coed, brother of Dafydd Gam, the legendary opponent of Owain Glyndwr.[55]

No clue has yet been found as to why Knapp chose to take his wife to Llanfoist, but he had spent a summer in Wales while collecting specimens for *Gramina* and maybe it was then that he fell in love with the place.[56] The section of his autobiographical poem, addressed to the goddess Flora, which covers this period, suggests that after the publication of *Gramina,* he simply wanted to give up the intensity of his research into grasses and revert to a general enjoyment of nature, in particular to his fascination for fungi. Monmouth ('Cambria's shades') was the place to do this:

> There, mark him now in Cambria's shades,
> Panting up steeps through forest glades,
> To woo thee in thy humblest seat,
> Rarely disturb'd by mortal feet,
> To view thee on thy mossy bed,

Where changeful agarics lift their head,
And riveted in rock, the oak
Scarcely has heard the woodman's stroke;
Whilst his grey lichen, pendant there,
Looks like some hoary peasant's hair.
The staring woodcock wakes in fright
From leafy bed on alpine height,
And flutt'ring from her foliage sere,
Steals to some silent valley near;
The squirrel peeps beside the tree,
Th'intruder on his haunts to see,
Then darts with agile leaps away
To watch him from some mossy spray.
The dark owl glares with moony eye,
As the lone wand'rer passes by,
And wonders what could bring him there,
To wake her in her beechen lair.

In a later prose account, referring again to his passion for fungi, he wrote lyrically 'no country is, I believe more favoured for the production of most of the kinds than Monmouth, with its deep dark woods, and alpine downs. A residence in that portion of the kingdom for some years introduced to my notice a larger portion of this singular race than every botanist is acquainted with. A sportsman then, but I fear I shall be called a recreant brother of the craft, when I own having more than once let my woodcock escape, to secure, and bear away some of these fair but perishable

6/4 St Faith's Church, Llanfoist, where six of John and Lydia Knapp's children were baptised and where their infant daughter Frances lies buried.

6/5 The stable yard at Middle Llanfoist, where Knapp kept the favourite pony that had carried him on his botanising expeditions, "a faithful and able performer of all the duties required of her," whose "fineness of limb, brilliancy of eye, and ardour of spirit" he greatly admired.

children of the groves. Travellers tell us of the splendour of this race in the jungles of Madagascar, but nothing can surely exceed the beauty of some old copse in Monmouthshire, deep in the valley, calm, serene, shaded by the pensile, elegant, autumnal-tinted sprays of the birch, the ground enamelled with every coloured agaric, from the deep scarlet to the pallid white, the gentle grey, and sober brown, and all their intermediate shadings. Fungi must be considered as an appendage and ornament of autumn; they are not generally in healthy splendour until fostered by the evening damps and dews of September, and in this season no part of the vegetable world can exceed them in elegance of form, and gentleness of fabrication; but these fragile-children of the earth are beauties of an hour:

> 'Transient as the morning dew,
> They glitter and exhale…"[57]

Just as there is no known record of the reason that John and Lydia Knapp chose to move to Middle Llanfoist, so there is no record of their reason for leaving, which they did between January 17th,1813, when their son William was baptised at Llanfoist and May 8th in the same year, when their the newly-born daughter Charlotte Henrietta was christened at St Helen's Church, Alveston, in southern Gloucestershire. If there was anything that might have spoilt their rustic dream, it was that high on the Blorenge above their house, the newly-built Brecknock and Abergavenny Canal had been sliced through the hillside. While today it is one of the prettiest and most romantic of all British canals, overhung with mature trees, then it must have been a brown scar across the countryside, freshly dug, freshly puddled and heavily embanked. It had just been completed when the Knapps arrived, but during their time, a second industrial scheme was proposed, to bring iron, coal and lime from Blaenavon to the canal's edge, by inclined planes and an iron-shod tramway. The junction was to be only some five hundred yards from their house and perhaps the threat of it became too much for the family to bear. Perhaps, it was the winter cold that drove them away, for wide and lush though the Usk floodplain is, the sun barely rises above the sheltering hills in the winter months and hard frosts remain unrelieved. Perhaps it was Knapp's health, which he frequently hinted had troubled him from his childhood, that precipitated the move. Perhaps it was the distance from city life. Whatever the reason, by May 1813, the family had crossed the River Severn and were now resident in Alveston.

7

THE PARISH OF Alveston is situated in southern Gloucestershire, on the last substantial ridge before the land falls away to former marshland that borders the River Severn. The first volume of Ralph Bigland's *Historical, Monumental and Genealogical Collections, relative to the County of Gloucester* was published in 1791 and is the closest in date of all the great surveys of the county, to Knapp's arrival there. Bigland explained that Alveston (then pronounced 'Alliston') is situated 'in the hundred of Langley and Swineshead, two miles south-west of Thornbury, six miles north-west from Sodbury, twenty-six miles south-west from Gloucester, and nine miles north-east from Bristol.' The parish lies, he said, 'high, and the soil is mostly clay, which however, in a good season, yields very good crops. A part of the soil is a coppery sand on a lime-stone, which produces a great deal of what is called Carnation-Grass. The western part forms a rock of limestone descending towards the Severn.' To this he added 'The great road from BRISTOL and the West and GLOUCESTER, to the North, passes through this place; at which an inn, known by the sign of The Ship, has lately been fitted up in a commodious manner for the convenience of travellers. A turnpike road is also making through it, leading to Sodbury through Iron Acton; and which, when completed from Aust, or the Old Passage, through Olveston, will be of singular advantage to travellers coming from the other side of the Severn, being the nearest road to LONDON from South Wales.'[58]

He explained that the parish then consisted of three hamlets, Alveston village itself, Earthcott (about one mile from the church) and Grovesend (more than two miles from the church), observing that 'about a mile and a quarter from the church, Henry King Esq. has a very good house and estate, with a fine plantation of pear trees, of which is made excellent perry.' This is the house that Knapp took on a lease from King's grandson-in-law, Rev. John Collinson (1782-1857).[59] Today it is divided into two, now known as 'Alveston Old House' and 'Dial House', the stables forming a third separate property. In the 1840 Electoral Register, against Knapp's name, the whole was simply recorded as 'Collinson's'.

7/1 "Collinson's," as seen from Shellard's Lane in Alveston, Gloucestershire. John and Lydia Knapp moved here in May 1813 and remained until their deaths in 1838 and 1845. Their farmland was to the left and right of the lane.

The fractured nature of the parish is undoubtedly accounted for by the fact that much of the open land which constitutes its centre, was once an enclosed mediaeval deer park. Habitation in those days had perforce to be around its edges and remarkably, it remains much in that configuration to the present day. 'The Street', at the end of which the former 'Collinson's' is situated, is a pretty lane, which may at one time have marked the park's northern boundary. In Knapp's day it was lined with some eight dwellings and numerous farm buildings and stables. Most of the houses were 17th century in origin. Many, including 'Collinson's, had been re-fronted in the Georgian style. When it was advertised to be let again, following Knapp's death in 1845, it was described as

> A CAPITAL MESSUAGE, suitable for the accommodation of a large family of respectability, with Greenhouse, ample Pleasure and Kitchen Gardens, Yards, and Offices, and Six Closes of excellent Pasture LAND, adjoining, within a ring Fence… and containing altogether by admeasurement, 40A. 1R. 2P.[60]
> The house is seated on a dry and healthy soil, amid highly ornamental Timber, with a southern aspect, commanding extensive and beautiful land and sea views. The supply of excellent spring water is inexhaustible. The air is remarkable for its tone and purity, and there is a cheerful repose pervading the spot very rarely to be met with.[61]

If Knapp used the Aust ferry to travel over the River Severn in his journeys between England and Wales, as he probably did, rather than taking the over-land route through Gloucester, then he would certainly have been familiar with the area. Even if he did not, it is clear from notes in his Herbarium that he had visited southern Gloucestershire in the late 18th century to collect grasses. His specimen of *Alopecurus bulbosus* is labelled from the 'sea marshes, Oldbury on Severn' as is his *Poa distans* or *Glyceria*. Another specimen of *Glyceria*, known to him as *Poa procumbens* is labelled 'from Rownham Ferry, Clifton,' which was also within the county of Gloucestershire at that date.[62] It may be that he actually decided to move to the limestone ridge above the marshes, after he had made a visit to Berkeley Castle with a friend. Though he does not mention Alveston in his account of that day's events, the journey from his friend's house near Bristol to the castle, meant that he travelled up what Bigland had called the 'great road from Bristol and the West', through Almondsbury, Rudgeway, Alveston and Thornbury before diverting, as he explained, to Tortworth. In doing so, he passed only six hundred yards from his future home. "Twas a lovely October morning,' he recalled, 'and fine threads of Gossamer floated lazily along, as some one quaintly observed, like 'scorched dew' – the maple, and the beech, just crested with their thin autumnal tints, stood eminent amidst the verdure of the woods, and all nature appeared in quiet, thoughtful, sober prime. Having passed about seven miles of most uninteresting scenery, a glorious prospect at Almondsbury hill, burst upon our view, the broad estuary of the Severn, its noble (though muddy) waters flowing majestically through the low lands of Gloucester, beneath us; with the high wooded rocks of Glamorgan, and Monmouth sloping into the stream, sprinkled with the peaceful habitations of rural life: this fine ridge with varied prospects continued for several miles, nearly until we turned off the main road at the little village of Falfield, to view the famous chestnut tree at Tortworth – about one and a half miles distant.'[63]

Once the family were established in Alveston, Knapp's busy life included serving as churchwarden at the ancient church of St Helen, just over a mile to the south of his house. Only the late 14th century chequered stone tower and a fragment of wall containing a Norman or perhaps late Saxon doorway, now remain, sold out of Church of England ownership some years ago.[64] St Helen's, which at that time was a chapel-of-ease to St Mary's in the adjoining parish of Olveston, was an uncomplicated building, with a nave and chancel only. It stood with its tower hard against the wall of the once grand Alveston Manor, which by then had become a tenanted farmhouse. Early English and Perpendicular windows pierced the church walls. A fine 14th century Sanctus bell-cot graced the ridge of the roof above the chancel arch. A low hagioscope,

7/2 *The old church of St Helen's, Alveston, recorded by Samuel Lysons for his Collection of Gloucestershire Antiquities in 1804. John Knapp served as churchwarden here and he and his family lie buried in its churchyard. The ancient yew tree still flourishes.*

7/3 The only known photograph of the chancel of Old St Helen's church, taken in 1948. The high box pews that filled the nave in Knapp's day had long since been cut up for firewood, but the 17th century Laudian altar rails still remained.

or squint, gave those who for health or other reasons could not be admitted to the church, a view of the altar. Inside, 17th century commandment boards framed the east window, and turned Laudian rails protected the altar from the ungodly. The pointed chancel arch sprang from earlier responds, each crowned with a naïve stiff-leaf capital. Fragments of wall-painting adorned the nave. The churchyard may well have been of pre-Christian origin. The ancient yew at the churchyard's centre, whose trunk was described in the 1930s as some thirty feet in circumference, almost certainly predates the Christian church.[65] Knapp must have known that tree well. It still thrives today. As the years passed, St Helen's must have become an increasingly poignant and important place for him. Between 1828 and 1833, three of his daughters, Frances Ann, aged fourteen, Charlotte Henrietta, aged twelve and Catherine Mary, aged eighteen died and were laid to rest in that tranquil place. In November 1838, his wife Lydia Frances died and was buried alongside her daughters, as, in due course, would be Knapp himself.[66]

The only known description of St Helen's when still in use and as Knapp would have known it, was written by Sarah Rugman of Alveston (c.1858-1942) and is quoted here, just as she wrote it:

The Alveston old church as it was when in use 60 years ago before the new one was built. As you enter the church you could go straight through into the vestary and to the left as you entered was two lots of childrens seats one side for girls and one side for the boys and the font in the midal and the gallery was up over them were the quire did sit and the chancel was at the top of the church and the Pews for the congregation to sit in was large Box seats nearly as big as horse boxes is today with two seats in them one each side and the one half of the congregation had to sit with there backs to the vicar and the other half facing him and when the peopel was standing up you could only just see there heads above the top of the pews and you could see all the working men goine to church on a Sunday morning with there clean smocks on and the little girls to Sunday school with there little linen bonnets on but that was nearly 80 years ago seeing them in there smocks… and the field that is called Bryars land used to be what was called the fee fees fund a part belonging to the church living and it was let out in allotments at that time and the old vicar used to go out and work on it like any other man and there was a cottage standing sixty years ago to the left of the church gate joining the church wall were the little plantason is. Under the church wall was a garden and that did belong to the church farm.[67]

How long Knapp served as churchwarden is not known, the only surviving evidence of his appointment was a single parish charity board, (recorded as surviving as late as 1958, but now missing), which bore his name and office. Serving the church must have been a responsibility which pleased him, as the Christian religion was of the foremost importance to him, though he did dislike the churchwarden's legal obligation to pay rewards to those who brought dead 'vermin' to his door, including the heads of tomtits. He regarded the unfortunate birds as wholly innocent.[68]

A letter in a private collection dated May 4th, 1833, suggests that as a parish official, Knapp did his best to assist neighbours in distress, and especially during the difficult and dangerous 1831-33 period of influenza:

My dear Madam,
Upon the receipt of your note, I immediately made enquiry into particulars, and regret to say they were too correct, as Bakers eldest girl is a most violent personage beyond the control of her Father, and laughs at the remonstrance of neighbours – I consequently sent over our assistant overseer to bring the child away from its Aunt's (Mrs. Pike) to keep in the poor-house, until something could be thought about it – but Mrs. Pike would not let him have the girl,

saying it was put there by Miss Bailey, and Mr. Holder,[69] and it was no affair of his – and did not appear to think that her niece had been unduly corrected – so that under these circumstances – nothing more can at present be done – I hope all your afflicted are recovering from their torments, which are said to be severe and lingering – Mrs. K. has for several days been a sad sufferer – does not recover as she should, and is now in bed – my poor half-alive child too is labouring under this additional complaint – but possibly not very severely – with all possible esteem believe me dear Madam truly yours, J.L. Knapp

The only other recorded instance of Knapp taking an official role in the community was in 1819, when he served as joint-collector of Land Tax in the parish, with the regular collector and assessor, Sampson Selman.[70] Although in twenty-first century terms, Knapp's personal attitude to the poor of his parish might be regarded as patronising, he was not out of step for his time, and he did care. Many references in his *Journal* show this to be the case. He took pleasure for example in allowing 'leasing' or gleaning in his fields after the harvest, not only because it benefited 'the needy and laborious community' in

7/4 *Old St Helen's church Alveston as it is today. A new church replaced this ancient building in 1885. Shamefully, its roof, chancel and nave walls were pulled down in 1960 as a Civil Defence exercise. The gravestones, including any marking the last resting place of the Knapp family, were moved to the churchyard edge and are presently inaccessible.*

financial terms, but because of the 'prattle, the gossip, the glee, the excitement' enjoyed by those who took part.[71] The Minute Book of the Gloucestershire Society, a charity which still flourishes and which at that time provided funds specifically for the apprenticeship of poor Gloucestershire-born boys and support for Gloucestershire-born women in childbirth ('the Gloucestershire Guinea'), shows that he was a contributor to its funds.[72]

As time passed, Knapp became known in Alveston as something of an eccentric collector. Villagers would bring him natural specimens of all kinds, no doubt hoping for reward. On occasions this led to the unnecessary capture and even death of rare birds, which like the thoughtless destruction of the poor tomtits, was a cause of deep regret to him. He was also brought historic artefacts, such as coins that had been found in the ground, and he was consulted or called in when evidence of archaeological interest was found. His observations on the Abbey Camp, a substantial Romano-British hillfort, situated a little over half a mile north-east of his house, can be found on pages 1, 58, 301 and 317 in part 2 of this book.

8

It has already been shown that Knapp was not only an experienced botanist and a fine botanical illustrator but also a lover of poetry. Twenty-four poems, written in Knapp's own hand and almost certainly composed by him, survive among his papers in the Buckinghamshire County Archive.[73] These have now been transcribed and are set out in Appendix 2 of this book. Of his poetic works, only three appeared in print in his lifetime. The most significant, an epic tale written in blank verse, was published by him anonymously, in 1818. The title page read

ARTHUR
OR THE
PASTOR OF THE VILLAGE

A POEM

NE NOSTER SERMO INDICET ALIQUID VITIUM IN NOSTRIS MORIBUS
INESSE[74]

I sing no high historic deeds
Of dying heros, bleeding steeds,
Nor wake of mould'ring ancestry the pride,
My muse a softer theme endites,
And roams where rural scenes invites,
With Science, Virtue: Innocence as guide

LONDON:
PRINTED BY BENSLEY AND SONS,
Bolt Court, Fleet Street,
SOLD BY MESSRS. PAYNE AND FOSS, PALL MALL

1818

For the most unfortunate of reasons explained later in this chapter, the poem brought neither fame nor fortune to Knapp, but it does reveal much about his character and beliefs as well as something of the circles in which he mixed.

Thomas Payne (senior), founder of the firm of Payne and Foss, and Knapp's chosen bookseller, had opened his business at Mew's Gate near Charing Cross around 1740. His ability, it is said, 'lay in fostering a lively social environment in his shop and in anticipating and shaping trends in bookselling.' *The Gentleman's Magazine* of 1799 described Payne's shop as 'the first that obtained the name of a literary coffee house in London, from the knot of literati that resorted to it', later describing it as 'the constant resort of men of rank and literature' and 'often mentioned in the correspondence of scholars and antiquaries as their daily resort for conversation, and their daily resource when in quest of books of rarity and value.'[75] So successful was it all that in 1808, Payne's nephew and successor Thomas, in partnership with his former apprentice Henry Foss, moved the business to the splendid Schomberg House, in Pall Mall.[76] There, Knapp no doubt mixed with the scholarly élite.

8/1 *The entrance to Schomberg House in Pall Mall, London. Knapp visited Thomas Payne and Henry Foss here in 1817, in the hope that they would publish his epic poem Arthur or the Pastor of the Village. This they did.*

Quite why Knapp allowed Payne & Foss to employ Thomas Bensley to print his new work, after the disastrous fire at Bensley's had destroyed the unbound copies of *Gramina Britannica*, is not known. It may have been a matter of cost, since after the fire, the entrepreneurial Bensley had invested heavily in an invention of the German, Friedrich Koenig (1744-1833), a revolutionary printing press that was powered by steam. *The Institutions of Physiology* written by J. Frederick Blumenback and John Elliotson and printed by Bensley, claimed to be the first complete book ever produced by this method. It was dated 1817 and stated in its introduction that it 'may be considered a typographical curiosity, being the first book ever printed by machinery. It is executed by Messrs. Bensley and Son's patent machine, which prints both sides of the sheet by one operation, at the rate of 900 an hour, and is the only one of the kind ever constructed.'[77] Knapp's book, though dated one year later, is very similar in design and typeface and is likely to have been printed on the same press.

It is almost as though printing John Knapp's work jinxed the Bensley premises. On July 3rd, 1819, only months after his new book had been completed and had been carefully stored in the printer's warehouse, disaster struck again. One of the most comprehensive reports of the event was carried by *The York Herald* for the following day. It read

DESTRUCTION OF MESSRS. BENSLEY'S PRINTING-OFFICE BY FIRE. Saturday about one o'clock, a fire broke out in the printing office of Messrs. Bensley and Son, Bolt-court, Fleet-street, London. Before the arrival of any of the fire engines, the flames had made rapid progress, and had communicated to a very large stock of paper, a considerable part of which was on lines hanging to dry. By two o'clock the whole range of printing-offices were ablaze and threatened destruction to a very close and crowded neighbourhood. By three o'clock, or very shortly after, not a vestige of these once-extensive premises remained, except the party walls and the basement floor. In the latter were the printing machines, but they, we hear, are greatly damaged. The very extensive printing materials of this establishment were entirely consumed.

The fire then communicated to the dwelling-house occupied by Mr. Bensley, sen. which lay to the south of the printing office, the back part of which, and the roof and upper room were destroyed. A considerable part of the furniture is burnt, and the remainder much damaged. The flames next caught the counting-house of the concern, which lay to the left of the dwelling house, and it was entirely destroyed. In a northerly direction, the fire extended to the back of the houses in Gough-square, occupied by Messrs. Smith and Co. Messrs. Elin and Co. (both furriers) and another large house occupied by Mrs. Salmon as a ladies' school; the latter was entirely destroyed, and the former very much damaged. – The house of a Mr. Smith, and that of a Mr. Fairbridge, were also greatly damaged. The property destroyed is said to be of the value of nearly £60,000. Mr. Bensley's premises were insured, but not to that extent which will in any degree cover his losses.[78]

Tragically, all the unbound copies of Knapp's *Arthur* were in store there, just as those of *Gramina Britannica* had been in 1807. Knapp later recalled 'Bensley was printing another work for me when his second fire took place,

8/2 Few illustrations of properties in Bolt Court were made during the 19th century. This watercolour by J.T. Smith (1766-1833) shows No. 8, recorded only because it was a former home of Dr. Samuel Johnson. It subsequently became part of Bensley's printing works and was badly damaged in the second disastrous fire.

and I was again unfortunate.'[79] This undoubtedly accounts for the fact that very few copies indeed of *Arthur* exist. Effectively, Knapp's work is only available today in facsimile.

The opening stanzas set out the background to the story, in which the gallant hero Raymond, falls in battle 'on Egypt's plain, in that ill-fated day /

Which gave a laurel to Britannia's wreath,' leaving his children as orphans. By God's good grace however, Raymond's brother, a kindly country priest named Arthur, took the orphans into his home.

If the plot is purely fictional and was not based on the death in battle of an actual relative or friend of Knapp's, then the choice of 'Egypt's plain' and not the traumatic field of Waterloo, suggests that the first lines were written after the Egyptian campaign of 1801 but before Waterloo in June 1816 and that Knapp began its composition in Monmouthshire, if not before.

Only a few lines later, Knapp reveals that his hero's purpose was to demonstrate to the orphans how morality and religious truth can be learned through the study of nature.

> Their early forms of education o'er
> Their pious teacher next, led the young mind
> Thro' Nature's paths, and journey'd upwards to
> That power which ordain'd them good, and fair;
> In many a ramble thro' the wilds, the
> Woods, and flow'ry fields, pour'd to their list'ning
> Ears a moral from the walks of truth, which
> Long experience gain'd, instructing whilst
> It made them good.[80]

Pausing then only to write several moving pages on the old age and gradual incapacity of Arthur's 'honest dog' Derry (surely a description of one of Knapp's own dogs), Knapp launches into the first of six 'rambles' that Arthur and the children undertake to local places of interest: 'The Copse,' 'The Abbey,' 'The Church-yard,' 'The Castle,' 'The Ocean' and 'The Ruined Cottage.' The style is the epitome of the romanticism of Knapp's period, but recognising the fragility of life at that time, makes occasional forays into the tragic, sometimes even the morbid. Always though, Knapp, through his modest hero Arthur, is educating, always moralising. He ends his tale:

> Much yet remains
> For me to tell, but from the chilly airs
> Of eve we now retirement seek. Yet come
> Again, sweet Summer, come, with all thy train
> Of wonder and delight, and linger with
> Us long; again, if Heaven approving
> Will permit, thy verdant fields perhaps we'll

Trace, and mark the florets of the op'ning
Spring; by chance some vagrant blossom find in
Fertile Vale, or on the mountain's dewy
Brow, whose richer bloom may grace, or fragrance
Scent the Vicar's humble manse. And you, sweet
Darlings of my life, whose feet have tripped with
Me this Summer round, dear to my heart's the
Wish, to make life's passing hours as welcome
And as happy here, as human frailness
And infirmity may know; to picture
Virtue to your view in all her brightest,
Truest colours drest, that you, enamour'd
With the form I draw, through life the lovely
Essence may regard, and love her dearest
When you know her best.[81]

9

KNAPP'S OBITUARY IN *The Gentleman's Magazine* records that in the years following *Arthur's* destruction, he contributed all the articles published under the title 'The Naturalist's Diary' in *The Time's Telescope*, between the years 1820 and 1830.[82] *The Time's Telescope* was an annual almanac published in London from 1813, whose regular format was to list month by month,

> Saints' Days and Holidays; with illustrations of British History and Antiquities; Existing and Obsolete Rites and Customs, Sketches of Comparative Chronology and Contemporary Biography. [Also] Astronomical Occurrences in every Month; comprising Remarks of the Phenomena of the Celestial Bodies; and The Naturalist's Diary explaining the various Appearances in the Animal and Vegetable Kingdoms.[83]

Poetry was included in almost every article, in place of pictorial illustrations, which did not appear in the almanac until 1830. '…As for Pictures – do not our numerous poetical sketches offer many a glowing scene, which, like that represented in Gray's Plano-Convex Mirror, were it realised on canvass, would make a fortune of the artist who should succeed in fixing the vivid colours of the poet's high imaginations?' the editor asked in 1827.[84]

Every new edition carried a list of reviews of previous editions. Most were extremely complimentary, and most were by respected literary reviewers.[85] The *Gentleman's Magazine* declared that,

> We recommend this volume as an excellent manual for young persons. It has the negative merit of being perfectly unexceptionable as to the information and entertainment it conveys… but it will also tend to cultivate those pure and simple pleasures which the God of Nature has so abundantly provided for inquiring minds.[86]

The reviewer for the *Literary Olio* of 1824, added light-heartedly that it is also deserving of a place in the libraries of 'grave and reverend seniors', as a book of reference, in Chronology, Biography, Antiquities, and obsolete Customs, and in almost every branch of Natural History: while he who has formerly delighted to climb the airy steep, or brush the dewy lawn, rejoicing in 'each rural sight, each rural sound,' now confined to his elbow chair, with his gouty foot resting on a cushioned stool, will wipe his spectacles, and in perusing the Naturalist's Diary, alternately smiling and sighing, will think of the joys and friendship of Auld Lang Syne; and like the Greenwich or Chelsea pensioner, reading a narrative of the campaigns in which he served, will, for a moment, live his youthful days again.[87]

It is impossible to know whether Knapp did indeed write every article, because each was published anonymously. If they were all Knapp's work, then from time to time, they offer a glimpse into his way of life. The article for December 1828 is a charming example:

A rainy day too has many pleasures peculiarly of its own, especially such a one as this on which I am writing, when the sky is robed in universal drab, and the rain knows no succession, and scarcely any variation of quality. Let him be melancholy who has no resources within his own mind, nor in books, the transcripts of the minds of others. I have had the summer ornaments removed from my brightened grate and a good fire kindled in their stead. I have refreshed my memory by reading the favourite passages of my favourite authors. I have turned over my whole collection of prints. I have assorted into bouquets the flowers which were gathered before the dappled sky of morning gave place to the heavy clouds or rather cloud which has succeeded it. And thus, having once more experienced the pleasures of a rainy day, I have drawn my elbow chair to the fire, and have put my ink-glass in requisition, in order to write a few lines on the subject; and although the wind, carrying the raindrops with it, ever and anon rattles a menace against the window, I throw my chair nearer to the fender, raising the glowing fire into a cheering blaze, and grasp my pen with all the resolution of a hero.[88]

Again, if every article is Knapp's work, then they show what an immensely wide knowledge of nature and of gardening that he had acquired, both indigenous and foreign and, in gathering quotes and poetry from seemingly endless books and articles by other authors, how astonishingly hard-working and well-read he must have been. Most of the poems quoted, acknowledge the names of their authors, but the remainder are anonymous.

Whether the unsigned verses were Knapp's own work or not, it is currently impossible to say. But it is not beyond the bounds of possibility that they were. As a record of animals, birds and insects commonly found in Great Britain in those years, his descriptions are remarkable. His occasional references to the climate and weather of his period too, can be astounding. An example published in the *Time's Telescope* for January 1820, is perhaps of particular interest in the present era of climate and weather concern. It begins with a poem describing the unseasonably warm winter of the previous year:

> Soft floated the clouds in the fields of blue ether;
> The earth with fresh flowers was still covered o'er;
> Like Egypt's enchantress, whom age could not wither,
> They bloomed, though the season of youth was no more.
>
> The voice of delight was long heard from the thorn,
> Its tenants no end to their happiness knew;
> The evening of love was as sweet as its morn,
> And the bird of the spring lingered all the year thro'.
>
> And even that day, whose return we revere,
> Though often it lours in our northern skies,
> Arose with mild aspect, unclouded and clear,
> As of old it appeared to the Bethlemites' eyes.

Such is the poetical description of the winter of 1818-19, as it was experienced in England. The singularly mild temperature of this winter, and the want of frost and snow, was not confined to our own island; it was equally observed in almost all parts of the European Continent. In Sweden, and most parts of Russia, they had, instead of the usual degree of cold, a temperature of several degrees above the freezing point. This was the case even in Lapland, to the north of Tornea, where, instead of the usual cold of 20o of Reaumur, they had 6o of warmth. This want of frost and snow proved a serious inconvenience in these northern regions, by preventing the conveyance of iron ore from the mines in Sweden to the smelting houses; and in Russia, the carriage of goods from the interior to the seaports for exportation, which is regularly done in winter, when the hard frozen ground, covered with snow many feet deep, affords a solid, even, and commodious road.

From the Meteorological Journal, kept at the Botanic Garden of Geneva, the same phenomenon, of want of snow, appears to have occurred on the Alps.

In the three months of October, November and December, there was only once so much as a hard frost. In the whole of November, says the Journal, 'the snow has not laid a single day on the mountains that surround our lake. This a phenomenon of which the oldest inhabitants can remember no previous instance. The wheat is remarkably beautiful; the cattle are still in the pasture as in the month of September.' The same journal, for December, says 'the continued fineness of the temperature, during this month, is without parallel in our country. Mount Jura, which is generally covered with snow in November, is still almost free from it today, the 31st December. There is none at all on the summit of La Dole, and very little on the summits near fort l'Ecluse. In consequence of the dryness of the temperature all the year, the springs are very low, and we begin to be uneasy for next year, seeing that the mountains have no snow at all on them.' Although, of late years, comparatively without snow, the month of January is not without its storms of wind and rain…'[89]

Two further climate or weather-related quotes, describe circumstances remarkably familiar to those in the twenty-first century. The first is serious.

In the years 1799, perpetual rain rendered the country in August as green as it usually is in May. Many thousand acres of wheat and other grain were covered with water. The rivers overflowed, and swept away the produce of whole farms; and a great scarcity of bread ensued.[90]

The second shows the light-hearted touch that Knapp was prepared to bring to leaven his otherwise studious work.

Horace Walpole, in one of his letters, dated the 15th June, 1768, says 'It rained near eight and forty hours without intermission. My poor hay has not a dry thread to its back. I have had a fire these three days. In short, every summer one lives in a state of mutiny and murmur, and I have found the reason: it is because we will affect to have a summer, and we have no title to any such thing. Our poets learnt their trade of the Romans, and so adopted the terms of their masters. They talk of shady groves, purling streams, and cooling breezes, and we get sore throats and agues with attempting to realise these visions. Master Damon writes a song, and invites Miss Chloe to enjoy the cool of the evening, and we have no such thing as the cool of the evening. Zephyr is a North-East-Wind that makes Damon button up to the chin, and pinches Chloe's nose until it is red and blue; and then they cry, this is a bad summer, as if we ever had any other. The best sun we have is made of Newcastle coal, and I am

9/1 The Ship Inn at Alveston gave Knapp easy access to long-distance transport. This early photograph shows the coaching inn as Knapp must have known it. Here local intellectuals met, including Dr. Edward Jenner's Convivio-Medical Society.

determined never to reckon on any other. We ruin ourselves by inviting over foreign trees, and make our houses clamber up hills to look at prospects. How our ancestors would laugh at us, who knew there was no being comfortable, unless you had a high hill before your nose, and a thick warm wood at your back! Taste is too freezing a commodity for us, and, depend upon it, it will go out of fashion again.[91]

To this amusing piece, Knapp adds an up-beat note

But in answer to our lively author's complaints, and to those of other complaining Englishmen, it may be observed, that although our vernal seasons are commonly rendered cold and uncomfortable by the long continuance of easterly winds, or by a superabundance of rain, we derive from these circumstances a beauty unknown in the gardens of a warmer country; the soft verdure of a fine well-kept lawn is a luxury not to be procured in more southern climates. If to this, and a thousand other advantages to be found in England, we add the blessing of a free government, and the impartial administration of the laws, we shall cheerfully unite with the poet, who, when speaking of Italy, says

'How has kind Heav'n adorned the happy land
And scattered blessings with a wasteful hand....'[92]

To achieve the varied content of twelve lengthy articles a year, for a span of ten years, Knapp would have needed access to a vast number of reference books, journals and other printed records and this was something that must surely have influenced his choice of Alveston as his new home. Within the parish were two coaching inns, *The Royal Oak* and *The Ship*, both of which were served by transport leading to and from Bristol and Gloucester. Just down the escarpment was the market town of Thornbury, with another substantial coaching inn, *The Swan*. Throughout the decade Knapp owned 'a favourite pony – for she has been a faithful and able performer of all the duties required of her in my service for upwards of two and twenty years – and, though now above five and twenty years of age, retains all her powers perfectly, without any diminution or symptoms of decrepitude; the fineness of limb, brilliancy of eye, and ardour of spirit, are those of a colt, and though treated with no remarkable care, she has never been disabled by the illness of a day, or sickened by the drench of a farrier.'[93] He also acquired a phaeton (a sporting and reputedly rather dangerous vehicle), harness, and an 'excellent hackney horse,' together with a lady's and a gentleman's saddle, so travel by horseback or phaeton to the nearest long-distance transport would only have taken a matter of minutes.[94] The Birmingham to Bristol mail coaches ran regularly through Alveston and waggons for the carriage of goods (and occasional passengers) ran daily from Thornbury to Bristol. Bristol's famous King Street Library was therefore within relatively easy reach of Alveston. Failing Bristol, coaches provided a regular service onwards to Bath and London: 'The Post to London from Bristol goes out every afternoon (except Saturday) at 4 o'clock – Arrives every morning, (except Monday) about noon. To Bath Goes out every morning between 9 or 10 in the evening.' [95] Knapp's solicitor and near neighbour Richard Crossman, wrote in his diary not long after this period, 'Left home by coach at 3 and arrived in London at 10... Journey to London £1.9.0. Packet of tobacco 9d.'[96]

10

ALL THE WHILE between his arrival in the parish and the year 1829 though, Knapp was engaged not only in cultivating his own garden, overseeing the farming of his land[97] and serving the church, but in keeping a detailed record of everything that he observed on his daily rambles in the countryside around him. For many years he maintained a friendship with other nearby amateur natural historians. These included the distinguished local doctor and pioneer of vaccination, Edward Jenner of Berkeley (1749 – 1823), with whom, he 'frequently had the honour of conversing, and always with profit.'[98] All this not only provided background copy for his articles in *The Time's Telescope* but also built up a treasure house of information for a far more important project, which would eventually earn him praise on both sides of the Atlantic: *The Journal of a Naturalist.*

If Knapp had a hero, it was the Rev. Gilbert White (1720-1793), author of *The Natural History of Selborne* (1789) and uncle of the London publisher of his ill-fated *Gramina Britannica*. White's book had become a best-seller and given that it was almost entirely taken up with reflections on the flora and fauna in and around the Hampshire village of Selborne, one might wonder why it was of interest to anybody living outside that immediate area. The answer quite simply was that it was the one of the first books to reject folk-lore and long-accepted views relating to plants and animals and replace them with personal observation of such things in the present and in context. White introduced his late 18th century readers to a way of looking at the world around them, which had not occurred to most of them before. He showed by example, that anyone with the time and the inclination to do so, could become part of the new and exciting science of natural history and ecology, that barely existed before he put pen to paper. And he showed that they could do this merely by walking outside their houses, looking and listening, no matter where they lived.

Knapp was captivated by *Selborne* and sought to become Gilbert White's successor. In the preface of *The Journal of a Naturalist,* he said so:

Many years have now passed away since we were presented with that very interesting and amusing book, the 'Natural History of Selborne': nor do I recollect any publication at all resembling it having since appeared. It early impressed on my mind an ardent love for all the ways and economy of nature, and I was thereby led to the constant observance of the rural objects around me. Accordingly, reflections have arisen, and notes have been made, such as the reader will find them. The two works do not, I apprehend, interfere with each other. The meditations of separate naturalists in fields, in wilds, in woods, may yield a similarity of ideas; yet the different aspects under which the same things are viewed, and characters considered, afford infinite variety of description and narrative: mine I confess are but brief and slight sketches; plain observations of nature, the produce often of intervals of leisure and shattered health, affording no history of the country; a mere outline of rural things; the journal of a traveller through the inexhaustible regions of nature.

Gilbert White had assembled his book from letters that he had previously written to his naturalist friends, and presented each chapter, just as the letters were written. Where letters did not exist, (for example in terms of general introduction), he was obliged to invent them.[99] Knapp, as his disciple, was spared this awkwardness, being able to plan his book from start to finish on the outline set by his hero, but as a single complete and polished work. It was in effect one long letter, almost (in 21st century parlance), a blog, taken from the private notes that he had been keeping. It was almost certainly modesty that drove him to launch his *Journal of a Naturalist* into the world anonymously and never once to reveal the name of the village he described, just as he had previously published *Arthur* and all the *Time's Telescope* articles, without claiming authorship. He confirmed his reluctance to be named in later years, stating that he was 'a person who avoids all publicity' and revealing that he had 'uniformly declined becoming the editor or reviewer of several works, flattering as the offer has been, – from a perfect sense of incapacity on the one hand and aversion to notoriety on the other.'[100]

Knapp began his manuscript as his hero Gilbert White had done forty years earlier, by briefly introducing his parish (though not by name) and the area in which he lived. He gave a brief insight into the area's history, geology, farming and land use, especially the growing of potatoes, teasels and dyer's broom, crops which provided the local people with their livelihood in the warmer months. Similarly, he described the profits made by adults and children in the winter, breaking the local limestone for use on the roads. 'A man or healthy woman can easily supply about a ton in the day; a child that

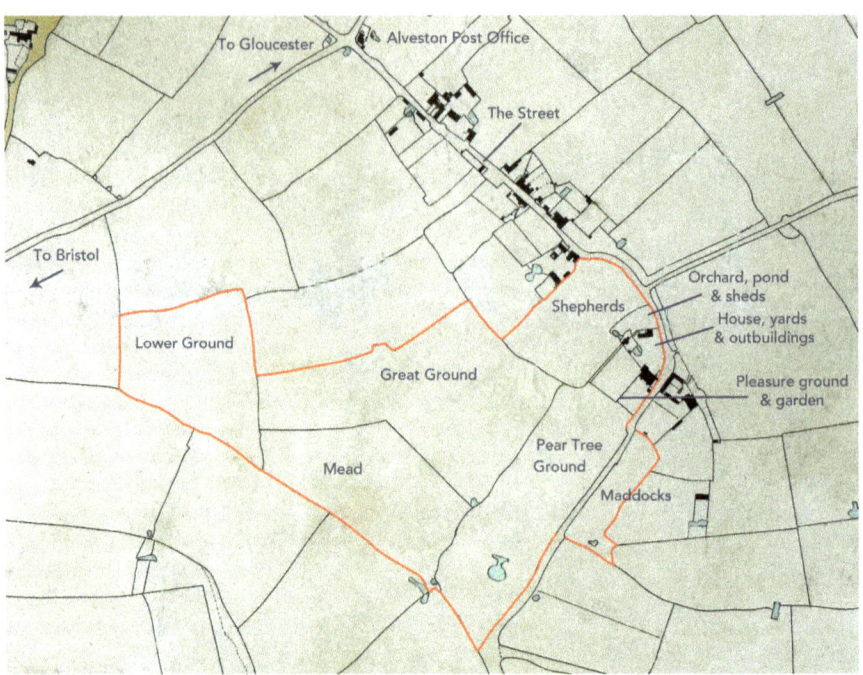

10/1 Part of a map of Alveston made by Henry Moore, surveyor, in 1829, annotated to show Knapp's house and land, as listed and described in the 1841 Tithe Apportionment for the parish.

goes on steadily, about one third of this quantity; and as we give one shilling for a ton, a man, his wife, and two tolerable-sized children, can obtain from 2s 8d to 3s per day by this employ for the greater part of the winter.'[101] As with every subject that he recorded, his enquiring mind led him to look far deeper into detail than any casual passer-by might have done.

In the case of limestone, he described how he had chemically analysed two samples from separate local quarries, after it had been burned in a local kiln to make fertilizer for the land. His purpose was to settle the origin of the rocks. He set out their composition as he found it, adding 'these two specimens so clearly prove that the original materials were derived from the deep, that no further arguments need be advanced.' The first sample he surmised was 'the mountain limestone of Werner; the latter a variety of dolomite.'[102]

It should not be thought from this that Knapp's manuscript consisted simply of dry factual analysis. Far from it. His observations throughout were accompanied by all manner of anecdotes. One of the most memorable was a story frequently quoted in newspaper reviews, after his book had been published. It was an account of an unfortunate travelling man, who settled down for the night near one of the local limekilns, to take advantage of the

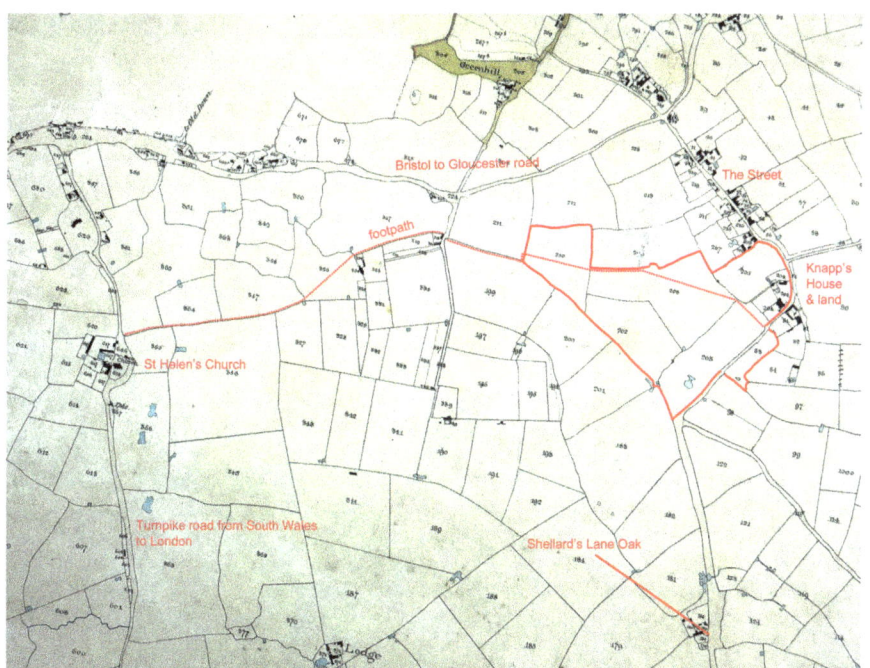

10/2 The footpath leading from Knapp's house to Old St Helen's church. The ancient "Shellard's Lane Oak", a favourite tree of his, stood to the south of his house beside the entrance to Hayward Farm.

warmth. He placed his feet on the stones that had been loaded onto the kiln to burn throughout the night and fell deeply asleep. As the night wore on, the fire rose, eventually consuming one of his feet completely and calcinating the bone of the leg above the ankle. He was found the following morning by the kiln-man, apparently suffering no pain, but concerned by the disappearance of his shoe. When he put his leg to the ground, his tibia 'crumbled into fragments, having been calcined into lime.' He did not survive. But it is typical of Knapp that he was not so much taken by the singularity of the story, but how 'perfectly insensible the human frame may be to pains and afflictions in peculiar circumstances; and that which would be torture if endured in general, may be experienced at other times without any sense of suffering.'[103]

Just as Knapp was able to analyse limestone chemically, so he was capable of analysing soil, though he could not see much advantage of doing so, because he found that ploughing changed the soil's constituent parts and each year yielded different results. Nevertheless, he was asked to do so by three of his neighbours and he duly obliged them. He described the practical difficulties that he himself faced, when attempting to alter the profitability of his own land, both in the improvement of the soil and in the introduction of new grasses. This last was a subject on which he was a considerable expert, but even so, he never succeeded in preventing the local grasses from re-establishing themselves, within a very short period. When he walked the long footpath from his house to the church, he passed a series of deep natural pits in the ground, some of which he said, 'present dark and frightful chasms.' The country people knew them as 'whirly pits,' but to Knapp inevitably, their interest lay in how they had been formed by nature and why.[104]

Knapp used his microscope for all manner of purposes, from determining the reason for decay in plants, to noting the beauty of pores in roots, stems and branches and examining the fur of animals and the scales of fish. Indeed, he had to restrain himself from listing the differences between each animal's fur that he had found. 'In a mere sketch like this,' he wrote 'it would conduct me infinitely beyond my intentions, to enumerate the many varieties of hair that are rendered manifest by the microscope; but three of four many be mentioned.'[105] His home became a miniature laboratory, where he could work on everything that caught his imagination. 'We investigators,' he wrote, 'endeavour to find a reason and a cause for all things.'[106] Collecting birds' eggs, for example, was not enough for him. 'The elegant blue that distinguishes the eggs of the firetail and the hedge sparrow, though corroded away, is not destroyed by… muriatic acid,' he recorded. 'The blue calcareous coating of the thrush's egg is consumed; but the dark spots, like the markings upon the eggs of the yellowhammer, house sparrow, magpie, &c., still preserve their stations on the film, though loosened and rendered mucilaginous… These circumstances seem to imply, that the colouring matter on the shells of eggs does not contribute to the various hues of the plumage.'[107] Fascinated by the quantity of egg-shell that a domestic chicken must produce, when laying some two hundred eggs a year, Knapp analysed a shell and concluded that 2 ¾ lbs of carbonate of lime per chicken was required.[108] Struggling to understand the life-cycle of the goat moth, and in particular to determine the reason for the unpleasant odour it exudes, again he used his microscope, noting that he found the only way to obtain adult moths was 'the often-tedious process of feeding the larva, and waiting for its change.'[109] He described, amongst many other experiments, one to measure the temperature of the soil at a depth of three inches, another to establish the amount of moisture exhaled by grass when compared with that of potato leaves or turnips, another microscopically to examine the woolly fibres in morells ('which in agarics would be flesh'), and finally one to establish which creatures ate the seeds of nidularia by microscopically examining their faeces.

Meteorology was another of Knapp's interests and he was at pains to record the unusual.

We are naturally solicitous [he wrote] to look back upon seasons remarkable for atmospheric phenomena and compare their results with that which is passing before us, though we may be fully sensible, that no conclusions can safely be drawn from them, a variety of circumstances not known, or not comprehended, combining to produce results beyond our means of calculation. There have been times when such recollections brought no pleasure with them by displaying the injuries and sufferings that hurricanes and floods have occasioned; and thus we who were witnesses of the distress occasioned by the lamentable rains of 1793, and the several successive years, when every wheat-sheaf presented a turf of verdant vegetation, cannot recollect it without sorrow, or ever forget that famine in our land. Yet it is amusing, on some occasions, to note the extremes of weather that our island has experienced; for though in general our seasons passed away without any very considerable dissimilitude, still we have known periods of great irregularity, drought or moisture, cold or heat. The freezing of great rivers, with the roasting of animals and passage of carriages upon the ice, our calendars and diaries relate; but instances of an opposite temperature, affording less striking events, are not so fully detailed as might be wished....

But perhaps the year 1825, taking all its circumstances, is the most extraordinary to be found in our annals. The winter of 1824-5 had been mild and wet; the ensuing spring dry, but with keen winds and frosty mornings, which greatly injured the fine blossoms, that appeared on our fruit trees; and the continued and profuse nightly fall of the honey-dew was quite unusual: the leaves of the oak, the cherry, and the plum, were constantly smeared and dropping with this clammy liquor, which, falling from the foliage on the ground, blackened it as if some dark fluid had been spilled upon it; the leaves of most of our stone fruits curled up, covered with aphids, and became deciduous; and the young shoots were destroyed by the punctures of the insects that clustered on them. This honey-dew continued to fall till about the middle of July, affording an abundant supply of food to multitudes of bees, moths, and other insects, which swarmed about the trees. We rarely begin cutting our grass before the first week in July; but in consequence of the heat of June in this year, it was so drawn up, that much hay was made and carried by the 20th of June, which commonly is not accomplished till August. Our crops on good ground were considered as fair, though in general the chilling season of May had occasions a deficiency; but all our clover crops and artificial grasses were harvested in the finest order, producing good size ricks and mows; yet their bulk was delusive, the provender cutting out light and strawy. The heat and drought continued with very partial and slight showers of rain all June and July; nor had we anything like serviceable rain till the end of August. In consequence our grasslands were burned up, and our fields parched, presenting deep fissures in whole parts. The heat was unusually distressing all day; and the evening brought us little or no relief, as every wall radiated throughout the night the heat it had imbibed from the torrid sun of the day. Our bedroom windows were kept constantly open, all apprehension from damps and night airs, which at other times were of the first consideration, being disregarded; a cooler temperature, however obtained, was alone required; and we lingered below, unwilling to encounter the tossings and restlessness that our heated beds occasioned. Our wainscots cracked, furniture contracted and gaped with seams; a sandalwood box, which had been in use for upwards of twenty years in dry rooms, shrunk and warped out of all form; a capsule of the sandbox tree (hura crepitans), which had remained in repose over a shelf above the fireplace for an unknown length of time, now first experienced an excess of dryness, and exploded in every direction; door frames contracted, windows sashes became fixed and immovable. These are trifles to relate, but yet they mark the very unusual dryness of the atmosphere....[110]

As for the usefulness of keeping detailed records, Knapp insisted that 'the recording of events is the province of the naturalist; and perhaps occasionally by comparing existing circumstances with past events, something approximating to probability may be obtained... but these things pass away, unless recorded; and though we may resort to the oldest memory for evidence, yet memory is oblivious, often exaggerative, and cannot safely be trusted.'[111]

The principal purpose of Knapp's book was of course to record the flora and fauna of his parish. And this he did, dealing section by section with wildflowers, trees, grass, birds, insects and indeed everything living or growing that came to his attention. And he missed very little. But his were not simple descriptions of species for identification purposes. What he recorded were the particular quirks and habits of each plant and living thing, based on familiarity gained by watching them over many years. His unshakeable Christian faith inevitably became part of the many conclusions he drew from his work and the descriptions that he gave. 'I would advocate the cause of all creatures, had I the privilege of knowing the excellency of them; not willingly assigning vague and fanciful claims to excite wonder, or manifest a base pride by any vaunt of superior observation; but when we see, blind as we are, that all things are formed in justice, mercy, truth, I would tell my tale as a man, glory as a Christian, and bless the gracious power, that permitted me to obtain this knowledge.'[112] He was almost certainly unaware that he was so well-respected for these views, that following the publication of his work, he came within a hair's breadth of being chosen by a sub-committee of the Royal Society, to become one of eight distinguished men to contribute to an eight-volume treatise under the will of Francis Henry Egerton, 8th Earl of Bridgewater.

Bridgewater, who was an eccentric clergyman living latterly in Paris, died in 1829. In his will, he left the astonishing sum of £8,000 to the Royal Society, to be used to 'write, print and publish' one thousand copies of a work 'on the power, wisdom and goodness of God, as manifested in the creation.' The Society appointed Davies Gilbert FRS (1767-1889) its President, Charles Blomfield (1786-1867), the Bishop of London, and William Howley (1766-1848) the Archbishop of Canterbury, to recommend suitable authors. It is recorded that all three initially favoured Knapp as one of the eight, but Blomfield 'upon looking again at Mr Knapp's book' at length decided against him. The reason, according to Gilbert's nephew John Enys, was that *The Journal* had been written for popular consumption, rather than as a scholarly work.[113]

Knapp can certainly be regarded as a pioneer conservationist. This is a theme which runs throughout his work. Of particular concern to him was the increase in population. 'Mankind appears to be progressively increasing,' he wrote; 'it was an original command of his Creator, and the animals domesticated by him, and fostered for his use, are probably multiplied in proportion to his requirements; but we have no reason to suppose, that this annual augmentation proceeds in a proportionate degree with the wild creatures upon the surface of the globe; and we know, that many of them are yearly decreasing, and very many that once existed have even become extinct.'[114] Again he said: 'Some of our birds are annually diminishing in numbers, others have been entirely destroyed, or no longer visit the shores of Britain. The increase of our population, enclosure, and clearage of rude and open places, and the drainage of marshy lands, added to the noise of our firearms, have driven them away, or rendered their former breeding and feeding stations no longer eligible to many, especially of the waders and aquatic birds. The great Swan-pool, near the city of Lincoln, on which I have seen at one time forty of these majestic creatures sailing in all their dignity, is, I am told, no longer a pool; the extensive marshes of Glastonbury, which have afforded me the finest snipe shooting, are now luxuriant corn farms; and multitudes of other cases of such subversions of harbour for birds are within memory. An ornithological list made no longer ago than the days of Elizabeth would present names of multitudes now aliens to our shores. The nightingale was common with us here a few years past, the rival songs of many were heard every evening during the season, and in most of our shady lanes we were saluted by the harsh warning note of the parent to its young; but from the assiduity of bird catchers, or some local change that we are not sensible of, a solitary vocalist or so now only delights our evening walk.'[115]

He especially disliked wanton killing and cruelty. Like so many today who have enjoyed field sports in their youth, but have turned conservationist in later years, he explained

> partial as I am to the habits and all the concerns of the country, I regret to say, that rural amusements, connected as they commonly are with the creatures about us, are frequently cruel; and that we often inconsiderately, in our sports, are the cause of misery and suffering to such as nestle around our dwellings, or frequent our fields, which, from some particular cause or motive, become the object of pursuit. I say nothing of the birds known as game, as perhaps we cannot obtain them by less painful means than we are accustomed to inflict, and the pursuit is frequently conducive to recreation and health; but the sportsman's essaying his skill on the swallow race, that 'skim the dimpled pool,' or harmless glide along the flowery mead, when, if successful, he consigns whole nests of infant broods to famine and to death, is pitiable indeed! No injury, no meditated crime, was ever imputed to these birds; they free our dwellings from multitudes of insects; their unsuspicious confidence and familiarity with men merit protection not punishment from him. The sufferings of their broods, when the parents are destroyed, should excite humanity, and demand our forbearance. But the wheatear, in an unfortunate hour, has been called the English ortolan; and is pursued as a delicate morsel, through all its inland haunts, when hatching and feeding its young, the only period in which it frequents our heaths. I execrate the practice as most cruel: their death evinces no skill in the gunner; their wretched bodies, when obtained, are useless, being embittered by the bruises of the shot, and unskilful operations of the picker and dresser. No, let the parental duties cease, and when the bird retires to its maritime downs, if doomed to suffer, the individual dies alone, and no starving broods perish with it. I supplicate from the youthful sportsman his consideration for those most innocent creatures, the summer wheatear and the swallow.'[116]

Of the steep decline of specific species, Knapp wrote movingly.

> That sober coloured, but beautifully feathered bird, the Water-Rail, is yet found with us, lingering in its ancient abodes, but in such diminished numbers, tho' always thinly scattered, that we must shortly expect it wholly to disappear and

no more seen with us: as man multiplies and advances, waste places are digged up, the marsh is drained, the water course is cleared, and that which secreted the wilding of the field, supplied them with food, or harbour for their young are removed as injurious to the requirements of the human race, and the few which are not destroyed are driven to other retirements: man and independent nature are ever hostile, warfare for sport or spoliation unremitting, his power and his arts become irresistible, flight alone remains, until they are no longer found.

To this Knapp added

killing everything, and cruelty are the common vices of the ignorant; and unresisting innocence becomes a ready victim to prejudice and power .... all created things were equally the objects of their Maker's benevolence and care; the worm that creepeth, the beast that perisheth, deserve our consideration, and claim, from human reason, mercy and compassion.[117]

11

With his lengthy manuscript mostly complete, Knapp decided to chance his arm. On 22nd July 1827 he approached John Murray (1778-1843), the celebrated London publisher of Lord Byron, Sir Walter Scott, Jane Austen and many other famous names. He wrote a polite unsolicited letter from Alveston, saying that he had nearly completed a work on a popular subject. He explained that he would be pleased to have an evaluation of it from someone less biased than himself. Not waiting for a reply, the following day he sent half of his draft manuscript to London, accompanying it with a note to the effect that he had an attractive illustration drawn by one of his family, that he believed would provide an ideal frontispiece.[118]

It took Murray almost two months to reply, which must have made for a tense wait. But reply Murray did, requesting sight of the second half on the manuscript. Knapp rather gingerly indicated that if Murray was thinking of publication, he would be happy to go ahead with it, but explained that in the light of the awful fate by fire of his two previous books, he could not take on the financial risk himself. He wondered if Murray might wish to buy the copyright outright, sell it to a bookseller or perhaps publish the work at his own expense and split any profits with the author.[119]

11/1 Engraving of John Murray (1778-1843) by E.F. Finden after H.W. Pickersgill. Murray published the works of Lord Byron, Sir Walter Scott and Jane Austin and many others. Between 1829 and 1838 he published four editions of John Knapp's Journal of a Naturalist.

Another tense month passed, during which Knapp wrote twice to Murray, but received no answer. In mid-October he wrote a third time, insisting that if the project was to come to nothing, then Murray must return his manuscript without delay. This precipitated a reply, but certainly not one that Knapp had expected. Murray, it seems, had sent the unfinished manuscript to the printers, even though it was missing text and was without Knapp's drawing for the frontispiece. Poor Knapp was appalled. He demanded that printing should stop immediately and insisted that the manuscript should be returned for completion and correction. If that was not upsetting enough, Knapp then saw a preliminary advertisement that Murray had placed in a London paper which described the size of the forthcoming book as octavo, when Knapp had specifically asked for it to be quarto, to match the few surviving copies of *Gramina Britannica* and of *Arthur*.[120]

Despite his objections, still the project did not run as smoothly as Knapp might have wished. Murray offered 100 guineas for the copyright, which Knapp declined, believing it to be derisory. Instead, he tentatively accepted an offer to share half of any profit. In his letter of acceptance, Knapp mentioned what seems to have been ongoing problems with his health but once again did not specify the cause.[121] When he was sent copies of the proof illustrations in late December 1827, he approved them, but expressed himself disappointed that in the letterpress proofs, the Latin names of the plants and animals had not been placed beside the text, rather than in it, as he had requested. He felt that there was insufficient space between each section and that the page size appeared to be small octavo, rather than the slightly larger Imperial, that he had been prepared to settle for.[122]

Nearly six months then passed with no news from London. Exasperated, Knapp dispatched a sharp reproof. 'What has become of our Journal of a Naturalist?' he wrote, 'having commenced the work so long before Christmas, I am at a loss to understand the cause of the delay in the publication – surely Sir, we ought not to keep back a work of this nature apposite to the season in which we have entered…'[123] But then, on or about July 23rd, 1828, his twelve-year-old daughter Charlotte Henrietta died. She was the second of his daughters to die in childhood and it took him some three months to face a return to his work and to send back the corrected letterpress proofs to London.

It was not until February 1829 therefore, eighteen months after the project began, that Knapp finally heard that Murray's chosen printer, Thomas Davison of Whitefriars (d. 1831), had completed his task. Knapp was delighted with the result, though he grumbled that its production had been unnecessarily slow. He requested that a copy each should be sent speedily to the two London learned societies of which he was a fellow, the Antiquaries and the Linnean, a copy to J. Millard of Camden Place, a copy to Inigo Thomas of Ratton,[124] a copy to Miss Knapp of Shenley[125] and four copies to his home at Alveston. He added hopefully that should a second edition be required, his journal was 'not exhausted of its contents.'[126]

On 12th February, John Murray published an advertisement in the London newspaper, *The Sun,* announcing 'In a few days will be published in post 8vo., with Plates, THE JOURNAL OF A NATURALIST – Plants, Trees, and Stones, we note, Birds, Insects, Beasts, and many rural things.' And then the excitement began.

In the next weeks the book received a series of very favourable reviews and was quoted in papers as far afield as Belfast.[127] *The Monthly Review* devoted seventeen pages to its assessment, beginning in the most flattering tones.

> This book reminds us forcibly of the line in Gray's Elegy – 'Some mute inglorious Milton may here rest,' inasmuch as it is evidently the production of a man of genius, who has spent the greater part of his life in a village 'situate,' as he tells us, 'upon a very ancient road, connecting the city of Bristol with that of Gloucester;' and many such men there are, we cannot doubt, in our villages, whose journals were never written, or at least, never published, though they might have done, as in the instance before us, high honour to the literary and scientific character of England.[128]

The Spectator in its review, before gently mocking contemporary education, began

> The country is so dull! There is so little society, and the walks soon grow tiresome; and when the weather is bad, how dreadful it is to be confined to the house without any amusements! Ignorance and blasphemy! Or if not blasphemy, only because ignorance is its excuse. To live in the midst of God's manufactures, and to find no object to rest the attention upon – to detect no lesson, no object of imitation, no infinite matter of perusal is blindness which, if it were wilful, would be sinful.

It concludes

> Such books as the Journal of a Naturalist show us what people ought to be taught, and how they might, by a little previous instruction, be enabled to pass their lives in contentment, usefulness and the practice of benevolence. If

the nature of the world and the creatures in it, the processes of the elements, the occupations of animals, and the adaptation of their forms to the several conditions of life, and a multitude of other branches of the study which it would be too tedious to enumerate, were made the subjects of early education, that world, now so dull and objectless to many individuals, would appear as it is, bursting with sources of delight, full to overflowing with interest and instruction. We should hear no more of the dullness of the country – no more repining for heated rooms and bare bosoms, the deafening noise of crashing concerts, or the lifeless joys of a cold or else suffocating playhouse...

...These few remarks will prove the hearty welcome we are prepared to give to all books worthy of bearing the title of the 'Journal of a Naturalist.' The most charming book of the kind we are acquainted with, is WHITE'S Natural History of Selborne; a book that Mr. MURRAY (or any other publisher) would do well to print in a neat and portable form, either to match with his beautiful edition of FRANKLIN'S Voyages, or this Journal of a Naturalist.

This work is avowedly suggested by the book we have spoken of; and we are not niggardly of our praises when we say that it is worthy to be joined with it. It is full of very curious and very pleasing information. We have read it with great satisfaction; and glad should we be if every district possessed a man capable of producing its fellow, and willing to do so. The spread of this work will contribute to the increase in numbers of such individuals: we recommend it with all our strength to all those who love the calm and peaceful contemplation of nature.'[129]

The Quarterly Review, a publication admittedly owned by John Murray, devoted twenty-five pages to its examination of the book, quoting large tracts from it and discussing each one in detail. It opened with the paragraphs

We believe very few books, on the subject of Natural History, have met with such unqualified praise from those whom the contemplation of the various objects of nature can afford rational amusement, as the 'Natural History of Selborne,' by the Rev. Gilbert White. The author of the little volume with the modest title now before us, admits that, in the collection of its own materials, he had this interesting book in his eye; that the perusal of it early impressed on his mind an ardent love for all the ways and economy of nature; and that he was thereby led to the constant observance of the various rural objects with which he was surrounded. But, as he observes, many years have passed away since the publication of Mr. White's amusing book, without its being followed up by any other bearing the least resemblance to it.... This is unquestionably true;

and we can assure him, that a close perusal of the two productions has satisfied us that they do not in the least interfere with each other. Both are well suited for a country library – both are almost sure to awake that degree of curiosity which promotes inquiry, and stimulates to the investigation of nature and the confirmation of truth. But the 'Journal of a Naturalist' is particularly calculated for this end. It brings home to its readers the structure, the splendour, and the utility of various species of the vegetable part of the creation. The peculiar habits and economy of different quadrupeds, birds, reptiles, and insects, that are found in and about a certain district of country, and the observations and reflections which the contemplation of the several objects gives rise to, are so just and admirable that they cannot fail to gratify the curiosity of the reader, and, at the same time, command his approbation and sympathy. In short, it is a book that ought to find its way into every rural drawing-room in the kingdom, and one that may be safely placed in every lady's boudoir, be her rank and station in life what they may; which is more than we can venture to say, with regard to Mr. White's volume. That good old clergyman, in the simplicity of his heart, sometimes spoke of matters in a way not exactly suited to female delicacy; but the most fastidious eye may, without fear of offence, consult the 'Journal of a Naturalist.'"[130]

Other equally positive reviews appeared in *The Athenaeum, The New Monthly Magazine* and *Blackwood's Magazine*.[131]

The result was a runaway success. Within days, another publisher had not only offered Knapp the opportunity to edit and revise Gilbert White's *Selborne*, but Thomas Pennant's son had offered Knapp the opportunity to edit and revise his father's celebrated *Zoology*. Knapp turned both down, he told Murray, because he feared that new editions of those works would affect the sales of his own.[132] Murray, realising that he had a modest winner on his hands, upped his offer for Knapp's copyright to three hundred guineas, but Knapp, sensing still greater success, stuck firmly to their original agreement to share profits.[133]

It is not known how many copies of the first edition Murray printed, but only eight weeks after its launch, a second edition was required. By the end of May 1829, Knapp had submitted one hundred and forty corrected pages and four additional drawings. He was overwhelmed with the congratulations he was receiving. 'You may smile,' he wrote to Murray, 'yet it is nevertheless perfectly true – that I hear so much said, which I never wish to hear, flattering as it may be, that it quite confuses a plain countryman like myself.'[134]

The second edition was completed by August, this time printed by William Clowes (1779-1847) of Duke Street, off Stamford Street, Blackfriars

11/2 First editions of The Journal of a Naturalist *include a fold-out frontispiece of the Shellard's Lane oak. Knapp attributed the drawing to a member of his family, but it was almost certainly his own work.*

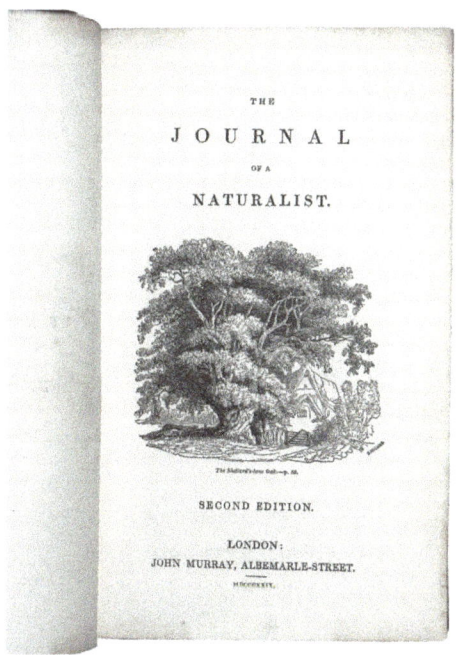

11/3 In second editions of The Journal *a small woodcut of the Shellard's Lane oak replaced the larger fold-out lithograph. It was engraved by the self-taught artist Samuel Williams.*

Road, London. It differed in content from the first edition not only because it included extra illustrations and added text, but because the frontispiece used in the first edition, the fold-out aquatint of the Shellard's Lane oak, taken from a drawing by one of Knapp's family, was omitted. It was replaced in the second edition by an elegant vignette of the same scene, engraved by the prolific self-taught artist and wood engraver Samuel Williams (1788 – 1853) and placed centrally on the title page. Knapp was delighted with the vignette as he had a particular fondness for the tree, which grew in the lane which runs beside his house. He described it later as 'a favourite tree – [which] has private associations connected with it.'[135] He asked Murray for six extra copies of the print and eight copies of the new book itself, specially printed on large paper, 'done up in boards with green calico.' He asked for one of these to be sent to John Walmesley, at the Circus, Bath and one to the Rev. William Kinsey at Trinity College, Oxford. The other six were to be delivered to him at Alveston.[136] One of these six extra-large books has recently been discovered in America and repatriated, inscribed on the front free end paper in Knapp's distinctive hand 'Presented by the author to his beloved and affectionate daughter Catherine Mary.'

Murray was an exceedingly busy man and hugely successful though *The Journal* was proving to be, it perhaps did not come all that high on his list of priorities. He was not good at keeping Knapp up to date with sales data and one senses that Knapp did everything he could to be patient, but from time to time could no longer bear the suspense. So it was that in November 1829, four months after the launch of the second edition, Knapp wrote to Murray saying that he had heard that sales had been very extensive and asked whether a third edition was planned. He was preparing further text and to avoid any delay, asked when it might be required. But then tragedy struck again. Only five or six days later, death claimed yet another of his daughters, eighteen-year-old Catherine Mary. Knapp wrote no more letters to London until January 1830. Then, perhaps because of a return of his chronic illness, perhaps because of the loss of his child or possibly both, he then sent two searing letters to Murray.

The first acknowledged receipt of one hundred and fifty three pounds, twelve shillings, but questioned the seventeen guinea charge against the profits, apparently paid to a professional indexer. Knapp had, he said, already provided a better index. He expressed dismay that less than two thousand copies had been printed, which left him 'but a small remuneration.' He softened his criticism by adding a grateful note, but with a sting in its tail: 'I trust Sir that you are in all respects satisfied, as I certainly am with your concern in this affair, but I cannot but think the author's profits trifling indeed.'[137]

Ten days later he had not received a reply, and so in a second letter, he let go all restraint. It was the kind of letter that should have been written to remove extreme frustration from its author's system, but the following day, torn up. He was writing, he said, on the advice of

> a literary friend, nearly the most approved, & most successful author of the day, perfectly intimate with all the arcana of the book trade – to request that you be so kind as to overlook the extraordinary return that has been sent to me regarding the publication, & receipts of the Journal of a Natt: – my own

sentiments are that some vast error has led to it and that it will be set right under your examination. I am told in every quarter that it has been the most decided hit of the season, but if so, the author has been most effectually missed! – the public are charged 15s and the author gains a trifle above 2!! Near a third of the impression disposed of the retail trade to allow it a profit of 5s 4d the volume – ! and the remainder given to the vendor at a price to afford him 4s 4d.! but the author is to have only less than one 7th a few pence like the pauper at the gate!! The charge for those simple outlines – excessive – that for the index I must remonstrate against, as a better was provided – a charge of £450 – for a small impression of a small 8vo work –

Indeed Sir, this will form a prominent example of unsuccessful authorship – £500 had been offered for the work – but it was valued at £1400 – Mr Hook obtains £1500 for every series of his 'sayings and doings' – and the unfortunate author of the most approved and lauded work of the season – £153.12s.9d – and even that paltry sum to be received in the distant perspective of Three Months! – about enough to pay the postage for the congratulations of success.'[138]

John Murray replied in an exemplary manner. He took the greatest trouble to lay out in simple terms the agreement that the notorious Theodore Hook had with his publisher[139] and show that in percentage terms, Knapp's remuneration should shortly match it. He set out the cost of advertising. He explained that the charge for the index included the charge for the editing and revision of Knapp's text by 'a person of great respectability & employed in this line for years by the first Authors,' who had asked £21 for his work, but, under pressure from Murray, had accepted £17. As for the engravings, he explained that they had been executed 'upon Steel, by which the expence of re-engraving them over again for every succeeding Edition will be saved.'

He ended his letter most elegantly.

I believe I have now replied to all your objections – my Accounts of the Items charged in the Statement are open to any one. Now I must add that the letter received from you is one of the rudest I ever received in the whole course of my business – but it is so completely at variance from the gentlemanly character of all your preceding communications – that it is merely the promptings of some injudicious friend – & conscious as I am of not deserving it – you will see that I am not in the slightest degree displeased at it. –

I should have written several days ago, but your Son called soon after the arrival of your Letter, and thinking he might desire to look into the Accounts, I made an appointment for him for the next day at 2, but he has never called.

I do not like to let another post off without writing to you. – If you will name any reasonable sum including the note I lately sent for the Copyright, to put an end to accounts, I shall have pleasure in entertaining it. In the meantime with all good humour, I remain, my dear Sir, Your faithful Serv. John Murray.[140]

Within days, a somewhat humbled Knapp replied that he was glad that his outburst had cleared the air and apologised for what he described as 'unintentional offence.' Thanks to Murray's decency, their relationship was repaired and within eight weeks or so, in March 1830, when the stocks of the second edition began to run low, work began on a third edition, with Knapp sending in yet more fresh material. Murray suggested that perhaps it would be more satisfactory to publish the new content in a second volume, but Knapp, in acknowledging a sizeable payment of £570.15s.3d which he had just received, regretted that this would be impossible. His notes would need a great deal of work to put in order, he said and 'my anxiety from domestic indisposition and my own state of health would prevent my attempting any arrangement.'[141] Though neither publisher nor author would have been aware of it, this was perhaps the highest point that *The Journal* would reach in its meteoric rise. Meanwhile Knapp's health was in decline. Even though references to *The Journal* and quotes from it appeared frequently in newspapers throughout the country, sales, in Great Britain at least, gradually began to slow.

The third edition was ready for sale by July 1830, again printed by William Clowes. Knapp chose to top and tail it with two newly written verses. The first, printed on the reverse of the title page, takes the form of an epigraph. On the one hand it betrays Knapp's nervousness at the launch of yet another edition, but on the other reveals his delight and surprise at the success of his work to that point:

> Go forth, my little bark, again, and risk
> Once more thy fragile form upon the world's
> Unsteady surge. Rude gales and currents may
> Be found to meet thee on thy way, and check
> Thy progress to a ready mart: yet steer,
> If happily though canst, thy course – light is
> Thy freight, nor rare; and few I deem'd would prize
> Such merchandize as thine, nor willing aid
> The foundering on the wave; but thou hast sail'd
> In tranquil seas – warm, sunny gleams have cheer'd
> Thee on; and friends – kind friends! – were seen,

> Who slighted not thy ware, all rustic as
> It was. Yet bear thee steady on thy course;
> And chance some wandering trafficker may come
> To seek a sample of thy stores, and find
> The lading to its invoice true.[142]

At the end of the edition, Knapp added the autobiographical poem that he had written, dedicated to the goddess Flora and named 'The Progress of a Naturalist.' Quotations from it have been freely used to illustrate the earlier chapters of this book and the entire poem, which runs into six pages in the third edition, may be found on pages 271 – 275, in the second half of this book. It is certainly symptomatic of the way that Knapp was feeling during the months leading up to the third edition. Indeed, the final verses might have been written by a man much older than Knapp's sixty-three years:

> But time a change to all must bring,
> And wear the form of mortal thing,-
> The ardour of the fire decay
> As its best fuel wastes away,
> And things are lost, and heeded not –
> This is forsaken, that forgot.
>
> And where is he, that infant, fled,
> Which wreath'd the wild flower round his head?
>
> That aged man reposing there,
> On cushion soft in elbow chair,
> Was he, the slender, musing boy,
> Which play'd with Flora, as his toy,
> And his early passion true,
> Fair nymph! Still often thinks of you!
> He sits in silent humble bower,
> And marks the border's varied flower;
> Exulting to his neighbours shows
> The bright Geranium's vivid glows;
> Some new Carnation from its bed
> Lifts by his care her speckled head;
> And other simple pretty things
> As harmless recreation brings;

Tells of the seasons that are flown,
And waits in patient hope – his own. –

Yet one spray more – one wreath would crave,
Such palmy boughs as angels wave;
In that fair pasture he would stray,
Where faith, where virtue, point the way,
With amaranthine crown to sing
Hosannas sweet to Flora's King.'[143]

This is not to suggest that Knapp had given up on his writing career, for along with the new additions to the third edition, he had sent Murray another manuscript of what he described as 'of entirely original matter.' When two months had passed and no comment on it had been received from Murray, Knapp wrote to Murray's son, asking if his father was away from home and if he was, whether he had shown any interest in publishing the new work. If not, he asked for his manuscript to be returned, because he was certain that he would find other publishers to take it on. He explained that it was named 'The Hostel &c.' It was in fact never published, though the manuscript still survives.[144] It is described in some detail in Appendix 2 of this book, where poems which form part of it, are transcribed.

It was at this period too that Knapp sent a brief piece that he had composed to the Bristol Horticultural and Botanical Society, to be read in his absence. Although in it he speaks specifically about horticulture, he reveals a truly modern understanding of the processes of nature, that foreshadows all the principles of 'rewilding' and ecology that are so much in vogue today. It was billed as being 'received from a friend and Well-Wisher… to whom (as we are to understand) the public are indebted for that interesting work, *The Journal of a Naturalist.*' It read:

> Perhaps, Gentlemen, you will permit me to dwell a little upon the results that may arise, not generally contemplated, from cultivation, more especially from Horticulture, which remarkably instance the universal tendency of nature to supply and occupy; and from what simple and natural causes life and being may originate. We may place ourselves upon a wild and open heath, where little can be seen to invite the settlement of man, nothing heard but the whistling of the passing wind, the murmur of some wondering bee; – we trench the soil, plant the rose, the bean, the apple, and they become immediately invested with numerous races of aphides, – then comes coccinellae or lady-birds, to consume

these creatures; tribes of little birds arrive to seek subsistence too: – we set the current, the raspberry – the black-bird, red-breast, and white-throat appear: the cabbage springs up, – now the butterfly with its larvae abound, troops of Wasps, flies, spiders, ichneumons, and others hover around, some to feed, others to nidify there. The Grove arises under our culture, warding the winds from our nurslings, and where at first crawled the beetle or the worm, we now find numerous races nourished, and nestling in the foliage. The woodland warblers come, the song of the Nightingale is heard – there is the moss, the lichen, and the fern, each with some creature pasturing upon it – a voice seems uttered, and creation obeys the mandates – the simple culture of man calling, from we know not where, the race of the earth – the heath becomes a paradise of flowers, fruits, and shades – harmony and animation abound – we hear no more the icy whistle of wind – the very atmosphere is changed – population follows – the entomologist – the botanist – the naturalist – human manners – human customs all succeed – and would have been unknown but for the wonder-working tool – the Spade. – The labour of man manifesting that universal link that binds all nature one interesting union, and presenting this fact, that as there is a law to regulate the individual, there is another providing for the succession and provision of the whole.[145]

12

WHEN THE THIRD edition of *The Journal* finally became available, the second edition had still not quite sold out. Knapp felt that it might be best to keep the new edition back until all copies of the second had gone. By October 1831, some two months after publication, sales proved steady but no longer the runaway success that they had been two years earlier. Knapp received a payment of £169.4s.9d which was good money, but he was disappointed. The problem it seems was firstly that the peak of enthusiasm for the book had passed and secondly that its remarkable success had spawned a plethora of similar works, rather in the way that the meteoric success of Dava Sobel's modest, pocket-sized book *Longitude* published in the 1990s, immediately inspired a whole new genre of 'popular science' lookalikes. In Knapp's case, Whittaker, Treacher & Co published Robert Mudie's *The British Naturalist* in 1830, in an almost identical format to *The Journal*. They then published *Researches in Natural History*, claiming in their advertisement that Knapp himself had recommended it as 'that most clever and ingenious little work.'[146] 'Howitt's *Book of the Seasons* appeared in 1831. It was praised as a 'pleasant rural companion' in the *London Morning Herald*, which added that 'since the publication of the 'Journal of a Naturalist', no work at once so interesting and instructive as 'The Book of the Seasons' has been submitted to the public.'[147] Even Knapp's own publishing house undermined *The Journal's* sales, by producing Edward Jesse's *Gleanings in Natural History with Local Recollections* in 1832. This began 'I can only hope that my little book may be received with indulgence. Almost every one now writes something: the scribimus indocti doctique was never in greater force:

> Those who cannot write, and those who can,
> All rhyme, and scrawl, and scribble to a man.'

This was a confession if ever there was one that amateur naturalists everywhere were scrambling to climb on Knapp's bandwagon. Jesse even included an epigraph in his work, which mirrored Knapp's. It begins 'Go, little book! And in those lists of Fame / Where strive the mightiest, run thy destined race…'[148] A second edition of *Gleanings* then included 'Some Extracts from the Unpublished MSS of the Late Mr White of Selborne.' Numerous other comparable works on natural history joined this break-neck race for sales. Maria Spratt, having opened and closed her anthology *The Language of Birds* by appropriating some fourteen pages of Knapp's work, at least had the courtesy of describing him as 'one of our most admired popular writers.'[149]

If that was not enough, for the first time, Knapp found his book facing adverse public criticism, combined with what today might be called the 'trolling' of the author. It began modestly in the letters section of the *Monmouth Merlin*, where Charles Sharp of Ross-on-Wye described Knapp's views on American blight in apples trees as 'egregiously mistaken.'[150] Sharp added that in other respects he admired *The Journal* very much, but his letter was probably the first to publish a negative comment of any kind.

In January 1832, Knapp then received a hoax letter suggesting that he had been named of in two scurrilous pamphlets called *Mad Houses* and *Harriette Wilson,* which had been published in London by the pornographer J.J. Stockdale. As it happens, neither pamphlet so much as mentioned him, but Knapp, living so far from the capital was not to know that. He wrote a rather desperate letter to Murray, asking if there was any truth in the report. He hoped against hope that 'a person who avoids all publicity as much as I do might have been exempted from the vulgarisms of the idle.'[151]

In March 1833 came the greatest blow of all to Knapp's self-respect. The well-known agitator, journalist and politician William Cobbett (1763-1835), author of the famous of *Rural Rides,* was apparently offended when he belatedly read the extremely favourable account of *The Journal* in the *Quarterly Review.* He made his feelings painfully clear over eleven pages in 1833 in his self-published *Cobbett's Political Register.*[152] There is a certain irony in the fact that Cobbett was based in Bolt Court, the scene of the fires that had destroyed Knapp's first two publications. It was an address that had brought little pleasure into Knapp's life and was now the source of yet more distress.

Cobbett had come from farming stock and had long campaigned for improvement in the lot of the labouring classes. To this end he published *Cottage Economy* in 1821. He was infuriated by Knapp's praise of an impoverished local labourer (Knapp had called him a 'worthy peasant') who had used the modest wages paid to him to such good effect, that at the time of his death six years later, he was found to have left his widow 'some money, though I know not the amount; two fine hogs, and a flock of 49 good sheep, many far advanced in lamb; and all this stock was acquired on ten shillings a week.'[153]

Cobbett simply did not believe that this was possible and devoted page after page to proving to his own satisfaction at least, that this story was a blatant lie. He hated the fact the labourer had been quoted as addressing his employer as 'his honour' and said so.[154] He hated what he saw as the patronising attitude of *The Journal's* author towards those in what Cobbett described as 'the lower walks of life.' He did not of course know who the author was, because Knapp's book had been published anonymously. This too he hated. But when he searched among the several hundred pages of his copy and managed to find thirteen similarities between descriptions in *The Journal* and those in Gilbert White's *Selborne*, he sprang to the conclusion that the book was not written by a naturalist at all but had been plagiarised from multiple other publications. He suspected, and hinted so, that *The Journal* (which he did not believe had ever been a journal), had been cobbled together by John Murray himself. His was an excoriating review.

12/1 *A likeness of William Cobbett, by Francesco Bartolozzi after a portrait by J.R. Smith. Cobbett's bad-tempered and ill-informed criticism of The Journal of a Naturalist caused Knapp deep distress.*

Knapp was deeply hurt by Cobbett's merciless attack. He did not hold back in a letter to Murray, describing Cobbett's opinion as 'exemplified by that sophistry, and false conclusion against facts, that he is so much in the habit of advancing.' So distressed was Knapp by what he saw as a complete 'lack of truth and respect', that he threw his copy of Cobbett's pamphlet into the fire.[155]

While all these curious events were occurring, another had taken place. It may well be that neither Knapp nor Murray knew anything about it. Certainly, neither mentioned it in their correspondence. In his epigraph, Knapp had portrayed his book as a bark, or three-masted ship, and wished it a fair wind. Fact then followed fiction, for the wind carried his bark across the ocean to the United States of America. There it came into the possession of Carey and Lea, publishers of Philadelphia, who replicated it exactly and offered it for sale.[156] At this period, no copyright laws existed in America that might restrict the republication of British authors. Henry Carey is recorded as having campaigned energetically to keep matters that way, as he and his fellow publishers made a good living from it. Charles Dickens was famously maddened some twelve years later, when he travelled to America to perform his novels, to find them freely available there in pirated editions. 'I am the greatest loser alive by the present law,' he complained.[157] If Knapp's *Journal* was pirated, as it seems it was, and if he knew that it had been, it would have been understandable if he felt similarly angry. But then again, he might have felt deeply proud that while sales in Britain were gently slowing, his book was on offer to a vast new audience on a vast and different continent.

The final curious event to have taken place at this time was the appearance of an advertisement in *The Saint James's Chronicle* of 10th January 1835 and repeated elsewhere. It had been placed by a bookseller named Davis in Chichester. It offered an octavo volume at a guinea, entitled *Reflections on the Bristol Riots in Twelve Sermons By the Reverends Charles and Henry Gray.* The sermons, the advertisement continued, had been 'preached at the parish churches of Almondsbury and Alveston.' The book itself, it informed readers, was 'dedicated to, and with a Steel Engraving, by Finden, from a fine portrait by R. West, of John Leonard Knapp, Esq., of Alveston; at whose request these Sermons have been published.' Knapp was astonished. He wrote to Murray at once explaining that 'the whole is one tissue of falsity, the object for so advertising and the person who is such a fool as to set about it incomprehensible! There is not an idea – nor ever was upon the subject.'[158] No copy of the book has been found, (if one ever existed), nor has a portrait of Knapp by either of the prolific line engravers Edward Francis Finden (1791-1857) or his brother William (1787-1852) yet been discovered.[159]

As the decade drew to a close, Knapp still hoped for a further enhanced edition of *The Journal* and he continued to offer Murray new text, even though towards the end of 1837 nearly two thousand copies of the previous editions remained in the publisher's hands. He was not disappointed, though a little taken aback, when he read an advertisement placed by Murray in May 1838, again in the *Saint James's Chronicle,* announcing 'in a few days, a New and Cheaper Edition, with Plates and Woodcuts, THE JOURNAL OF A NATURALIST.' He wrote at once to London asking for assurances that this was not simply a third edition in disguise. He had described himself in earlier days as 'a plain man and wishing to proceed in all my doings by an open road, free from the least contention or misunderstanding upon any matter' and he did not want the public misled.[160] Murray's reply does not survive and Knapp

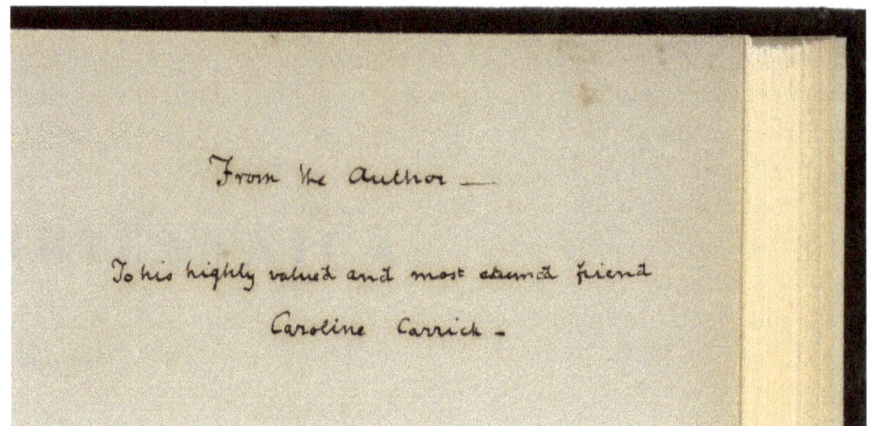

12/2 The inscription in Knapp's hand of the copy of Gramina Britannica that he gave his near neighbour, Caroline Carrick.

made no further comment on the matter, beyond agreeing that a reduced price to 'get rid of any stock in hand by any reasonable sacrifice' was worthwhile.[161] It is no great surprise therefore that copies dated 1838 seem to match the third edition in all but name, the title pages being tipped in, rather than bound with the rest of the text.

By 1840, Knapp had received a total of at least £1,555 in royalties, which is said to be the equivalent of nearly £200,000 in terms of today's money.[162] By any measure he had written a remarkably successful book and had greatly encouraged mid-19th century popular enthusiasm for natural history and pioneering ecology. It is noteworthy that almost every copy that comes on the market today includes pencilled annotations by successive owners, who have enthusiastically added their own observations to Knapp's. *The Minute Book of the Ephemeral Book Society of Bristol,* reveals that at a meeting on March 14th, 1829, at 10 Richmond Terrace, a copy of *The Journal* was ordered for members to pass round and read, on the suggestion of Mr. Ward. On 11th July 1829, Mr Bigg ordered a copy of *The Natural History of Selborne.* Again, this suggests that Knapp's book was not only enjoyed, but was re-awakening interest in the earlier work.[163] He was however seventy-three years old, and life had begun to close in on him. His older brother Primatt had died at Shenley in January 1838 and his wife, Lydia had died in October of the same year at Alveston. He buried her next to their daughters in St. Helen's churchyard. His household was now reduced to Knapp himself, Lydia Margaret (his only surviving unmarried daughter) and three servants.[164] He may have found some solace in his friendship with a near neighbour, Caroline Carrick of Marlwood Grange (1787-1872), who had lost her husband Andrew, a celebrated Bristol surgeon, in the previous year. On the front free end paper of a copy of *Gramina Britannica* that Knapp gave her, probably in the summer of 1842, he wrote 'From the Author – To his highly valued and most esteemed friend Caroline Carrick.'[165] Perhaps it was Caroline who encouraged him to keep writing.

13

IN AUGUST 1839, John Knapp was approached by an unnamed provincial publisher, who suggested that a second edition of *Gramina Britannica* ought to be commissioned. This is likely to have been William Strong (d.1846) of Clare Street, Bristol and Fore Street, Exeter, whose attention may have been drawn to Knapp by the newly formed Bristol Philo-Botanical Society. The Society had perhaps inadvertently revealed Knapp's name as the author of the *Journal of a Naturalist* in a letter to *The Bristol Mercury*.[166] Knapp still had the valuable engraved copper plates from which the original illustrations for *Gramina* had been made, but not wishing to change publishing houses, he wrote to Murray asking if he would be interested in taking the project on. He attached a 'sadly soiled' specimen of one of the original prints (*Festuca Sylvatica*), that had been rescued from the fire that had destroyed the greater part of the original edition. Murray did not reply immediately, as was his wont, but in January 1840, asked Knapp to send him a complete book. Knapp nervously sent one of the only two copies that he had in his possession by coach via Bristol to London and was greatly alarmed when Murray did not immediately confirm that it had arrived.

13/1 Wax bust of John Leonard Knapp in his seventieth year, modelled by G.G. Parker of Bath and dated August 1836. The bust is now in the care of the Royal Botanic Gardens, Kew.

However, just as he thought his precious book was lost beyond recall, Murray sent a kindly reply, explaining that *Gramina Britannica* was not the kind of book in which he specialised and recommended that he should approach a Mr. Ridgway. This was probably James Ridgway (fl.1784-1847), a distinguished

London publisher of botanical books and illustrations. Ridgway however turned the offer down, on the grounds that he was overwhelmed with work. Knapp fell into a gloom, believing that a version published in the provinces could 'hardly but be attended with disappointment.' He decided to abandon the whole idea.[167]

Strong however did not give up. He was extremely well respected in both the cities in which he ran his business and frequently collaborated with Longman and Company of Paternoster Row, London. So it was that just before Christmas 1841, Knapp was persuaded to pen a new introduction to his original work, dating it December 1st. In it he explained that he had been induced to present a new edition to the public after much solicitation, and had altered little from the first edition, beyond taking into account some of the changes of

13/2 *Festuca Sylvatica. A copy of the plate from Gramina Britannica that Knapp sent to John Murray.*

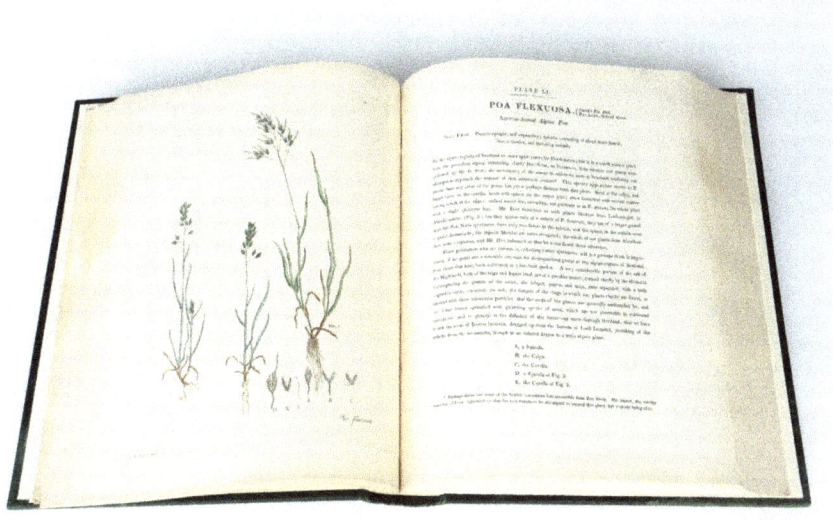

13/3 *The second edition of Gramina Britannica, published in 1842 by William Strong of Bristol and Longman & Co of London.*

generic names and genera that had become current in the intervening period. The copper plates and the text were then put into the hands of John Chilcott (d.1856) printer, stationer, and bookseller, of Wine Street, Bristol.

The finished book, quarto, bound with boards and selling at three pounds sixteen shillings, was first advertised in *The Bristol Mirror* on July 23rd, 1842. This was followed by a remarkable burst of publicity arranged by Longmans, who placed advertisements in a wide range of newspapers from *The London Morning Herald* to *The Limerick Chronicle*, *The Liverpool Mail* and *The Hull Advertiser*. No record is known of the number of copies printed or sold. Suffice it to say that copies of the second edition are far more frequently found in antiquarian bookshops today, than the painfully scarce first editions. Knapp must have been delighted that finally, in his seventy-sixth year, his phoenix had risen triumphantly from the ashes.

Only three months after the new *Gramina* had been published, Knapp had the additional pleasure of seeing his eldest son, Arthur John Knapp married at Willingdon in Sussex. The bride was Knapp's niece-by-marriage, Louisa, the youngest daughter of Inigo Thomas of Ratton.[168] Arthur had

13/4 The historic Lauderdale House on Highgate Hill, London. Knapp inherited two fourths of the ownership of the house and estate in 1841. His nephews Matthew Knapp of Little Linford and Dr. John Knapp of Edinburgh inherited a fourth each. It is now an arts and education centre.

13/5 The three notebooks containing the additional text for John Knapp's much hoped-for 5th edition of The Journal of a Naturalist.

been articled to John Daniel of Bristol in 1824 and was by 1842 a prosperous solicitor in the same city, in partnership with Robert Osborne, and Francis and Charles Ward. No doubt his legal skills had been much to the fore in 1841, when his father's first cousin, Miss Harriett Knapp (1760-1841) died in her 80th year, and ownership of the historic Lauderdale House and its estate on Highgate Hill, London, which she had inherited in 1795, was divided between her co-heirs. John Leonard Knapp inherited two fourths and Matthew Knapp of Little Linford and Dr. John Knapp of Edinburgh inherited a fourth each. When Dr. Knapp sold his share to Arthur in 1845, the naturalist passed his two shares to Arthur also, leaving Arthur and his cousin Matthew Knapp as joint owners.[169]

Knapp continued to write sporadically to John Murray as the months passed, asking how sales of his book were going, and whether an expanded edition of *The Journal* might still be published. He had prepared three notebooks containing the additional text for a fifth edition, together with detailed instructions as to how each part might be dovetailed into the text of the fourth. But as Murray remained hesitant, Knapp wrote a note to be kept with his notebooks to explain their purpose to his heirs, should death

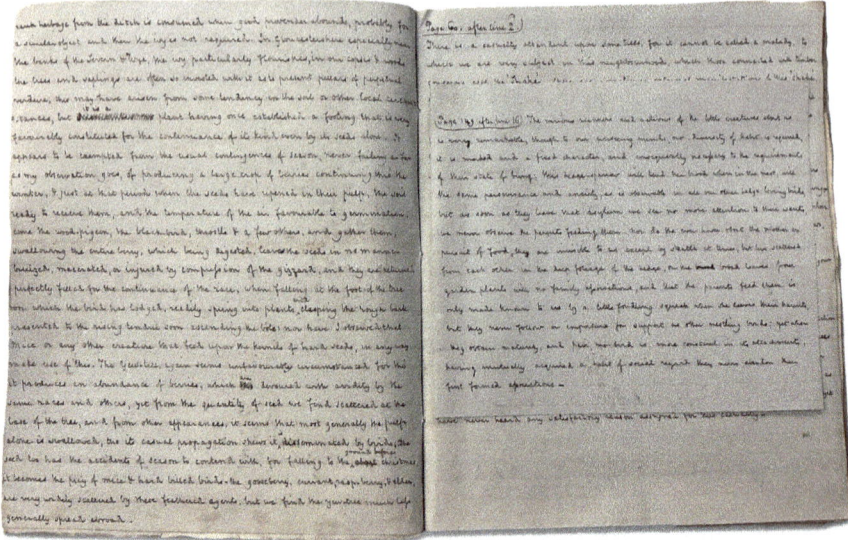

13/6 Typical pages from Knapp's notebooks, written in his own neat hand.

overtake him. By now his handwriting had become noticeably less legible, and he had begun to write on scraps of paper rather than the larger sheets he had used consistently before. A lengthy article published in *The Bristol Mercury* in 1890, quotes from several letters that Knapp had written to his friend John Elton of Redland, Bristol in his final days.[170] They reveal that he was having increasing difficulty with his sight. He was now devoting many hours to taxidermy, partly for building his study collection, but largely because 'it was not trying to his eyes, which had grown weak.' He certainly did not advocate the killing of animals for this purpose unless it was for strictly scientific reasons. 'We do not destroy in wantonness, or unnecessarily;' he had written earlier,

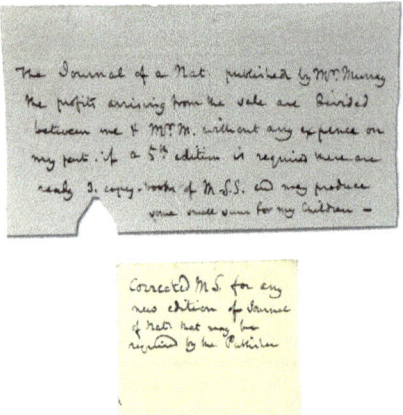

13/7 The memoranda on scraps of paper that Knapp left with his notebooks, describing their content and purpose: "…if a 5th edition is required there are ready 3. copy-books of M.S.S.…"

'and that life of which it is expedient to deprive a creature, is taken by the most speedy, and in the least painful manner known.'[171]

He and John Elton frequently sent each other specimens, Elton on one occasion sending him a white rat, a cream-coloured mole and an otter. On another occasion Elton sent him whimbrels and other sea birds from the island of Steep Holm, a present which pleased him greatly, as marine birds were not his usual subject. Many of the creatures he received brought back memories of the pleasures of the past. Of the specimen of a great bat (*Vespertilus noctule*) that he received, he wrote nostalgically 'I always associate with this creature fine summer days and the enjoyments of that brief period of our year, when we see him sailing alone in hasty flight amidst the saffron rays of twilight and squeaking his pleasure when but scarcely visible from his distance.'

In a letter that speaks of the unusual quantities of wild fowl that year in the marshes of the Severn, when as many as seven thousand wild geese had been seen in one meadow at Slimbridge,[172] Knapp noted that he employed the village carpenter to bring him fresh additions to his 'infantine menagerie.' Not every attempt to add to his collection, he told Elton, was successful though, and this amused him. 'Within these few days a bittern was killed near us,' he wrote, 'a very rare bird in these parts – and being designed for me had arrived almost, when the bearer was met by an idle, mindless young fellow and induced to give it up for a roast! Think of that!' On another occasion he described how 'a farming man this winter refused a guinea for a Canada goose – not tendered by me; but in the sequel, when he had been at the expense of having it preserved, he offered it to me for ten shillings, and had it declined, it not being the animal so much wanted as the occupation of doing it that is desirable for my weak eyes.'

When Murray once again held out some hope of a fifth edition of *The Journal* in November 1844, Knapp replied ruefully 'from my advance in life there can be no expectation that I shall survive to see it – however I cannot reasonably lament that event but those who come after me can supply what might be required.'[173] On April 29th, 1845, he died, aged seventy-eight. He was buried beside his wife and daughters in the tranquillity of St Helen's churchyard. The Rev. Cadell Holder, the future brother-in-law of the novelist Thomas Hardy, conducted the service. His obituarist in *The Gentleman's Magazine* wrote 'his later years were passed almost entirely in the pursuit of his favourite study, which continued even up to the day preceding his death. His garden afforded him one of his principal amusements.'[174]

14

IT HAS BEEN said of Rev. Gilbert White that 'his blend of evidence and empathy, science and emotion, is probably his most important and enduring legacy.'[175] The same can be said with complete accuracy of Knapp. He was the first to re-ignite the fire originally lit by White in his *Natural History of Selborne*, and such was the popularity of *The Journal* that look-alike popular natural history books of similar style and content continued to flood onto the market, despite his death. Then, remarkably, in 1853 another edition of *The Journal* itself appeared, boasting 'notes and additions' and edited by the American, Susan Augusta Fenimore Cooper (1813-1894).

Susan Fenimore Cooper was born in Scarsdale, New York. She was the daughter of the celebrated American novelist James Fenimore Cooper (1789-1851) and his wife Susan Augusta DeLancey. She was said to be especially devoted to her father and acted as his secretary and amanuensis. She travelled in Europe with her family and whether it was in Europe that she was introduced

14/1 Susan Fenimore Cooper (1813-1894), celebrated as America's first female naturalist, whose early work may well have been inspired by Knapp.

to *The Journal of a Naturalist*, or whether she read one of the pirated copies that had been published in the United States in 1831 is not known. It is also not known whether it was Knapp's example that inspired her to begin a natural history diary of her own from 1848 onwards.

Whether it was or was not, in 1850 she published her American diary, which like Knapp's consisted of personal observations of flora and fauna, made in her own locality. It was produced under the title *Rural Hours, a Nature Diary of Cooperstown*. Just as Knapp's published work had done twenty-one years earlier, it became a considerable success. It is now regarded as one of the earliest and most admirable examples of popular natural history writing in the United States and certainly the first by a woman. After it was republished in England in 1855, (confusingly but perhaps significantly under the title *The Journal of a Naturalist*), the celebrated Charles Darwin (1809-1882) wrote '…talking of Books, I am in the middle of one which pleases me, though it is very innocent food, viz 'Miss Cooper's Journal of a Naturalist'. Who is she? She seems a very clever woman…'[176]

To follow-up to this remarkable success, Cooper (with Phinney & Co. of Buffalo) then republished an edition of Knapp's book as *Country Rambles in England; or Journal of an English Naturalist with notes and additions by the Author of Rural Hours*. This was effectively a reprint of Knapp's 3rd edition, including his 'Go forth, my little bark' epigraph, but not his lengthy poem, *The Progress of a Naturalist*. Cooper's introduction to the book began

> It is now nearly twenty-and-five-years since the 'Journal of a Naturalist' first appeared in England. The author, Mr. Knapp, has told us himself that the book owes its origins to the 'Natural History of Selborne,' a work of the last century, which it is quite needless to say has become one of the standards of English literature; and the reader is probably also aware that the honors acceded to the disciple, are in this instance, scarcely less than those of his master – The Journal of a Naturalist, and Selborne, stand side by side, on the same shelf, in the better libraries of England.

Since no correspondence can be found in the John Murray Archive between John Murray in London and Phinney & Co. in Buffalo, it would appear that *Country Rambles in England* was again a pirated edition. If it was, it was one produced with great respect and admiration.

Cooper was a competent artist and although there is no proof that she embellished *Country Rambles* herself, the first pages are charmingly illustrated in colour with a selection of English birds. These are reputed to have been

 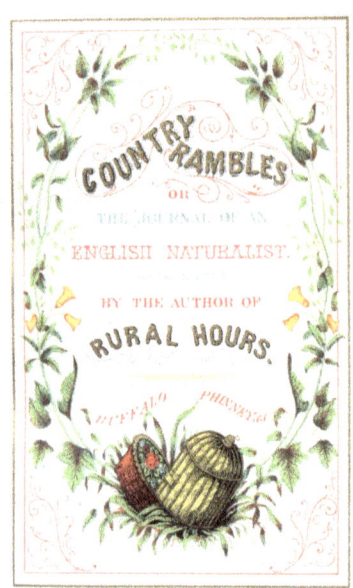

14/2, 14/3 The frontispiece and title page of Country Rambles in England, *Susan Fenimore Cooper's American edition of Knapp's* Journal of a Naturalist, *published in 1853. The drawings are thought to be her work.*

her work. The book's title is also encircled with a prettily drawn rustic wreath and printed in colour. Although Cooper modestly claimed not to be any kind of expert in the field of natural history and regretted 'that the task of editing the 'Journal of a Naturalist' should not have fallen into hands better able to render the author full justice,' she consulted what she believed were the best books on the subject and then took advice from Professor S.F. Baird, Major Le Conte, and Mr M.A. Curtis. These were significant American authorities in their day. Professor Baird (1823-1887) was appointed Assistant Secretary of the Smithsonian Institution in 1850 and was still in post when consulted by Cooper. He was a prolific writer on natural history subjects, especially ornithology and over the years was showered with international awards and honours. Major Le Conte was John Eaton Le Conte (1784-1860) an explorer of the American continent. His principal interest was zoology, though he was a fellow of the Linnean Society of London. Rev. Moses Ashley Curtis (1808-1872) was an American botanist of renown. That these men came to know Knapp's work through Cooper is remarkable. The scholarly appendices that they helped her provide, explained the flora and fauna that would have been familiar to British readers, but not an American audience.

As the years have passed, and following the success of her natural history books, Cooper has become highly regarded as a pioneering advocate of American ecology and the preservation of forests. Perhaps she was reacting to Knapp's warning:

> Where shall we look for the remnants of that mighty wood filled with boars, bulls, and savage beasts, that surrounded London? Even in our own days, heaths, moors, and wilds, have disappeared, so as to leave no indications of their former state but the name. Woods and forests seem to be the original productions of most soils and countries favourable for the abode of mankind, as if inviting a settlement, and offering materials for its use. As colonies increase, wants are augmented; the woods are consumed; the plough is introduced, division of property follows; a total change and obliteration ensues…[177]

The *Journal of a Naturalist* also came into the hands of Henry David Thoreau (1817-1862), the American essayist, poet and philosopher. It is said to have had some influence on his famous work *Waldon, or Life in the Woods*, published in 1844. Again, whether it did or did not, Thoreau certainly quoted Knapp as a reference work in his correspondence with other naturalists and referred to Knapp's writing in his journal.[178] So it was that Knapp's lifetime's work became familiar to many of the most serious students of natural history in America.

In terms of Knapp's legacy elsewhere, *Gramina Britannica*, *Arthur the Pastor of the Village* and *The Journal of a Naturalist* all remain obtainable today through 'print on demand' services on the internet. The first four editions of *The Journal of a Naturalist* and Susan Fenimore Cooper's version, *Country Rambles in England* can be read in online editions and many original copies of both books can be found in antiquarian bookshops. His curious unpublished manuscript *The Hostel* is held by Buckinghamshire Archives. His unpublished notes and instructions for a further edition of *The Journal*, are also held by Buckinghamshire Archives. These have been transcribed and used to create the previously unpublished fifth edition, that forms the second part of this book. One hundred original drawings by Knapp for *Gramina Britannica* were sold at Christie's London saleroom on 31st May 1996. It is to be hoped that they have been kept together, but their fate is presently unknown. His herbarium, containing the pressed grasses illustrated and described in *Gramina Britannica,* is in the care of Bristol Museum.[179] He had presented it to the Bristol Philosophical Institution, a forerunner of the present Bristol Museum, inscribing it on the title page 'These SPECIMENS of BRITISH GRASSES Collected in their NATIVE SITUATIONS are respectfully presented To the BRISTOL PHILOSOPHICAL INSTITUTION By their obliged Honorary Member J.L. Knapp – 1824'. The five volumes of his drawings of fungi are reputed to be in the hands of the British Museum, but to date have not been found. The oval oil portrait of Knapp, probably by Smith, that was photographed at his family home c. 1911 no doubt still exists, but again has not yet been found. The small wax bust of him by Parker of Bath is in the care of the Royal Botanic Gardens at Kew.

None of Knapp's children made natural history their career, as he had done, but his nephew John Knapp of Edinburgh (1800-1857), the third son of his brother Primatt and a respected doctor of medicine, became a serious collector of foreign and indigenous seashells. His collection ultimately consisted of twenty-five thousand examples contained in five mahogany and rosewood cabinets. These were sold at auction after his death and carried the substantial reserve price of five hundred guineas. Also offered was a copy of Lovell Reeves' *Conchologia Iconica*, 'splendidly bound', with a reserve of fifty guineas.[180] It seems likely that the plant specimens held today in the Royal Botanic Garden Edinburgh, often attributed to John Leonard Knapp, were in fact collected by Dr. John Knapp, since the signed and dated descriptive labels on them are not in his uncle's distinctive handwriting. A pressed fern (*Polystichum lonchitis*) now in Bristol Museum, which clearly came from the same original collection, is labelled 'Clova Mountains Augst / 36 Dr J Knapp.'[181]

14/4 Portrait of Arthur John Knapp (1808 – 1883), eldest son of John and Lydia Knapp. The original watercolour was photographed in black and white at Little Linford Hall in 1911 and this has been recently coloured. A.J. Knapp, a solicitor, served as Treasurer of the Clifton, Bristol and West of England Zoological Society, from 1858 until 1873.

Knapp's eldest son Arthur John, the Bristol solicitor, who lived at Llanfoist, a substantial Victorian house in Clifton that he named after his birthplace, may not have been a professional botanist or natural historian, but even so become Treasurer of the Clifton and Bristol and West of England Zoological Society in 1858[182]. The Society's gardens and zoological collection, which were opened to the public in 1836, were within sight of his house. 'Treasurer' at that time is said to have been a role half-way between today's Director and President, so his position was an active one. He retired in 1873, leaving the Society 'stronger than it had ever been.'[183] As it happens, the Zoological Society, which still flourishes, has recently moved its collections out of Bristol and is now situated at the Hollywood Tower estate, Cribbs Causeway, only five miles from John Knapp's house at Alveston.[184] Its stated purpose is to work for the conservation of endangered species worldwide, and to study and conserve the flora and fauna of its magnificent South Gloucestershire site. Arthur Knapp would surely have been delighted that his legacy continues so close to his father's home, though not perhaps as pleased as his father would have been, to know that a fifth edition of his *Journal of a Naturalist*, which he lobbied so hard for, would finally be assembled and published in this volume, albeit one hundred and eighty years after his death. He had set out in his youth to prove that he was the intellectual equal of his more highly educated older brothers. There can be little doubt that in that, he succeeded.

APPENDIX I

Knapp only recorded the names of about half the places he visited to collect specimen grasses. They are listed below. Some were noted in *Gramina Britannica,* some in his Herbarium now in Bristol Museum and some on the labelling of the specimens that he sent to Sir James Edward Smith at the Linnean Society.

Aldborough, Suffolk
Ambleside, Westmoreland
Battersea-fields 'opposite the linseed mill', Surrey
Ben Lawers, Perthshire
Ben Ledi, Callander, Stirlingshire
Ben Nevis Inverness-shire
Blackheath 'from fields', Kent
Buckingham, Buckinghamshire
Caerphilly, Glamorgan
Callander, Perthshire
Carew Castle, Pembrokeshire
Castle Eden Dene, Durham
Cirencester, Gloucestershire
Corby Craggs, (probably Corbies Craig, Edinburgh, as Knapp rather vaguely suggests 'near Forfar')
Craig Hallech (probably Craig Hulich), Perth
Dawlish, Devon
Deptford, Surrey (now greater London)
Erith, Kent
Forfar Heath, Angus
Glastonbury, Tor Hill, Somerset
Hartlepool, Durham
Hotwells, Gloucestershire (now Bristol)
Ingleton, Helks-wood, Yorkshire
Isle of Anglesey

Isle of Jersey
Isle of Sheppey, Kent
Killin 'upon Loch Tay', Perthshire
Lamorna Cove, Cornwall
Lamphey, Pembroke
Lincoln, Lincolshire
Loch Lomond (Now W. Dunbarton)
Loch-ne-gar, (Lochnagar) Aberdeenshire
Lowdore (Lodore) Falls, Cumberland
Marazion, 'on the sands' Cornwall
Margate, Kent
Matlock, 'wood opposite the old Baths', Derbyshire
Maudsley, (Mauldslie) Lanarkshire
Melford Bridge, Nr Woodbridge, (perhaps Knapp meant Wilford Bridge?) Suffolk.
Melton, 'the Walks' sandy fields, Suffolk
Merthyr, Glamorgan
Middleton Stoney, Oxfordshire
Moffat Hills, Dumfries
Narborough, 'sandy fields,' Norfolk
Oldbury on Severn, 'sea marshes' Gloucestershire
Orcheston St Mary, Wiltshire
Oxford, Oxfordshire
Penzance, Cornwall
Pill, Somerset
Purfleet, Essex
Rownham Ferry, Clifton, Gloucestershire (now Bristol)
St Vincent's Rocks, Gloucestershire (now Bristol)
Seaton, Durham
Seaton Snook, Durham
Settle, Yorkshire, (now North Yorkshire)
Snowdon, Caernarfonshire
Studley Woods, Rippon, Yorkshire
Sunderland 'ballast hills', Durham (now Tyne & Wear)
Swaffham 'by the road from Narborough Hall to Marram' Norfolk
Tees, 'the mouth of', Durham
Teignmouth, Devon
Tenby Warren, Pembrokeshire
Thorp Arch, Yorkshire
Yarmouth, 'the Dunes', Norfolk

APPENDIX 2
JOHN KNAPP, AS POET.

JOHN KNAPP PUBLISHED three poems. The first was his epic and lengthy tale in blank verse, *Arthur, or the Pastor of the Village*, printed in London in 1818 by Bensley and Sons. Most copies were lost in Bensley's second fire of 1819. The second was his biographical poem, *The Progress of a Naturalist*, dedicated to the goddess Flora, which first appeared as an appendix to the third edition of his *Journal of a Naturalist* in 1830, was repeated in the fourth edition of 1838 and can be found at the close of the new fifth edition which forms the second part of this book (pages 271 – 275). His poem *Go forth my little bark*, also appeared first in the third edition and can be found following the title page of the new fifth edition. It may be that some or all the poems published anonymously in the articles attributed to Knapp in *The Times Telescope* between 1820 and 1830, were his, but it is now impossible to be sure. In the same way, it is impossible to tell whether Knapp wrote any of the brief pieces quoted in his *Journal of a Naturalist*. If some or most were his, then the description of the medicinal value of the foxglove, 'summoned to aid us in our distresses' must be one of the most moving pieces that he ever wrote, born perhaps of the agonies that he experienced in the early deaths of four of his daughters:

> —Thy last, sole aid (which art can give),
> The woe-worn parent seeks, and, hoping, clings
> In tearless wretchedness to thee; watches
> With anxious heart thy subtle progress through
> The day, and of thee fitful dreams through all
> The night —
> — spare, if thou
> Canst, his hopeless grief; save worth, save beauty,
> From an early grave.

The remainder of Knapp's poetical works, or poems attributable to him, exist either in an unpublished sixty-six-page manuscript in Knapp's hand, held by the Buckinghamshire County Archive and called *The Hostel*, or on numerous individual scraps of paper, again in his distinctive handwriting, in the same collection. None however are signed. Considerable time and effort have been spent attempting to ensure that these are not by other authors. Only two in the collection, *The Magpie and her Brood* and a fragment of the traditional folk-song *The Carrion Crow* have so far fallen into this category. It is possible that others will be identified, but this does not seem likely.

The Hostel or the Records of my Landlord is a fictional work of prose that Knapp sent to John Murray in March 1830 in the hope of publication, but it was returned unpublished. It consists of a collection of curious stories, some light-hearted, some historical, some ponderous, set out along the lines of Chaucer's Canterbury Tales. The introduction, clearly weaving Knapp's actual experience with make-believe, tells how the author was staying in a remote inn at Ingleton in Yorkshire, in order to 'botanize' or collect plants and minerals in the forest and mountain of Ingleborough.

One day, so the story goes, the landlord of the inn happened to mention that some years before, an association of friends had formed at his hostelry, to meet each month for mutual amusement and good fellowship. Each in turn was expected to tell a tale, which the landlord (an educated man who had fallen on hard times) assiduously wrote down in a battered, brown-covered book. As all the members had long since died, the landlord supposedly gave Knapp this book, which had been annotated 'by the unmannered frequenters of the tap-room' with such remarks as 'a long story about nothing,' 'that ain't true,' and 'a humbug.'

Those whose tales were recorded were the Vicar, the Captain, the Squire, the Banker, the Quiet Gentleman, the Curate, the Bookseller, the

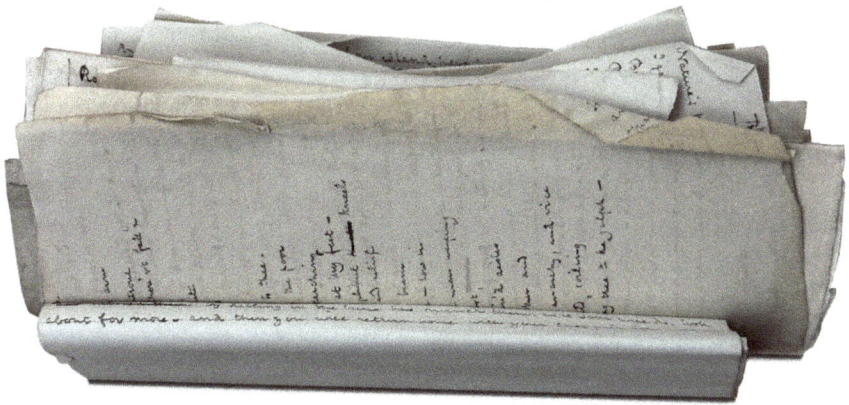

A2/1 The tightly rolled bundle of John Knapp's poems, as held by the Buckinghamshire Archive.

Merchant, the Doctor and the Antiquary. Poetry forms part of the Vicar's tale (numbered 2 below) and is entitled 'vice – virtue – and penitence.' It tells the story of a desperate character, saved through marriage to a righteous wife. The Bookseller, (one Abel Drewe), included three poems in his tale (3, 5 and 6 below). No 3 tells the story of an orphaned child, with a passion for wildflowers, who sold posies around the village. No 5 reflects Drewe's ability to write moving epitaphs and No 6 describes Tintern Abbey in Monmouthshire. No 4 forms part of the Merchant's tale. The lengthy poem relating the story Mary Queen of Scots at Lochleven Castle (No. 1 below) is included in *The Hostel* for no apparent reason. An earlier much corrected version on a paper scrap is headed in Knapp's hand 'Written at Lock Leven near Kinross July 1802', exactly the date at which he was botanizing there with George Don, though this information is not divulged in *The Hostel* text. Given that Knapp intended his collection of stories to be published, there seems little doubt that the verses which form part of them were his own work.

The remainder (numbered 8 to 22 below) are written very much in the Knapp style and there is little reason to doubt that he wrote them. Many give the impression of having been written to illustrate some further prose stories of *The Hostel* type. This would account for the unusually named characters that appear in them. If that was their purpose, then the prose text for which they were intended, is now lost.

The poems appear below exactly as written by Knapp, with no changes to his spelling, punctuation, capitalisation or lay-out.

(1)
Tell me Lockleven, placid Lake!
Did e'er thy lovely scenes a Native view,
With heart so cold that could not feel,
Or heave a sigh to injur'd beauty due? –

Then let not one, tho' Britain's son,
And all her glories to his bosom dear,
Forget thy Naiads rights to pay
And Mary's woes their tributary tear –

Here twelve long months did Scotland's pride
Her beauteous Mary pine in lonely care
The roses in her visage died
And sorrow spread her pallid Lilies there –

Oh minutes, how ye creep along
If care or sorrow but thy burden swells!
Then what are days, when all alone
Their tardy hours the fetter'd captive tells –

P'rhaps on yon turrets ruin'd brow
Where creeping Ivy wraps its mantle round
The beauteous captive nightly wept
Or listen'd to the passing billows sound –

And as the Stockdove o'er the wave
Would homeward silent wing her weary flight
She'd watch the wand'rer thro' the skies
Till all was lost in closing shades of night.

Poor bird! She'd say, thy plumes e're long
Will bear thee to thy mate and peaceful nest.
I would for once those wings were mine
That I might flee from hence and be at rest –

But where could Scotlands daughter roam
Where find one faithful friend her griefs to share?
Where press a bosom warm as hers
Or sighing tell her varied sorrows there? –

Ah no! the few my childhood gave
Misfortunes with'ring blighting frowns have flown –
Then bosom take thy tears again
For all its sorrows are alone thine own –

Near where yon riven oak is seen
She'd constant pass her melancholy eve,
Low to the virgin chaunt her prayer
Or some soft mournful harmony would weave-

The plaintive music o'er the wave
The passing wing of breezy Zephyr tost,
And now you'd hear it murm'ring sweet,
And now again was hush'd, in distance lost –

The nightly trav'ler scarce could deem
Such strains to issue from a mortal lyre,
But awful list'ning pause to hear
The dulcet warb'lings of celestial quire –

On Gallia then she fondly mus'd
And those few halc'on hours she only knew,
Whilst mem'ry ah too faithful here,
Her blissful infant days would oft renew –

Oh Mary! Could thy soul have seen
But half the mis'ry of thy future hours,
Thou'adst bless the dusky robe of night,
That safely hid thee in Locklevens towers –

Thou trust Eliza's treach'rous heart
Which weaves for thee a dark and fatal snare,
Ambit'on sway'd her scepter'd pow'r
And stifl'd infant pity whisp'ring there

Hope with delusive phantom train
Seductive smil'd to lure thee to thy doom,
And Douglas plum'd with love's light wing
Bare thee with speed to Britain, and thy tomb–

The ling'ring peasant loves to shew
Her walks, and still will oft her fate bewail –
And parents musing o'er her ills
Tell to their list'ning child the mournful tale-

And strangers here will often trace
(Their minds by Fancy's erring finger drest)
The ruined chapels weedy aisles
And ev'ry spot the feet of Mary prest-

Oh ill star'd Queen in distant days
We note thy wrongs and suff'rings with regret
Light o'er thy errors pass the eye,
And tho' we can't forgive them – must forget.

(2)
-Yes there he rests –
Nor e'er will time of man, his history reveal –

I knew him once, an idle brawling boy,
Content'eous, and in awe of nought – ling'ring
Ere his book he found – the first to shut and
Scamper out – to plan illegal deeds, and
Riot in forbidden ground – a reckless
Youth, that scorn'd the shackles of control –

Careless of fortune and of fame, he rac'd
His wild career on passion's steed –

Our paths of life divided soon – I sought
A quiet and a pleasant track – he rush'd
Amidst the waves and hurricanes of the
Western climes – he heard the strife and gave his
Aid to break the gyves from old Vespucios
Southern realms – and by the whirlwind of the
Contest cast beyond all knowledge of his
Deeds, or fate –

But years rolled on, and summer suns had warm'd
And Winters frosts had chill'd me in the calm
Seclusion of domestic life – the change
That days must bring on all I knew – and felt.
But grieved not – disease had slackened
Not the sinews of my arm – my pate tho'
Blanch'd was by the hand of time – not sorrows
Gift – one ev'ning when the stars were sparkling
In Novembers rimy air, and the dry
Log bright glowing of the hearth – a stranger
Ask'd admittance to the manse – he came –
Was courteously receiv'd and seated –
Warm'd o'er the burning brands a withered
Hand – boney, & gaunt – and then in silence

Bent a look intent on me – as mine had
Been on him – he seem'd a tall dark man – and
Like a rock calcin'd by frost, the softer
Parts all worn away, the flinty core deep
Scor'd, and seem'd by time was spar'd, and now in
Rough defiance stood – black pass'ons ridg'd a
Fearless front – yet amidst the intrepid
Daring of his mien, at times some soft'ned
Lines of sadness might be seen crouching to
View – then as asham'd some stern reproachful
Lineament would come and cover all –

At length he told his errand and his tale –
And with the brawling boy of distant time,
I trac'd some semblance to the vet'ran man –
I must not speak what then was told – nor think
Upon the crimes – the horrors then disclos'd –
Nor say how red the withered hand that
Trembled o'er the blaze – my wonder that a
Human heart so utterly remorseless
Could exist – a being with a manly
Form – a mind – capacity of reason –
Could walk the earth without the knowledge of
An after state! – dismal, and cold, a life
Like this must be – yet still a spark surviv'd –

Fancying the beauty of Miranda's
Form, he wedded – a british fair who liv'd
Domesticated in the land –

Her ever blameless life – her constant love,
Was as a reclaiming angel sent to
Enlighten and to save – admiring her
He listened to her words – knowing that
She thought them true – would read the unknown page
She pointed out – musing in seeming thoughts –
Then hasty clasp the book and flee – as if
His rugged mind was all ashamed of
The weakness and infirmity he felt –

Pained in absence from her side, he would
Return again to meet Miranda's smile,
Dote on the gentleness, the justness of
Her mind, the calm serenity of brow –
And think that some deep seated truth alone
Could yield tranquillity like hers –

She left him
When his love was warm in union with
Esteem – the fiery tumult of his
Days had ceas'd – and when Miranda died
The old excitements of his mind had lost
Their spring – sad in memory of her he'd
Turn the page she shew'd – and muse upon her
Words till they became his faith –

We cannot tell the endless ways which he
Who wills that none should die, may call from death
To life – we know that some have been so good,
So gentle, and so pure – so slightly bound
To earthly things, they seemed scarcely form'd
Of mortal soil – but as the spirits of
The realms of truth, walking in guise unknown
To lead the sinner to his peace –

To Europe he return'd
And chance directed him to these our fields,
From whence he never wandered more –

The wild relentless man was gone – and I
Have seen his tear, fall on the sacred leaf –
Seen him bend submissive on the altar's
Steps – he'd listen with intense delight to
Our revealed truths, abstracted from all
Other things – 'tell me such words again, for
Such Miranda told ! tell them once more, I
Cannot weary but believe them true –

My office was to soothe

His terminating hours – but so calmly
Did he rest – so gently had the spirit
Flown, that we knew not when it pass'd –
Peace to his earth – we trust atonement made -

(3)
May-flower – the wanderer of the village

Nature's pure child, a guiless lass,
I gather pleasures others pass,
But ah! They say my brain is weak
Because I do such trifles seek –
If I in woodlands happy stray
And with sweet flowers pass my day
My joys to these wild things confined,
Why do they say I'm weak of mind?
I love my flowers to me they bring
Renewed hopes in ev'ry spring,
These hopes, these joys of summer give
With them my pretty flowers I live –
I grieve when autumn takes away
These pleasure of my sunny day
And when they're gone in wintry hours
Fancy still brings my pretty flowers -

The yellow wren was in the tree,
That summer bird so dear to me –
The little vagrant came to tell,
Of the wild blossoms in the dell,
The Primrose from her mossy bed
Heard the glad tale and rais'd her head,
The grove, the mead too, heard the call
And wak'd their slumb'ring florets all,
And I have wand'red round and round
And many a pretty flow'r I've found –

Here lady, here's the azure bell,
That loves in woodland shades to dwell,
A bell, that gives nor joy or sorrow

A laugh today – a tear tomorrow –
But hangs its pretty head to earth
Nor sounds one note of grief or mirth –
And here's lads-love! ah me! ah me!
That such a silly name should be
And foolish maids to prize it so
And let it in their gardens grow! –
And here's hearts-ease – so seldom found,
To flourish in the self-same ground –
And see another beauty here
We only find in early year,
A pensive maid that looks so pale
If march but breathe an icy gale,
And yet within her leafy bow'r –
Will blush at Aprils tepid show'r
And one would almost think it true,
That flowers human follies knew –

But if I once begin to tell,
Of these wild things I love so well,
Fancy away will wand'ring rove
The heath, the meadow and the grove,
Till ev'ry floret of the year,
Will vis'ond in my mind appear –
And village clocks – the Daisy here,
Bids me go home, for eve is near –

(4)
Give me again my native hill
The whisp'ring Pine, the mossy rill,
The scenes of artless infant joy
When Donald, was a mountain boy –

The bliss, my fancy paints, 'tis true
Time sober'd hours may ne'er renew,
Affection wishes still to trace
Where first I ran my childhoods race,
With pious grief once more – to lave
With grateful tears my parents grave –

(5)
Yes my dear babe! With tearless eye,
I see thee on thy pallet lie –
Oh if that eye would yield its grief.
My loaded heart might know relief –
Our master, and redeemer dear
Drop'd for his friends a heavenly tear,
Nor would he now rebuke the dew,
A mothers love might shed for you –

My babe! My babe will take its flight,
And rise to realms of glorious light! –
Then shall it be my constant care
Rightly to live, & join him there –

(6)
The Abbey of St. Mary – Tintern

In days remote beyond the ken of man,
Long ere the Normans' rigid reign began,
This region round was all one tangl'd wood
Where mighty Oaks decayed as they stood,
And the rude Hind, but just above the beast,
Shook to his bristly herd their acorn feast;
A slave in body and a brute in mind –
A base example of the human kind,
Mean to his lord, he'd abject kiss the earth
Yet act the tyrant to a lower birth –

There the wild Stag, just stretching from his bed
Shaking the moisture from his antler'd head,
Listen'd, if chance the breeze of op'ning day,
Brought to his timid ear the Buck-hounds bay –
'Twas silence all, save from the crag, the thrush,
Chanting his mellow note in haw-thorn bush –
Pawing the moss, and bounding from the shade,
He sought the herbage of some fav'rite glade –
Such was this site in some far distant hour,
Ere art call'd nature to obey her pow'r –

By Biscay's rock-bound shore, the briny main
Bare the mail'd chieftain to his home again,
From Palestine he came, where frantick war
Had drench'd in human blood her ruthless car,
Calm was the hour – but soon the billows growl'd
Dash'd their white foam, and loud the tempest howl'd,
Heav'd o'er the deck a broad resistless spray,
And delug'd half her mariners away –
The master's left – but he with pallid fear
Calls to the Priest, and whispers 'death is near' –

The rugged chief who knew no martial awe
Whose pow'r was right, whose will to all a law,
Whose joy and pastime was the war-crys crash,
Trembling now view'd the elemental flash -
Meek to the Virgin pray'd, if life she'd spare,
He'd build a sumpt'ous fane, endow it fair,
Where stol'd priest should chaunt her endless praise,
Before her shrine perenn'al lustres blaze –

The storm has ceas'd – the Pinnace gaily drest
Bears the mail'd chieftain to the havens rest –
Hark to the axe's ring, the mallets dash -
The groaning of the Beech, the Oaks loud crash –
The woodmans laugh – the creaking of the wain,
Bearing its pond'rous load to build the fane –

Some learned churchman with his rule in hand,
Squares the free stone – and shapes the slender band,
Bends the high arch – the fine pilasters trace,
Lifts the light roof – to guard the sacred place –
From the rude mine his zigzag wonders call,
And spreads his florid trac'ry round the wall –
And thus was form'd this mystic, votive pyre,
Which a more polish'd age does yet admire –

When the first order'd course of time began,
All things were meted by a destin'd span,

The highest, and the lowest had a space,
A goal where all must end the stated race –

Grizzled by time, and gaunt, the chieftains head,
Reclines from weakness on a restless bed,
War, and the chace were weight enough to bear
He scowl'd religion to the churchmans care –
In languor now, the Baron has a fear,
There may be truths he once had smil'd to hear –

The bold Cistert'an comes with martial tread,
Rousing the feeble chieftain on his bed –
Fram'd to conduct the revel at the board,
He took the crozier – but preferred the sword –
And better knew the merits of his steed,
Than canons, pater nosters, or the creed –
Swings his large form along in scornful pride,
And rudely seats him by the sick-mans side –

'My son thy many sins are in that book,
Where saints above but tremble when they look,
For crimes like thine no mercy now can flow,
But flames are roaring for thy soul below –
Think of thy laugh at shaven Priests! – thy slights! –
The niggard hand that dol'd our vested rights –
Nor will our sacred mother now incline,
To pardon unrepented sins like thine' -

'Oh tell me father in this hour of grief,
What can I do – how hope some slight relief'-
'Give of thy self my son, some worthless share,
And holy church will give for thee – its pray'r -
Then bending grateful o'er the Abbots hand,
The dying man surrenders half his land,
Whilst his stern heir with dark revengeful lour,
Curst the stol'd priest – yet humbled at his pow'r –

And that grim belted figure, cross legg'd there
Is the sole remnant of the great De Clare!

Here on a shield, all heralded by rules
His arms – field Or – and then three chevr'rons gules

Grey in the vest of time, the noble fane,
Long years in haughty splendour did remain,
Fearing no monarchs threat, no laymans ill,
E'en hardly bending to the Pontiffs will –

But truth began her banners now to wave,
Sweeping the cobwebs from religions slave;
Taught men to try the fact of what they heard
Nor rest salvation on a sinners word –
To see in faith and love, more real good
Than all the glitt'ring trump'ry of the rood –
The film that vail'd the word, grown thin and spent
Parted in twain, and light stream'd thro' the rent –

And then young Henry, mounted on the throne,
Sternly forbade all freedom – but his own –
Urg'd on by pride, by lust, by love of gold,
He burst the very bonds he would uphold,
A priest in soul – a tyrant at the heart,
Rage 'gainst the men who would not take his part,
He drave the old monastic fathers hence,
Wealth was the object – crime was the pretence –
And when he'd done – with papist heart he burn'd
To reinstate the faith he overturn'd –
Despoil'd the fanes lest chance these scatter'd men,
Might raise to power, and collect again –

And now all rest, and open to the sky,
Shatter'd and lorn the noble remnants lie –
And ah! How vain the ancient fathers toil,
Cemented masses cumb'ring all the soil –
O'er labour'd trac'ry hangs the spiders web,
And florid sculpture shields the swallows shed –
And in that place where holy songs were heard,
Its choiristers are now the cackling bird,
There by the shrine where warm tapers beam'd

Unscath'd for ages has the Martin scream'd -
And in the aisle our fathers trod with fear,
The peasant whistling litters down his steer –

Crime might be theirs – a faith perverted too,
Still they upheld the best – that then we knew,
And tho' I glory in the light it brought,
I feel for those who sang – as they were taught –
'Twas holy once, and we the site revere,
Whatever follies might be sanction'd here.
Pagan, or false, 'twas once a house of pray'r,
And solemn vows and off'rings tender'd there-

But Time, which often plays a wantons part,
Pilf'ring the finest feelings of the heart,
Sneering away long wish'd faiths of mind,
Leaving us nothing half so good behind,
Will long spare thee – and whilst one stone remain,
One arch alone thy tow'ring wall sustain,
Hundreds will come, and sketch thee in the book,
Stroll o'er thy site, upon thy fragments look,
And seated in the shade from mid-day sun,
Marvel and meditate – as I have done –

----

(7)
The Christian's Dirge

'Tis past – 'tis past – 'tis past –
All earthly thoughts are fled –
And human cares all now repose –
Repose – in man's last bed –

Rest Christian Rest
Rest Christian Rest –
On him who call'd that dust to life
Then bid its being cease –
Receive – Receive for Jesu's sake

The spirit to thy peace
Rest Christian Rest
Rest Christian Rest

(8)
Ah Severns banks, and Severns sea –
Thou art not what thou us'd to be –
No nothing that I now can see
Is as it once appear'd to me –
For He's away upon the tide*
Far Far from me and Severns side -

When Apples bloom'd in lovely May –
With him so lov'd I oft would stray
And happy happy were the hours
We rambled out to gather flowers.

But flowers no more for me will spring
Nor youthful days their pleasure bring
Time calls the changes of the year
But all to me is winter drear
For he's away upon the tide
Far-far from me and Severns side -

* Knapp first wrote 'for my dear Lad' instead of 'He' and later 'him'

(9)
An infant sleeping

Sleep – Sleep, my lovely infant sleep!
A guiless heart is now thine own –
If years in sin that heart should steep
No more this tranquil rest is known –

Sleep – sleep my babe and take thy rest
Thy wants are all supplied by me –
No sorrows now disturb thy breast
In coming days may't be as free! –

Yes my dear babe an anxious heart
Bends in thy Couch and weeps for thee –
A few short years – tis then thy part
To bend in mine – and weep for me –

(10)
The Forget me not

How few who've pac'd a rough and lengthened way –
With human follies – passions as their lot –
Could parting wish, or truly serious say –
To those they leave behind – 'forget me not' –

The youth who deeply graves on beechen tree
The last memorials of his schoolboys day –
May wish perhaps, or fondly think to be,
Remember'd often when he's far away –

When manhood comes, and adverse fortunes fret,
With hosts of worldly ills to fight, & toil,
How little can he love – how much regret,
The deeds, or objects of his mortal coil? –

Memory at best is shaded dark with grief,
And half lifes actions better be forgot,
Should one kind deed be done – one wish, tho' brief
Perhaps I'd say for that – 'forget me not' –

(11)
The Fishers chapel Tenby

– On that far coast where antient Pembroke's
Rugged cliffs for ages has withstood the
Raging of the furious main, and bravely
Battled with the mining surge, tho' shrinking
Yearly from contending might – a ruin
Stands – neglected and alone – bare on the
Jutting crags that breast the harbours wave – the
Sculptors patient hand ne'er laboured there – no

Fretted roof from reeded pillars rose – nor
Tinted windows gleamed with orient hues –

A simple fabric built by artless man –
But long abandon'd, now the wild herb comes
And waves in ev'nings breeze – Sedum – Thrift
The Wall-flower and the Fern – all desolate
As it is, a gentle voice is heard that
Whispers still of piety and Faith – a
Rustic chapel 'twas – and one erected
On a purer motive hardly can be
Found – raised in confidence to him whose
Mercy only can restrain the rav'ning
Main – here in the twilight of the early
Morn, the ocean fishers came imploring
Heavens blessing, on their coming toil – to
Pray for succour and defence from all the
Perils of the sea in their long hours of
Absence on the distant wave – then launch'd their
Crazy barks and sail'd – Heaven only knows
The secret heart of man – but rising from
An act of suppliance and trust like this,
Their progress thro' the day must surely be
In fearless confidence – and where the gale
Came on, calm resignation nerv'd the arm
To bravely bide the storm – to rest on truth
That cannot fail – to conquer or to die –

But time that wears the granite rock – spares not
The ordinance of man – but innovates
On all – till thoughts and manners pass away –
And now all waste the Fishers chapel rests –
No prayers are tender'd there – no heart bends down –
With thankfulness to him whose pity spared-
Alone we hear the beating of the surge
The sea-gulls melancholy scream –
Nor let the thoughtless smile contemptuous at
The old mens tale, as basking in the sun
They linger o'er their distant fathers days

Telling of oceans wealth – what store of prey
Was theirs – the captives that obedient came –
No toiling trade was then the Fishers craft –

One would not cherish such false views of fact
As theirs might be – but if a faith like that
Will give a reverence for abandon'd rights
And vivify a lively principle
Of gratitude and trust in him who gives
And takes from all – 'tis good that such should live –

(12)
Summer is a coming

Summer surely is a coming
Look the trees there all are greening –
Bees are active cheerful humming
Birds their ruffled plumes are preening –

In furzy field the linnet sings
Doves in Beechen copse are cooing,
And all the little world of wings
Their various pleasures shewing –

Cowslips in the mead are yellow
Blue bells now in the copse are seen
Larks they spring from off the fallow
And daisies deck the village green

Perch'd on the Haw-thorns upmost spray
Hark, hark the Cuckoo's voice I hear –
Oh how I love his simple lay
This herald of the summer year –

Little maids in fields are rambling
Gathering flowers here and there
Now for this, for that one scrambling
The very least to snatch its share –

Haste my Ellen! Fetch thy bonnet –
Up! Up! And let us speed away
Time they say won't wait – fie on it! –
Let us enjoy this sunny day –

(13)
Haste my Dorinda let us stray
Ere this calm scene shall fade away –
The silver planet of the night
Now sheds o'er all her placid light
Her tender beams so softly fall
Glitt'ring with gems the ivy'd wall
We'll listen to the distant lay
Of ev'nings bird on Hawthorn spray
The murm'ring stream will glide along
And babble to his plaintive song
We'll see in sleep each floret rest
Reposing on earth's verdant breast
With guiless hearts we'll tell our say
And silent tread our lovely way
So led by fancy – joys so dear
Forget that time ne'er lingers here –

(14)
Oh tell me not of doubts and fears –
My hearts my love to thee devoted -
Yon summer fly that wanders by
May gaily sip, at ev'ry trip –
The honeyed sweets of opening roses –
On many a flow'r in beauties bower
With flutt'ring wing – regard may fling.
On one, with perfect love reposes
Tell me not then of doubts and fears
My hearts my love to thee devoted –

(15)
The Spring Swallow –

The Swallow! The Swallow! There see her on wing,
Attended by Zephyrs and flow'rs of spring –
Oh welcome sweet swallow I'm pleased with thy sight!
Now twitter and frolick from morning till night
There's flys by the may-bush, there's flys at the pool –
The children are here – but will soon go to school –
My cat will not scare thee, nor gunner annoy
So build in my windows in freedom and joy –
But pleasures are brief both to bird and to man
Thy life is but transient, and his a mere span –
Cold winter will drive thee far far o'er the main
But haste my sweet swallow and come back again.

(16)
The Garniture of Time –

Once, on a distant, soft and sunny hour
A flaxen-headed child in parent bow'r
Played – or leaning laugh'd in thoughtless glee
Against his fond, his widow'd mothers knee –
Her slender fingers wanton'd in his hair
Or rang'd with blameless pride those ringlets fair-
He feels amidst the reck of time, e'en now
The graceful – gentle hand that smoothed his brow-
His days, like other infants days were spent,
His sleep like theirs – was calm & sweet content –
As time mov'd on these trifles were forsook
And flaxen locks assumed a darker look
And as the manly Auburn curl'd around
New thoughts – new wishes to his fancy bound.
Idle and crude ideas form his mind
The thoughts of childhood still – but more refin'd-
But reason comes, and brings a soberer plan,
And childhoods fancy fades, as grows the man–
In that same bow'r, midst the full flow of joy,
Laughs the fine youth – where play'd the gleeful boy –

A widow'd hand once curl'd his flaxen hair,
A gentle damsel comes, as lov'd, as fair,
Parts the brown ringlets on his manly brow,
And gives a kiss of deep affection now –
But years roll on – tho' gentle be their pace
Yet still their furrows we can plainly trace –
And each tho' mingled with its joys – its care
Shades in its turn again, his auburn hair,
And frosts of ev'ning unregretted spread
A graceful token on the elders head,
Softly, scarce noticed, they lighted down,
Mingling the tints of age, with autumns brown -
Winter at length comes graceful on & slow
Hiding the auburn 'neath a veil of snow –
The damsel, now no more – but cherish'd mate
Kisses her partners bald but honor'd pate,
Pats the thin wool upon the polish'd crown
And looks to heav'n to send its blessings down–
Serene the winters mantle's on his head
The summers ardour – autumns glows are fled,
He would not yet recal the roseate hours,
He laugh'd with happiness in beauties bow'rs,
Grateful as every colour came, then pass'd,
Content with each – resigned to the last –

(17)
The transports child –

For my far distant home Oh how often I sigh
And muse o'er the pastimes of infancy's days
I still hear the Field Lark singing blithe in the sky
And wake at the blush of the mornings first rays –

The soft moan of the Stock-dove yet dwells in my ears
The thrill of the Throstle perch'd high on the tree
The voice and the breath of the new opening year
All – all are still present – remembered by me–

Those affections which nature supplied at my birth –
Those feeling which cling both to health and decay –
The scant fare – yet the warmth round my poor father's hearth
Unbidden remind me I'm far far away –

In my innocence taken from sin and from shame –
And saved in my childhood from harm and from wrong
Yet my love for my home is a long lasting flame –
And burns undiminished yet ardent and strong.

Oh my mother! Can she ever think on her child ! –
She knows not my welfare – I know not her lot–
Yet fondly would trust as she roams o'er the wild
Her lost one – her last one – was not quite forgot.

I am grateful for all the distress I am spared
The kindness that saved a poor child from her death
Yet I cannot forget what my infancy shared
But grieve for those parents who gave me my breath

(18)
A Winters Evening in the Country

Winter now comes with cheerless mien
Wrap'd in his snowy mantle round –
The Pine-tree bends his head to earth –
And hidden all the verdant ground –

The sparrows cluster at the barn
To seek the grains which we reject
The Red-breast too he has been here
And quietly at the window peck'd

With feeble note has beg'd our aid
To help him in his sad distress.
With suppl'ant eye he seems to know
His trifling want: we should redress –

The Song thrush too on leafy spray
His whitethorn berry now does claim
His summer's active hops not seen,
Subdued by want he's pinch'd and tame –

The Black bird seeks the laurel hedge
If chance he may some fragment find
And all things pit'ous look to man
And pat'ent wait his bounty kind –

Haste my good John unto the Rick
And cut some Hay for that poor Cow
Last night she had no want of food
Be sure to give her plenty now

And those poor sheep that scratch the ground
To clear away the harden'd snow –
Scatter some fodder by the hedge
And see how quickly it will go

Old Betty Grevel in the Lane
Must have direct this shilling – here –
And let her have that faggot too
Her wood-stacks very low I fear –

Now let us shut our windows up –
Have tea – our eve'ning book begin –
And since 'tis comfortless without
Let's find our comforts all within –

(19)
The family – Renown

---- 'twas a noble sight to see those
Fine and healthy boys, and laughing girls – step
Above step – the father's pride, the mothers
Joy – pacing along the village path with
Agile feet upon the sabbath morn – to
Hear the shout, the mirth that ev'ry pastime

Brought – the glee exuberant that could not
Be suppres'd, when school-day hours were o'er, with
Friends – the Hall – and happiness in view –

It was an ancient Hall;
Where once had stood a Saxon's massive hold,
But Will'am had bestow'd it to reward
The prowess of a Norman chief – time had
Now moulder'd much – and scars, and rifts were seen
To tell of hostile violence in war –
Fancy and shifting manners of the
Day, at various periods too, had changes
Wrought – still 'twas a venerable hall – one
Tow'r there was, a strong cemented mass, which
Stood unscath'd, hallowed by time and man –
The reverenced treasure house of years –
Here was suspended from the walls, the Mail
The red-crossed hero won on Syria's
Arid sands – the Hauberk, and the glave – the
Falchion which had reek'd at Cressy and
Poitiers – the batter'd helm, the banner and
The spear – the Bow that drove unerring death,
Thro' Barnett's field – the Sword the conquering
Edward in the hour of Victory gave –
The match-lock, and the leathern doublet, of
A later day – all here were found – all view'd
With transports here – with ecstacy of heart
Far nobler in the main, but mingled p'rhaps
With human, pardonable pride –

But days pass on, the flowery and the
Leafy hours – & fields are sheeted with the
Winters snow – the youth are bronz'd with summers
Sun – their manly honours curling round their
Brow – the pastime of the boy no more its
Transports give – the leap – the race – is now a
Steady walk erect – new feelings – wishes
Hopes arise – the page of history is
Read – and the old record tells of barbed

Steeds, and arms – the Tournament, and joust – the
Feats of Gilbert, and Sir Hugh their sires – the
Scarf the high-born beauty gave her Knight, the
Glance, that stole his heart – and hours are past in
Thoughtful, sober, mood, where hang the trophies
Of their ancient race – and as the wind of
Autumn sighs along the Lance, and waves the
Mouldering banner on the wall, 'tis as
A voice that calls to fame, a signal to
Achieve the like –

How willing is the human heart to be
Deceiv'd – to trust the shadowy phantoms
Of the brain, the meteor, or the falling
Leaf, when inclinations prompt – and where's
The British heart that hears not clear the still
Small voice that calls to honour and to fame?

Away are these brave boys
Breasting the battles wave, or sick'ning in
An adverse clime – the one in Indian
Warfare full, and ere three years were past the
Muffl'd drum the vollied peal proclaim the
End of his career – another now with
Fever pin'd on Affricks sands – no mother
There to bathe the burning brow – to kiss the
Parched lip, that ap'd to utter but her
Name – at length delirium came and all
The boy return'd – his pastimes and his loves –
'Twas said that in the deep still hour of night
A low, faint voice was heard to call for 'Home,
Home, Home', – then all was still.

Reginald too heard the call – chiding
The tardy steed that bore him to the port,
The breeze that play'd upon the rivers wave,
And as the Bark mov'd slowly down the stream
He wav'd with joyous heart his last farewell –
The Bark sail'd on – a vessel homeward bound

Told of his health – his prosperous state – these
Tidings were the last – an unknown fate on
Some far distant sea was fear'd – hope, without
Hope, liv'd long – and ceased –

Edgar was wreck'd upon the strand –
That gallant boy! – the riot of the Hall –
The fearless, faithful, and the kind – that held
In thraldom ev'ry heart – and gave his own
To all – ne'er say his fathers land again –

One stripling,
Yet survived – a slender thread to bear the
Weighty records of his house – the gather'd
Glories of eight hundred years! – a youth with
College honours wreath'd around his brow – who
Lov'd the woodland walk to listen to the
Bird of eve – noting the silver planet
Gliding thro' the rack of night – to woo the
Maid of Poesy, burning the midnight
Taper at her shine – he saw no banner
Waving on the lance – nor heard the breeze of
Ev'ning chiming on the rusted helm – another
Vision yet devour'd his peace, as baseless
And as fatal too – his mind with restless
Ardour toiled to excel – to pass the
Intellectual bounds that nature had
Assign'd – along the pictur'd wall he'd look
Desponding, at his old forefather's Coif,
The ermine – and the Lawn – pining away
The very sustenance of life – then pass'd
A shadow to his tomb.

Stout was his heart, and bravely
For a time – the old man combated the
Adversities of life – vain of his high
Norman race – his sons – his trophied and
His pictured line, he saw in ages
Yet remote his name survive wreath'd midst the

Records of his country's deeds, with honour
And renown – but when he heard of Edgar's
Death, he gave the struggle o'er – to him had
Been consign'd this long descent, centuries
Of Fame – and he had lost it – life became
A blank – infirmity prevail'd and bare
His mutual energies away –

But then was beautiful to see
The temper'd calmness of Matilda's mind!
Nature at first had sway – permitting of
A mothers grief – the strife was bitter but
It pass'd away – no pride of heritage
Subdued the better regulated
Feelings of her breast – she could have borne to
Hear her offsprings name resound with praise, to
See them foremost midst the glorious and the
Great – but the ascendant bias of her
Mind was quiet fame – it might be female
Weakness p'rhaps – to hear their actions oft the
Theme of virtue and of truth – to see their
Lives deserve the meed of moral goodness,
Pure from the semblance of imagin'd wrong –
But vice had not been theirs – nor shame – this was
Her Pride – her lofty boast, when thoughts came o'er-
She wound the mantle resignation gave
Around her heart, shut out the icyness,
The shiverings of mortal grief, and in
The temperate glow of confidence and
Faith, reposed thro' all her after life -

(20)
The Village Elm –

We live amidst the things of earth,
And they obtain a being with us – have
Their progress, and decay – a measured
Period with man – familiar to
Our sight we almost form affection for

Them – for those who feel not, or return our
Fondness ever – Those aged Thorns which oft
In infant days the earliest May-flower gave,
The Red-starts constant haunt in summer hours,
The Throstles in the winter morn, have pass'd
Their days with me – time and the sweeping storm
Have rudely torn them – but still their relicks
Kindle something like regard, bringing to
View the lost, past deeds and periods of Life –
The scenes which childhoods frolic days unite
With manhoods meditative mind, frequent
A melancholy pleasure will recal –
We sigh not for departed hours perhaps
For health – for times unfetter'd with a care –
We'ave had our world, and that's enough –
Old friendships, pastimes, merriments appear –
But oh! How oft does mem'ry of other
Things intrude – waking reflections that might
Have rested else –
The venerable Elm, whose branching
Limbs had shelter'd once my school-dames cot,
Where fancy almost hears the wailing and the
Mirth of thoughtless youth – yet lives – an hollow
Shell, the owlets haunt – a few grey-headed
Men, and sober matrons linger too, the
Babblers of a distant day – good Martha
Long has pass'd – her dwelling place a grassy
Knoll – all else is chang'd – all new – the stranger
Paces o'er the soil – what hours return at
Sight of that old stump! The fates, the fortunes
Of us idlers round his twisted roots! The
Swift career, the pains, the pleasures, contests
Of our lives – its summers day a dwarf and
Scanty foliage gives – the autumns gleam
But poorly gilds his leaves – the Red-breast sings
Upon the bleached spray – and ragged, rough
And grey's its ancient vest – yet still we view
Its relicks with regard, and feel a pain
For ev'ry fragment as it falls – it falls

And jars a chord within – 'tis parting from
A venerable friend, associate
With years and scenes that visit us no more –

(21)
Fragmentary Records of an Antient Family –

'tis a tale so old, so utter faint that
Memory retains it not – traditions dead –
And history, has nothing told -
There was a feeble woman
It is said, who lived beside the Woodman's
Cottage in vale – and did some fifty
Years agone, or more – who heard in youth the
Tale – but talk'd not of it much – misfortunes
Came, distress, and care, had other things to
Muse upon, than these old-world tales – and age
Had craz'd her brain – and all was gone –
In the pale ray of some mild autumn's eve
She'd sit upon a time worn stool before
Her cottage door – maund'ring out about some
White scarv'd Knight – and ladies fair wandering
Beneath the moon-lit rocks, and woody shades –
And blood stain'd arrows – then shriek and laugh – and
Doze – and memory of all connecting tale
Had pass'd –

(22)
The Exile – with prospect of the year at the close of 1831

Land of my fathers must I leave thee! – home
Of the dearest and the best! The treasure
House of noble and of virtuous hearts – and
Must I leave behind thy deathless fame, the
Sage – the Hero – Patriot – and Divine,
Crowding thro' ev'ry period and space –
Yes! Land of the mighty ones adieu – thou
Buckler of the helpless and the brave – the
Asylum of the world – I leave thee –

But Oh my country would that heaven might
Spare thee – as oceans gem – the star that lit
The world – but days of darkness come, and thou
Wilt fall in plenitude of wealth, and fame –
Vigourous in strength in science and in
Art – thou hast not heard the war-horse snorting
In his pride – the foot of baseness has not
Touch'd thee – all – all have fallen – and thy hour
Must come – flower of the earth! There is a
Worm mining his secret way – sapping the
Rills of life – heedless of the fate that waits
His labours end – thy sustenance will fail –
And he will die – and misery – deeper
Than misery ere was known, o'ershadow
All – Oh for that arm and mind, which heaven
Gave us once to save, would come again and
Crush the reptile in his lair

I leave thee not my country,
But in bitterness of heart – grateful for
Thy boons – thy comforts and defence
Thro' long and happy years – my feeble arm
Has no support – nor loudest voice controul
The uproar to destroy – and when thou'rt fall'n
No history can tell – no time restore
The like of thee – nor ere again unite
The gathered wisdom of a thousand years
To succour and to bless –

Oh I would raise a monument to thee-
Surrounded by the world in tears – the poor
The widow, and the orphan there searching
Thy stones for bread – then famish at thy feet –
Ruin'd by power or fraud a suppliant kneels
And spreads his hands for aidance and relief
Bearing thy even balance and thy beam
To mete him justice and his right – low in
The Dust should bend poor sinful man weeping

O'er the book of life, bewildered, lost,
In terror pale, seeking thro' ruin'd aisles
And prostrate fanes, his pastor, father and
His friend – whilst raging anarchy, and vice
And envy come with uproar wild, coiling
Their chain round all, and drag thee to the gulph -

APPENDIX 4

THIS PIECE ENTITLED 'Rapacious animals destroying the innocent & feeble,' was written by Knapp on three sides of a large, folded sheet of paper which now constitutes part of the Knapp family records, held by the Buckinghamshire Archive. He left no clue as to whether it was intended for publication in a further edition of *The Journal of a Naturalist*, or not. Had he been ordained, like his father, grandfather and brother, his words might have made a thought-provoking sermon.

> Human reason, is at all times a most deceitful guide, but its directions are least to be depended upon, when it attempts to illucidate the destinations of providence; among the appointments of nature, there are few ordinances more inscrutable than the invariable permission that is given to the strong and bold to prey upon and plunder the weak and timid; such is the condition of the four orders of creation – Quadrupeds, Birds – Fishes, Insects, the last order, being the most feeble, becomes not only the prey of its own mighty ones, but of the other three likewise. It has been attempted to account for this state of things, from the disobedience of man, and the evil that flowed into the world in consequence; but Scripture, which can be our only informant, does not apparently justify this influence... Revelation has declared to us the creation of the animal and vegetable worlds, the origin of man, and all the glorious system of the heavens. Man being so created, in a state of innocence was placed in a situation abundantly filled with everything agreeable to his nature, the free use of all (with one exception) confirmed to him, blessed spirits surrounded his dwelling, & his creator held converse with him – Having disobeyed the only command that was given him, his nature became changed, and susceptible of all the passions, and consequent evils attendant on them: what the condition of man would have been had he not sinned, otherwise than happy, his duration, or his termination, we are not told, and all conjecture is idle. The consequences of sin began immediately to appear, and its darkness, variations & shades, were, from what we find in scripture, as manifest and as numerous in the early ages, as at the present day. The mild dispensations of Christianity where its influence

has been felt, has softened in most, and subdued in some, the darker tints of sin, and passion, but pagan cruelty and malevolence has never been lost, and probably will remain on earth, till we are Christians indeed, and He, the sun of righteousness shall come – Tho' man fell in consequence of his disobedience from happiness to sorrow, and tho' the earth was cursed for his sake, and thorns, and weeds obstructed his way, and perplexed his labours, yet we are nowhere told in scripture that the animal world underwent any change, or were punished for his crimes – The creation and appointment of the rapacious orders and the unremitting system of destruction, persecution, and suffering of the weaker tribes, are not perhaps reconcilable to our ideas of a beneficent providence, limited as our vision, and perceptions are, yet that these rapacious animals were so destined from their first creation without any reference to sin of man seems conclusive from reason, for had they partaken of the malediction that ensued, the whole internal construction, and external formation must have been very different from what we now find them, and we cannot in defiance of reason, and without revelation to aid us suppose a reorganization after the alteration, of man's condition and rejection from Paradise – The construction of the teeth of beast of prey – their jaws – their strength, senses, powers of abstinence – and the shortness of their intestines, are peculiar to them, and admirably adapted to their habits of life, but perfectly unfitted, or useless as graminivorous animals. The whole internal and external structure of birds of prey are in like manner constituted solely for their mode of life, their power of vision is so strong, and so essentially different from that of birds whose food consists of seeds, that clearly manifests a designed conformation, and appointed destination in life – all the harmless creatures, or those applicable to food, are produced in greater portions than the noxious, or those not edible; in the Quadrupeds this produce is perhaps not so obvious, but as food is more easily obtained by graminivorous animals, than can be by rapacious ones, more of the former are bred and brought up especially in the Deer, and Antilope race, we find prodigious herds of these creatures in some parts of Africa and tho' inhabiting the same regions, as the lion, and other carnivorous beasts, yet we find them only singly, or in small parties. In Birds, and Fish, this increase is very remarkable, The Eagle, and other rapacious birds bring up one, or at most only two young ones, the Gallinaceous birds, and all the Duck tribe, hatch perhaps twenty at a time, The Shark, and other rapacious Fish devour their own young ones – and few are brought up – the Cod, herring, macrill, salmon, Pilchard – &c – produce in incredible swarms – The myriads of insects, in some seasons are greater than in others, but the increase in insectivorous birds are in general proportionate; the swarms of fly's and gnats that are generated by the humidity and putrescence of some rivers, particularly near great cities, are incredible, yet those who have seen the multitudes of Hirundines at the beginning of August traversing the river Thames, about Fulham will admit they abound in numbers, equal to their destruction – If, as has been said, the strong animals feed on the others, and thus keep creation in due bounds – the question is asked, but why this extraordinary increase to necessitate reduction? – and we get entangled in inextricable difficulties. Whilst Adam was in Paradise every created bird, and quadruped, was by an especial ordination brought before him to give names to; we are not told that they were inhabitants of paradise, and probably were not, because they were 'brought to him', These blessed gardens were regions, separated from the world, and all their products peculiar, as we know, not only from their producing the tree of life, but by one of the punishments of man's sin being to eat the 'herb of the field,' that is to live on fruits, seeds, roots &c, vegetables, the common products of the earth, (for man was not permitted the use of animal food until after the deluge) and not the particular production of paradise – possibly then in this separated spot the rapacious tribes did not inhabit. Man being now thrust from out this heavenly boundary, and in a depraved state, went into the world abounding with all the beings that were created, a world now suited to his condition, and altered nature, a world as we now find it inhabited by all those creatures which infinite wisdom has formed, manifesting that there was a sufficient cause, and that the cause yet remains, by their being fed and protected by a preserving providence – who having given man powers over them, which he has exercised, and it has been his great employ and pleasure in every period of the world, yet prohibits their destruction.

APPENDIX 5

JOHN KNAPP'S PRAYERS for morning and evening use, preserved in his own handwriting in the Buckinghamshire Archive. They are composed of comfortable phrases borrowed from the 1662 revision of the Book of Common Prayer, combined with Knapp's own thoughts, needs and desires. They reveal his complete trust in the Christian faith.

Morning

O eternal being, father of all mercies, accept the acknowledgement of a poor and sinful creature for all thy bounties towards me, for thy especial protection of me this night; it was thy power O lord that brought me into this world, and it hath pleased thee to support me in it, thro' all the perils and dangers which by my helpless nature I have been hourly exposed to – O my father still continue thy mercy towards me, bless me with a full conviction of thy all-seeing providence, replenish my soul with a grateful sense of thy bounties towards me, and make me earnestly endeavour to deserve them better, correct me in that which is amiss, strengthen and support me to accomplish that which is pleasing unto thee, and in thy especial goodness, through the merits of my saviour and redeemer Jesus Christ, so let thy holy spirit rest with me now, that my conduct thro' this day, may afford me comfort in the hour of death, and in the day of judgement – for his sake O father, pardon our sins, and accept our prayers –

Evening

O merciful father, who despiseth not the sighing of a contrite heart, but suffereth the prayers of those who with true sorrow turn unto thee, accept now thro' the merits of my redeemer, my sincere repentance for the evils I this day may have committed by thought, word, or deed against thy divine majesty; I do earnestly repent, and am heartily sorry for these my misdoings; forgive me father for what is past; and so let thy blessed spirit rest with me, that I may strive to please thee better for the hours that thou permittest to me – And accept O supreme being! the acknowledgements of a poor and sinful creature, for all thy merciful protections to me thro' the various perils of this day: and now O father as it has been thy good will thus long, so signally to preserve me, still let thy blessed angels watch over me and defend me thro' the hours of darkness, and thro' the shadows of death – refresh me with a sweet and comfortable rest: and when the light of day returns, let my soul waken with a grateful sense of thy mercies unto me, and with an earnest wish to pass the coming hours in doing justice, loving mercy, and walking humbly with my god; that when this transitory life shall have passed away, I may receive the blessings thou hast promised me thro' the merits of our redeemer, and rest with thee in everlasting tranquillity and peace – Grant this we beseech thee, O merciful father for Jesus Christ's sake our mediator and our judge -

APPENDIX 6

"The naturalist, from his habit of observing, sees many things not obvious to all persons; his province is to investigate all the operations of nature, and if he record them truly, he has done his duty..." (John Knapp, 5th edition p. 64.)

"We know that many of them are yearly decreasing, and very many that once existed have even become extinct..." (John Knapp, 5th edition, p. 85.)

LISTED HERE ARE the flora and fauna discussed by John Knapp in his *Journal of a Naturalist*, which existed in and around the parish of Alveston in his day. The list presents the opportunity of comparing what species could be found in southern Gloucestershire then, with what might be found there today. Omitted are plants and species mentioned in his book, that he almost certainly saw in other parts of Great Britain. The descriptions used are those given by Knapp himself.

Grasses, meadow and woodland plants

Agrostis, Autumn violet, Bent grass (*Agrostis vulgaris*), Bramble (*Rubus caesius & fruticosus*), Briony (*Tamus communis*), Buttercup, Carding teasle (probably an introduction,*Carduus veneris*), Carnation grass, Cherlock, Cinquefoil, Clematis, Clover, Common Burnet (*Poterium sanguisorba*), Common field scabious (*Scabiosa succisa*), Coarse grass (*Aira caespitosa*), Cockfoot, Cornel (*Cornus sanguinea*), Corn rose (*Rosa arvensis*), Cornflower, Cowslip, Crowfeet (*Ranunculus acris & bulbosus*), Crocus, Daisy, Daffodil, Damask rose, Dandelion, Dewberry (*Rubus caesius*), Dock, Dog rose, Dogstail (*Cynosaurus cristatus*), Dyers' Broom (*Genista tinctoria*), Dyers weed, yellow weed, weld or wold (*Reseda luteola*), Fern, *Festuca sylvatica*, Fetid Hellebore (*Helleborus foetidus*), Fetid iris (*Iris foetidissima*), Foxglove (*Digitalis purpurea*), Foxtail (*Alopecurus pratensis*), Globeflower, Goosegrass, Groundsel (*Senecio vulgaris*), Great ox-eye daisy (*Chrysanthemum lucanthemum*), Holcus, Holly, Horsetail, Ivy (*Hedera helix*), Lichen farinaceous, Lichen prunastri, Lilly, large white (*Lilium candidum*), Logger-heads (*Centauria nigra*), Mat grass, Meadow-fescue, Meadow poa (*P.*

pratensis), Mistletoe, Orchises [*sic*], Ox-eye daisy, Pasture scabious (*Scabiosa succisa*), Phleum, Pilewort, Pimpernel (*Anagallis arvensis*), Plantain, Primrose, Quaking grass, Ray grass (*Lolium perenne*) with white trefoil and hop clover– planted by Knapp, Red archangel (*Lamium purpureum*), Red poppy, Redshanks (*Geranium Robertyianum*), Rough stalked poa (*P. trivialis*), Scabious, Snapdragon, great (*Antirrhinum majus* – white, pink and common), Snowdrop (*Galanthus nivalis*), Soft grass (*Holcus lanatus*), Spring violet, Thistle, Turfey hair grass (*Aira caespitosa*), Vetch, Vervain (*Verbena officinalis*), Water hemlock, Water parsnip, Wild clematis or travellers'-joy (*Clematis vitalba*), Wild mustard, Wild rose, Woody night shade, Yellow oat grass (*Avena flavescens*).

[N.B. Knapp lists many herbs and medicinal plants (5th edition, p. 53 onwards) but does not make it clear whether they may be found locally or not.]

Trees
Apple, Ash, Aspen (*Populus tremula*), Beech, Blackthorn, Bullace plum, Cherry, Crab apple, Crofton pippin apple, Elm, Gansel Bergamot pear, Hazel, Holly, Horse chestnut, Larch, Laurel (*Prunus laurocerasus*), Maple (*Acer campestre*), Oak, Pear, Poplar, Russet apple, Sallow (*Salix caprea*), Sycamore (*Acer pseudoplatanus*), Walnut, White thorn or May (*Mespilus oxycanthus*), Wheeler's Russet apple, Willow, Wych or broad-leafed elm, Yew.

Fungi
Agaric, *Agaricus caseus*, *Agaric surrectus*, Fairy rings, Fingered clavaria (*Clavaria hypoxylon*). Grey puff-ball, Hydnum fungus, Mitred helvella (*Helvella mitra*), Morell (*Morchella esculenta*) 'very rare', Nidularia (*Nidularia campanulata*), *Nidularia striata*, Oderus agaric (*Agaricus odorus*), Pale grey fungus (*Agaricus fimiputris*), Puffball (*Lycoperdon cinereum*), Scented agaric (*Agaricus fragrans*), Starry puff (*Lycoperdon stellatum*), Stinking morell, Stinking phallus (*Phallus impudicus*), Turetted puff (*Lycoperdon fornicatum*), Verdigris agaric (*Agaricus aeruginosus*).Floriform hydnum (*Hydnum floriforme*).

Birds
Blackbird, Blackcap (*Motacilla atracailla*), Buzzard, Bullfinch (*Loxia pyrrhula*, known locally as 'pick-a-bud'), Butcher bird or Great Shrike, Chaffinch, 'twink' (*Fringilla caelebs*), Cock, Cuckoo, Colemouse (*Parus ater*), Common bunting (*Emberiza miliaris*), Common Thrush, Common Wagtail, Crossbill (*Loxia curvirostra*) – 'an occasional visitor', Crow, Daw, Fern Owl, Fieldfare (*Turdus pilaris*), Firetail, Golden-crested wren (*Motacilla regulus*), Goldfinch (*Fringilla*

carduelis), Grey flycatcher (*Muscicapa grisola*), Grey Wagtail, Great Shrike, 'butcher bird' (*Lanius excubitor*), Greenfinch, Green legged sandpiper, Hawk owl, Hedge sparrow or shufflewing (*Motacilla modularis*), Heron, House sparrow, Jack runner (or Water-rail), Jay, common jay (*Corvus glandarius*), Kestrel (*Falco tinnunculus*), Kite (*Fulco milvus*) seen passing overhead, 'but never visits us,' Land-rail, Linnet (*Fringilla linota*), Long-tailed titmouse, long-tailed tomtit, long tom, poke pudding (*Parus caudatus*), Magpie, 'pie', 'magget' (*Corvus pica*), Martin, Meadow lark, Missel thrush (*Turdus viscivorus*), Nightjar, Nightingale, Nuthatch (*Sitta europaea*), Peewit (*Tringa vanellus*), Raven (*Corvus corax*), Red start, Redwing (*Turdus iliacus*), Robin (*Motacilla rubecola*), Rock pigeon, Rook (*Corvus frugilegus*), Sea Swallow (once swept inland in great numbers), Shrike, Shuffle-wing, Skylark, Snipe or jack snipe, judcock, half snipe (*Scolopax gallinula*), Sparrow (*Fringilla domestica*), Sparrow Hawk, Stormy petrel, Mother Cary's chicken, (*Procellaria pelagica*, brought by a hurricane in 1824), Starling (*Sturnus vulgaris*), Swallow, Swift, Thrush, Tit-lark, Titmouse (*Parus*), Tomtit (*Parus caeruleus*), Tomtit (*Parus major*), Tree creeper (*Certhia familiaris*) 'becoming a scarcer bird', Water hen, Water-rail, Wheatear (*Sylvia oenanthe*), Whitethroat, White wagtail, Willow wren (*Motacilla trochilus*), Woodlark (*Alauda arborea*), Woodpecker, Wood pigeon, Wren, Wryneck (*Jynx torquilla*) 'visits annually… in yearly diminishing quantities,' Yellow hammer, Yellow wagtail.

'The world of sensitive nature'
Badger, Common Shrew, Fox, Great bat, Hamster mouse, Harvest mouse (*Mus messorius*), Hedgehog (*Crinaceus europaeus*), Leveret, Long-tailed garden mouse, Marten (*Mustela martes*), Meadow mouse, Mole want, mouldwarper or mould-turner (*Talpa europaea*), Pale blue shrew (*Sorex Daubentonii*? [sic] *Cuvier*), Polecat, Rabbit, Rat (*Mus rattus*), Short-tailed meadow mouse, Squirrel, Stoat, Water Shrew (*Sorex diens*), Weasel.

Insects excluding lepidoptera
American blight, Aphids, Beetle, Black ant, Black pismire, Bonbylius (*Bombylius medius*), Bookworm, Carpenter bee, Cockchafer (*Melolantha vulgaris*), Common hill ant, Common wasp (*Vespa vulgaris*), Dragon fly, four-spotted (*Libellula quadrimaculata*), Fly (*Stomoxys calcitrans*), Glow worm (*Lampyris noctiluca*), Grasshopper, Great black ant (*Formica fuliginosa*), Great black fly (*Musca grossa*), Great Dorr beetle or clock, (*Scarabaeus stercorarius*), Great water beetle (*Ditiscus marginalis*), House fly (*Musca carnaria*), Hornet (*Vespa crabro*), Humblebee, Ichneumon wasp, *Ips niger*, Ladybird (*Coccinella septem punctata*), Mineuses, Red ant (*Formica rubea*), Rose beetle (*Cetonia aurata*), Solitary wasp, solitary

(*Vespa campanaria*), Weevil, White blight (*Aphis lanata*), Winter gnat (*Tipula hiemalis*), Wood louse (*Oniscus armadillo*), Yellow ant (*Formica flava*).

Butterflies.
Admiral (*Vanessa atalanta*), Blue argus (*Papilio argus*), Cabbage white, Common white, Large Admiral (Vanessa atalanta), Little blue argus, Marble (*Papilio galanthea*), Meadow brown (*Papilio janira*), Melitæa artemis, Peacock (*Vanessa Iö*), Painted lady (*Papilio cardui*), *Papilio phlaes*, Spring Azure (*Papilio argiolus*), Sulpher (*Gonepteryx rhamni, Papilio rhamni*).

Moths
Alder moth, Buff-tip, Dartford emerald (*Phalaena lucidata*), Death's-head sphinx (*Acherontia atropos*), Emperor (*Phalaena pavonia*? [*sic*]), Ermine (*Phalaena evonymella & P. padella*), Gamma (*Phalaena gamma*), Ghost moth (*Hepialus humuli*), Goat moth (*Phalaena cossus*), Green housewife moth (*Phalaena vernaria*), Hummingbird hawkmoth (*Sphinx stellatarum*), *Phalaena potamogeta*, Reaumur, Scarlet tiger (*Callimorpha caja*), Small oak (*Phalaena viridana*), Tiger, Verdigris, Yellow underwing (*Phalaena pronuba*).

Arachnids
Spider (unknown type).

Crustacean
Water flea.

Reptiles
Blindworm, Coluber natrix, Common newt (*Lacertus aquaticus*), Grass snake, Slowworm, Viper.

Molluscs
Common snail (*Helix aspera*), *Helix hortensis, Helix nemoralis*, Little banded (*Helix virgata*), Wreathed, slug, *Tellina cornea*.

Amphibians
Toad.

Worms
Auger, Clay hairworm (*Gordius argillaceus*), Common earth worm (*Lumbricus terrestris*), Dew-worm, Wire-worm.

BIBLIOGRAPHY

Alveston Women's Institute: *Alveston, our village within living memory* [Alveston 1957].
Barker: W.R. Barker, *The Bristol Museum and Art Gallery 1772-1906* [J.W. Arrowsmith, Bristol 1906].
Bigland: Ralph Bigland, *Historical, Monumental and Genealogical Collections, relative to the County of Gloucester, Vol. 1.* [London, 1791], reprinted by The Bristol & Gloucestershire Archaeological Society, 1989.
Blackler: Helen Blackler Ph.D., *The Herbarium of Thomas Velley.* Online at www.northwesternnaturalistsunion.org.uk
Blumenbach: J. F. Blumenbach, trans. John Elliotson *Institutions of Physiology,* 2nd edition [London 1817].
Bristol Mercury: *The Bristol Mercury* [Bristol, April 20th, 1839].
Bristol Times and Mirror: *Bristol Times and Mirror* [Bristol, March 10th, 1832].
Brock: W.H. Brock, *The selection of Authors of the Bridgewater Treatises,* [Notes & Records of the Royal Society of London, December 1966, Vol. 21 No. 2].
Cameron: Kenneth Walter Cameron, *Companion to Thoreau's Correspondence,* [Transcendental Books, Hartford, Connecticut, 1964].
Cobbett: William Cobbett, *Cobbett's Political Register Vol LXXVIII* [London, 1832].
Darwin 1: Charles Darwin, *Letter to Asa Grey* [6th November 1862]. Online: https://www.biodiversitylibrary.org/item/225921#page/312/mode/1up.
Darwin 2: Charles Darwin *Reading List.* Online: https://www.darwinproject.ac.uk/people/about-darwin/what-darwin-read/darwin-s-reading-notebooks
Enys: John D. Enys, Davies Gilbert & John F.W. Herschel, *Correspondence regarding the Appointment of the Writers of the Bridgewater Treatises between Davies Gilbert and Others.* [Printed by J. Gill and Son, Penryn, 1877].
Fallon: Dr. David Fallon, *'Stuffd up with books': the bookshops and business of Thomas Payne and Son, 1740-1831.* [Roehampton University 2019]. Online: https://doi.org/10.1080/2373518X.2019.1703322
Fenimore Cooper: Susan Augusta Fenimore Cooper, *Country Rambles of The Journal of a Naturalist* [Phinney & Co., Buffalo, 1853].
Fox Talbot: Henry Fox Talbot, *The Pencil of Nature,* [Longman, Brown, Green & Longmans, London, 1844].
Gentleman's Magazine: Sylvanus Urban, *The Gentleman's Magazine and Historical Chronicle* [London, 1799].
Gloucestershire Society: *The Minute Book of the Gloucestershire Society,* beginning in 1657, Bristol Archives Ref: 40558, NRA 7563 RO Misc.
Green-Armitage: A.H.N. Green Armitage, *Bristol Zoo,* [J.W. Arrowsmith, Bristol,

1964].

Grimaldi: **Stacey Grimaldi F.S.A**, *Miscellaneous Writings from Printed & Manuscript Sources Part III* [One hundred copies posthumously and privately printed, London 1881].

Howard: Joseph Jackson Howard, LL.D., F.S.A, *Miscellanea Genealogica et Heraldica,* Vol 3, [Hamilton, Adams & Co, London 1880].

Howitt: William Howitt, *Book of the Seasons* [Henry Colburn and Richard Bentley, London 1831].

Jardine: Sir W. Jardine and others, *The Annals & Magazine of Natural History, volume 16 "Proceedings of the Linnean Society,"* [R & J.E. Taylor, London 1845].

Jesse: Edward Jesse, *Gleanings in Natural History with Local Recollections,* [John Murray, London 1832].

Knapp: Arthur John, *Roots and Ramifications or extracts from various books explanatory of the derivation or meaning of words* [John Murray, London 1857].

Knapp 1: John Leonard Knapp F.L.S, A.S., *Gramina Britannica,* [Printed for John White, London, 1804].

Knapp 2: John Leonard Knapp, *Arthur, or, the Pastor of the Village; a poem* published anonymously, [Payne & Foss, London 1818].

Knapp 3: John Leonard Knapp, *The Journal of a Naturalist* 1st edition, published anonymously, [John Murray, London, 1829].

Knapp 4: John Leonard Knapp, *The Journal of a Naturalist* 2nd edition published anonymously, [John Murray, London, 1829].

Knapp 5: John Leonard Knapp, *The Journal of a Naturalist* 3rd Edition published anonymously, [John Murray, London, 1830].

Knapp 6: John Leonard Knapp, *The Journal of a Naturalist* 4th Edition published anonymously, [John Murray, London, 1838].

Knapp 7: *John Knapp's letters to John Murray, London,* [National Library of Scotland, MS 40657 Folios 1-94 + 1, July 22nd, 1827 – November 2nd, 1844].

Knapp 8: John Leonard Knapp, *The Journal of a Naturalist,* (almost certainly a pirated edition), published anonymously, [Carey & Lea, Philadelphia 1831].

Knapp 9: John Leonard Knapp, *Herbarium; Specimens of British Grasses Collected in their Native Situations.* Bristol Museum Registered number: Ba1405-1523.

Knapp 10: John Leonard Knapp F.L.S & A.S., *Gramina Britannica,* 2nd Edition, [Longman & Co, London and W. Strong, Bristol 1842].

Knapp 11: John Leonard Knapp, *The Hostel or the Records of my Landlord.* Unpublished manuscript, Buckinghamshire Archives, Ref D-KN/Box 13.

Knapp 12: John Leonard Knapp, an unpublished part manuscript, with its first pages torn out, labelled in Knapp's hand on an attached fragment of paper *Berkeley Castle.* Buckinghamshire Archives, Ref D-KN/Box 13.

Knapp 13: John Leonard Knapp, *The Journal of a Naturalist* 5th Edition, Ed. Sir George White Bt., FSA., [Hobnob Press, 2025]. First printed as an integral part of this book.

Knapp L: Leonora Knapp, *The Moravian Memoir of Leonora Knapp.* Presented to the Moravian Church in Bristol and dated October 17th, 1820. Online at http://moravianlives.org/scripto/?scripto_action=transcribe&scripto_doc_id=300030&scripto_doc_page_id=4000415

Knapp O: Oswald Greenwaye Knapp M.A., *A History of the Chief English Families bearing the name of Knapp* [St. Catherine Press, Strand, London, 1911].
Land Tax: *Land Tax Records for the Parish of Alveston, Gloucestershire.* Online at www.ancestry.co.uk.
Linnean Society 1: *Proceedings of the Linnean Society of London, Vol. 1* [R & J.E. Taylor, London, 1849].
Linnean Society 2: *Letters from J.L. Knapp to James Edward Smith.* The Linnean Collections. Available online through www.linnean-online.org
Linnean Society 3: *Letter from Hugh Davies to J.E. Smith.* The Linnean Collections. Available online through www.linnean-online.org.
London Gazette: *The London Gazette,* Issues 13498 & 15016 [London 1793 & 1798].
Matthews: William Matthews, *The New History, Survey and Description of the City and Suburbs Bristol, or Complete Guide and Bristol Directory for the year 1793-4.* [Bristol 1794].
Mee: Arthur Mee, *The King's England "Gloucestershire"* [London 1938].
Monmouthshire Merlin: *Monmouthshire Merlin,* [Monmouth, January 23rd, 1830].
Mudie: Robert Mudie, *The British Naturalist,* [Whittaker, Treacher & Co, London 1830].
Murray: John Murray, *Letters to John Knapp,* [National Library of Scotland, MS 41910, pp. 53,70 & 71. May 19th, 1829, and January 22nd, 1830].
Randall: Paul Randall, forthcoming *History of St. Mary's Church, Shenley Church End.*
Riddelsdell, Hedley & Price: Rev. H.J. Riddelsdell, G.W. Hedley & W.R. Price, *Flora of Gloucestershire,* [The Cotteswold Naturalists' Field Club, Cheltenham, 1948].
Seaward: Mark R.D. Seaward, *Joseph Banks, James Edward Smith and the Linnean Society.* [Annual Lecture of the Sir Joseph Banks Society 2010.] Online. Available through www.joseph-banks.org.uk
Society of Thornbury Folk: *The Bulletin of the Society of Thornbury Folk,* Series 2, No. 12 [Thornbury c.1955].
Spratt: Mrs G, *The Language of Birds, comprising Poetic and Prose illustrations of the Most Favourite Cage Birds with twelve highly coloured plates.* [Saunders and Otley, Conduit Street, London, 1837].
Stockdale: J.J. Stockdale, *Harriette Wilson* [London 1825].
Stockdale: J.J. Stockdale, *Mad Houses* [London 1831].
Swayne: Rev. George Swayne, *Gramina Pascua: or, A Collection of Specimens of the Common Pasture Grasses, arranged in the Order of their Flowering.* [S. Bonner, Castle Green, Bristol, sold by W. Richardson, Royal Exchange, London, 1790].
Tallis: John Tallis, *London Street Views,* [John Tallis, London] published in booklet form, from 1838-45.
The Monthly Review: *The Monthly Review,* No. XLIV, Vol. X [G. Henderson, London 1829].
The Time's Telescope: *The Time's Telescope,* [Sherwood, London, published annually from 1813]. Online: through https://archive.org
The Spectator: *The Spectator, No. 40 "Literary Spectator"* [London 4th April 1829]. Online: through http://archive.spectator.co.uk/article/4th-april-1829/11/literary-spectator

Thornbury Roots: www.thornburyroots.co.uk

Wax Chandlers: *Apprenticeship Records of the Worshipful Company of Wax Chandlers,* Available online through www.findmypast.co.uk

White: The Revd. Gilbert White, Ed. Grant Allen, *The Natural History of Selborne,* [John Lane, The Bodley Head, London 1900] First published 1789.

Will of John Knapp: *Will of John Knapp proved 1710/11.* [Prerog Court of Canterbury 1384-1858 Prob. 11 1705-1712, Piece 519: Young, Quire Nos 1-47.] Available online through www.ancestry.co.uk

Will of John Leonard Knapp: *Will of John Leonard Knapp,* [Prob 11: Will Registers 1845-1847, Piece 2019: Vol 10, Quire Nos. 451-500 (1845).] Online. Available through www.ancestry.co.uk

Will of Nathaniel Matthew Knapp: *Will of Nathaniel Matthew Knapp,* [PRO Prerogative Court of Canterbury Prob 11/1264 2nd June 1795.] Available online through www.ancestry.co.uk

York Herald: *York Herald* [York, North Yorkshire] Online: https://www.newspapers.com/paper/the-yorkshire-herald-and-the-york-herald/7868/

ENDNOTES

1. There is much confusion over the place, date and parentage of John Knapp. Howard p. 262, states that John Knapp was baptised at Harwell, Berkshire in 1638 and that his brothers and sisters were baptised at Blewberry and Harwell. A detailed Genealogical table of the family of Knapp of Little Linford, Buck, offered on eBay in May 2024 states that he was baptised at Tilehurst, Berkshire. Grimaldi p.329 suggests that John Knapp and his siblings were baptised at Chilton, John in 1646. This last is confirmed as correct by the Chilton Parish Registers. They were the children of Richard Knapp and Elizabeth his wife. The story is now further confused by the fact that since 1974, Chilton and Harwell were transferred from the historic county of Berkshire to Oxfordshire and since 1976, the ecclesiastical parishes of Harwell and Chilton have been combined.
2. The Will of Johannis Knapp unmistakably begins "This is the last will and testament of me John Knapp Citizen and wax chandler of London". Grimaldi wrongly states (p. 330) that John Knapp was a "citizen and Wool Chandler."
3. Grimaldi p.329 onwards
4. The Will of Johannis Knapp, Prerogative Court of Canterbury, Wills 1384-1858 Prob 11: 1705-12, piece 519, Quire numbers 1-47 (1711) Ancestry.co.uk
5. Shenley Baptismal Register May 9th 1767 reads "John Leonard Knapp Son of Primatt & Keturah was born and privately baptised"
6. From the Moravian Memoir of Leonora Knapp, dated Bristol October 17th, 1850, http://moravianlives.org/memoirs/Leonora-Knapp/.
7. Knapp 2, pp. 37-8.
8. Knapp 5, pp. 427-432. See also part 2 of this book, pp. 271-275
9. Knapp included two further memories from his early and very early childhood in *The Journal of a Naturalist* 1st edition (Knapp 3, p. 231). See also part 2 of this book, pp. 102-3 and 141-142.
10. *Gentleman's Magazine*, "J.L. Knapp, Esq., F.S.A., and F.L.S." Vol. 23, pp 653-4.
11. Ibid.
12. Buckinghamshire Archive D-KN/13
13. Jardine, vol. 16 pp. 419-20. Problems with his health are a recurring theme in Knapp's writing, but what afflicted him is never revealed. In his 1837 poem, he describes himself as "a pallid youth". In the preface of his *Journal of a Naturalist* (op. cit) he writes of "intervals of … shattered health" and on page 270 he refers to the "oblivion of pain" that the study of natural history afforded him.
14. The Americans today see John Paul Jones, a Scot by birth, as the father of their navy. The British, in Knapp's day, saw him as a pirate.

15   Gentleman's Magazine, op. cit., pp 653-4.
16   Mary Pierce, née Knapp, was baptised at Shenley on 3rd August 1770. She married John Pierce of Hertford Street, Fitzroy Square at All Saints, Worcester in 1796. He died in 1803. Mary died at Wollstein, (now Wolsztyn, Poland) without issue, on 18th June 1860. She was buried there in the Catholic Ground. (Vol 3 Miscellanea Genealogica, 1880). She had set up and funded a hospice in her house, a pretty 'English Gothic' building in Biala Góra street (now Pod Roberta Kocha), which continued after her death. It became the home of the Nobel prizewinner Dr. Robert Koch. (Lata Mlodzieńcze: Robert Koch (1843-1910), *Kosmos Problemy nauk Biologicznych*, Tom 59 p.18.) It is now the Muzeum dr Roberta Kocha. A group of Mary Pierce's fascinating letters to her nephew Arthur Knapp survive in the Buckinghamshire Archive, Box D-KN/13.
17   *London Gazette,* Issue 15016, p 411.
18   Will of Nathaniel Matthew Knapp, 2nd June 1795.
19   Knapp 11, p.1
20   *Gentleman's Magazine op. cit.*
21   https://www.aucklandmuseum.com/collection/object/am_library-catalogq40-10961
22   One hundred drawings of grasses by Knapp were sold at Christie's on 31st May 1996. The watermark is described in the sale catalogue. [Online] https://www.christies.com/en/lot/lot-1019385
23   Knapp 1, preface.
24   Royal quarto measures 10" x 12 1/2" (250 x 320mm)
25   George Don was the great uncle of the broadcaster writer and celebrity gardener, Monty Don OBE.
26   In 1864 a massive explosion occurred in the gunpowder magazine at Erith. *The Illustrated London News* reported that "not a single stone remained upon another, the very foundations being torn up…. the site was marked with huge fissures and chasms in the earth, immense lumps of which had been scooped out and hurled into the adjacent fields." Several deaths occurred. Some of the bodies were never found. Whether the fields containing Knapp's grasses were destroyed, is not known.
27   LINN-HS 1708.22.1 *Physica indet. (Herb Smith)*
28   Knapp 5, op. cit. See also part 2 of this book, p. 273
29   Carl Linnaeus, *Systema Naturae*, first published in 1735.
30   Seaward.
31   Now usually known as *Mibora minima.*
32   Letter to James Edward Smith dated December 19th 1801, The Linnean Collections GB-110./JES/COR/6/8. Online, available through www.linnean-online.org
33   Letter to James Edward Smith, ibid.
34   Will of John Leonard Knapp.
35   Sir Noah Thomas FRS (1720-1792), born in Neath, became physician to George III. He died in Bath.
36   Letter from Knapp to J.E. Smith, The Linnean Collections GB-110/JES/COR/6/12 July 30th1809. Velley's Herbarium is now in the Collection of The

37 Letter from Knapp to J. E. Smith, The Linnean Collections GB-110/JES/COR/6/7 Dec. 19th, 1801. Online www.linnean-online.org
38 Letter from Hugh Davies to J.E. Smith, The Linnean Collections GB-110/JES/COR/4/19 Jan 11th, 1802. Online www.linnean-online.org
39 Letter from Knapp to J. E. Smith, The Linnean Collections GB-110/JES/COR/6/12 July 30th, 1809. Online www.linnean-online.org
40 Letter from Thomas Velley to James Edward Smith The Linnean Collections GB-110/JES/COR/26/35 Dec. 14th 1802. Online www.linnean-online.org
41 Seaward, op. cit.
42 Letter from Knapp to J.E. Smith, The Linnean Collections GB-110/JES/COR/6/11 28th November 1802, Online www.linnean-online.org
43 Records of the Society of Antiquaries of London SAL/02/030/001 23rd June 1803. Online.
44 Ibid. SAL/02/030/006 Dec. 1st, 1803. Online. Candidates for election to the fellowship of the Society of Antiquaries c.1800 were formally described in the Society's Minute Book as "a Gentleman conversant in the History & Antiquities of this Country." Knapp was described in that way, but additionally as "an accurate observer of the natural History of this Country." He was proposed by Walden Henry Hanmer FSA, Robert Smith FSA and Edmund Turner FSA. He was admitted a Fellow at Somerset House, London on Thursday December 8th, 1803.
45 *A Dictionary of Printers and Printing Obituary* for Thomas Bensley 1889 and *Dictionary of National Biography* 1900 both quoted online at www.paekakarikipress.com
46 Knapp later described his own attempts at reseeding fields in Knapp 3, pp 25-27. See also part 2 of this book: p.13
47 Knapp had inherited £1,000, some furniture, a silver waiter, a marrow spoon and a silver pint mug on the death of his father.
48 Georges-Louis Leclerc, Comte de Buffon (1707-1788), one of the first naturalists to recognise ecological succession.
49 The Star, Saturday 7th November 1807
50 St James's Chronicle, Saturday 7th November 1807.
51 Vulcan was the Greek god of the fire and the forge.
52 A miniature portrait of Arthur Freeman (d.1779) by John Smart (1743-1811) is held in the Museum of Fine Arts, Houston, U.S.A. Freeman owned considerable estates in Antigua, and while in his forties, eloped with the 19-year-old Margaret Thomas, daughter of Sir George Thomas, 1st Baronet, Governor of the Leeward Islands. He left his estates in trust to secure £2,000 for each of his children, the residue to his eldest son Inigo Freeman, who on inheriting his maternal grandfather's estates, changed his surname to Freeman Thomas.
53 St Mary's Chapel (now demolished) was designed and built by John Wood and was opened on Christmas Day 1734. In the mid-1820s, a legal dispute was instigated by the Rector of St Swithin's concerning the legal basis of the Chapel and so whether marriages that had taken place there were valid or not. Where

(Liverpool Botanic Gardens. See Helen Blackler Ph.D., *The Herbarium of Thomas Velley.* Online www.northwesternnaturalistsunion.org.uk — continuation above item 37)

that left Mr and Mrs Knapp is not recorded. See https://www.batharchives.co.uk.
54 Gentleman's Magazine and Linnean Society 1 *op. cit.*
55 https://coflein.gov.uk/en/site/37007/
56 Letter from Knapp to J.E. Smith, The Linnean Collections GB-110/JES/COR/6/10 14th November 1802: "I spent one summer in Wales, and this last was devoted to rambling amidst the Caledonian Alps…". Online www.linnean-online.org
57 Knapp 3, p.116. See also part 2 of this book: p.70
58 Bigland, Vol 1 p.50.
59 John Collinson was ordained in Bristol in 1803. He was collated to the north-eastern rectory of Gateshead in 1810, a post he held for 29 years. In 1839 he removed to the rectory of Bolden, while also serving as perpetual curate of Lamesly, rector and honorary canon of Durham and rector of Mortlake. He was elected to the exclusive Literary Society in London in 1807. He published a number of religious works and contributed to the *Gentleman's Magazine* in 1821 and to the influential London magazine the *Quarterly Review*. Through this he was associated with John Murray, later publisher of Knapp's most successful work. Online https://www.wikitree.com/wiki/Collinson-184
60 40 acres, 1 rood and 2 perches equals just over 16 hectares.
61 Bristol Times and Mirror, Saturday 12th July 1845.
62 Knapp 9, pp. 17, 47 & 49.
63 Knapp 12, p 1. This description bears a marked similarity to the opening sentences of Knapp's *Journal of a Naturalist* and was perhaps their inspiration. See part 2 of this book, p.1
64 A replacement church, designed by the Bristol architect Henry Lloyd (1812-1887) but of modest architectural merit, was built supposedly nearer to the centre of population in 1885, leading to the abandonment of the mediaeval building. In 1960, the roof of the old church was removed and all the walls, except part of the north nave wall, were demolished. This remarkable act of vandalism was carried out as a Civil Defence exercise.
65 Mee p. 23
66 Charlotte Henrietta Knapp was buried on July 30th, 1828, Revd. Dr. Charleton officiated. Catherine Mary Knapp (wrongly entered in the register as "Catherine Ann") was buried on 27th November 1829, Revd. Mr Holder officiated. Frances Ann Knapp was buried on 30th October 1833, Caddell Holder, curate, officiated. Lydia Frances Knapp was buried on 29th October 1838, the Revd Caddell Holder, curate, officiated. John Leonard Knapp was buried on 5th May 1845, the Revd. Caddell Holder officiated.
67 Society of Thornbury Folk, *Bulletin*, Series 2, No. 12, quoted in *Alveston, our village within living memory,* Appendix VIII.
68 Knapp 3, pp 163-6. See also part 2 of this book: pp. 99-101
69 Rev. Caddell Holder (1803-1882) served as a curate at Alveston. Later, as Rector of St. Juliot in Cornwall, he employed Thomas Hardy (1840-1828), then an aspiring architect, but later the celebrated author, to refurbish his church. Hardy met Holder's sister-in-law Emma Gifford while there. He subsequently married her.

70 Land Tax 1819.
71 Knapp 3, pp 349-353
72 Bristol Archives, Ref: 40558, NRA 7563 RO Misc. p. 422. During the period of Knapp's residence in Gloucestershire, lists of subscribers who did not attend the Society's annual fund-raising dinners are only recorded for the year 1815.
73 Buckinghamshire Archives, Ref: D-KN/Box13
74 'Let nothing we say reveal any failing in our character.'
75 *The Gentleman's Magazine*, Vol. LXIX 1st part, 1799 pp.171-2
76 Fallon.
77 Blumenbach, p. iii.
78 *York Herald* 4th July 1819.
79 Letter from Knapp to John Murray Sept 3rd 1827
80 Knapp 2, p. 8.
81 Knapp 2, p. 144
82 *The Gentleman's Magazine*, April 29th, 1845
83 *The Times Telescope*, title page 1828.
84 *Time's Telescope*, introductory "Advertisement" 1827.
85 This was following an unfortunate start, albeit one fairly typical among such publications of the period. An article in *The Morning Post for* 6th December 1813 under the heading "Court of Chancery, Saturday December 4. Literary Property, Brady v. Sherwood &c" reveals that the first edition of *Time's Telescope*, (seven years before Knapp's involvement with it), had blatantly plagiarised the complainant John Brady's Clavis Calendria. Brady's death in 1814 perhaps eased the problem.
86 Quoted in *The Time's Telescope* "Notices" 1827.
87 Quoted in *Time's Telescope* 1824 "Notices."
88 *Time's Telescope*, "The Naturalist's Diary" 1828 p.372.
89 *Time's Telescope* 1820 pp. 2-3.
90 *Time's Telescope* 1820 footnote 1.
91 *Time's Telescope* 1820 pp.164-5.
92 Ibid. p.165.
93 Knapp 3, pp. 180-181. See also part 2 of this book: p.108
94 *Bristol Mercury*, Saturday 17th May 1845: "ALL the Live and Dead FARMING STOCK, PHAETON, neat HOUSEHOLD FURNITURE, and other effects, late property of John Leonard Knapp, Esq."
95 Matthews, p. 89.
96 Alveston Women's Institute, p.89.
97 Knapp's sister, Mary Pearce, who lived in Wollsztyn, on the Polish / German border, wrote to him on July 22nd 1842, "The Countess I Plater has given me a Cow which I nourish with the Grass and products of my Ground a little excepted. They have quite other ways of feeding the Cattle here, than we have in England. It is so much more expensive, and the Animals neither give so much Milk nor so good as yours." (Buckinghamshire Archive D-KN/13). In the same Archive, is an undated note written by Knapp to his son Arthur which reads: "I have been very fortunate with my hay having secured the whole without rain – a large crop – and long about from the impossibility of obtaining mowers – all being occupied

about their Potato crop."
98 Knapp 13 p.161
99 White: p. xxix.
100 Letter to John Murray April 16th, 1833.
101 Knapp 3, p.18. See also part 2 of this book: p.8
102 Ibid. p.13. See also part 2 of this book: p.6
103 Ibid. p.15. See also part 2 of this book: p.7
104 Ibid. pp.358-9. See also part 2 of this book: pp.242-243
105 Ibid. pp. 147-8. See also part 2 of this book: pp.89-90
106 Ibid. p. 358. See also part 2 of this book: p.238
107 Ibid. pp. 229-30. See also part 2 of this book: p.138
108 Knapp 13 p.139
109 Ibid. pp.187-188
110 Ibid. p.249
111 Ibid. p.228.
112 Ibid. p.219
113 Brock pp. 162-179.
114 Ibid. p. 137.
115 Ibid. pp.197-8.
116 Ibid. pp. 226-7.
117 Ibid. p. 226-7.
118 Knapp 7, folios 1 & 3
119 Ibid, folio 4.
120 Ibid, folio 9.
121 Ibid, folio 13.
122 Ibid, folio 11.
123 Ibid, folio 17.
124 Knapp's brother-in-law.
125 Probably his niece, Mary, who within a month married The Hon. and Rev. C.G. Perceval.
126 Knapp 7, folio 21.
127 *The Belfast Commercial Chronicle,* March 2nd, 1829
128 *Monthly Review,* 1829.
129 *The Spectator,* 4th April 1829, pp. 11-12.
130 *The Quarterly Review,* Vol 39, Art V, pp. 406-431.
131 All three quoted in "Cabinet Library No. II", an appendix to *The Journal of a Naturalist* published by Carey & Lea, Philadelphia 1831.
132 Knapp 7, folio 23.
133 Knapp 7, folio 25.
134 Knapp 7, folios 27-28.
135 Knapp 7, folio 39. The tree no longer exists. It was almost certainly a relic from the mediaeval Royal Park of Alveston. Although Knapp believed it to be the only survivor, one other hollow oak of similar age is known, 400m north of Old St Helen's churchyard, in a field adjoining the footpath to Knapp's home.
136 Ibid, folio31.
137 Ibid, folio 35.

138 Ibid, folio 38.
139 Theodore Hook (1788-1824) was a writer, wit, one-time civil servant, jailbird and famous practical joker, who while in confinement for debt wrote nine volumes of stories collected under the title *Sayings and Doings*.
140 Murray, January 22nd, 1830.
141 Knapp 7, folio 42
142 Knapp 5, opposite the title page. See also Knapp 13, opposite the title page
143 Ibid, pp. 431-2. See also Knapp 13 pp.274-275
144 The original manuscript of *The Hostel or the Records of my Landlord* is kept with the Knapp Family Deeds and Papers (ref: D-KN/BOX 13) in the Buckinghamshire County Archives. The Siege of Calais, a tragedy also forms part of this collection and despite being catalogued as part of John Knapp's papers, is not in his handwriting and is most likely the work of his elder brother.
145 *Bristol Times and Mirror,* Sunday 10th March 1832. See also Knapp 13, pp.61-62, where this piece is woven into the text.
146 *Morning Herald* September 18th, 1830.
147 *Morning Herald,* London, 9th June 1831.
148 Jesse, 1832, p. vii. Perhaps the first words of both Knapp and Jesse's epigraphs were a nod towards Chaucer's "Go litel boke" at the end of his epic poem *Troilus and Criseyde*. Jesse however wrote of his: "I will only add, in the last two lines of a stanza which one of our poets, later than Chaucer, offers me.".
149 Spratt, pp. 19–31, 332-335. Maria Spratt was wife of the "surgeon-accoucheur", artist, caricaturist and printmaker George Spratt, author of *Obstetric Tables*.
150 *Monmouthshire Merlin,* January 23rd, 1830.
151 Knapp 7, folio 51.
152 Cobbett, "Journal of a Naturalist" 2nd March 1833, columns 542-564.
153 Knapp 3, p. 20. See also Knapp 13, p.9
154 Knapp 3, pp. 19- 20. Cobbett's criticism clearly struck home though. In the additional piece that he wrote for "page 207, after line 14" of the 3rd edition, he finished "I am very well aware that an anecdote relating to the same bird similar to what is stated above is in 'Whites Selborne' – but I have not suppressed my statement, it being a full corroboration of that account given by my great and amiable predecessor-"
155 Knapp 7, folio 55.
156 In the same year, the same Philadelphia company published a replica of William Howitt's *The Book of the Seasons*. Presumably this had been pirated also.
157 Online: https://theprivateerclause.com/charles-dickens-1842-tour-of-america/
158 Knapp 7, folio 58.
159 In an undated note to his son Arthur, now included in the file relating to Mary Pearce in the Buckinghamshire Archive, he wrote "I have this morning given Mr. Smith a long sitting, and a most frightful Phiz – he has produced – to be amended it is to be hoped – he talks of five more sittings." This presumably was a painted portrait, which to date has not been found.
160 Ibid, folio 25.
161 Ibid, folio 70.
162 https://www.in2013dollars.com/uk/inflation/1840, consulted in February 2024.

163 *Minutes of the Ephemeral Books Society of Bristol,* a society which still flourishes today. Bristol Records Office MS 46524 p. 048.
164 The 1841 Census Return lists Knapp's servants as Ann Bulgin, aged sixty-five, Mary Binden, aged twenty-five and Susannah Saniger, aged twenty. It may be that Lydia Margaret, his daughter was unwell and needed care, for Knapp's sister Mary had written to him from Wollsztyn on 22nd July 1842 "embrace Margaret for me, & tell her how much I wish she had a better state of body – it is cruel for so young a person to be either quite, or almost a cripple." (Buckinghamshire Archive D-KN/13.
165 The inscribed book was sold by the auctioneers Flints of Thatcham, on May 1st, 2024. A description of the Carrick family can be found on https://www.thornburyroots.co.uk/outside-thornbury/poultrybrook-farm/ Caroline Carrick was born Caroline Tudway in Wells on 16th March 1787. She was the daughter of Robert Tudway J.P. and his wife Mary. She was granddaughter of Rev. Thomas Paine, a canon of Wells Cathedral.
166 *Bristol Mercury* April 20th, 1839.
167 Knapp 7, folios 78 and 87.
168 The marriage took place on October 6th, 1842, and was reported on 8th in the *London Evening Standard.* Knapp's second son William Thomas Knapp had married Sarah Mary Marsh at Walcott in Bath, Somerset on 19th September 1836. After her death, William Knapp married Philadelphia, fifth daughter of Sir Percival Hart Dyke and widow of Rev. James Mackenzie.
169 https://www.british-history.ac.uk/survey-london/vol17/pt1/pp.7-18
170 *Bristol Mercury* Monday, March 17th, 1890. The article was by "Thomas Wright, author of "The Town of Cowper," "The Chalice of Carden," &c." from his address at Cowper School, Olney. Wright noted that he was indebted to "one of his [Knapp's] sons, [who] is still living" for the information quoted in his article.
171 Knapp 3, p. 374. He continues by giving a long description of what he believed was the most instantaneous and humane method of killing insects for preservation. See also Knapp 13, p.252
172 Slimbridge, now home to the Wildfowl and Wetlands Trust, is thirteen miles to the north of Knapp's home at Alveston.
173 Knapp 7, folio 93.
174 *The Gentleman's Magazine,* Vol. 23, pp. 653-4.
175 Pallant House Gallery website: https://pallant.org.uk/who-was-gilbert-white-and-why-is-he-important/
176 Darwin 1, November 6th, 1862.
177 Knapp 3, pp 5-6. See also Knapp 13 p.3
178 Reported in the unusual website www.kouroo.info. Cameron confirms that Thoreau owned a copy of the 1831 Philadelphia edition of *The Journal of a Naturalist* and that his wife, Sophia, gave it to the Concord Free Public Library after his death.
179 Knapp 9.
180 Advertisement by the auctioneers, Dowells and Lyon, placed in the *Edinburgh Evening Courant,* 26th June 1858.
181 Information kindly supplied by Rhian Rowson, Natural History Curator, Bristol

Museums: e-mail February 2025.
182 Arthur John Knapp was also an etymologist. In 1856, John Murray, his father's publisher, published his *Roots and Ramifications or extracts from various books explanatory of the derivation or meaning of words,* in a similar format to *The Journal of a Naturalist.* A.J. Knapp wrote it originally to raise funds for a school at Pickwick in Wiltshire, which had a large population of labourers and quarrymen.
183 Green-Armitage, pp. 25-5.
184 It is conceivable that John Knapp knew the owner of the Hollywood estate (then known as the Compton House estate), and so perhaps he knew the estate itself. Knapp was made an Honorary Member of the Bristol Philosophical Institution in February 1824. Henry Sheppard of Compton House became a member in June 1832.

INDEX

Alveston, Gloucestershire, 37-45, — St Helen's Church, 37, 41-44, 84, 148n64 — Earthcott, 38, Grovesend, ibid. — Collinson's (Knapp's home), 38-39 — The Ship Inn, 38, 56-57 — Shellards Lane, 39, 68, 73-74, 150n135— Mediaeval deer park, 39 — The Street, 40 — Land Tax at, 44 — Abbey Camp (Romano British hillfort) at, 45 — Royal Oak, 57 — Maps of Knapp's farmland, 60-61, Unfortunate travelling man at, 60-61 — Limestone at, 60-61 — Whirly pits at 62.

Arthur, or the Pastor of the Village, 46-49, plot of, 49-51, 59, 69.

Aspley Guise Classical Academy, 7-8.

Aust, Old Passage, 38, ferry, 40.

Avery, Dighton, apprentice to John Knapp I, 1.

Baird, Prof. S.F. (1823-1887, Assistant Secretary of the Smithsonian Institute), preface, 94.

Banks, Sir Joseph (1743-1820), 16.

Barrington, Hon. Daines, (1728-1800), 26.

Bath, Somerset, 23, 32 — St Mary's Chapel at, 32.

Bensley, Thomas (d.1835), printer, 26, 29-30 — Bensley & Sons, 46 — Fires at Bensley's warehouse, 29-31, 48-49.

Berkeley Castle, Gloucestershire, 40.

Bigland, Ralph (1712-1784, Gloucestershire historian), 38.

Blomfield, Charles (1786-1867, Bishop of London), 65.

Bolt Court, Fleet Street, London, 26 — First fire in, 29-31, second fire in, 48-49, 81, reference to, 68.

Bridgewater, 8th Earl of, see Egerton.

Bristol, 40, Rownham ferry, Clifton, ibid. — King Street Library at, 57, Ephemeral Book Society at, 84 — Philo-Botanical Society at 86 — Philosophical Institution at, 95-6 — Bristol Museum, 21, 96, 98, — Llanfoist, Clifton, Bristol, 96, home of Arthur Knapp.

Brockford, Suffolk, 32.

Caldecot (Buckinghamshire), 9.

Carey & Lea, publishers, Philadelphia, 82.

Carteret, Capt. Philip (d. 1796), 7-8.

Carrick, Caroline of Marlwood Grange, Thornbury, 84, 152n165.

Chilcot, John, printer, stationer and bookseller, Bristol, 88.

Clifton and Bristol and West of England Zoological Society, 96-97.

Clowes, Willian, printer, 72.

Cobbett, William (1763-1835), 81-82.

Collinson, Rev. John (1782-1857, Knapp's landlord), 38, 148n59.

Cooper, Susan Augusta Fenimore (1813-1894), preface, 92-95.

Craven the, 11.

Crossman, Richard, solicitor, 57.

Curtis, Rev. Moses Ashley (1808-1872, botanist), 94.

Darwin, Charles, preface, 93.

Davies, Rev. Hugh (1739-1821), 16, 18, 20.

Davison, Thomas, printer, 69.

Deptford, 1.

Dobson, James, apprentice to John Knapp I, 1.

Don, George (1764-1814), 14, 16.

Drapers, Worshipful Company of, 1.

Dyrham, Gloucestershire, 12.

East Harptree, Somerset, 12.

Egerton, Francis Henry, 8th Earl of Bridgewater, 65.

Elton, John, of Redland, Bristol, 90-91.

Enys, John, 65.

Experiments by Knapp, with grass 61-62, fish scales 62, animal fur, ibid, birds' eggs, ibid., goat moth, ibid., soil temperature, ibid., moisture exhaled by plants, ibid., woolly fibres in morels, ibid.

Fires, see Bensley and Bolt Court.

Freeman, Lydia Frances (1772-1838), see Lydia Frances Knapp.

Freeman, Arthur, of Ratton, Sussex, 32, 147n52.

Fungi, 35-36

Gilbert, Davies FRS (1767-1889, President of the Royal Society), preface, 65.

Gloucestershire Society, the, 45.

Gramina Britannica, 12-15, 23-31, 34, 58, 69 — Contemporary review of, 27-29 — Second edition of, 86-88 — original drawings for, 95.

Gramina Pascula, 12.

Grasses, Panicum Crusgalli, 14, Alopecurus Bulbosa, ibid., 40, Phleum Bochmeri, ibid., Polypogon Littoralis, ibid., Aster Tripollum, ibid., Cyprinedium, ibid., Knappia Agrostidea, 17, Agrostis semi-nuda, 24, Festuca repens, 25, Carnation-grass, 38, Poa distans or Glyceria, 40, Poa procumbens, 40, Festuca sylvatica, 86-87.

Haberdashers, Worshipful Company of, 1.

Hereford Militia, 6.

Hibbert, John, engraver, (1768-c.1840), 23, 32.

Holder, Rev. Cadell, 91, 148nn66, n69.

Hollywood Tower estate, Cribbs Causeway, 97.

Howley, William (1766-1848, Archbishop of Canterbury), preface, 65.

Ingleton, 11.

Jenner, Dr. Edward of Berkeley, 56, 58.

Jones, John Paul, pirate, 7-8.

Journal of a Naturalist, first published 1829, 69, Second edition published 1829, 72-4, — Contemporary reviews of, 70-72 — Eight extra-large versions of the 2nd edition printed, 74 — Third edition, 1830, 74, 76-78, 80, 83-84 — Pirated in America, 82, preparation for fifth edition, 89-91, 95.

Kinsey, Rev. William, of Trinity College, Oxford, 74.

Knapp, Arthur John (1808-1883, Knapp's eldest son), preface, 75, 88-89, 96, 153n182.

Knapp, Catherine (Knapp's great grandmother), 1.

Knapp, John (1646-1710), Knapp's great grandfather), 1, 145n1.

Knapp, Dr. John, of Edinburgh, Knapp's nephew, 89, 96.

Knapp, John Leonard (1767 – 1845), preface — Poetry by, 3, 5, 13-14, 16, 34-5, 50-51, 76-78, 100-131 — Period biographies of, 6-7 — Education at Thame, 6, at Aspley Guise, 7-8, at Kensington, 8 — Service in the Navy, 6-8, 10, in the Worcester Militia, 8-9, in the Northampton Supplementary Militia, 8-10— Poor health of, 8, 37, 76, 90-91 — Botanizing at Ingleborough, 10-11, other places visited, 14, 98-99, 142n26 — Favourite pony, 11, 36, 57 — Collecting ferns, 12 — Drawing fungi, 12 — Snipe shooting, 12-13, 65 — Collecting grasses, 12-15 — Publication of, *Gramina Britannica*, 12-15, 23-31, 34, 58, 69, destruction by fire of, 29-31 — Second edition of, 86-88 — Third edition, 74, 76-78, 80, 83-84 — *Arthur or the Pastor of the Village*, 46 -51, 59, 69, destruction of by fire, 48-49 — *The Hostel or the Records of my Landlord*, 78, 95 — Payment for books sold, 74-6, 80, 84 — *Journal of a Naturalist*, first published 1829, 69, second edition published 1829, 72-4, third edition, 74, 76-78, 80, 83, supposed fourth edition, 83, preparation for fifth edition, 89-91, 95 — Fifth edition finally published, 97 — Contemporary reviews of first edition, 70-71 — Extra-large versions of the 2nd edition, 74 — Pirated in America, 82-83 — Rival works, 80-81 — Essay on rapacious animals by, 132-134 — Christian faith, 14-15, 18, 64-65, 67, 135-136.— Knappia Agrostidea, grass named in Knapp's honour, 17, 21 — Books read by, 17-19 — Herbaria studied by, 19-20 — Correspondence with the Linnean Society, 23, with John Elton, 90-91 — Lodging in Bath, 23, 32 — Learned societies, Linnean, 16, 70, Antiquaries, 23, 70, 147n44, — Marriage to Lydia Frances Freeman, 32, 147n53 — Children of, Lydia Margaret, (b. 1806), 33, 84, 152n164, Frances (b. 1807), 33, 35, Arthur John, (b. 1808), Frances Ann (b.1809), 33, 42, Catherine Mary (b. 1811), 33, 42, 74, William Thomas (b. 1813), 33, 152n168, Charlotte Henrietta, (b.1816), 37, 42, 69 — Residence at Llanfoist, 32-37 — Woodcock shooting, 35 — Herbarium created by, 21, 40 — Service as churchwarden, 41-43 — Collector of Land Tax, 44 — Interest in horticulture and agriculture, 58-59, 61-62, 77-79 — Views on conservation and decline of habitat, 65-67 — Dislike of wanton cruelty, 66-67 — Supposed portrait by Finden after R. West, 83 — Death of, 91 — Lost portrait of, by Smith, title page, 96, 151n159 — Wax bust of by Parker of Bath, 86, 96.

Knapp, Harriett, of Highgate, (1760-1841), Knapp's first cousin once removed. 89.

Knapp, Leonora, (1774-1851, Knapp's youngest sister) 2-3.

Knapp, Louisa, (née Thomas), wife of Arthur Knapp, 88.

Knapp, Lydia Margaret, (b.1806) Knapp's daughter, 33, 84, 152n164.

Knapp, Lydia Frances, née Freeman (1772-1838, Knapp's wife), 32, 42, 44, 84.

Knapp, Mary, (see Pearce).

Knapp, Rev. Matthew (1684-1752, Knapp's grandfather), 1-2.

Knapp, Matthew of Little Linford, Knapp's nephew, 88-89.

Knapp, Miss, of Shenley, 70.

Knapp, Nathaniel Matthew (1760-1795, Knapp's eldest brother), 9.

Knapp, Rev. Primatt, (1730-1793, Knapp's father), 2-3, death at Powick, 9.

Knapp, Rev. Primatt (1764-1838, Knapp's elder brother), 5, 84.

Knapp, William Thomas, (b. 1813) 33, 152n168.

Lauderdale House, Highgate, 88-89.

Le Conte, Major John Eaton (1784-1860, explorer), 94

Linnæus, Carl (1707-1778), 16, 20, 146n29.

Linnean Society of London, 6, 9, 14, 16, 17, 20, 23, 29, 32, 70, 94, 98.

Linford, Great, 1 — Little, 1.

Llanfoist, Abergavenny, Knapp's residence at, 32 -37 — St Faith's Church, 33, 35 — Llanfoist, Clifton, Bristol, 96, home of Arthur Knapp — Middle Llanfoist, 33-34.

Longman and Sons, Paternoster Square, publishers, 87-88.

Loughton, 1.

Meteorology, 53-56. 62-64 — the Year 1825, 63-64.

Millard, J. of Camden Place, 70.

Murray, John (1778-1843), publisher, 30, 68-72, 74-76, 78, 81-83, 86-87, 89, 91, 93, 101, 148n59, 153n182.

Naturalist's Diary, see *Time's Telescope*.

Nodder, Richard Polydore, (fl. 1790-1820) engraver, 23, 25.

Northampton, Lord, 8-9.

Northampton Militia, 6, 9-10.

Payne & Foss, Pall Mall, publishers, 46-47.

Pearce, Mary (née Knapp, 1770-1860), Knapp's sister), 6-7, 9, 146n16, 149n97, 151n159.

Pennant, Thomas (1726-1798), 19, 26, 72.

Phinney & Co, of Buffalo, USA, 93.

Pony, Knapp's faithful, 11, 36, 57.

Powick, Worcestershire, 9.

Pratten, Lydia Margaret, see Lydia Margaret Knapp.

Pucklechurch, Gloucestershire, 12.

Ridgway, Mr (probably James Ridgway (fl. 1784-1847) publisher, 86-87.

Riots, at Bristol, 7, 9, 83.

Romano British hillfort at Alveston, 45.

Royal Society, the, preface, 16, 65.

Rugman, Sarah of Alveston, 42-43.

Selborne, Natural History of, preface, see White, Rev. Gilbert

Selman, Sampson, collector of Land Tax, 44

Severn, the River, 37-38, 40-41, 91, 99, 115.

Shellard's Lane, 39 — oak tree in, 61, 68, 73-4, 150n135.

Shenley Church End, 1-5 — St. Mary's Church, 3 — Rectory, 4 — Brook End, 9.

Slimbridge, Gloucestershire, 91, 152n172.

Smith, Dr. (later Sir) James Edward, 16-17, 18, 19, 20-22, 98, 148n56.

Smith, Mr, artist, title page, 96

Society of Antiquaries of London, 23, 70, 147n44.

Sole, William, (1741-1802), 20-22.

Stockdale, J.J. pornographer, 81.

Strong, William (d. 1846), of Bristol and Exeter, publisher, 86-87.

Swayne, Rev George, (c.1746-1827), 12 — *Gramina Pascula*, ibid, 27.

Thame, Grammar School at, 6.

Thomas, Inigo, 70, 88, 147n52.

Thoreau, David, preface., 95, 152n178.

Thornbury, S. Gloucestershire, 38, 40, Swan Inn at, 57.

Time's Telescope, the Naturalist's diary, 52 – 57, 58, 59, 100, 149n85.

Tomtits, regarded as vermin, 43, 45.

Tonge, Cmdr. William Norris, preface.

Tortworth Chestnut, the, 40-41.

Transport, Mail coach from Bristol to Birmingham, 57, Bristol to London via Bath, ibid.

United States of America, preface, 74, 82-83, 92-93.

Velley, Thomas, (1748-1806), 19-22.

Vitriol, green, 1.

Walmesley, John, of the Circus, Bath, 74.

Walpole, Horace, 55-6.

Wax Chandlers, Worshipful Company of, 1.

Weather, see Meteorology.

Whaddon, 1, 9.

White, Benjamin, (1725-1794), publisher & bookseller, 24-25.

White, John (fl. 1785-1816), publisher & bookseller, 24, 26-27, 30-31.

White, Rev. Gilbert, (1720-1793), preface, 19, 24, 26 58-59, 71-72, 81, 82, 84, 92-93, 151n154.

Williams, Samuel, wood engraver, 73.

Wolstein, (now Wolsztyn), 7, 146n16.

Worcester Militia, 8-9.

An 'AI' generated portrait of John Leonard Knapp (1767-1845), derived from a poor black and white photograph of the lost original. (courtesy of Paul Randall.)

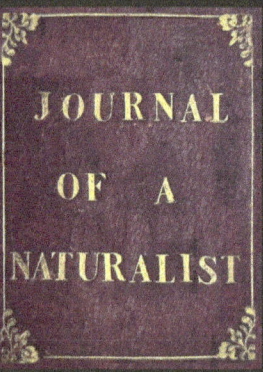

JOURNAL

OF A NATURALIST.

———

— "Plants, trees, and stones, we note,
Birds, insects, beasts, and many rural things."

　　　　Go forth, my little bark, again, and risk
　　　　Once more thy fragile form upon the world's
　　　　Unsteady surge. Rude gales and currents may
　　　　Be found to meet thee on thy way, and check
　　　　　　Thy progress to a ready mart: yet steer,
　　　　　　If happily thou canst, thy course — light is
　　　　Thy freight, nor rare; and few I deem'd would prize
　　　　　　Such merchandize as thine, nor willing aid
　　　　Thee foundering in the wave; but thou hast sail'd
　　　　In tranquil seas — warm, sunny gleams have cheer'd
　　　　Thee on; and friends— kind friends!— were seen,
　　　　　　Who slighted not thy ware, all rustic as
　　　　　　It was. Yet bear thee steady on thy course;
　　　　And chance some wandering trafficker may come
　　　　　　To seek a sample of thy stores, and find
　　　　　　　　The lading to its invoice true.

THE JOURNAL OF A NATURALIST

The Shellard's-lane Oak— p. 29.

JOHN LEONARD KNAPP F.S.A., F.L.S
Hon. Member, the Bristol Philosophical Institution.

Previously unpublished
FIFTH EDITION

Edited by Sir GEORGE S.J. WHITE Bt., F.S.A.

THE HOBNOB PRESS
GLOUCESTER
MMXXV.

First edition, published by John Murray, London, February 1829

Second edition with added text, John Murray, July 1829

Third edition, with added text, John Murray, July 1830

Fourth edition, John Murray June 1838

Unauthorised American editions:

Carey & Lea, Philadelphia 1831

"Country Rambles or the Journal of an English Naturalist" with notes by Susan Fenimore Cooper,
Phinney & Co., Buffalo, 1853

5th edition, including John Knapp's unpublished additions, Hobnob Press, Gloucester, 2025

Introduction to the Fifth Edition

THE FIRST EDITION of John Leonard Knapp's *Journal of a Naturalist*, which the author described as "the observation and application of years," was published by John Murray in London in February 1829. It was the first book of its kind since the Revd. Gilbert White's best-selling *Natural History of Selborne* of 1789. It was critically acclaimed, though its author hid from personal recognition, by choosing to have his work published anonymously. He went to some lengths to avoid naming Alveston, in southern Gloucestershire as the village in which he lived and compiled his many notes, explaining later that he was "a person who avoids all publicity."

So successful was *The Journal*, that in July of the same year, a second edition of some two thousand copies was printed, with added text and extra illustrations. A third edition, also updated, was produced in July 1830. Almost certainly unknown to Knapp and his publisher, the book was pirated and published in America in 1831, by Carey and Lea of Philadelphia.

In due course, the run-away sales began to slow. Knapp himself modestly believed that the novelty of his book had been the chief cause of its success, and that it was inevitable that sales would taper off. In this he was partially right. But great numbers of 'look-alike' books had begun to flood the market, rapidly produced by authors keen to share in Knapp's extraordinary success. Knapp had single-handedly reignited popular enthusiasm for natural history and enthusiastic competition was the price that he had to pay for it.

None of this prevented him from hoping that John Murray would publish a fourth, updated edition and he made all the necessary preparations for it. He was pleased, although taken aback, when in June 1838, he read in a newspaper that "a New and Cheaper Edition, with Plates and Woodcuts" had been issued, which he had known nothing about. He wrote at once to Murray expressing his surprise and hoping that this fourth edition was not the unsold copies of the third, in disguise. Murray's reply has not survived, but the evidence of existing fourth editions, whose text matches the third edition and whose title pages appear to be tipped in, rather than bound, suggests that they are indeed what Knapp had suspected.

Still Knapp did not give up hope of a genuinely new edition and continued to make fresh observations for it. These he recorded in three marbled exercise books, against the day that they would be needed. He wrote to Murray on the subject regularly. This correspondence, over several years, kept his hopes alive, either of a revised fifth edition, or a supplementary volume containing his additional notes alone. But by then, scientific discoveries were beginning to undermine the widely accepted theory of Creationism, the bedrock of Knapp's work. Given that Murray would eventually become the publisher of Darwin's *Origin of the Species*, it is conceivable that his desire to risk his capital for a fifth time on Knapp's *Journal*, was in the very early stages of slipping away.

But old age and infirmity eventually brought the matter to a close. Knapp's final letter to Murray on the subject was sent in November 1844, only six months before he died. He wrote "I am quite of your opinion concerning a new edition – but from my advance in life there can be no expectation that I shall survive to see it – however I cannot reasonably lament that event but those who come after me can supply what may be required."

Following Knapp's death in April 1845, Murray returned the three exercise books of additional notes that had been sent to him, the new information within them still unpublished. Today, all three reside in a cardboard box held by the Buckinghamshire County Archive in Aylesbury. The Archive generously gave permission for the present writer to transcribe and reproduce their contents as part of his newly researched biography of John Knapp. John Chandler and Louise Ryland-Epton, publishers of this work, encouraged him not to include the notes simply as a series of miscellaneous essays, but to follow the detailed instructions that Knapp had left with them and to create the new updated edition of the *Journal* that he had longed for. This he has done. Only those essays without specific instructions for positioning within the previous text and those few where the instructions are unclear, remain as an addendum, their order matching that in Knapp's notebooks. As the drawings that Knapp made to accompany his new text were never made into woodcuts, they have been included as he drew them. So it is that exactly one hundred and eighty years after Knapp's death, a wholly authentic fifth English edition of his best-selling *Journal of a Naturalist* has been born.

Just as John Murray had found it necessary to have Knapp's sometimes breathless style tidied up by an experienced editor, so very modest changes have been made to his hand-written additions, mostly in punctuation and capitalisation, to break up sentences which in his notebooks run on unchecked. Knapp himself feared that his notes would require "much rearrangement" and that "my anxiety from domestic indisposition and my own state of health would prevent my attempting any…" As it turns out, they have required minimal rearrangement, and the present writer hopes most sincerely that both John Knapp and John Murray would have approved of what he has done.

At the end of all this, a reader might reasonably ask "what is the point of publishing a new edition John Knapp's work today?" The answer is simple. His book is an invaluable record of the natural world that existed in his day and the threats that it faced. It describes the extremes of climate that he experienced, even containing what may be the earliest use of the words "climate" and "change" in the same sentence. Although knowledge of natural history has moved on since his time, proving areas of his speculation wrong, his enthusiastic observation of the world that he knew is pure gold. As a result, almost every page of his Journal is an inspiration to its readers to go out, observe, understand and take pleasure in the world around them, just as John Knapp did, so many years before

George S.J. White Bt., FSA., 2025.

Preface

MANY YEARS HAVE now passed away since we were presented with that very interesting and amusing book, the "Natural History of Selborne:" nor do I recollect any publication at all resembling it having since appeared. It early impressed on my mind an ardent love for all the ways and economy of nature, and I was thereby led to the constant observance of the rural objects around me. Accordingly, reflections have arisen, and notes been made, such as the reader will find them. The two works do not, I apprehend, interfere with each other. The meditations of separate naturalists in fields, in wilds, in woods, may yield a similarity of ideas; yet the different aspects under which the same things are viewed, and characters considered, afford infinite variety of description and narrative: mine, I confess, are but brief and slight sketches; plain observations of nature, the produce often of intervals of leisure and shattered health, affording no history of the country; a mere outline of rural things; the journal of a traveller through the inexhaustible regions of Nature.

Contents

RESIDENCE of the Author — Extensive prospect on the banks of the Severn — Welsh mountains, and passages of the river — Roman encampment upon a British site — Roman roads — Coins — Skeletons of men and horses — Traces of a royal forest — Soils of the Author's parish — Limestone — its power of absorbing moisture, and acquiring carbonic acid — its various uses — its great abundance — its existence and origin pre-eminent amongst the wonders of the creation — Rocks formed in the parish by the coral polypi — analysis of — Rocks of deposit — analysis of — Lead ore — Carbonate of strontian — A traveller's foot burned off by his sleeping on a lime-kiln — Residences upon limestone soils supposed healthy — The village affords employment for labourers — Amount of limestone disposed of — Praiseworthy conduct of a peasant.........   Page 1 to 9

Analysis of soils considered as fallacious — in what the real goodness of a soil consists — Dairy processes in the Author's parish — Cheese-making — its agriculture — nature of its grass lands — Wild plants — Predominating plants in corn-fields — Soils will produce particular herbage — Mode of saving hay in bad seasons — Saving wheat ....   Page 9 to 15

The potato — culture of — different sorts of — planted by dibble and spade — expense and profit of cultivating — its effect upon the soil — not considered as injurious — sketch of the history of this extraordinary root — idle stories respecting its introduction — Lord Bacon recommends the use of it with grain in the brewing of ale — some soils not favourable for the root — Cultivated for the London market — introduced later than tobacco — its great value as an article of food — our ignorance of the first habitats of the Cerealia — Tendency of plants to revert to their original creation — the original species of the potato cannot now be ascertained — component parts of some varieties of — the most valuable production that Europe has received from America —

Bishop Heber says it is esteemed in the East as the greatest benefit ever received from Europe — will in time become naturalized to every region, and circulate its benefits round the globe.......... Page 15 to 23

The teazle. plant — its introduction into England — cultivation of — mode of gathering — its value — not injurious to the soil — its variety of names — its uses — Bad custom in farming — Clatters......... Page 23 to 27

Study of natural history — too much neglected — dismissed as the employment of children and inferior capacities — one of the most delightful occupations that can employ the attention of rational beings — occupies and elevates the mind — communicates an interest to every rural walk — ought to be made an object in youth — A beautiful Oak tree — Age of trees — prodigious size of several trees — the oak uncertain in producing acorns — barking of the oak — an interesting "History of the Oak" might be written — all its products valuable — Britain's guardian tree — The Wych, or broad leaved elm — its character — uses — magnitude— arrows made from the wood of— name — etymology of — suffers in early frosts — not beautiful in autumn — The buff-tip moth — Trees attractors of humidity— condense fogs — Air under trees — cause of the verdure beneath — a casualty attendant on some trees......... Page 27 to 35

Foliage of Trees and Plants — their great utility and agency — Peculiar propensities bestowed upon the vegetable world — Preference given by plants to particular soils — The fetid hellebore — its medicinal uses — The village doctress — Blossoms of plants — their utility not obvious — The carpenter bee — What flowers most abundant — design of flowers — application of flowers — natural love of flowers — emblems of innocence and purity — the natural playthings of children — universal ornament— cultivation of flowers — the bouquet — introduction of alien species......... Page 35 to 42

The aspen or poplar tree — construction of its foliage — formation of foot-stalks—curious construction of the leaf-stalks of plants — tendency of the poplar to extend its root and throw out suckers —The dyers' broom — gathering of— its medicinal virtues not resorted to— dishonest practice in the sale of — its uses confined to the dyer — Certain natural productions receive and retain artificial colourings — others reject, or soon part with it — Difference in the fibre of flax and silk — Nature of colour — effects of the rays of the sun on colours — Peculiarities of the snapdragon — a perfect insect trap — extraordinary formation of its corolla — The dogsbane — cruelly destructive of animal life— the object mysterious — The glaucous birthwort — The fluids secreted by the nectaries of plants the food of insects — Taste, the most variable of our senses — Sugar, in accordance with the likings of all orders of creation — The book-worm — its delight in sweets — The saccharine principle — Sugar contains above half its weight of oxalic acid — The very air a compounded poison, in the proportion of seventy-eight parts of deleterious matter, in a mixture of one hundred — The snapdragon vegetates in great droughts — Power of vegetation in plants — Effect of evaporation from the earth......... Page 42 to 49

The ivy — its shelter and food for birds and insects — love of ivy — ornament to ruins — its effect — medicinal value to animals — distribution of seeds by birds and other animals —The foxglove — grows only in particular soils — medicinal uses — uncertain application — its name — Vindication of old epithets— Ancient and modern remedies — The snowdrop — a native plant — remains long in neglected places — character of the snowdrop — a melancholy flower — Yellow oat grass — affected by drought — The vervain — estimation in which it was formerly held — supposed qualities of — its ancient and modern virtues — 'red water' in cattle perhaps caused by louse wort — The dyer's weed — its value and uses — its culture recommended — Yellow colour — most permanent and common — The brimstone butterfly — The day's eye — The dandelion—remarkable consequences of cultivation — Singular appearance of a grass — Common brambles - insect path on its leaves — its uses......... Page 49 to 64

The maple tree — an early autumn beau — the fashion followed by others — the maple wood, a beautiful microscopic object — medicinal properties ascribed to it by Pliny — hoary appearance in 1830 — its leaves punctured by insects — The wild clematis, or traveller's joy — grows in limestone soils by preference — its uses — pores of the wood in the microscope — Vessels of plants — uninjured by dry seasons — seeds of the clematis..... Page 65 to 68

The Naturalist's autumnal walk — beautiful, and full of variety — The agaric — its beauty and variety — plentiful in Monmouthshire— Beauties of the fungi race — uncertain appearance of the fungi — mutations of the fungi — Decomposition of plants — Laurel — Holly — two-fronted sphæria — Elm leaves — Sycamore leaves — Bark of plants — The nut — Beech — Odorous agaric — Fragrant agaric — "Stainer agaric" — Stinking phallus — Mode of propagation— Dispersion of the seeds of plants — The turreted puff — Starry puff — Morell — Bell-shaped nidularia — Food for mice..... Page 68 to 77

Quadrupeds — The marten cat — his capture — well adapted for a predatory life — its skin — stoats and weasels —The hedgehog — its mode of life—always destroyed — prejudices against — Cruelty of man towards animals — hedgehogs formerly an article of food — sensibility of the spines — behaviour in captivity — value in bakehouses — The harvest mouse — where found — its character — its strength and perseverance — Increase and decrease of animals — Migration of rats — The water shrew — its residence and habits — The common shrew mouse — The pale blue shrew — The mole — its actions — character— abundance of — easily discovers its food - structure of its body — The hair and fur of animals — flesh of the moles — killed by weazels .... Page 77 to 91

Birds — the Author's admiration of — The rook — The hedge-sparrow — contingencies of its life — its song — example of a domestic character — aliment provided for all creatures in hour of want — The willow wren — her early appearance and departure — her nest — object of her migration — Difficulty of rearing young birds — The golden-crested wren — The linnet — their song and habits — The bull-finch — its character — injurious to trees — preference of food — no destroyer of insects — The robin — its character — always found — The song of birds — The chaffinch — beautifully feathered — habits of the female — country epithets — conduct in spring — moisten their eggs in hot weather — fondness for the tender shoots of peas — Parish rewards for vermin — The blue tomtit — perishes in winter — mode of obtaining food......... Page 91 to 101

The wiles and stratagems of birds — Distinguished by voice — Note of the cole-mouse — The white throat — The long tailed titmouse — their nests—

journeys—eggs— labour to feed their young— Winter food of birds— Great variety of the nests of birds — Beautiful nests of the goldfinch— Sufferings of the swallow — Maternal care of a little blue tomtit — industry — The raven — scared from its nest — faculty of discovering its food — universally found — duration of life — superstitions wearing out — duration of animal life — aided or injured by man — An old horse — Life of man — examples of longevity — The crossbill — breeds in England — many wantonly killed at Backwell — The rook — suffers in cold and dry seasons — his life in the year 1825 — various habits of— detects grubs in the earth — his habits in the spring — associations — sense of smelling — migration of birds.........   Page 101—113

The magpie — its nest — its habits — a plunderer of the farm-yard — its natural affection — The jay — conduct of the old bird — winter habits — its feathers — its predaceous nature — The plundering of birds' eggs during incubation —The shrike, or butcher-bird — its nest — its young — kills other birds — a sentinel — its mischievous disposition — its use of a shamble-hook — The stormy petrel — its habits — held in reverence by seamen — The wryneck — its habits — Disappearance of certain birds — their numbers annually diminishing — The nightingale — migrating birds — Rooks love long avenues — The starling, or stare — great flights — social habits — breeding — a stray bird — actions before roosting — congregate — very attentive to their young — journeyings — Laborious life of birds — The redstart— The fire-tail — theft of its eggs by sparrows — The swift — The swallow tribe — seen in winter at Totnes —The sparrow-hawk and kestrel — capture by intimidation — single out individuals — Variableness of seasons — Early seasons bring rain — Blooming of the white thorn — Conduct of migrating birds — The butcher bird — The grey flycatcher— care of its young in extreme temperature — 1833 a year deficient in migrating birds —The thrush — instance of affection — motives of action — utility in a garden — The sparrow — domestic habits — manners — increase — destruction — great consumers of insects — accommodating appetite — Roosting of small birds.........   Page 113 — 133

Creatures associating with man — The common mouse — The rat — The house-fly — Usefulness of animals to man — Conduct of man towards animals — The wheatear — its habits — conceals its nest with great artifice — Country amusements often cruel — Supplication for pity — The eggs of birds — their markings — numbers of eggs laid — Foolish superstitions—

The kite — his habits — great capture of — The blackcap — its habits — song — nest — food — shyness — Habits of our occasional visiting birds — The pettychaps — White throats— Willow wren — Fear of man in animals — Stratagem of a wren — Instinct — Early awakening of birds— The early morning— Morning in autumn — The goldfinch — captured — die in the winter — soon reconciled to captivity — The tree creeper — winters in England — not an increasing bird — The yellow wagtail — The pied wagtail — unusual method of catching its prey......... Page 133—151

Rapacious birds — Passerine birds — The common bunting — unthatch corn-ricks — The green finch — extraordinary mating ritual — Old tokens of wind and weather — The white lily — The pimpernel — The missel-thrush — his note — breeds near the dwellings of man — Change of character in birds — Love of offspring in birds — Ordinations of Providence— The jack snipe — its solitary habits — The Christmas shooter — Association of birds — The green legged sand piper — The peewit — its habits — eggs — An albino white throat — Destruction among the rooks — Second establishment of birds — Prognostications — Hedge fruit — Haws and heps — The fieldfares — Redwings — feeding in the lowlands — uplands — Egg of the fieldfare—Rural sounds — voices of birds — Plumage of birds — Song of birds — The woodlark — its habits — voice — capture — its hinder claw — Language of man — of birds — Note significant of danger — Singing a spontaneous effusion — Variety of note in some species — 'Lady-bird' note of a song thrush — Croaking of the nightingale — Crowing of the cock — Admiration of birds — The winter gnat — Birds, cleanly and innocent creatures......... Page 151 — 170

Knowledge slow of attainment — Entomology a difficult study — Wonders around us — The objects of many insects unknown — limbs of small insects beyond comprehension — Chrysalis of a moth — Four-spotted dragon fly— The ghost moth — soon destroyed — Specimens of plumage of butterflies — The scales of fish — Habits of insects — The blue argus butterfly — a pugnacious insect — combats — The azure butterfly — seldom seen — The humming bird hawk-moth — habits — wildness — tamed by familiarity — feigns death — The painted lady butterfly — uncertain in appearing — The marble butterfly — The common wasp — retires over winter with folded wings — The meadow brown butterfly — The yellow winged moth— The admirable butterfly — drawn to decaying fruit — Foreign plants ignored by indigenous insects — The gamma moth — Sudden appearance of Melitæa artemis in

1834 — The tiger moth — Provisions of creatures for their young — The goat moth — their numbers — odour — power of destruction — tenacious of life — Larvae of phalæna cossus, where plentifully found...... Page 170—188

Designs of Supreme Intelligence in the creation and preservation of insects — Summer evening's ramble — Insects abounding — ignorance of their objects — The glow-worm — curious contrivance about their eyes — light — migration — The strength of insects — Eggs of the snake— destruction — harmless in England — general aversion to snakes — Paucity of noxious creatures inhabiting with us — The slow-worm — The small bombyx — vigilance — animation — quarrels — The black ant— combats of strength — Red ants mortality — Yellow ants — winter nests— ants not injured by cold — millipedes — support a great degree of cold — The purpose of insects — Insects in stagnant water — The newt — its voracity — The water flea — an amusing insect — The dorr-beetle — their numbers — feign death to avoid injury — Cleanliness of animals in health......... Page 188 — 203

Recurrence to causes — Changes that take place in nature — Cockchaffers — The death's-head moth — chrysalides — superstitions regarding the insect — voice — The great water-beetle — its habits and voracity — Nest of an unknown spider — The hair-worm — its object — Nests of a solitary wasp — Nest of the carpenter bee — Hornets — their abundance at times, and voracity — kill each other — The garden snail — its injuries — generally secure from destruction — snails double-glazed in the severe winters of 1829 and 1830 — faculties — Small banded snails — their numbers — superstitions concerning them — Earthworms — numbers of — the prey of all creatures — utility of — drain watery soils......... Page 203 — 219

Inattention to the works of Providence — Empiricism — Apple tree blight— progress — injury — remedy for blight — Aphides — food for the long-tailed titmouse and golden crested wren — Effects of season on vegetation — Destruction of grass-roots — vegetation — The squirrel — damage caused by — Emission of scents from plants — Honey-dew — a close hot night in 1829 — Destruction to foliage by small moths — Salt winds — Leasing — its profits — an innocent occupation — ordained by the Almighty — Old holiday amusements — wearing out — Maypoles — Christmassing — Kitchen bushes — young holly-trees— Superstitions relating to plants —

Village method of recording impounded cattle —Anecdote of the finding of a guinea — Influence of electric atmosphere on vegetation — Hummings in the air — possible cause — Fairy rings — Spring changes — Periodical winds — The absorption of minerals by different plants — Whirly pits — Sinkings in the earth — Lichen fascicularis — Village manufacturer of gut for musical instruments — Withering and searing of leaves — Salt winds destructive of vegetation— Spottings on apples — Spottings on strawberry leaves — Curious agaricus surrectus — Curious analogy between plants and animated beings — Roots of an ash tree......... Page 219 — 248

The year 1825 — its peculiarities and influences — A speedy method of killing insects — Preservation of insects — Pollarding of trees — most injurious — Insects that destroy the ash — The willow rarely seen as a tree — a remarkable one near Gloucester — Foggy morning — Reeking of the earth — the cause — and utility — Winter of the year — Ice in pools — Law of nature—Winter called a dull season — Nature actively employed — Exhausting power observed in air — A minute vegetable product — Influence of green on the senses — Winter subsistence of birds — Admiration of Nature — Contemplation of the stars — their lustre — Celestial intelligences — Heavenly bodies — Atmospheric experiments — Effect of the approach of autumn on the mind— An early frost — Spring returns — Extraordinary tendencies of nature to produce — Capabilities of man — The Naturalist's concluding reflections Page 248— 269

THE PROGRESS OF A NATURALIST; TO FLORA.   Page 271

BRIEF ESSAYS

The Wren — The Ivy — The Land-rail — Sea swallows swept inland by gales — Rooks — Apple trees infected by aphids — Strong blight — The Blue Argus — The preservation of fruit — The effects of drought — Ancient vegetation preserved beneath a Roman path — The village well — Autumnal sensations — Growths beneath oak leaves — The plumage of owls — The migration of rats — The spade — Instinct — The danger of black-thorn — Climate change — The Water-rail — Feathers of birds — The Great Bat — The Abbey Camp — Changes in plumage — Larvae of Lepidoptera — Vulnerability of the

Robin and Wren — High winds providing food for birds — The destruction of birds' eggs — The capture of Woodpeckers — Large gatherings of creatures — Cuckoos' eggs — Fossil remains of plants and animals — The Gansel Bergamot pear......... Page 290 — 325

Explanation of the Plates

PLATE I.

A. Limestone from the down, enlarged.
2. The same in decay, representing the empty cells.
3. Ditto, enlarged.
B. C. D. The forms of the insect, which have remained.
E. The stony matter remaining at the mouth of the cells.
4. Dipsacus fullonum, the fullers' teazle.
5. Apocynum androsæmifolium, or flower of the dogsbane.
A. Flower enlarged.
B. The pointal, with its seed at the base.
C. Two expanded anthers, ready for capturing a fly.
D. The anthers closed, and the insect captured.

PLATE II.

Plant of a grass, from Malvern hills.

PLATE III.

Fig. 1. A section of the wood, from a young spray of the maple.
2. Ditto, from the clematis vitalba.
3. A rose leaf, bearing the mark of an insect's path.
4. A. The fur of a mole.
B. Ditto of the bat.
C. Ditto of the hamster mouse.
D. Hair of callimorpha caja.
5. Egg of the nightingale.

PLATE IV.

Fig. 1. Sphæria on the loaf of an elm.
A. A portion enlarged, and the cuticle parting.
B. The same enlarged, representing the capsules. Fig. 2. Sphæria bifrons, on a laurel leaf.
C. The front, and dorsal parts.
D. Imbedded capsules.
Fig. 3. Sphæria coryli, on a nut branch.
E. The tubercle enlarged, bordered with the epidermis.
F. A section of the capsules at the base. Fig. 4. Sphæria faginea, on a beech stick.
G. Section of a tube, with the capsules at the base.
H. Group of the tubes detached from the bough, with their capsules.
I. A tube detached.

PLATE V.

Fig. 1. A chrysalis of an insect.
B. The inner hood.

Fig. 2. The male glow-worm.
D. Under side of the head, representing the hood-winkers and eyes.

Fig. 3. The branch of an apple tree, infested with the aphis lanata.
B. The aphis enlarged, with the globules, and the cotton that surrounds them.
D. The early appearance of the insect with its terminating bristle.
E. Appearance of the creature in winter.

Fig. 4. Eggs of the snake.

PLATE VI.

Fig. 1. Nest of vespa campanaria.
Fig. 2. Gordius argillaceus.
2. The same enlarged.
3. Ditto, and fully extended.

PLATE VII.

The Agaricus Infector.

To THE BINDER — Place the Plates I. to VII. in order after page 275, and before the Index.

WOODCUTS, DRAWINGS & WATERCOLOUR.

VIGNETTE. — The Shellard's-lane oak, title page
Spines and tubes of the hedgehog, enlarged, p. 81.
Harvest mouse and nest, p. 82.
Plumage of lepidopterous insects, p. 174.
Scales of fish, p. 177
Nest of an unknown spider, p. 208
Nest of the solitary wasp, p. 210
Nest of the carpenter bee, p. 211.
Agaricus surrectus, p. 246
Roots of an ash, p. 247.
Feather of the short-eared owl, p.307
The Alder moth, p. 320

JOURNAL
OF
A NATURALIST.

THE VILLAGE IN which I reside is situated upon a very ancient road, connecting the city of Bristol with that of Gloucester, and thus with all the great towns in the North of England. This road runs for the chief part upon a high limestone ridge, from which we obtain a very beautiful and extensive prospect: the broad estuary of the river Severn, the mountains of Glamorgan, Monmouth, and Brecon, with their peaceful vales, and cheerful looking white cottages, form the distant view: beneath it lies a vast extent of arable and pasture land, gained originally by the power of man from this great river, and preserved now from her incursions by a considerable annual expenditure, testifying his industry and perseverance, and exhibiting his reward. The Aust ferry, supposed to be the "trajectus," or place where the Romans were accustomed to pass the Severn, is visible, with several stations of that people and the ancient British, being a part of that great chain of forts originally maintained to restrain the plundering inroads of the restless inhabitants of the other bank of the river: Thornbury, with its fine cathedral-like church and castle, the opposite red cliffs of the Severn, and the stream itself, are fine and interesting features.

An encampment of some people, probably Romans, occupies a rather elevated part of the parish, consisting of perhaps three acres of ground, surrounded by a high agger, with no ditch, or a very imperfect one, and probably was never designed for protracted resistance: it appears to form one of the above-mentioned series of forts erected by Ostorius, commencing at Weston, in Somersetshire, and terminating at Bredon in the county of Worcester — ours was probably a specula, or watch-hill, of the larger kind. We can yet trace, though at places but obscurely, the roads that connected this encampment with other posts in adjoining villages. A few years sweep away commonly all traces of roads of later periods, and the testimony of some old man is often required to substantiate that one had ever been in existence within the memory of life: yet these uniting roads, which, as works, must have been originally insignificant, little more than by-ways, after disuse for above fourteen hundred years, and encountering all the erasements of time, enclosures, and the plough, are yet manifest, and an evidence of that wonderful people, thieves and ruffians though they were, who constructed them. There is probably no region on the face of the globe ever colonised, or long possessed, by this nation, which does not yet afford some testimony of their having had a footing on it; this people, who, so long before their power existed, it was predicted, should be of "a fierce countenance, dreadful, terrible, strong exceedingly, with great iron teeth that devoured and broke in pieces:"

— where'er thy legions camp'd,
Stern sons of conquest, still is known,
By many a grassy mound, by many a sculptured stone.

Almost every Roman road that I have observed appears to have been considerably elevated above the surrounding soil, and hence more likely to remain apparent for a length of time than any of those of modern construction, which are flat, or with a slight central convexity; the turf, that in time by disuse would be formed over them, would in one case present a grassy ridge, in the other be confounded with the adjoining land.

Coins of an ancient date, I think, have not been found here[1]; nor do we possess any remains of warlike edifices, or religious endowments. Our labourers have at various times dug up by the road side several skeletons of human beings, and of horses; they were in general but slightly covered with earth; and though the bones were much decayed, yet the teeth were sound, and appeared most commonly to have belonged to young persons, and probably had been deposited in their present situations at no very distant period of time. With the bones of a horse so found there remained the iron head of a lance, about a foot long, corroded, but not greatly decayed. Unable better to account for these skeletons, we suppose that they constituted, when alive, part of the forces of General Fairfax, and that they fell in some partial encounters with the peasantry when defending their property about to be plundered by the foragers of his army in 1645, at the time he was besieging the castle of Bristol. The siege lasted sixteen or seventeen days; many parties during that time must have been sent out by him to plunder us cavaliers, and contention would take place.

1   Some money was found in one of our fields a few years past, which fame, as in all such cases, without perhaps any foundation, enlarged to a considerable sum. The nature of the coin I know not. A few old guineas were admitted; but from fear of that spectre "tresor trove," the whole was concealed, whatever it might be.

It is foreign to my plan to enumerate, and it might be difficult to discover, all the changes and revolutions which have taken place here; and I shall merely mention, that this district formerly constituted a regal forest, and we find Robert Fitzharding holding it by grant in the time of King John. We have a "lodge farm," it is true, and the adjoining grange, the "conygar," i.e. coneygard, the rabbit-keeper's dwelling, may, perhaps, have been the situation of the sylvan warren; but there are no remains, or any other indications, of a forest ever having been in existence. Names and traditional tales are all that remain in most places now to remind us of the ancient state of England, or to make credible the narratives of our old historians, who lived when Britain was a forest. Where shall we look for the remnants of that mighty wood filled with boars, bulls, and savage beasts, that surrounded London? Even in our own days, heaths, moors, and wilds have disappeared, so as to leave no indications of their former state but the name. Woods and forests seem to be the original productions of most soils and countries favourable for the abode of mankind, as if inviting a settlement and offering materials for its use. As colonies increase, wants are augmented; the woods are consumed; the plough is introduced, division of property follows; a total change and obliteration ensues, though the ancient appellation by which the district was known yet continues.

The parish consists in parts of a poor, shattery grey clay, beneath which we find, in some places, a coarse lias; in others a spongy, rough, impure limestone; in other parts a thin stratum of soil is spread over an immense and irregular rock of carbonate of lime, running to an unknown depth: this in many cases protrudes in great blocks through the thin skin of earth. The rock, though usually stratified, has no uniform dip, but tends to different directions; in some places it appears as if immense sheets of semifluid matter had been pushed out of the station it had settled in, by some other or later-formed heavy-moving mass, or met with an impediment, and so rolled up: that these sheets had not fully hardened at the time of being moved is yet made probable by the whole crystallization of the mass being interrupted; so that no part adheres firmly, but separates into small shattery fragments when struck. This substance we burn in very large quantities for building purposes, and for manure, which, by the facility which we have of obtaining small coal, is rendered at the low rate of three-pence a bushel at the kiln. Our farmers, availing themselves of this cheap article, use considerable quantities, composted with earth, for their different crops, at the rate of not less than a hundred bushels to the acre. This is a favourite substance for their potato land. The return in general is not so large as when grown in manure from the yard: but the root is said to be more mealy, and better flavoured.

The utility of lime as manure consists in loosening the tenacious nature of some soils; rendering them more friable and receptive of vegetable fibres: it especially facilitates the dissolution and putrefaction of animal and vegetable substances, which are thus more readily received and circulated in the growing plant; and it has the power of acquiring and long retaining moisture; thus rendering a soil cool and nutritive to the plants that vegetate in it. The power that lime has of absorbing moisture will be better understood, when we say, that one hundred weight will, in five or six days, when fresh, absorb five pounds of water, and that it will retain in the shape of powder, when slakened, loosened, as is commonly said, nearly one-fourth of its weight[2].

That lime rehardens after being made soft, as in mortar, is owing to the power which it has of acquiring carbonic acid — the fixed air of Dr. Black —from the atmosphere; when the stone is burned, it loses this principle, but re-absorbs it, though slowly, yet in time, and it thus becomes as hard as stone again: we unite it with sand to promote the crystallization and hardening. The utility of lime in various arts, agriculture, manufactories, and medicine, is very extensive, and in many cases indispensable; and the abundance of it spread through the world seems designed as a particular provision of Providence for the various ends of creation. Lime, and siliceous substances, compose a very large portion of the dense matter of our earth; the shells of marine animals contain it abundantly; our bones have eighty parts in one hundred of it; the egg-shells of birds above nine parts in ten — during incubation, it is received by the embryo of the bird, indurating the cartilages, and forming the bones. But the existence and origin of limestone are preeminent amongst the wonders of creation; nor should we have been able, rationally, to account for the great diffusion of this substance throughout the globe, however we might have conjectured the formation, without the Mosaical revelation. It may startle, perhaps, the belief of some, who have never considered the subject, to assert, what is apparently a fact, that a considerable portion of those prodigious cliffs of chalk and calcareous stone, that in many places control the advance of the ocean, protrude in rocks through its waters, or incrust such large portions of the globe, are of animal origin — the exuviae of marine substances, or the labours of minute insects, which once inhabited the deep. In this conclusion

---

2   The weight of lime is very variable, differing in different places; but taking our lime at the average of eighty pounds to the bushel, some idea may be conceived of the cooling nature of this substance. Lime, to be used as manure, must be in a pulverised state; and by drawing on the land the quantity that we do, we convey to every acre so dressed equivalent to two hundred and fifty gallons of water, not to be evaporated, but retained in the soil as a refrigerant to the fibres of vegetation.

now chemists and philosophers seem in great measure to coincide. Fourcroy observed, forty years ago, that "it could not be denied, that the strata of calcareous matter, which constitute, as it were, the bark or external covering of our globe, in a great part of its extent, are owing to the remains of the skeletons of sea animals, more or less broken down by the waters; that these beds have been deposited at the bottom of the sea, immense masses of chalk, deposited on its bottom, absorb or fix the waters, or convert into a solid substance part of the liquid which fills its vast basin" — *Supplement to Chemistry*, p. 263. Such are the conclusions of philosophical investigation; and the discoveries of all our circumnavigators fully corroborate these decisions as to formation. Revelation in part accounts for the removal of these stupendous masses; though, probably, unrecorded concussions since the great subversion of our planet have, in remote periods, effected many of the removals of these deposits. We find the basement of many of the South Sea Islands, some of which are twenty miles long, formed of this matter. Captain Flinders, in the gulf of Carpentaria, held his course by the sides of limestone reefs, five hundred miles in extent, with a depth irregular and uncertain; and still more recently Captain King, seven hundred miles, almost a continent, of rock, increasing, and visibly forming: — all drawn from the waters of the ocean by a minute creature, that wonderful agent in the hands of Providence, the coral insect. This brief account of the origin of calcareous rocks was, perhaps, necessary before mentioning an extraordinary fact, that, after the lapse of so vast a portion of time since the basement of the mighty deep was heaved on high, existing proofs of this event should remain in our obscure village.

The limestone rocks here are differently composed, but are principally of four kinds — a pale gray, hard and compact; a pale cream-coloured, fine-grained and sonorous; these form the upper stratum of stone on our down, a recent deposit, or more probably a mass heaved up from its original station. The whole of this mass, running nearly half a mile long, is obviously of animal formation, a coral rock; a compounded body of minute cylindrical columns, the cells of the animals which constructed the material, the mouths of which are all manifest by a magnifier. The stop in the progress of the work is even visible; soft; stony matter having arisen from one of the tubes, and become indurated there in a convex form; in others the creatures have perished, but their forms or moulds remain, though obscure, yet sufficiently perfect to manifest the fact; these tubes, by exposure to the air for any length of time, have the internal or softer parts decomposed, and the stone becomes cellular. A. (Plate 1, fig. 1) represents an enlarged fragment of the down limestone, with the mouths of the cells; B. C. D. are the appearance of some of them, with the forms of the insect which constructed them; E. represents the stony matter by which the work is continued fixed at the mouth of the cell; Fig. 2, the same when decomposed, the animal matter having quite perished in some cases, and partially in others; Fig. 3, enlarged. This stone burns to a fine white lime, and is very free from impurities, containing in a hundred parts —

Carbonate of lime …	88
Magnesia…	8
Silex…	1
Alumine[3], coloured with iron…	3
	100

Another quarry presents, likewise, unquestionable evidence of animal origin, veins of it being composed of shattered parts of shells, and marine substances, greatly consumed and imperfect, embedded in a coarse, gray, sparry compound; an ocean deposit, not a fabrication, and consequently has more impurities in its substance than that of insect formation: it contains about—

Carbonate of lime…	73
Magnesia…	11
Clay…	14
Silex…	2
	100

These two specimens so clearly prove that the original materials of their substance were derived from the deep, that no further arguments need be advanced to support this fact as to our limestone. The former is, perhaps, the mountain limestone of Werner; the latter a variety of dolomite. Our other quarries, as well as the lower strata of the above, present no such indications of animal formation, and they are probably sediment arising from a minute division of shelly bodies, now indurated by time and superincumbent pressure, and become a coarse-grained marble. Our limestone thus appearing not to be contaminated with any great portion of magnesian earth, it may be used for all agricultural purposes with advantage. Many detached blocks of limestone are

---

3   I have called this alumine, stained with oxide of iron; but it seems more like vegetable or animal remains, adhering to the filter like a fine peaty deposit, and is lost in combustion.

found about us, having broken shelly remains, and the joints of the encrinite, greatly mutilated, embedded in them. Irregularly wandering near the lime-ridge is a vein of impure sandy soil, covering a coarse-grained siliceous stone; sand agglutinated, and coloured by oxide of iron, resisting heat, and used in the construction of our lime-kilns; the labourers call it "fire-stone."

We occasionally, though sparingly, find, in a few places on our downs, nodules of lead ore, which induced persons in years past to seek for mineral riches; but the trial being soon abandoned, the result, I suppose, afforded no reasonable ground for success. We likewise find thin veins of carbonate of strontian, but make no use of it; nor is it noted by us different from common rubbish; nor do I know any purpose to which it is peculiarly applicable, but in pyrotechnics. Spirit of wine, in which nitrate of strontian has been mixed, will burn with a beautiful bright-red flame; barytes, which approaches near to strontia, affords a fine green: nitrates of both, compounded with other matters, are used in theatrical representations. Strontian exists in many places, and plentifully; some future wants or experiments will probably bring it into notice, and indicate the latent virtues of this mineral.

Perhaps I may here mention an incident, that occurred a few years past at one of our lime-kilns, because it manifests how perfectly insensible the human frame may be to pains and afflictions in peculiar circumstances, and that which would be torture if endured in general, may be experienced at other times without any sense of suffering. A travelling man, one winter's evening, laid himself down upon the platform of a lime-kiln, placing his feet, probably numbed with cold, upon the heap of stones, newly put on to burn through the night. Sleep overcame him in this situation; the fire gradually rising and increasing until it ignited the stones upon which his feet were placed. Lulled by the warmth, the man slept on; the fire increased until it burned one foot (which probably was extended over a vent hole) and part of the leg above the ancle entirely off, consuming that part so effectually, that a cinder-like fragment was alone remaining— and still the wretch slept on! and in this state was found by the kiln-man in the morning. Insensible to any pain, and ignorant of his misfortune, he attempted to rise and pursue his journey, but missing his shoe, requested to have it found; and when he was raised, putting his burnt limb to the ground to support his body, the extremity of his leg-bone, the tibia, crumbled into fragments, having been calcined into lime. Still he expressed no sense of pain, and probably experienced none, from the gradual operation of the fire, and his own torpidity during the hours his foot was consuming. This poor drover survived his misfortunes in the hospital about a fortnight; but the fire having extended to other parts of his body, recovery was hopeless.

Residences upon limestone soils have generally been considered as less liable than other situations to infectious and epidemic disorders; and such places being usually more elevated, they become better ventilated, and freed from stagnated and unwholesome airs, and by the absorbing principle of the soil are kept constantly dry. All this seems to favour the supposition that they are healthy; but if exempted from ailments arising from mal-aria, inflammatory complaints do not seem excluded from such situations. When the typhus fever prevailed in the country, we were by no means exempted from its effects; the severe coughs attending the spring of 1826 afflicted grievously most individuals in every house; and the measles, which prevailed so greatly at the same season, visited every cottage, though built upon the very limestone rock.

This village and its neighbouring parishes, by reason of the peculiar culture carried on in them, and the natural production of the district, afford the most ample employment for their labouring inhabitants; nor perhaps could any portion of the kingdom, neither possessing mineral riches, manufactories, or mills, nor situate in the immediate vicinity of a great town, be found to afford superior demand for the labour, healthy employment, and reasonable toil of its population. Our lime-kilns engage throughout the year several persons; this is, perhaps, our most laborious employ, though its returns are considered as fair. In our culture, after all the various business of the farms, comes the potato setting; nor is this finished wholly before hay-making commences. Teazleing succeeds; the corn harvest comes on, followed shortly by the requirements of the potato again, and the digging out and securing this requires the labour of multitudes until the very verge of winter. Then comes our employment for this dark season of the year, the breaking of our limestone for the use of the roads, of which we afford a large supply to less favoured districts. This material is not to be sought for in distant places, or of difficult attainment, but to be found almost at the very doors of the cottages; and old men, women, and children can obtain a comfortable maintenance by it without any great exertion of strength, or protraction of labour. The rough material costs nothing, a short pickaxe to detach the stone, and a hammer to break it, are all the tools required. A man or healthy woman can easily supply about a ton in the day, a child that goes on steadily, about one-third of this quantity; and as we give one shilling for a ton, a man, his wife, and two tolerable-sized children, can obtain from two shillings and eightpence to three shillings per day by this employ the greater part of the winter; and should the weather be bad, they can work at intervals, and various broken hours, and obtain something — and there is a constant demand for the article. The winter accumulation is carted away as the frost occurs, or the spring repair comes on. Our labourers, their children and cottages, I think, present a testimony of their well-doing, by the orderly, decent conduct of the former, and the comforts of the latter. There are years when we have disposed of about three thousand tons of stone, chiefly broken up for use by a few of our village poor; if we say by twenty families, it will have produced perhaps seven pounds to each, — a most comfortable addition to their means, when we consider that this has been obtained by the weak and infirm, at intervals of time without more than the cost of labour, when employment elsewhere was in no request.

I may perhaps be pardoned in relating here the good conduct of a villager, deserving more approbation than my simple record will bestow; and it affords an eminent example of what may be accomplished by industry and economy, and a manifestation that high wages are not always essential, or solely contributive to the welfare of the labourer. — When I first knew A. B., he was in a state of poverty, possessing, it is true, a cottage of his own, with a very small garden; but his constitution being delicate, and health precarious, so that he was not a profitable labourer, the farmers were unwilling to employ him. In this condition he came into my service: his wife at that time having a young child, contributed very little to the general maintenance of the family: his wages were ten shillings per week, dieting himself, and with little besides that could be considered as profitable. We soon perceived that the clothing of the family became more neat and improved; certain gradations of bodily health appeared; the cottage was white-washed, and enclosed with a rough wall and gate; the rose and the corchorus began to blossom about it; the pig became two; and a few sheep marked A. B. were running about the lanes: then his wife had a little cow, which it was "hoped his honour would let eat some of the rough grass in the upper field;" but this was not entirely given: this cow, in spring, was joined by a better; but finding such cattle difficult to maintain through the winter, they were disposed of, and the sheep augmented. After about six years' service, my honest, quiet, sober labourer died, leaving his wife and two children surviving: a third had recently died. We found him possessed of some money, though I know not the amount; two fine hogs, and a flock of forty-nine good sheep, many far advanced in lamb; and all this stock was acquired solely with the regular wages of ten shillings a week, in conjunction with the simple aids of rigid sobriety and economy, without a murmur, a complaint, or a grievance!

I REPORT NOTHING concerning our variously constituted soil, thinking that no correct statement can be given by any detail of a local district under cultivation, beyond generally observing its tendency, as every soil under tillage must be factitious and changeable. As a mere matter of curiosity, I might easily find out the proportions of lime, sand, clay, and vegetable earth, &c., that a given quantity of a certain field contained; but the very next ploughing would perhaps move a substratum, and alter the proportions; or a subsequent dressing change the analysis: the adjoining field would be differently treated, and yield a different result. I do not comprehend what general practical benefit can arise from chemical analysis of soils; but as eminent persons maintain the great advantages of it, I suppose they are right, and regret my ignorance. That the component parts of certain lands can easily be detected, and the virtues or deficiencies of them for particular crops be pointed out, I readily admit; but when known, how rarely can the remedy be applied! I have three correspondents, who send me samples of their several farms, and request to know by what means they can meliorate the soil. I find that B. is deficient in lime; but understand in reply, that this earth is distant from his residence, and too costly to be applied. D. wants clay; E. is too retentive and cold, and requires silex or sand; but both are so circumstanced, that they cannot afford to supply the article required. Indeed it is difficult to say what ought to be the component parts of a soil, unless the production of one article or grain is made the standard; for differently constituted soil will produce different crops advantageously: one farm produces fine wheat, another barley; others again the finest oats and beans in the parish. To compound a soil of exact chemical parts, so as to afford permanent fertility, is a mere theory. Nature and circumstances may produce a piece of land that will yield unremitting crops of grass, and we call it a permanently good soil; but art cannot effect this upon a great scale. A small field in this parish always produces good crops; not in consequence of any treatment it receives, but by its natural composition; consisting principally of finely pulverised clay, stained with red oxide of iron, a considerable portion of sand, and vegetable earth: but though I know the probable cause of this field bearing such good wheat, I cannot bring the surrounding and inferior ones into a like constitution, the expense far exceeding any hope of remuneration. Rudolph Glauber obtained gold from common sand, but it was an expensive article! Temporary food for a crop may be found in animal, vegetable, or earthy manures, but these are exhaustible; and when aliment ceases, the crop proportionably diminishes. In one respect, chemical investigation may importantly aid the agriculturist, by pointing out the proportion of magnesian earth in certain limes used for manure, and thus indicate its beneficial or injurious effects on vegetation. I should not like lime containing twenty per cent, of this earth; but when it contains a much smaller proportion, I should not think it very deleterious. This earth acts as a caustic to vegetation, and, neither being soluble in water, nor possessing the other virtues of lime, diminishes the number of bushels used according to its existence, and thus deprives the crop of that portion of benefit: but after all, as Kirwan says, the secret processes of vegetation take place in the dark, exposed to the various and indeterminable influences of the atmosphere; and hence the difficulty of determining on what peculiar circumstance success or failure depends, for the diversified experience of years alone can afford a rational foundation for solid and specific conclusions.

The real goodness of a soil consists principally, perhaps, in the power it possesses of maintaining a certain degree of moisture; for without this, the plant can have no power of deriving nutriment from any aliment: it might be planted on a dunghill; but if this had no moisture in it, no nutriment would be yielded; but as long as the soil preserves a moisture, either by its own constituent parts, or by means of a retentive substratum, vegetation goes on. Continue the moisture, and increase the aliment, and the plant will flourish in proportion; but let the moisture be denied by soil, substratum, or manure, and vegetation ceases; for, though certain plants will long subsist by moisture obtained from the air, yet, generally speaking, without a supply by the root, they will languish and fade.

Our dairy processes, I believe, present nothing deserving of particular notice. From our milk, after being skimmed for butter, we make a thin, poor cheese, rendered at a low price, but for which there is a constant demand. Some of our cold lands, too, yield a kind greatly esteemed for toasting; and we likewise manufacture a thicker and better sort, though we do not contend in the market with the productions of North Wilts, or the deeper pastures of Cheshire or Huntingdon.

Our dairy farmers complain occasionally of a circumstance attendant upon cheese-making, by which they become seriously injured, called the "heaving of the curd", that substance when recently turned from the vat fermenting, swelling and warping out of all form, becoming spongy & unsaleable: this heaving will at times occur in the press, but usually only when turned from the mould: all farms are not subject to this warping, & it seems that the curd from cows pastured in deep rich land, or on very poor soils are less subject to it than when the animal has pastured on moderately good herbage. That some part of their provender occasions it is thus manifest, and suspicion has been directed to the vernal grass, (*anthoxanthum*) with some apparent reason; the evil chiefly happens when cows are first turned to grass, and now it is that this plant most abounds. As the season advances the vernal grass withers away, the "heaving" in great measure ceasing; deep rich soils except when boggy, or reclaimed are not favourable to its growth, and when produced other counteracting species predominate and neutralize its effects, and the specimens on the poor lands though abundant, are dwarfish and less succulent than in moderately good pastures: the plant however when produced contains a larger portion of essential oil than its congeners, as the odour arising from the dryed herbage manifests. One of our cheese farmers had entertained the notion, or more probably had heard it from another person, that this "heaving" might be caused by the woody night shade (*solanum dul*.) and brought an handful of the black briony (*tamus*) to me to investigate for him, being supplied with the required plant he made use of it, and the females concerned in the dairy were so satisfied with its efficacy that a fresh supply was requested. I understand that small bundles of the solanum are placed in the tub containing the milk whilst it undergoes the process of curdling with the rennet, and remains for successive curdlings until its virtues are supposed to be exhausted, care being taken that its strong flavour should not be communicated to the milk by an overlarge quantity: this testimony is not I think satisfactory, or of much value, as the trial was made when the heave is always upon the wane, and as the woody nightshade is not a plant that vegetates early enough commonly to be used when the vernal grass abounds and most of the heave is particularly injurious, its virtues can but be seldom put to the test; however when practicable it may deserve a trial.[4]

The agriculture of a small district like ours affords no great scope to expatiate upon: great deviations from general practice we do not aim at; experimental husbandry is beyond our means, perhaps our faculties. Local habits, though often the subject of censure, are frequently such as the "genius of the soil" and situation render necessary, and the experience of years has proved most advantageous.

Our grass in the pastures of the clay lands, in the mowing season, which, from late feeding in the spring and coldness in the soil, is always late[5], presents a curious appearance; and I should apprehend, that a truss of our hay from these districts, brought into the London market, or exhibited as a new article of provender at a Smithfield cattle show, would occasion

4   Editor's note: Knapp's suggestion for curing 'heaving' in cheese, using woody night shade, is probably poisonous, and so should not be attempted.

5   In 1826, the herbage on some of our clay lands designed for mowing was, by reason of its tardy growth and the dry ness of the season, in such small quantities, that the owners let it grow untouched until after the corn harvest, in order to obtain some bottom grass, and, in consequence, our hay-making, as it was called, was not over until the last week in September.

conversation and comment. The crop consists almost entirely of the common field scabious (*scabiosa succisa*), loggerheads (*centauria nigra*), and the great ox-eye daisy (*chrysanthemum lucanthemum*). There is a scattering of bent (*agrostis vulgaris*), and here and there a specimen of the better grasses; but the predominant portion, the staple of the crop, is scabious— it is emphatically a promiscuous herbage; yet on this rubbish do the cattle thrive, and from their milk is produced a cheese greatly esteemed for toasting — melting, fat, and good flavoured, and, perhaps, inferior to none used for this purpose. The best grasses, indeed, with the exception of the dogstail (*cynosurus cristatus*), do not delight in our soil: the meadow poa (*p. pratensis*), and the rough stalked poa (*p. trivialis*), when found, are dwarfish; and having once occasion for a few specimens of the foxtail (*alopecurus pratensis*), I found it a scarce and a local plant: but I am convinced, from much observation, that certain species of plants, and grasses in particular, are indigenous to some soils, and that they will vegetate and ultimately predominate over others that may be introduced. In my own very small practice, a field of exceedingly indifferent herbage was broken up, underwent many ploughings, was exposed to the roastings of successive suns, and alternations of the year under various crops; amongst others that of potatoes; the requisite hackings, hoeings, and diggings of which alone were sufficient to eradicate any original fibrous, rooted herbage. This field was laid down with clean ray grass (*lolium perenne*), white trefoil, and hop clover, and did tolerably well for one year: and then the original softgrass (*holcus lanatus*) appeared, overpowered the crop, and repossessed the field; and yet the seed of this holcus could not have lain inert in the soil all this time, as it is a grass that rarely or never perfects its seed, but propagates by its root. The only grass that is purposely sown — trefoils are not grasses — is, I believe, the ray, or rye, no others being obtainable from the seedsman: this we consider as perennial; yet, let us lay down two pieces of land with seeds, from the same sack, the one a low, moist, deep soil, the other a dry upland, and in three or four years we shall find the natural herbage of the country spring up, dispute, and acquire in part possession of the soil, in despite of the raygrass sown: in the deep soil, the predominant crop will probably consist of poæ, cockfoot, meadow-fescue, holcus, phleum, foxtail, &c.; in the dry soil it will be dogstail, quaking grass, agrostis, &c., not one species of which was ever sown by us. It appears that the herbage of our poor, thin claylands is the natural produce of the soil, for every fixed soil will produce something, and would without care always exclude better herbage. Attention and manures, a kind of armed force, would certainly support other vegetation, alien introductions, for a time, but the profit would not always be adequate. In a piece of land of this nature I have suppressed the natural produce, by altering the soil with draining, sheep-feeding, stocking up, and composting: and scabious, carnation grass, mat grass, and their companions, no longer thrive; but if I should remit this treatment, they would again predominate, and constitute the crop.

Most counties seem to have some individual or species of wild plants predominating in their soil, which may be scarce, or only locally found in another; this is chiefly manifested in the corn-lands — for aquatic or alpine districts, and some other peculiarities, must form exceptions. This may be in some measure occasioned by treatment or manure, but commonly must be attributed to the chemical composition of the soil, as most plants have organs particularly adapted for imbibing certain substances from the earth, which may be rejected or not sought after by the fibrous or penetrating roots of another. *Festuca sylvatica* abounds in every soil without an apparent predilection for any one: *F. uniglumis*, only where it can imbibe marine salt: *F. pinnata*, is found vegetating upon calcareous soils alone, and I have known it appear immediately as the limestone inclined to the surface, as if all other soils were deficient in the requisite nutriment. Many of the maiden-hairs and ferns, pellitory, cotyledon, &c., are attached in the crevices of old walls, seeking as it were for the calcareous nitrate found there, this saltpetre appearing essential to their vigour and health. The predominating plants in some corn-fields is the red poppy, cherlock (*sinapis arvensis*), mustard (*sin. nigra.*), wild oat, corn-flower (*cyanus*); but in some adjoining parish we shall only sparingly find them. With us, in our cold clay-lands we find the slender foxtail grass (*alopecurus agr.*) abounding like a cultivated plant; when growing in clover, or the ray-grass, the whole are cut together, and though not a desirable addition, is not essentially injurious; but vegetating in the corn, it is a very pernicious weed, drawing nutriment from the crop, and overpowering it by its more early growth, at times so impoverishing the barley or the oats, as to render them comparatively of little value. The upright brome grass (*bromus erectus*) is a pest in our grass lands, giving the semblance of a crop in a most unproductive soil; hard and wiry, it possesses no virtue as food, and is useless as a grass: this bromus inclines to the limestone, the lias, or claystone, as if alumine was required, to effect some essential purpose in its nature; but this is a plant not found universally.

We have in use generally here a very prudential method of saving our crops in bad and catching seasons, by securing the hay in windcocks, and wheat in pooks. As soon as a portion of our grass becomes sufficiently dry, we do not wait for the whole crop being in the same state, but, collecting together about a good waggon-load of it, we make a large cock in the field, and as soon as a like quantity is ready we stack that likewise, until the whole field is successively finished, and on the first fine day unite the whole in a mow. Some farmers, in very precarious seasons, only cut enough to make one of these cocks, and having secured this, cut again for another. Should we be necessitated, from the state of the weather, to let these parcels remain long on the ground, or be a little dilatory, which I believe we sometimes are, before they are carried, or as we say, hawled (haled), the cocks are apt to get a little warm, and only partially heat in the mow, the hay cutting out streaky, and not perhaps so bright or fragrant as when uniformly heated in body: but I am acquainted with no other disadvantage from this practice, and it is assuredly the least expensive, and most ready way of saving a crop in a moist and uncertain season.

For wheat it is a very efficacious plan, as these stacks or pooks, (a corruption perhaps of packs,) when properly made, resist long and heavy rains; the sheaves not being simply piled together, but the heads gradually elevated to a certain degree in the centre, and the butt end then shoots off the water, the summit being lightly thatched. An objection has been raised to this custom, from the idea that the mice in the field take refuge in the pooks, and are thus carried home; but mice will resort to the sheaves as well when drying, and be conveyed in like manner to the barn: we have certainly no equally efficacious or speedy plan for securing a crop of wheat, and thousands of loads are thus commonly saved, which would otherwise be endangered, or lost by vegetating in the sheaf.

We will admit that grain, hardened by exposure to the sun and air in the sheaf, is sooner ready for the miller, and is generally a brighter article than that which has been hastily heaped up in the pook; but when the season does not allow of this exposure, but obliges us to prevent the germinating of the grain by any means, I know no practice, as an expedient, rather than a recommendation in all cases, more prompt and efficacious than this.

TWO OF OUR crops not being of universal culture are entitled to a brief mention. We grow the potato extensively in our fields, a root which must be considered, after bread-corn and rice, the kindest vegetable gift of Providence to mankind. This root forms the chief support of our population as their food, and affords them a healthful employment for three months in the year, during the various stages of planting, hacking, hoeing, harvesting. Every labourer rents of the farmer some portion of his land, to the amount of a rood or more, for this culture, the profits of which enable him frequently to build a cottage, and, with the aid of a little bread, furnishes a regular, plentiful, nutritious food for himself, his wife, and children within, and his pig without doors; and they all grow fat and healthy upon this diet, and use has rendered it essential to their being. The population of England, Europe perhaps, would never have been numerous as it is, without this vegetable; and if the human race continue increasing, the cultivation of it may be extended to meet every demand, which no other earthly product could scarcely be found to admit of. The increase of mankind throughout Europe, within the last forty years, has been most remarkable, as every census informs us, notwithstanding the havoc and waste of continual warfare, and most extensive emigration; and as it seems to be an established maxim, that population will increase according to the means of supply, so, if a northern hive should swarm again, or

> Blue-eyed myriads from the Baltic shore

once more arise, future historians will probably attribute this excess of population, and the revolutions it may effect, to the introduction of vaccination on the one part, and the cultivation of the potato on the other.

The varieties of this tuber, like apples, seem annually extending, and every village has its own approved sorts and names, different soils being found preferable for particular kinds, and local treatment advantageous. We plant both by the dibble[6] and the spade: our chief sorts are pink eyes, princess beauty, magpies, and china oranges, for our first crop; blacks, roughs, and reds, for the latter crop; and horses' legs, for cattle. We have a new sort under trial, with rather an extraordinary name, which I must here call "femora dominarum!" But we find here, as is usual with other vegetable varieties, that after a few years' cultivation the sorts lose their original characters, or, as the men say, "the land gets sick of them," and they cease to produce as at first, and new sets are resorted to. We have no vegetable under cultivation more probably remunerative than this, or more certain of being in demand sooner or later; it consequently becomes an article of speculation, but not to such an injurious extent as some others are: it gives a sufficient profit to the farmer and his subrenter. Our land is variously rented for this culture; but perhaps eight pounds per acre is a general standard: the farmer gives it two ploughings, finds manure,

---

6 But dibbling is not held in esteem by us: we think that in wet seasons the holes retain the moisture and the sets perish; and that in dry weather, being less covered than when planted by the spade, they are more obnoxious to injury by birds and mice, become affected by droughts, are longer in shooting out, and produce, in most cases, inferior crops. In a lighter soil these objections, perhaps, would not be found reasonable.

and pays the tithe; the seed is found, and all the labour in and out is performed by the renter; or the farmer, in lieu of any rent, receives half the crop. The farmer's expenses may be rated at —

	£.	s.	d.
Rent to his landlord ...	1	10	0
Two ploughings...	1	6	0
Twelve loads of manure...	1	16	0
Tithe...	0	10	0
Rates...	0	3	0
	£5	5.	0

leaving him a clear profit of 2*l*. 15*s*. per acre. The sub-renter's expenditure and profit will be —

	£.	s.	d.
Rent...	8	0	0
Labour in and out...	3	0	0
Five sacks of seed...	1	12	6
	12	12	6
	£.	s.	d.
Produce 50 sacks, at 6s. 6d...	16	5	0
Trash, or small pigs...	1	0	0
	£17	5	0

leaving a profit of £4. 12*s*. 6*d*. per acre. The produce will vary greatly at times, and then the price of the article varies too. The returns to the labourer are always ample, when conducted with any thing like discretion; and the emolument to the farmer is also quite sufficient, as, beside the rent, he is paid for the manuring of his land for a succeeding crop, be it wheat or barley; hence land is always to be obtained by the cotter, upon application. We have a marked instance in the year 1825 how little we can predict what the product of this crop will be, or the change that alteration of weather may effect; for after the drought of the summer, after our apprehensions, our dismay (for the loss of this root is a very serious calamity), the produce of potatoes was generally fair, in places abundant; many acres yielding full eighty sacks, which, at the digging out price of 6s. the sack, gave a clear profit to the labourer of £11. 7s. 6d. per acre ! But at any rate it gives infinite comfort to the poor man, which no other article can equally do, and a plentiful subsistence, when grain would be poverty and want. The injudicious manner in which some farmers have let their land has certainly, under old acts of parliament, brought many families into a parish; but we have very few instances where a potato-land renter to any extent is supported by the parish. In this village a very large portion of our peasantry inhabit their own cottages, the greater number of which have been obtained by their industry, and the successful culture of this root. The getting in and out of the crop is solely performed by the cotter and his family; a child drops a set in the dibble-hole or the trench made by the father, the wife with her hoe covering it up; and in harvesting all the family are in action; the baby is wrapped up when asleep in its mother's cloak, and laid under the shelter of some hedge, and the digging, picking, and conveying to the great store-heap, commences; a primitive occupation and community of labour, that I believe no other article admits of or affords.

It has been said that the culture of the potato is injurious to the farm in general, and I know landlords who restrict the growth of it; but, perhaps, the extent of injury has been greatly overrated. The potato, it is true, makes no return to the land in straw for manure, and a large portion of that which is made in the barton is occasionally required for its cultivation; and thus it is said to consume without any repayment what is equally due to other crops: but the cultivation of this tuber requires that the soil should be moved and turned repeatedly; it is generally twice at least ploughed, trenched by the spade for sets, and hacked when the plant is above ground, then hoed into ridges, and finally, the whole turned over again when the crop is got out: thus is the soil six times turned and exposed to the sun and air; and it is kept perfectly free from weeds of all kinds — both of which circumstances are essentially beneficially to the soil. If the potato must have manure, it does not exhaust all the virtues of it, as the crop which succeeds it, be it wheat or barley, sufficiently manifests: there are, besides, exertions made by the renter to obtain this profitable crop, that greatly improve the farm, and which a less promising one would not always stimulate him to attempt — he will cut up his ditch banks, collect the waste soil of his fields, composting it with lime and other matters, as a dressing for the potato crop, and it answers well; the usual returns from corn, and fluctuations in the price, will not often induce him to make such exertions. All this is no robbery of the farmyard, but solely a profitable reward and premium to industry.

Since the time that the Potato became extensively cultivated, the corn crops which have succeeded, have manifestly been more free from predominating weeds than they formerly were; *Cyanus* (corn flower) is scarce with us, but the glow of the poppy, the gay yellow of the gold of pleasure (*Alyssum sat:*) and such plants have been sparingly observed, with the exception of charlock, which yet keeps its ground, and weeds in general diminished contention with the crop; we may be a little more attentive to cleanliness, and eradication of intrusive herbage perhaps, but we must rather conclude that the repeated hacking & hoeing of the potato crop, and exposure of the earth, at first, and overshadowing foliage afterwards, by destroying the annual weed have accomplished this; but however reasonable or defective the cause assigned may be, it is a fact that the grain crops of late years have been much less than formerly intruded upon by tall and conspicuous weeds

Much has been said and written about the potato, but as some erroneous ideas have been received concerning its early introduction into Europe, perhaps a slight sketch of the history of this extraordinary root may not be uninteresting — a summary of the perusal of multitudes of volumes, papers, treatises!

The sweet Spanish potato (*convolvulus batatus*), a native of the East, was very early dispersed throughout the continent of Europe; and all the ancient accounts, in which the name of potato is mentioned, relate exclusively to this plant, a convolvulus; but our inquiry at present regards that root now in such extensive cultivation with us, which is an American plant (*solanum tuberosum*). Perhaps the first mention that is known concerning the root is that of the great German botanist, Clusius, in 1588, who received a present of two of the tubers in that year from Flanders, and there is a plate of it among his rare plants. The first certain account which I know of by any English writer is in Gerard, who mentions, in his herbal, receiving some roots from Virginia, and planting them in his garden near London as a curiosity, in the year 1597. All the multiform tales which we have of its introduction by Hawkins, shipwrecked vessels, Raleigh, and his boiling the apples instead of the roots, are merely traditional fancies, or modern inventions, with little or no probability for support. There is some possibility that Sir Walter Raleigh might have introduced the potato into Ireland from America, when he returned in 1584, or rather after his last voyage, eleven years later; but if so, it was much confined in its culture, and slowly acquired estimation even in that island, for Dr. Campbell does not admit that it was known there before the year 1610, fifteen years after Sir Walter's final return. In England it seems to have been yet more tardy in obtaining notice, for the first mention which I can find wherein this tuber is regarded as possessing any virtue, is by that great man, Sir Francis Bacon, who investigated nature from the "cedar that is in Lebanon even unto the hyssop that springeth out of the wall: he spake also of beasts, and of fowls, and of fishes, and of creeping things," in his history of "Life and Death," written, probably, in retirement after his disgrace. He observes, that "if ale was brewed with one-fourth part of some fat root, such as the potato, to three-fourths of grain, it would be more conducive to longevity than with grain alone." It was thus full twenty-four years after its being planted by Gerard that the nutritive virtues of this root appear to have been understood; but with us there seems to have been almost an antipathy against this root, as an article of food, which can scarcely excite surprise, when we consider what a wretched sort must have been grown, which one writer tells us was very near the nature of Jerusalem artichokes, but not so good or wholesome; and that they were to be roasted and sliced, and eaten with a sauce composed of wine and sugar! Even Philip Miller, who wrote his account not quite seventy years ago, says, "they were despised by the rich, and deemed only proper food for the meaner sorts of persons;" and this at a time when that sorry root, the underground or Jerusalem artichoke (*helianthus tuberosus*) was in great esteem, and extensively cultivated. And we must bear in mind the disinclination, the prejudice I might almost call it, that this root manifests for particular soils. Most of our esculent vegetables thrive better — are better flavoured, when growing in certain soils, and under different influences; but the potato becomes actually deteriorated in some land. And every cultivator knows from experience, that the much-admired product of some friend's domain or garden becomes, when introduced into his own, a very inferior, or even an unpalatable root. Potatoes will grow in certain parishes and districts, and even remain unvitiated; but the product will be scanty, as if they tolerated the culture only, and produced by favour; whereas in an adjoining station, possessing some different admixture of soil, some change of aspect, the crop will be highly remunerative. These circumstances, in earlier days, when their value, and the necessity of possessing them, were not felt, counteracted any attempt for extensive cultivation, or, probably, influenced the dislike to their use.

However locally this solanum might have been planted, yet it appears, after consulting a variety of agricultural reports, garden books, husbandmen's directions, &c., down to the statements of Arthur Young, that the potato has not been grown in gardens in England more than one hundred and seventy years; or to any extent in the field above seventy-five. At length, however, as better sorts were introduced, and better modes of dressing found out, it became esteemed; and the value of this most inestimable root was so rapidly manifested, and the demand for it so great, that we find by a survey made about thirty years ago, that the county of Essex alone cultivated about seventeen hundred acres for the London market. I know not the extent of land now required for the supply of our metropolis, but it must be prodigious.

Amidst the numerous remarkable productions ushered into the old continent from the new world, there are two which stand pre-eminently conspicuous from their general adoption; unlike in their natures, both have been received as extensive blessings— the one by its nutritive powers tends to support, the other by its narcotic virtues to soothe and comfort the human frame — the potato and tobacco; but very different was the favour with which these plants were viewed: the one, long rejected, by the slow operation of time, and perhaps of necessity, was at length cherished, and has become the support of millions; but nearly one hundred and twenty years passed away before even a trial of its merits was attempted: whereas the tobacco from Yuccatan, in less than seventy years after the discovery, appears to have been extensively cultivated in Portugal, and is, perhaps, the most generally adopted superfluous vegetable product known; for sugar and opium are not in such common use. Luxuries, usually, are expensive pleasures, and hence confined to few: but this sedative herb, from its cheapness, is accessible to almost every one, and is the favourite indulgence of a large portion of mankind. Food and rest are the great requirements of mortal life: the potato, by its starch, satisfies the demands of hunger; the tobacco, by its morphin, calms the turbulence of the mind: the former becomes a necessity required; the latter a gratification sought for.

Many as the uses are to which this root is applicable— and it will be annually applied to more; if we consider it merely as an article of food, though subject to occasional partial failures, yet exempted from the blights, the mildews, the wire-worms, the germinatings of corn, which have often filled our land with wailings and with death, we will hail the individual, whoever he might be, who brought it to us, as one of the greatest benefactors to the human race, and with grateful hearts thank the bountiful giver of all good things for this most extensive blessing.

It is a well-known fact, that we are perfectly ignorant of the native sites of nearly all those gramineous plants, distinguished by Linnaeus as Cerealia, whose seeds have from the earliest periods of time served for the food of man, such as wheat, rye, barley, rice, maize, oats: perhaps we must except the two last, as the oat was discovered by Bruce growing under the culture of Nature alone; and he was too good a botanist to have mistaken the identity of *Avena sativa* — and Indian corn may have been found. That some of them were produced in those regions first inhabited by mankind, we have every reason to believe, and the warrant of something like obscure tradition; but our ignorance of the first habitats of these plants is the less to be wondered at, when we consider that it is more than probable that culture and the arts of man have so infinitely changed the form, improved the nature, and obscured the original species, that it is no longer traceable in any existent state. There appears to be a permission from Nature to effect certain changes in vegetables, yet she retains an inherent propensity in the plant to revert to its original creation, which is very manifest in this particular race; for the sorts which we now make use of will not endure the thraldom of our perversion without the artifices, the restraints of man, but have a constant tendency to return to some other nature, or to run wild, as we call it. Man bears them with him in all his wanderings; by his treatment they remain obedient to his desires, and are identified with colonization; but as soon as he remits his attentions, the seeds perish in the soil, or their offspring dwindle in the earth, and are lost. Or we may say, that Nature, having created these things, permits him, in the sweat of his brow, to effect an improvement, and consigns the custody of them to his care, satisfied that he will preserve them for his own benefit as long as required; when his occasion for them ceases, or when by sloth he neglects them, they return to their original creation: the earth might be cursed to bring forth thorns and thistles, but an attendant blessing and mercy was reserved of permitting them to be cultivated, producing healthful recreation and grateful food. If these are plants of immemorial antiquity, the potato is yet of comparatively modern introduction, but the original species from whence all our endless varieties have emanated cannot probably now be ascertained, man having, as observed above, almost created an essential article of food; and it is not unimportant to note the great difference that subsists in the component parts of these varieties — for though, in common estimation, a potato may be a potato, yet we find them very differently compounded. The influence of different temperatures and years may cause these proportions to vary, but I give them as observed in 1828.

Black or purple	Fibre 9¾	Fecula... 9¾	Water 80½	= 100
Prince's beauty	Ditto 15	Ditto... 11¾	Ditto 70¼	do.
Horses' legs....	Ditto 13	Ditto... 15	Ditto 72	do.

The proportion of fecula varies greatly, and as the principle of nutriment is supposed to exist in this matter, the value of each sort, if mere nutriment is required, is indicated by this analysis.

The potato may be considered as the most valuable production that Europe has received from the continent of America, and is now, as Bishop

Heber informs us, much esteemed in the East, and regarded as the greatest benefit the country ever received from its European masters. A plant that can so climatise and preserve its valuable properties in such different temperatures as northern Europe and Bengal, where the thermometer ranges up to 90 or 100 degrees of heat, must be particularly endowed, and in time will probably become naturalized to every region, and circulate its benefits round the globe. The strenuous manner in which I have lauded this root may, perhaps, excite a smile in some, who only know it as a table viand; but those who have witnessed the blessings which this tuber confers, by affording a sufficiency of food to man and beast, will not be disposed to regard lightly such comforts obtainable by their poorer neighbours.

OUR SECOND CROP to which I alluded, and which some years we grow largely, is the teazle (*dipsacus fullonum*), a plant which is probably no native of this country, but, like woad; canary-grass, &c., originally introduced by some of the numerous foreign artisans, who have at various times sought refuge here, or been encouraged to settle in England. Our woollen manufactory could hardly have made any progress without this plant: the constant continental wars in the earlier part of our monarchy, and the rival jealousies of foreign nations, would have impeded, or prohibited, the necessary supply of teazles, and thus rendered the domestic cultivation of this indispensable plant a primary object. The manufactory of cloth was certainly carried on in England during the reign of Richard I., perhaps in his father's reign; but it was probably not until after the tenth of Edward III., that the teazle was cultivated to any extent with us; for about that time the exportation of English wool was prohibited, and the wearing of foreign cloth opposed by government. Flemish artisans were encouraged to settle in this country, and carry on their trade, with every liberty and protection; a regular mart was established; and the tuckers, or woollen weavers, became an incorporated body; particular towns began to furnish peculiar colours — Kendal its green, Coventry its blue, Bristol its red, &c.; and from this period, I think, we may date the cultivation of the teazle in England.

Hudson, in considering this species as indigenous, directs us to hedges for our specimens; but, though the teazle is certainly found a wilding in some places astray from cultivation, yet it is singular that with us it does not wander from culture: though the seeds are scattered about and swept from the barns where the heads are dried, into the yard, and vegetate in profusion on the dung-heaps and the by-ways where dropped, yet I have never observed it growing in the surrounding hedges.

Teazles are cultivated in some of the strong clay lands of Wilts, Essex, Gloucester, and Somerset. The latter county is supposed to have grown them earliest. The manufacturers rather give the preference to those of Gloucester, as lands repeatedly cropped are thought not to produce them so good in some respects. Strong land, thrown up as for wheat, and kept dry, affords the best teazles. Weeding, draining, and other requisites, demand a constant labour through great part of the year; and hence a certain expense is incurred: but remuneration, loss, or great profit, circumstances must determine; nor, perhaps, is there any article grown more precarious or mutable in its returns.

The teazle throws up its heads in July and August; and these are cut from the plant by hand, with a knife particularly formed, and then fastened to poles for drying: the terminating heads are ready first, and called "kings:" they are larger and coarser than the others, and fitted only for the strongest kinds of cloth, and are about half the value of the best. The collateral heads then succeed, and receive the name of "middlings," and are the prime teazles. Should the season prove moist, great injury ensues; but exposure to wet for any length of time ruins the head, which, by its peculiar construction, retains the moisture, and it decays. We cannot stack them like corn, as pressure destroys the spines, and a free circulation of air is required to dry them thoroughly; and we seek for barns, sheds, and shelter of any kind, crowd the very bed-rooms of our cottages with them in dripping seasons, and bask them in every sunny gleam that breaks out: this is attended with infinite trouble; and as few farmers, who have so many other concerns on their hands, like to encounter it, they become the speculation of the most opulent class of cottagers. When dry, they are picked and sorted into bundles for sale, ten thousand best and small middlings making a pack; nine thousand constitute the pack of kings. If there be a stock on hand, and the season favourable, there is a sufficiency for the demand, and the price low: if adverse weather ensue, the price becomes greatly advanced, and we have known them in the course of a few months vary from £4 to £22 the pack! but from £5 to £7 is perhaps the average price of this article. This variation in value affords the growers a subject for constant speculation — a source of rapid wealth to some, and injury to others — and we most emphatically call teazles a "casualty crop." Our manufacturers occasionally import teazles from Holland and France, when the price is high in England: this they can do when the home price exceeds £8.

In letting teazle land, various agreements are made, not necessary to mention in a note like this; but it is usually taken for two years, it requiring much of this time from sowing the seed to cutting the heads for sale. In rating the expenses, we will say —

	£.	s.	d.
One acre, at £2 per ann. (for two years)...	4	0	0
Expense of culture, £3 per ann. per acre...	6	0	0
Tithe...	0	8	0
Cutting the heads, per acre...	0	6	0
Sorting and packing, at 6s. for seven packs, average Crop...	2	2	0
Miscellaneous expenses, polls, sticks, &c...	1	0.	0
	13	16	0
Average crop brought to market, seven packs, at £6	42	0	0
Leaving a profit for the 2 years, upon an acre, of	£28	4	3

As the teazle man seldom rents less than four or six acres, which he can very well attend to, it may produce at the two years' end a return upon the six acres of £169, if all circumstances should be favourable — a tempting inducement to speculation, when a labourer, by regular daily pay, cannot earn above £32 per annum. But it requires some ready money to support the family during this period of expectation — and if a bad season occur, all the labour is lost, the profit destroyed, the anxiety of months ends in disappointment, and debt only remains. This is most truly a casualty crop; and the manufacturers are so sensible of the risk and trouble attending the cultivation of this plant, that they prefer purchasing to growing it for their own use; and I know one who has declared his loss in the attempt to exceed £500.

It has been thought that the cultivation of teazles exhausted the land, and some landlords in consequence have forbidden the growth of them in their agreements; perhaps I can be no sufficient judge of the accuracy of this idea, from our limited growth; but speaking locally, such land as we make use of for their culture is of so inferior a nature, that little deterioration can ensue from any crop. The teazle, having a tap root, does not exhaust the superficial soil as a fibrous-rooted plant would do; the ground on which they grow is hoed, and turned by the spade repeatedly, and up to a certain period kept free from weeds; but as the plant is forming heads, little attention seems given to the eradication of intrusive rubbish, and, consequently, after gathering the crop, the soil is frequently in a very foul state, and from hence the chief injury to the land may arise, rather than from the teazle plant. Though this crop requires no manure, nor affords any to the soil, yet the removal of the earth so repeatedly by the hoe and spade becomes equivalent to a fallow: with us a wheat crop often succeeds the teazle, and I have observed in this case as good a return of that grain as is produced by the adjoining fields where teazles had not been grown.

This plant seems to be known in many countries by a name expressive of its use. Old Gerard has recorded several of these names. Its old English name was the carding teazle; the Latin name *carduus veneris*; the French call it *chardon de foullon;* the Danes and Swedes, *karde tidsel*; the Flemings, *karden distel*; the Hollanders, *kaarden*; Italy and Portugal, *cardo*; the Spaniards, *cardencha*, &c.

I believe that the teazle affords a solitary instance of a natural production being applied to mechanical purposes in the state in which it is produced[7]. It appears, from many attempts, that the object designed to be effected by the teazle cannot be supplied by any contrivance — successive inventions having been abandoned as defective or injurious. The use of the teazle is to draw out the ends of the wool from the manufactured cloth, so as to bring a regular pile or nap upon the surface, free from twistings and knottings, and to comb off the coarse and loose parts of the wool. The head of the true teazle is composed of incorporated flowers, each separated by a long, rigid, chaffy substance, the terminating point of which is furnished with a fine hook. (See Plate 1, Fig. 4.) Many of these heads are fixed in a frame; and with this the surface of the cloth is teased, or brushed, until all the ends are drawn out, the loose parts combed off, and the cloth ceases to yield impediments to the free passage of the wheel, or frame, of teazles. Should the hook of the chaff, when in use, become fixed in a knot, or find sufficient resistance, it breaks without injuring or contending with the cloth, and care is taken by successive applications to draw the impediment out: but all mechanical inventions hitherto made use of offer resistance to the knot; and, instead of yielding and breaking as the teazle does, resist and tear it out, making a hole, or injuring the surface. The dressing of a piece of cloth consumes a great multitude of teazles — it requiring from fifteen hundred to two thousand heads to accomplish the work properly. They are used repeatedly in the different stages of the process; but a piece of fine cloth generally breaks this number before it is finished, or we may say that there is a consumption answering to the proposed fineness — pieces of the best kinds requiring one hundred and fifty or two hundred runnings up, according to circumstances.

Our small farmers here have a vile practice of picking from their turf, in the spring of the year, all the droppings of their autumn and winter fed cattle to

---

7 *Equisetum hyemale*, the Dutch rush, or shave grass, is yet used in its natural state for finishing fine models in wood, and in removing roughnesses in plaster casts.

carry on their arable land for the potato, or some grain crop: this affords no great supply to ploughed land, and is very injurious to their grazing grounds; but the answer generally is, "that the corn must have manure, and the beast can take care of itself;" and in many cases, I fear, from the starved appearance of the young cattle, that their best endeavours have afforded a very inadequate supply.

This picking of the field was formerly very generally resorted to in the midland counties; but the farmer at that time had a sufficient excuse in the scarcity of common fuel. The droppings of the cows were collected in heaps, and beaten into a mass with water; then pressed by the feet into moulds like bricks, by regular professional persons, called clatters (clodders); then dried in the sun, and stacked like peat, and a dry March for the clat harvest was considered as very desirable. These answered very well for heating water for the dairy and uses of the farm back-kitchen, giving a steady, dull heat, without flame; but navigable canals, and other conveniences of a similar nature, have rendered the practice now unnecessary. With us this bad custom is declining, and probably in time will cease altogether.

IT IS RATHER a subject of surprise, that in our general associations and commixtures in life, in times so highly enlightened as the present, when many ancient prejudices are gradually flitting away, as reason and science dawn on mankind, we should meet with so few, comparatively speaking, who have any knowledge of, or take the least interest in natural history; or if the subject obtain a moment's consideration, it has no abiding place in the mind, being dismissed as the fitting employ of children and inferior capacities. But the natural historian is required to attend to something more than the vagaries of butterflies, and the spinnings of caterpillars; his study, considered abstractedly from the various branches of science which it embraces, is one of the most delightful occupations that can employ the attention of reasoning beings; a beautiful landscape, grateful objects, pleasures received by the eye or the senses, become the common property of all who can enjoy them, being in some measure obvious to every one; but the naturalist must reflect upon hidden things, investigate by comparison, and testify by experience, and living amidst the wonders of creation, it becomes his occupation to note and proclaim such manifestations of wisdom or goodness as may be perceived by him. And perhaps none of the amusements of human life are more satisfactory and dignified, than the investigation and survey of the workings and ways of Providence in this created world of wonders, filled with his never-absent power; it occupies and elevates the mind, is inexhaustible in supply, and, while it furnishes meditation for the closet of the studious, gives to the reflections of the moralizing rambler admiration and delight, and is an engaging companion, that will communicate an interest to every rural walk. We need not live with the humble denizens of the air, the tenants of the woods and hedges, or the grasses of the field; but to pass them by in utter disregard, is to neglect a large portion of rational pleasure open to our view, which may edify and employ many a passing hour, and by easy gradations will often become the source whence flow contemplations of the highest order. Young minds cannot, I should conceive, be too strongly impressed with the simple wonders of creation by which they are surrounded: in the race of life they may be passed by, the occupation of existence may not admit attention to them, or the unceasing cares of the world may smother early attainments — but they can never be injurious — will give a bias to a reasoning mind, and tend, in some after, thoughtful, sobered hour, to comfort and to soothe. The little insights that we have obtained into Nature's works are many of them the offspring of scientific research; and partial and uncertain as our labours are, yet a brief gleam will occasionally lighten the darksome path of the humble inquirer, and give him a momentary glimpse of hidden truths: let not, then, the idle and the ignorant scoff at him who devotes an unemployed hour, —

No calling left, no duty broke,

to investigate a moss, a fungus, a beetle, or a shell, in "ways of pleasantness, and in paths of peace." They are all the formation of Supreme intelligence, for a wise and a worthy end, and may lead us by gentle gradations to a faint conception of the powers of infinite wisdom. They have calmed and amused some of us worms and reptiles, and possibly bettered us for our change to a new and more perfect order of being.

When we investigate the works of nature our business is not simply to admire this, to define and arrange that, its mere reference to man, but endeavour to trace the object, to indicate the goodness of the creator in his works, with all humbleness of heart be grateful for any permission that may be afforded us of perceiving his power and universal regard for his creatures; not to let our operations be a simple form of the head or hand, but covert them to a practical religion of mind, a kneeling down of the heart to benevolence and perfection. Nor do I know any employment that can so well exercise our affections and produce such a natural love of our maker as the contemplation of these his works on earth. The heavenly bodies perhaps may appear more wonderful, but we do not dwell with them, know their influences less, and find such mystery surrounding them, such awful grandeur, that tho' they elevate the mind, yet depress it with a sense of insignificance, with something like fear of approaching too closely the heavenly sanctity; as in reading the Revelations of S^t. John, we seem to be involved with mighty powers and things requiring a reverence and honour unglorified beings know not how worthily to bestow; but in these lowly works, companions of man, we dare to admire, sometimes fancy we comprehend, and always may presume exempted from offence to search out, and they will lead to faith & love.

We yet possess two forest-trees, beautiful and unmutilated! An oak in Shellard's-lane has escaped the woodman's axe, the hedger's bill: it stands on the side of the waste, and has long afforded shade and shelter to an adjoining farm-house. These circumstances, and not being valuable as a timber tree, may have contributed to its preservation: its hamadryad is left alone in the land to mourn her lost companions. This tree is not mentioned as being at all comparable with the gigantic productions of the kind that we have accounts of, and perhaps by many would be passed by unnoticed; yet it is deserving of some regard, from the vegetable powers that have existed, and still continue in its trunk, The bole, at some very distant period, by accident or design, appears to have lost its leading shoot, and in consequence has thrown out several collateral branches; three remain, which have now grown into trees themselves, existing in full vigour and constituting a whole of much beauty. It is a characteristic specimen of an oak, with all the corrugations, twistings, furrows, and irregularities, which this tree with a free growth generally exhibits; expanding its three vigorous arms to the Sun of Heaven with a pendent, easy dignity, that seems like an enjoyment of unrestrained liberty. We have no good criterion to regulate our judgment with regard to the age of trees of considerable antiquity. In young ones the rings of the wood will often afford a reasonable ground for opinion; but in old trees these marks are absorbed, obscured, or uncertainly formed, so as to be no sufficient guide. In particular cases, such as inclosure of waste or other lands, formation of parks and plantations, the times of planting are sufficiently recorded; but generally speaking, neither oral tradition, nor written testimony, remains to indicate the period when a tree sprang up. This oak, however, from all the signs of age that it retains, must have existed as a sapling at some very distant day, and is the most undoubted relic of antiquity in the vegetable world that we possess.

The elm and the beech, in age, frequently present very decided vestiges of a former day; but the oak of centuries has impressed upon it indelible characters of antiquity, and is a visible *vetustum monumentum*. The wreathings and contortions of its bark, even its once vigorous, but now sapless limbs, with their bare and bleached summits, stagheaded and erect, maintain a regality of character which perfectly indicates the monarch of the forest, and which no other tree assumes. We have many accounts in different authors of the prodigious size which the oak has attained in England; but most of the trees that have arrived at any vast circumference, seem, like this our village oak, to have lost their leaders when young, and hence are short in the butt: yet we have records of aspiring timber trees of this species, of astonishing magnitude, though perhaps none of them exceed those mentioned by Evelyn, cut down near Newberry, in Berkshire, one of which ran fifty feet clear without a knot, and cut clean timber five feet square at the base; its consort gave forty feet of clear, straight timber, squaring four feet at its base, and nearly a yard at the top. The "lady oak," mentioned by Sir E. Harley, produced a butt of forty feet, and squared five feet throughout its whole length, thus producing twenty tons of timber — a mass of surprising grandeur! But the most magnificent oak ever known to have grown in England was probably that dug out of Hatfield Bog: it was a hundred and twenty feet in length, twelve in diameter at the base, ten in the middle, and six at the smaller end where broken off; so that the butt for sixty feet squared seven feet of timber, and four its entire length. Twenty pounds were offered for this tree[8]. This extraordinary vegetable should have been preserved in some museum, as unequalled in ancient, unapproachable in modern days; exceeding in magnitude even that famous larch brought to Rome in the reign of Tiberius[9], and reserved as a curiosity for many years, which was one hundred and twenty feet long, and two feet in diameter its whole length.

Indigenous, flourishing, and inured to all the caprices of our climate as the oak is, yet it produces its fruit very precariously, and at times sparingly, like a plant of exotic origin; which does not appear to have been the case formerly, when such herds of swine were maintained by the produce of our woods alone, and grants from manorial lords for permission thus to feed them were recorded with care as valuable obtainments.

The cause of infertility in indigenous trees can arise from no defect of construction in the organs of fructification, but from some obstruction, perversion, redundancy, or vitiation of the natural powers; which is particularly manifested by the faculty which they possess at one period of producing fruit, and their impotency at another. This imbecility from one cause or another probably influences at periods every tree or herb that springs from the earth;

8   Philosoph. Trans., as quoted in the Sylva.
9   Pliny's Natural History.

but in regard to the oak, the most general and probable cause of its sterility is suspended circulation. This is more immediately brought to notice from our custom of barking the timber of this tree in the spring. At times our barkers go on rapidly with their work; yet in a few hours a frost, or a sharp wind, will put an entire stop to their operations, in consequence of the cessation of the flow of sap, which is followed by the adhesion of the bark to the wood. Whenever this nutriment ceases to be supplied, the immature and tender germen must languish; and if the supply be long suspended, it must perish from deficiency of food. That such is the natural effect of spring frosts and sudden chills, more injurious probably to the fruit in this immature state, from its greater delicacy, than when it is more developed, is reasonable to suppose: how far a change of seasons may have taken place to accomplish the injury alluded to, more commonly now than in former periods, we have no criterion for proving; but if failures of the acorn crop took place as frequently in times when swine's flesh was mostly the diet of the middle and lower classes of people as they do now, the privations of our forefathers were severe indeed.

An interesting volume might be formed, entitled the "History of the Oak." The first mention that we know of this tree is that ancient of days, the "oak of Mamre," under which Abraham sat in the heat of the day; and that it was an oak, one of the fathers, Eusebius, tells us, as it remained an object of veneration even in the time of Constantine. We would note all the celebrated querci of antiquity; the use, value, strength, duration, &c., of its timber; the infinite variety of purposes to which its various parts are applied by the mechanic, the dyer, the artisan; the insects, which amount to hundreds of species, that live and have their being on the oak; the vegetables it nourishes, ferns, lichens, mosses, agarics, boleti, &c.; the sawdust, apples, gallnuts, acorns, leaves, and innumerable et cetera of Britain's guardian tree. However highly the Druids might venerate the oak, and make it the emblem and residence of their deity, yet the intrinsic value of this tree was unknown to our remote forefathers. All their knowledge of its virtues was probably included in its uses for building, its acorns for their swine, and, perhaps, its bark for preserving the skins which they used. Modern ingenuity and necessity have brought its various qualities into notice, or our oak would have received such honours as in days of darkness were conferred upon inanimate things. Attica considered the olive as the gift of her tutelary goddess, and some benevolent saint would have been lauded and hymned, for having endowed the oak of Britain with such extensive virtues for the good of mankind.

The other tree that I mentioned above as one of our boasts, is a wych or broad-leaved elm (*ulmus montana*), standing near the turnpike-road.

This very fine and stately tree was saved, when the merciless axe levelled all its companions, at the solicitation of a lady now no more, and remains a testimony of her good taste, the civility of the agent, and the ornament of our village. When in youth, this species presents a character decidedly different from the common elm (*ulmus campestris*). Its branches at times are so strong as to be nearly equal in size with the main stem that supports them, and loaded with such a profusion of foliage, that the sprays become pendent, and give the idea of luxuriance with weakness, of a growth beyond strength: advancing in age, its arms and sprays become less pensile, as the leaves are smaller and less burdensome; yet they hang commonly in large heavy masses, like what we formerly were accustomed to see in the aquatintas of Jukes, and the prints of that period. It can, however, occasionally assume the appearance of elegance and lightness, and is usually less aspiring and more branching than the common elm; its dense foliage yields a fine shade for cattle, and it deserves, even on this account, if it possessed no other merit, a more general cultivation. The wych elm, though a rare tree in some counties, seems more extensively spread over England than the other species, and adventures farther to the north. Ray tells us, on the authority of Aubrey, that the common elm, so called, is scarcely found indigenous northward of Lincolnshire, whereas this species is found even in Scotland. Our soil is very favourable to the growth of both species. The wych elm affords a tough and valuable wood for the wheeler and the millwright; the bark from the young limbs is stripped off in long ribands, and often used, especially in Wales, for securing thatch, and for various bindings and tyings, to which purpose its flexible and tough nature renders it well adapted. Gerard says, that arrows were made from the wood of this tree, and he lived at a period when he could well ascertain the fact, during the reign of Elizabeth and her predecessor, before fire-arms had superseded this truly British weapon: he was, in the younger part of his life, gardener to the great Lord Burleigh. That the wych elm, when permitted, will attain large dimensions, is manifest by the size of several we have observed in many places; but that gigantic one which grew in Staffordshire exceeds in magnitude any other of this species which we ever heard of. It required the labour of two men for five days to fell it; it was forty yards in length, with a diameter of seventeen feet at the butt; yielding eight pair of naves, and eight thousand, six hundred, and sixty feet of boards, the sawing of which cost 10*l.* 17*s.* It contained ninety-seven tons of timber. As Evelyn says, "this was certainly a goodly tree!" The etymology of this tree seems to be unknown, and different authors, who mention it, spell it, accordingly, various ways: – Evelyn calls it, wich, and witch; Gilpin, wich; others, wych; Bacon, weech. The foliage of the young trees of this elm are the favourite food of the larvae of the buff-tip-moth (*Phal. Bucephala*), for though they likewise feed upon the young leaves of the oak and the lime, yet they give the preference to those of this tree: when so feeding it will always be known by their rejectments on the earth beneath, which, when the larvae are in any number, may be noticed by very unattentive persons. This caterpillar, when nearly fed for its change, becomes heavy, and commonly falls to the earth from the spray, and we can see them crawling along the paths, or even upon the clothes of persons that have walked under the trees where they have fed: though this creature is very often found in considerable numbers throughout the summer and autumn, yet, by reason of some fatality, the moth is by no means so common an insect as might be expected from the profusion of its larvae.

We have no indigenous tree that suffers from the advance of the winter season so early as the wych elm. A few others may manifest its approach nearly as soon, but they become augmented in splendour by a touch of the frosty air, not ruined and denuded like our elm, which contributes no grandeur, no beauty, to our autumnal scenery, as its leaves curl up, become brown, and flutter from their sprays, when growing in exposed situations, as early often as the middle of September, by constitutional mechanism alone, even before the beech or the maple seems sensibly affected by the cold. This character of itself marks a difference from the common elm, which preserves its verdure, except from accidental causes, long after this period; and then, when its season arrives, the foliage becomes tinged with a fine, mellow, yellow hue, contributing a full share with other trees to the character and splendour of autumn. The wych elm may occasionally be desirable in the few days that our northern summer requires its deep shades, but will not otherwise afford pleasure or beauty in the shrubbery or the park as an ornamental tree, as its leafless sprays announce too early the unwelcome termination of our floral year, and its sober russet foliage is scattered at our feet without preparation or a parting smile.

Trees in full foliage have long been noted as great attractors of humidity, and a young wych elm in full leaf affords a good example of this supposed power; but in the winter of the year, when trees are perfectly denuded, this faculty of creating moisture about them is equally obvious, though not so profusely. A strongly-marked instance of this was witnessed by me, when ascending a hill in the month of March. The weather had previously been very fine and dry, and the road in a dusty state; but a fog coming on, an ash tree hanging over the road was dripping with water so copiously, that the road beneath was in a puddle, when the other parts continued dry, and manifested no appearance of humidity. That leaves imbibe moisture by one set of vessels, and discharge them by another, is well known; but these imbibings are never discharged in falling drops: the real mystery was, the fog in its progress was impeded by the boughs of the tree, and gradually collected on the exposed side of them, until it became drops of water, whereas the surrounding country had only a mist flying over it. Thus, in fact, the tree was no attractor, but a condenser; the gate of a field will in the same manner run down with water on the one side, and be dry on the other; as will a stick, or a post, from the same cause. It is upon this principle that currents of air will be found under trees in summer, when little is perceived in open places; and the under leaves and sprays will be curled and scorched at times, when the parts above are uninjured. The air in its passage being stopped and condensed against the foliage of the tree, it accordingly descends along its surface or front, and escapes at the bottom, where there are no branches or leaves to interrupt its progress. In winter there is little to impede the breeze in its course, and it passes through; consequently, at this season the air under a tree is scarcely more sensibly felt than in the adjoining field.

There is a casualty attendant upon some trees, for it cannot be called a malady, to which we are very subject in this neighbourhood, which those connected with timber concerns call the "shake" – there are sometimes external manifestations of this "shake", but generally it is only obvious when the tree is cut down, and then the centre of the timber appears cracked, and the adhesion of the fibre so defective as to present fissures and rifts from the root to the summit; in very bad cases the heart of the tree seems separated from the surrounding parts as if produced by some violent agitation, consequently, when cut into boards or quarterings, portions of the sawings are found injured by the layers being shattery and loosely adhering – The oak, & the elm, are principally subject to this and I have seen pine timber from British America sadly shaken; Norway fir I think less so. Trees in hedge-rows, or single growers, are said to be more frequently affected than wood or coppice timber: the slow growing plant long lingering in the soil before it starts into vigour is more commonly subject to this injury than those which thrive and get forward early: with us some places are exempt, others always have a tendency to it, which I think the cold and deep clayey lands have, yet not universally. Agitation by winds, or exposure to periodical tempests cannot occasion this, as the pollarded trees which no storms could move, are fully as much internally soft as that which remains unmutilated and some sheltered places present as many instances, as open ones do. Though I have conversed much with those familiar with this subject, yet have never heard any satisfactory reason assigned for this casualty -

It may be observed, that in the spring of the year the herbage under trees is generally more vivid and luxuriant than that which is beyond the spread of the branches; this may be occasioned, in some instances, by cattle having harboured there, and the ground becoming in consequence more manured; but it will be found likewise manifestly verdant and flourishing where no such accessory could have enriched it, and is, I apprehend, in general, chiefly owing to the effects of the driving fogs and mists, which cause a frequent drip beneath the tree, not experienced in other places, and thus in a manner keep up a perpetual irrigation and refreshment of the soil, and promote the decomposition of the foliage beneath, which, being drawn into the earth by worms, contributes to the verdure by the nutriment they yield.

THE FOLIAGE OF trees and plants, by its amazing profusion, variety, and beauty, must ever have been, as it is now, a subject of admiration and delight, is perhaps full as deserving of notice, and at times even more to be regarded, than the blossoms which accompany it. Let us take only one yard square upon the first verdant ditch-bank in spring, and the variation of form and character which will there be presented may probably exceed general imagination; but the object of all this extraordinary diversity is concealed, with the many other mysteries of creation; yet we have such an ascendant thirst for information upon the causes and nature of the things about us, as to render it an apparent inherent principle of the mind, inducing it to gratitude and love. From information in all the works of Providence arises, as a necessary consequence, admiration, and an exalted sense of Supreme intelligence and goodness. Without the desire of knowing the designs and processes of things, no investigation would be bestowed, and we should remain in ignorance of all but the bare facts and gross perceptions of creation; nor can it be questioned but that the more extensive our acquaintance is with the objects of Providence, in such proportions must our convictions be of his justice, wisdom, and power.

The great utility of foliage, and its agency in accomplishing the requirements of the plant and its products, are well known; and we can form some comprehension of the vast supply that is required by a tree when we view its foliage, each leaf being employed in receiving and transmitting gases from the air in certain proportions to the plant: these great operations having been effected during the summer months, and this agency of the leaves finished, they fall to the ground, not as an useless incumbrance, but to convey a large portion of fresh soil peculiarly fitted for the nutriment of vegetation. Should they remain in any quantity beneath the tree, they appear to be injurious to the smaller herbage, but they are more generally dispersed as they part from the sprays by the gales of autumn, which whirl them along in crowds to the hedges, trenches, and ditches around: here they accumulate and decay, furnishing, in conjunction with other vegetable decompositions, a very nutritive earth, as is manifest by the wild plants growing in those situations, for notwithstanding all the obstructions of shade, thorns, and briers, they are generally found in great luxuriance or health. This earth, in time crumbled by frosts, and washed by rains into the ditches from the banks, becomes accumulated there, and we collect it, compost it with other matters, and use it as a beneficial dressing for our cultivated lands: many of these leaves, however, remain near the tree, and soon communicate their virtues to the herbage; some are consumed by natural consequences, others are attacked by small fungi, which break their surfaces, admit moisture, and facilitate decay; the worm now seizes them as his portion, and, having fed upon a part, draws the remainder into the earth, where a rapid separation of the parts takes place, and they are received through the roots into vegetable circulation anew; and thus the beautiful foliage which has been so pleasing during our summer months, supplied the tree with sustenance to increase its magnitude, and all the requisites demanded by its fruits and products— has glowed perhaps with splendour, and been our admiration in the decline of the year, now returns to the soil, not to encumber it, but to administer health and vigour to a new series of vegetation, and circulate in combinations hidden from any human perception.

By a very wise appointment, peculiar propensities have been bestowed upon the vegetable world, greatly assimilating to the tastes and inclinations of the animated tribes. Beasts and insects feed on particular plants, and reject others, and the delight of one is disgusting to another. So some plants, not having the power of locomotion, will thrive only in certain compounded soils, aspects, and situations, evincing a similar tendency to preference of nourishment as do the sensitive tribes; and some districts, that vary a little in their component parts or position from those adjoining, will present an individual or a race that is not found in another: the common product of the North or of the East is treasured in the Herbarium of the southern or western botanist; we can boast but few, yet we have some of these capricious children of the soil.

The fetid hellebore (*helleborus foetidus*) is not a common plant with us, but we find it sparingly in one or two places; and though a plant indigenous to Britain, yet it is not improbable that it has strayed from cultivation, and become naturalized in many of the places in which we now find it. Its uses as a herb of celebrity for some complaints of cattle occasioned its being fostered in many a cottage-garden long since erased, where the good wife was the simple doctress of the village, when perhaps mortality was not more extensive than in these days of greater pretension and display. Modern practice yet retains preparations of this herb, but it appears that, from the powerful manner in which they act, great discretion is necessary in their administration. This hellebore is one of our few plants that present us with a dull, unsightly, unpleasing blossom. We have many with a corolla so small as to be little noticed; but this plant, and the fetid iris (iris foetidissima), produce blossoms that would generally be considered as darksome and cheerless. There is no part of a vegetable which we usually admire more than its flowers, for that endless variety of colours, shades, forms, and odours, with which they are endowed; yet the utility of the blossom is by no means obvious. Linnaeus calls the corolla the arras, the tapestry of the plant; and we are perfectly sensible that the blossom in very many instances is essential in various ways to securing and perfecting the germen; that it often contains the food of multitudes of insects, which feed on the pollen, the honey, or the germen; and that the odour emitted by it leads frequently various creatures to the object in request, and, by their agency, the fecundation and perfecting of the seeds are often effected: but we are astonished at the elaborate mechanism and splendour of some species, and see the whole race of creation, with the exception of man, utterly regardless of them. Butterflies and other insects will bask on expanded flowers, and frequent their disks, but it is in wantonness, or to feed on the sweet liquors they contain. The carpenter-bee, that every summer cuts its little circular patches in such quantities from my roses to line its nest in the old garden-door, selects the green leaves only, chiefly from the China, Provence, and damask kinds[10], passing over the petals of their blossoms as useless. That splendid insect, the rose beetle (*cetonia aurata*), that beds and bathes in sweetness, will partially eat the flowers of some species of roses, and "lap the nectar they produce;" and a few others nibble a little; but the liliaceous tribes, and other glorious flowers, as far as we know, furnish to insects no supply, but expand, wither, and die, unnoticed but by the eye of man alone. Flowers that are grand, gay, cheerful or beautiful, predominate infinitely over those that are of a sombre hue or gloomy aspect. Employment and occupation were as much the design, as they are found to be essential to the happiness of human life: we are not all constituted to soar in the higher regions of scientific research; our dispositions are as various as our intellects. Horticulture was the first occupation instituted for man, and he cannot pursue a more innocent and harmless employ: we were given "every herb, and every tree upon the face of the earth." For food, or raiment, the immediate necessities of man, a very few

---

10   This bee does not exclusively make use of the leaves of rose for its purposes, as I have known it in some seasons cut away the young foliage of *cytisus laburnum*, even when growing in company with its favourite rose.

of them are applicable; but we can collect them for amusement, in admiration of their beauty. Without this beauty, they would be no object of research; and man, who is exclusively sensible of its existence, can alone find pleasure in viewing it. The mind that is delighted with such admiration, must be almost insensibly led to an attendant pleasure, the contemplation, the perception of infinite wisdom and power, manifested in the adornment, splendour, and formation, of even the simplest flower of the field. I would not arrogate for man an exclusive right, or make him generally the sole consideration of the beneficence of Providence; but there are influences, which his reason can alone perceive, incitements to good thoughts and worthy actions.

Flowers, in all ages, have been made the representatives of innocence and purity. We decorate the bride, and strew her path with flowers: we present the undefiled blossoms, as a similitude of her beauty and untainted mind; trusting that her destiny through life will be like theirs, grateful and pleasing to all. We scatter them over the shell, the bier, and the earth, when we consign our mortal blossoms to the dust, as emblems of transient joy, fading pleasures, withered hopes; yet rest in sure and certain trust that each in due season will be renewed again. All the writers of antiquity make mention of their uses and application in heathen and pagan ceremonies, whether of the temple, the banquet, or the tomb — the rites, the pleasures, or the sorrows of man; and in concord with the usages of the period, the author of the "Book of Wisdom" says, "Let us crown ourselves with rose-buds and flowers before they wither." All orders of creation, "every form of creeping things and abominable beasts," have been, perhaps, at one time or another, by some nation or sect, either the objects of direct worship, or emblem of an invisible sanctity; but though individuals of the vegetable world may have veiled the mysteries, and been rendered sacred to particular deities and purposes, yet in very few instances, we believe, were they made the representatives of a deified object, or bowed down to with divine honours. The worship of the one true Being could never have been polluted by any symbol suggested by the open flowers and lily-work of the temple.

The love of flowers seems a naturally implanted passion, without any alloy or debasing object as a motive: the cottage has its pink, its rose, its polyanthus; the villa, its geranium, its dahlia, and its clematis: we cherish them in youth, we admire them in declining days; but, perhaps, it is the early flowers of spring that always bring with them the greatest degree of pleasure, and our affections seem immediately to expand at the sight of the first opening blossom under the sunny wall, or sheltered bank, however humble its race may be. In the long and sombre months of winter our love of nature, like the buds of vegetation, seems closed and torpid; but, like them, it unfolds and reanimates with the opening year, and we welcome our long-lost associates with a cordiality, that no other season can excite, as friends in a foreign clime. The violet of autumn is greeted with none of the love with which we hail the violet of spring; it is unseasonable, perhaps it brings with it rather a thought of melancholy than of joy; we view it with curiosity, not affection: and thus the late is not like the early rose. It is not intrinsic beauty or splendour that so charms us, for the fair maids of spring cannot compete with the grander matrons of the advanced year; they would be unheeded, perhaps lost, in the rosy bowers of summer and of autumn; no, it is our first meeting with a long lost friend, the reviving glow of a natural affection, that so warms us at this season: to maturity they give pleasure, as a harbinger of the renewal of life, a signal of awakening nature, or of a higher promise; to youth, they are expanding being, opening years, hilarity, and joy; and the child, let loose from the house, riots in the flowery mead, and is

Monarch of all he surveys,

There is not a prettier emblem of spring than an infant sporting in the sunny field, with its osier basket wreathed with butter-cups, orchises, and daisies. With summer flowers we seem to live as with our neighbours, in harmony and good-will; but spring flowers are cherished as private friendships.

The amusements and fancies of children, when connected with flowers, are always pleasing, being generally the conceptions of innocent minds unbiassed by artifice or pretence; and their love of them seems to spring from a genuine feeling and admiration, a kind of sympathy with objects as fair as their own untainted minds: and I think that it is early flowers which constitute their first natural playthings; though summer presents a greater number and variety, they are not so fondly selected. We have our daisies strung and wreathed about our dress; our coronals of orchises and primroses; our cowslip balls, &c.; and one application of flowers at this season I have noticed, which, though perhaps it is local, yet it has a remarkably pretty effect, forming for the time one of the gayest little shrubs that can be seen. A small branch or long spray of the white-thorn, with all its spines uninjured, is selected; and on these its alternate thorns, a white and a blue violet, plucked from their stalks, are stuck upright in succession, until the thorns are covered, and when placed in a flower-pot of moss, has perfectly the appearance of a beautiful vernal flowering dwarf shrub, and as long as it remains fresh is an object of surprise and delight.

No portion of creation has been resorted to by mankind with more success for the ornament and decoration of their labours than the vegetable world. The rites, emblems, and mysteries of religion; national achievements, eccentric masks, and the capricious visions of fancy, have all been wrought by the hand of the sculptor, on the temple, the altar, or the tomb; but plants, their foliage, flowers, or fruits, as the most graceful, varied, and pleasing objects that meet our view, have been more universally the object of design, and have supplied the most beautiful, and perhaps the earliest, embellishments of art. The pomegranate, the almond, and flowers, were selected, even in the wilderness, by divine appointment, to give form to the sacred utensils; the rewards of merit, the wreath of the victor, were arboraceous; in later periods, the acanthus, the ivy, the lotus, the vine, the palm, and the oak, flourished under the chisel, or in the loom of the artist; and in modern days, the vegetable world affords the almost exclusive decorations of ingenuity and art. The cultivation of flowers is of all the amusements of mankind the one to be selected and approved as the most innocent in itself, and most perfectly devoid of injury or annoyance to others; the employment is not only conducive to health and peace of mind, but probably more good-will has arisen, and friendships been founded by the intercourse and communication connected with this pursuit, than from any other whatsoever: the pleasures, the ectasies of the horticulturist are harmless and pure; a streak, a tint, a shade, becomes his triumph, which, though often obtained by chance, are secured alone by morning care, by evening caution, and the vigilance of days: an employ which, in its various grades, excludes neither the opulent nor the indigent, and, teeming with boundless variety, affords an unceasing excitement to emulation without contention or ill-will.

The bouquet may be an exile now; but the revolutions of fashion will surely return this beautiful ornament to favour again. With us the nosegay yet retains its station as a decoration to our Sunday beaux; but at our spring clubs and associations it becomes an essential, indispensable appointment; a little of the spirit of rivalry seeming to animate our youths in the choice and magnitude of this adornment. The superb spike of a Brompton, or a ten weeks' stock, long cherished in some sheltered corner for the occasion, surrounded by all the gaiety the garden can afford, till it presents a very bush of flowers, forms the appendage of their bosoms, and, with the gay knots in their hats, their best garments, and the sprightly hilarity of their looks, constitutes a pleasing village scene, and gives an hour of unencumbered felicity to common man and rural life, not yet disturbed by refinement and taste.

One of the great amusements of my life has been Horticulture, the simple cultivation of Flowers and esculents, an unskilful labourer, yet persevering notwithstanding failures and accidents of all kinds, seeking recreation and employment; for the mind of man in an healthy state seems so constituted as always to require occupation and exertion of some nature, and dominion was given to man, to man only, over all the works of creation, not only that he might use them, but by continual consideration of them increase his intellect and elevate his condition, by calling to his aid every perfection of nature, for whatever our acquirements may be, arts or learning. The laws of nature are almost universally the source and basement of all – various as the amusements the business of man may have been, we have the highest authority, that of Scripture, for believing that the cultivation of plants was a very early occupation of mankind, from the tillage of Cain, the vineyard of Noah, the garden of herbs of king Ahab. We can frequently trace the progress of nations in this employ, and the sword of conquest whilst it exterminated regions without pity, planted the most desired vegetables of the subjugated in the gardens of the victors: The Olive of Minerva, the Vine of Bacchus, the Corn of Ceres, are probably merely fabulous notations of particular individuals who brought these plants from other regions. The Roman armies though their career was so often stained with cruelty and wrong, introduced many plants on their return into the gardens of Italy, & we see the vegetable trophy and the captive noble entering in the same triumphal procession through the gates of Rome: Lucullus brought the cherry in full fruit from the conquered country of Armenia – Tiberius the horse-chestnut – Pompey the ebony, a plant which was said to vegetate only in the shade, imbibing the darkness of the night, probably from the collection of that celebrated preparer of poisons and antidotes, Mithridates. Those famous gardens which he established on his return from the East, were adorned with many of the productions from the regions he subdued. To Vespasian, Rome was indebted for the Syrian balm, and their armies for 400 years resident in Britain introduced many medicinal plants and fruits from the east – the Paladins brought home the damson; the shallot - scallion - ascalon garlick and the officinal rose yet retains a name intimating its native soil. Perhaps the most ancient remnant of cultivation was observed by Rich, near the site of old Babylon, in an antient tree so evidently of exotic origin, as to be considered a descendant from those planted in the hanging gardens of the Median queen 2400 years ago, or part of the collection of that great cultivator Cyrus, who had a noble garden there in which he often worked to recruit his mind disturbed by the cares of government. If we direct our eyes to the map of Europe it will afford us pleasure to observe in more modern periods, wherever wealth had taken up her abode, science was invited to visit too, and though England was tardy in these invitations, yet many instances might be advanced of her noble and opulent citizens becoming the patrons and cultivators of science. The garden of Gerard in Holborn was sufficiently extensive to produce a printed catalogue 230 years past, and even a little before this period, the collection of the two Tradescants was held in great esteem, and numbers of individuals are recorded as the early introducers of plants. Busbec brought the lilac from Asia into England 380 years past; the peony, great sun flower, white rose, the hollyhock, and the lily were very early cultivated here. But it is to the extensive commerce of Britain that we are indebted for that great accession of beauty & worth which we possess, and the love of novelty, ever so ardent a principle in the mind of man has covered our borders, and filled our conservatories with splendours, and our minds with delight; the wild, the mountain, the marsh, all accessible places and regions have been traversed to gratify this desire, and what a magnificent, curious collection has been the result! Drawn from the den of the tiger, the lair of the crocodile, the sting of the mosquito, the death path of the rattlesnake and the asp – and thus we note the silent advance of science through realms and time, we mark the trophies of man's perseverance amidst our garden labours, we see the earth covered with vegetable wonders, the lily of the field, the grass that fadeth away, arrayed with that splendour which adorned them in their first creation, the odour of the violet and the rose unchanged and the same, telling us of a wisdom so very good that need not be amended, of a power controllable by nothing.

> Oh, who can hold a fire in his hand
> By thinking on the frosty Caucasus?

AND YET THE shivering of the aspen, or poplar tree (*populus tremula*) in the breeze will give us the sensation of coldness, and communicate an involuntary shuddering. The construction of the foliage of this tree is peculiarly adapted for motion; a broad leaf placed upon a long footstalk, so flexile as scarcely to be able to support the leaf in an upright posture: the upper part of this stalk, on which the play or action seems mainly to depend, is contrary to the nature of footstalks in general, being perfectly flattened, and, as an eminent botanist and esteemed gentleman, Dr. I. Stokes observes, is placed at a right angle with the leaf, being thus peculiarly fitted to receive the impulse of every wind that blows. This stalk is furnished with three strong nerves, placed parallel, and acting in unison with each other; but towards the base the stalk becomes round, and then the nerves assume a triangular form, and constitute three distinct supports and counteractions to each other's motions. I know no petiole with a similar conformation, or better calculated for the vibration of a leaf. The leaf-stalks of plants are very curious constructions; and the nerves and vessels contained in them, which are the vehicles of a large portion of that nourishment which plants receive through their foliage from the air, seem in general differently placed, and fitted for variety of operation. The poplar is a tree that occasions at times a good deal of trouble in our pasture lands, by the tendency which it has to extend its roots and throw out suckers. Three or four of this species in a hedge-row, bounding a meadow in my occupation, oblige me every year that the field is mowed, by their prolificacy, to send a man with his stockaxe to remove their numerous offspring, a mere temporary expedient, tending rather to increase the complaint, as eradication by trenching with the spade can alone effectually check the encroachments of runners so tenacious of life, and rapid in growth.

The dyer's broom (*genista tinctoria*) abounds with us, and becomes a perfect incumbrance in our clay land pastures. It is seldom eaten by cattle, except in cases of great necessity, and remains untouched, if other food be obtainable, giving a deceitful appearance of verdure to a naked pasture. It yet retains a place in some of our dispensatories; but its medicinal virtues are probably never made trial of in modern practice, the lenient assuasives of our forefathers seeming unequal to contention with the constitutions of these days. I know not any use to which it is applicable but for the dyer. Our poorer people a few years ago used to collect it by cart-loads about the month of July; and the season of "woodwaxen" was a little harvest to them: but it interfered greatly with our haymaking. Women could gain each about two shillings a day, clear of all expenses, by gathering it; but they complained that it was a very hard and laborious occupation, the plant being drawn up by the roots, which are strongly interwoven in the soil. The dyer gave them eightpence for a hundred weight; but I fear the amount was greatly enhanced by the dishonest practice of watering the load, for the specious purpose of keeping it green; and the old woodwaxers tell me, that, without the increase of weight which the water gave the article, they should have had but little reward for their labour. Greediness here, however, as in most other cases, ruined the trade, the plant becoming so injured and stinted by repeated pullings, as to be in these parts no longer an object worth seeking for; and our farmers rather discountenance the custom, as the "greenweed" preserves and shelters at its roots a considerable quantity of coarse herbage, which in the winter and spring months is of great importance to the young cattle browsing in the pastures. The use of this dyer's broom is to prepare woollen cloths for the reception of another colour. It communicates to the article a dull yellow, which will then, by being dipped in another liquor, or composition, according to the shade required, receives a green hue. Vegetable filaments, cotton, flax, &c., are very differently formed from those threads afforded by animals, as silk and wool, and are differently disposed to receive colours. The dye that will give a fine colour to the one, is perhaps rejected by the other; and this plant is rarely or never used by the dyer for cotton articles. That certain natural productions receive and retain, and others reject, or soon part with, artificial colourings, are in some cases in consequence of the nature of the substance, and in others by reason of the conformation of the fibre; but any examination of this kind would only occasion a tedious discussion, and remain very obscure at last. We find certain effects produced and reason upon them, but so small are the parts operated upon, minute the agents, and equivocal the connexion, that we can do little more than theorize upon the subject; but, perhaps, I may slightly instance the difference existing in the fibre of flax and silk. The parts which compose the filaments of the former are generally considered as being flat and flaky, whereas those of the latter are tubular and round: this conformation renders silk so soft to the touch, and refracting more perfectly the rays of light, occasions much of its lustre, and the brilliancy of its hues. Perhaps we have no art or trade less confined within the trammels of formulae than that of the dyer; every professor appearing to have his own methods of acquiring particular tints and shades, guided often in his proportions by that mutable sense, the taste, and regulating the temperature of his compositions, not by the thermometer, but by the feeling of the hand. So capricious are these tests, so different the sensations of the operator, or the variable influences of solar light, that success on one day does not ensure a similar result on another.

Colour is, probably, only reflected light; but by what means the absorption of oxygen increases the lustre is not quite obvious — yet the power of the sun's rays, in augmenting the intensity of the hues of many things, is well known: there is an admirable green colour for foliage, to be obtained by the union of the light Prussian blue with the dark gamboge; but I could never acquire this clear and lustrous, without compounding it in the light of the sun. As the young artist will find this a most useful pigment, I may in addition say, that a small bit of the light Prussian, with three or four times the quantity of gamboge, must be laid upon the pallet, or in the saucer, and with a drop or two of water, only enough to make it work easily, be most thoroughly united and incorporated by the finger, with the sun shining upon the mixture, adding more gamboge repeatedly during the operation, until the blue is subdued and a clear green produced; but, if a tedious operation, yet perseverance will ultimately produce a very brilliant permanent green.

If the rays of the sun augment the intensity of colour in many instances, they will likewise destroy the hues that have been acquired by many articles; for the solar beams consist of three distinctly acting rays, by one of which oxygen is conveyed, by another it is taken away: or we may say that part of the solar beam, by reason of its affinity for oxygen, extracts it, and consequently the shades of colour, entirely or in part, the solid body of the oxygen becoming converted into gas and flying off; and thus we say the article has faded.

We have our walls in many places here decorated with most of the varieties of the great snapdragon (*antirrhinum majus*); the white, the pink, and the common: and that beautiful deviation, with a white tube and crimson termination, is slowly wandering from the garden, and mixing with its congeners. It has not, perhaps, been generally observed, that the flowers of this plant, "bull-dogs," as the boys call them, are perfect insect traps; multitudes of small creatures seek an entrance into the corolla through the closed lips, which upon a slight pressure yield a passage, attracted by the sweet liquor that is found at the base of the germen; but when so admitted, there is no return, the lips are closed, and all advance to them is impeded by a dense thicket of woolly matter, which invests the mouth of the lower jaw.

Smooth lies the road to Pluto's gloomy shade;
But 'tis a long, unconquerable pain,
To climb to these aetherial realms again.

But this snapdragon is more merciful than most of our insect traps. The creature receives no injury when in confinement; but, having consumed the nectareous liquor, and finding no egress, breaks from its dungeon by gnawing a hole at the base of the tube, and returns to liberty and light. The extraordinary manner in which the corolla of this plant is formed, the elastic force with which the lower limb closes and fits upon the projection of the upper, manifest the obvious design in the great Architect, "whose hands bended the rainbow;" and the insects are probably the destined agents whereby the germen is impregnated, for as soon as this is effected, the limbs become flaccid, lose their elasticity, are no longer a place of confinement, but open for the escape of any thing that might have entered. The little black pismire is a common plunderer of this honey.

It is a perplexing matter to reconcile our feelings to the rigour, and our reason to the necessity, of some plants being made the instrument of destruction to the insect world. Of British plants we have only a few so constructed, which, having clammy joints and calyxes, entangle them to death. The sun dew (*droserae*) destroys in a different manner, yet kills them without torture. But we have one plant in our gardens, a native of North America, than which none can be more cruelly destructive of animal life, the dogsbane (*apocynum androsaemifolium*) which is generally conducive to the death of every fly that settles upon it. Allured by the honey on the nectary of the expanded blossom, the instant the trunk is protruded to feed on it, the filaments close, and, catching the fly by the extremity of its proboscis, detain the poor prisoner writhing in protracted struggles till released by death, a death apparently occasioned by exhaustion alone; the filaments then relax, and the body falls to the ground. The plant will at times be dusky from the numbers of imprisoned wretches. This elastic action of the filaments may be conducive to the fertilizing of the seed by scattering the pollen from the anthers, as is the case with the berberry; but we are not sensible that the destruction of the creatures which excite the action is in any way essential to the wants or perfection of the plant, and our ignorance favours the idea of a wanton cruelty in the herb; but how little of the causes and motives of action of created things do we know! and it must be unlimitable arrogance alone that could question the wisdom of the mechanism of him "that judgeth rightly;" the operations of a simple plant confound and humble us, and, like the handwriting on the wall, though seen by many, can be explained but by ONE.

A. (Plate 1, Fig. 5,) a flower of the dogsbane enlarged; B. the pointal with seed at its base; C. two expanded anthers ready for capture; D. the anthers closed over the pointal, and the prey captured.

The different manner in which vegetables exert their organic powers to effect the destruction of insects is not, perhaps, unworthy of a brief notice; some, as those above mentioned, accomplish it by means of elastic or irritable actions, adhesive substances, and so forth; but we have another plant in our greenhouses, the glaucous birthwort (*aristol. glauca*), that effects these purposes without any of these means, but principally by conformation. The whole internal surface of the tubular flower is beset with minute strong spines, pointing downwards; these present no impediment to the descent of the animal which may seek for the sweet liquor lodged upon the nectarium at the base of the blossom, nor is there any obstruction provided for its return by means of valves or contractions, the tube remaining open; but the creature cannot crawl up by reason of the inverted spines; and to prevent its escape by flying up the tube, the flower makes an extraordinary curve, bending up like a horn, so that any winged creature must be beaten back by striking against the roof of this neck as often as it attempts to mount, and falling back to the bulbous prison at the base of the flower, dies by confinement and starvation, and there we find them: a certain number of these perishing, the blossom fades and drops off.

The fluids secreted by the nectaries of plants, being commonly sweet, are much sought after by insects for the purposes of food, and in some few cases, we find, with consequences fatal to them: it is the vegetable world which supplies the principal support to all animated things, and sweets are almost the sole production of this class of creation: air and water are probably the natural aliment of plants, but through the agency of light they imbibe oxygen, which produces the saccharine principle; and sugar, which is a vegetable oxide, is found to contain considerably above half its weight of oxygen — almost all vegetables possess this matter, though undetected by common observation; the young joints of the common grasses are sensibly replete with it; the very liquor of an onion abounds with sugar; and there seems sufficient reason to conjecture, that if vegetables were deprived of this principle, they would lose the most essential portion of their virtues; as, probably, no other constituent part contributes so much to nutriment as this does, which might be instanced by numerous examples. Taste, generally speaking, is the most variable sense which we are acquainted with, the likes and dislikes of sensitive beings being so often dissimilar, and even perfectly at variance; an ordination of supreme wisdom, and in perfect uniformity with the extensive scheme of nature — yet this one saccharine substance, whether in a compounded state in herbs, roots, or fruits, or uncombined in sweet exudations and excretions of plants, seems nearly in universal accordance with the likes of the highest, and most inferior orders of creation, from man through beasts and insects to the very slug, which eats the paper from our books and our walls, to feed upon the sugar in the starch and gluten of the paste which cements it; that little silvery creature, the book-worm, (*lepisma saccharina*[11],) detects it in the same substances, and mines his circular perforations through the bindings of our volumes to obtain it; and the weevil that so injuriously perforates my beechen table, is upon a similar pursuit. Nature has appointed for the nourishment of all mammiferous creatures a milky fluid as an intermediate substance between the animal and the vegetable kingdoms: this is replete with saccharine matter, and almost the only

---

11   This little creature has so great a delight in sweets, even in an uncombined state, as its name intimates, that it is frequently found inhabiting damp closets and such places, where crumbs of sugar may have been scattered.

animal product known to contain it[12]; and we must consequently conclude that it was placed there to furnish their principal support. Though this fluid in separate animals is supplied with different portions of this substance, and one vegetable affords a sensible evidence of its existence by its juices or its fruit, and another none whatsoever, yet that is no proof of a deficiency for some things of the nutritive principle, as the organs of one creature can and do convert a matter to subsistence which another has not the means of effecting. In cold and gloomy summers, we often complain that the fruit to our taste is wanting in its usual sweetness and flavour; this fact may be probably attributed to the deficiency of solar light diminishing the supply of oxygen to the plant, and thus the saccharine matter produced will be proportionably defective. And what a mysterious disclosure of the component parts of a natural body is here presented to us! This very sugar, which I have been lauding as the basis of animal nutriment, and in accordance with the gratification of so many creatures, yet contains above half its weight of that very pernicious poison, the oxalic acid[13], yet is so combined with other matters as to be salutary and pleasant. Even the very air we breathe is a compounded poison, in the immense proportion of seventy-eight parts of deleterious matter in a mixture of one hundred! This we know is eminently necessary; nor can any supposition at all have place but that of unlimited wisdom and providence — the creation of man and his continuance in life is all wonderful, his call, and his departure inaccessible to the understanding.

All the varieties of this snapdragon have the power of maintaining a state of vegetation in great droughts, when most other plants yield to the influence of the weather; and it is the more remarkable in these plants, as the places in which they chiefly delight to vegetate are particularly exposed to the influence of the sun. In that hot dry summer of 1825, when vegetation was in general burnt up and withered away, yet did the snapdragon continue to exist on parched walls, and draw nutriment from sources apparently unable to afford it; not in full vigour certainly, but in a state of verdure beyond any of its associates. The common burnet (*poterium sanguisorba*) of our pastures, in a remarkable degree, likewise possesses this faculty of preserving its verdure, and flourishing amid surrounding aridity and exhaustion. It is probable that these plants, and some others, have the power of imbibing that insensible moisture, which arises from the earth even in the driest weather, or from the

---

12 Wool contains this matter, and it is the cause probably why all vestments formed of it are so preyed upon by various little moths, and other creatures.

13 Henry.

air which passes over them. The immense evaporation proceeding from the earth, even in the hottest season, supplies the air constantly with moisture; and as every square foot of this element can sustain eleven grains of water, an abundant provision is made for every demand. We can do little more than note these facts: to attempt to reason upon the causes, why particular plants are endowed with peculiar faculties, would be mere idleness; yet, in remarking this, we cannot pass over the conviction, that the continual escape of moisture from one body, and its imbibition by another, this unremitting motion and circulation of matter, are parts of that wonderful ordination, whereby the beneficence and wisdom of Providence are manifested: without the agency of evaporation, not dwelling on the infinitude of effects and results, no vegetation could exist, no animal life continue.

THE IVY (*hedera helix*), the dark-looking ivy, almost covers with its thick foliage the pollards in our hedgerows; and, creeping up the sides of the old barn and chimney of the cottage, nearly hides them from our sight, affording a sheltered roosting-place to many poor birds, and is almost their only refuge in the cold season of the year. But the ivy can boast of much more extensive service to the poor wayfaring beings of creation, than the merely affording them a covering from the winds of winter. Those two extreme quarters of our year, autumn and spring, yield to most animals but a very slender and precarious supply of food; but the ivy, in those periods, saves many from want and death; and the peculiar situations, in which it prefers to flourish, are essential to the preservation of this supply, as in less sheltered ones it would be destroyed. In the month of October the ivy blooms in profusion, and, spreading over the warm side of some neglected wall, or the sunny bark of the broad ash on the bank, its flowers become a universal banquet to the insect race. The great black fly (*musca grossa*), and its numerous tribe, with multitudes of small winged creatures, resort to them; and there we see those beautiful animals, the latest birth of the year, the admiral (*vanessa atalanta*) and peacock (*vanessa Io*) butterflies, hanging with expanded wings, like open flowers themselves, enjoying the sunny gleam, and feeding on the sweet liquor that distils from the nectary of this plant. As this honey is produced in succession by the early or later expansion of the bud, it yields a constant supply of food, till the frosts of November destroy the insects, or drive them to their winter retreats. Spring arrives; and in the bitter months of March, April, and even May at times, when the wild products of the field are nearly consumed, the ivy ripens its berries, and then almost entirely constitutes the food of the missel thrush, wood-pigeon, and some other birds; and now these shy and wary birds, that commonly avoid the haunts of man, constrained by hunger, will approach our dwellings, to feed upon the ripe berries of the ivy. Now, too, the blackbird and the thrush resort to its cover, to conceal their nests. These early-building birds find little foliage at this period sufficient to hide their habitations; and did not the ivy lend its aid to preserve them, and no great number are preserved, perhaps few nests would be hidden from the young eyes that seek them. The early expansion of the catkins of the sallow (*salix caprea*), and others of the willow tribe, whence the bee extracts its first food, and the late blooming of this ivy, are indispensable provisions for the existence of many of the insect race; the "young raven does not cry in vain," nor is any thing abandoned by that Power which called it into being. We all seem to love the ivy —

The wanton ivy, wreath'd in amorous twines,

more than any other uncultured evergreen that we possess; yet it is difficult satisfactorily to answer why we have this regard for it. As a lover of the lone, the ivy-mantled ruin, I have often questioned with myself the cause and basis of my regard for that which was but a fragment of what might have been formerly splendid, and intrinsically possessed but little to engage admiration, yet, wreathed in the verdure of the ivy, was admired; but was never satisfied, perhaps unwilling to admit the answer that my mind seemed to give. The ivy is a dependant plant, and delights in waste and ruin. We do not often tolerate its growth when the building is in repair and perfect; but if time dilapidate the edifice, the ivy takes possession of the fragment, and we call it beautiful; it adorns the castle, but is an indispensable requisite to the remains of the monastic pile. There is an abbey in the north of England, which has been venerated by all its late possessors. It is trimmed, made neat, and looks, perhaps, much as it did formerly, except being in ruins. The situation is exquisite, the remains are splendid, yet with many it fails to excite such interest as it should do. It is a bare reality. A ruin in the west of England once interested me greatly. The design of revisiting and drawing it was expressed at the time. A few days only elapsed, but the inhabitant of a neighbouring cottage had most kindly laboured hard in the interval, and pulled down "all the nasty ivy, that the gentleman might see the ruin." He did see it, but every charm had departed. These two instances, from many that might be advanced, manifest that ivy most frequently gives to these ancient edifices the idea of beauty, and contributes chiefly to influence our feelings when viewing them. The ruins of a fortress, or warlike tower, may often historically interest us, from the renown of its founder or its possessor, some scene transacted, some villain punished, hero triumphant, or cause promoted, to which we wished success: but the quiet, secluded, monastic cell, or chapel, has no tale to tell; history hardly stays to note even its founder's name, and all the rest is doubt and darkness; yet, shrouded in its ivied folds, we reverence the remains, we call it picturesque, we draw, we engrave, we lithograph the ruin. We do not regard this ivy as a relic of ancient days — as having shadowed the religious reel use, and with it often, doubtless, piety and faith, for it did not hang around the building in old time, but is comparatively a modern upstart, a sharer of monastic spoils, a usurper of that which has been abandoned by another. The tendril pendant from the orient window, lightly defined in the ray which it excludes, twining with graceful ease round some slender shaft, or woven amid the tracery of the florid arch, is elegantly ornamental, and gives embellishment to beauty; but the main body of the ivy is dark, sombre, massy; yet, strip it from the pile, and we call it sacrilege, the interest of the whole is at an end, the effect ceases,

A moment seen, then lost for ever.

Yet what did the ivy effect? what has departed with it? This evanescent charm perhaps consists in the obscurity, in the sobriety of light it occasioned, in hiding the bare reality, and giving to fancy and imagination room to expand, a plaything to amuse them.

The Ivy is one of those plants having strong and rank leaves apparently possessing medicinal virtues, and is accordingly sought after by some animals with avidity. As the winter approaches termination we see the sheep thrusting their heads into the bushes of the hedge for the ivy found running in those places, or seeking what they can obtain from the lower part of the trees, the larger cattle standing on the hedge-bank stretching to their utmost up the pollards to gather the foliage of this plant, or cropping that which vegetates on the old wall, and we may generally find the fence trodden down by them beneath a tree that produces a good supply of it. At this season and in the early spring they seem alone anxious about it, nor does it appear to be sought for merely as green herbage for often plenty of grass may be in the pasture at the time, but seemingly they repair to it from some instinctive motive, to allay uneasy sensations or correct bodily ailments; in summer much rank herbage from the ditch is consumed when good provender abounds, probably for a similar object and then ivy is not required. In Gloucestershire especially near the banks of the Severn and Wye, the ivy particularly flourishes, in our copses

& woods the trees and saplings are often so invested with it as to present pillars of perpetual verdure; this may have arisen from some tendency in the soil or other local circumstances, but it is a plant having once established a footing that is very favourably constituted for the continuance of its kind even by its seeds alone. It appears to be exempted from the usual contingences of the season, never failing as far as my observation goes, of producing a large crop of berries continuing through the winter, and just at that period when the seeds have ripened in their pulp, the soil ready to receive them, and the temperature of the air favourable to germination, come the wood-pigeon, the black-bird, throstle and a few others, and gather them, swallowing the entire berry, which being digested, leaves the seeds in no manner bruised, macerated or injured by compression of the gizzard, and they are returned perfectly fitted for the continuation of the race, when falling at the foot of the tree on which the bird has lodged, readily spring into plants and clasping the rough bark presented to the rising tendril soon ascending the bole: nor have I observed that mice or any other creature that feed upon the kernils of hard seeds, in any way make use of this. The yew-tree, again seems unfavourably circumstanced for though it produces an abundance of berries, which are devoured with avidity by the same races and others, yet from the quantity of seed we find scattered at the base of the tree, and from other appearances, it seems that most generally the pulp alone is swallowed, though its casual propagation shews it is disseminated by birds; The seed too has the accidents of season to contend with, for falling to the ground before Christmas, it becomes the prey of mice and hard billed birds. The gooseberry, currant, rasp-berry, and elder are very widely scattered by these feathered agents, but we find the yew-tree much less generally spread abroad.

We still retain the name of the ivy as given by Pliny, though we know no reason why it was so called; but the word "helix," winding about, or twisting, is sufficiently apposite.

The foxglove (*digitalis purpurea*) is found with us in one or two places only, rather existing than flourishing, manifesting, like many other plants, a marked partiality to particular soils. It produces an abundance of seed, yet seems to wander little from the station its progenitors had fixed on, as if that alone was congenial to its habits; but with us the soil varies greatly. In the West of England, it thrives and increases with particular luxuriance; but many counties may be searched in vain for a single specimen. It seems to prefer a sandy, gravelly, or loose drained soil; not I think vegetating in strong retentive earths. We have few indigenous plants, not one, perhaps, which we have so often summoned to aid us in our distresses as the foxglove: no plant, not even the colchicum, has been more the object of our fears, our hopes, our trust, and disappointment, than this: we have been grateful for the relief it has afforded, and we have mourned the insufficiency of its powers: —

— Thy last, sole aid (which art can give),
    The woe-worn parent seeks, and, hoping, clings
In tearless wretchedness to thee; watches
With anxious heart thy subtle progress through
The day, and of thee fitful dreams through all
The night —
— spare, if thou
    Canst, his hopeless grief; save worth, save beauty,
From an early grave.

As a mere flower, the digitalis is a very handsome plant; and could we rely upon its yielding the virtues it is considered to possess, or could we regulate or control its influence, it would exist unrivalled for beauty and worth amidst our island plants. Why such a name as "foxesgloves" was bestowed upon this plant it is difficult to say, perhaps from the bare resemblance to finger-cases presented by its flowers: but I am not one of those who cavil or jeer at the common, or "vulgar names," as we are in the habit of denominating the unscientific appellations of plants; for we must remember that the culling of herbs and simples, and compounding preparations from them, to relieve the sufferings of nature, were the first rudiments of all our knowledge, the most grateful exertion of human talent, and, after food and clothing, the most necessary objects of life. In ages of simplicity, when every man was the usual dispenser of good or bad, benefit or injury, to his household or his cattle, ere the veterinary art was known, or the drugs of other regions introduced, necessity looked up to the products of our own clime, and the real or fanciful virtues of them were called to the trial, and manifests the reasonableness of bestowing upon plants and herbs such names as might immediately indicate their several uses, or fitness for application; when distinctive characters, had they been given, would have been little attended to; and hence, the numbers found favourable to the cure of particular complaints, the ailments of domestic creatures, or deemed injurious to them. Modern science may wrap up the meaning of its epithets in Greek and Latin terms; but in very many cases they are the mere translations of these despised, "old, vulgar names." What pleasure it must have afforded the poor sufferer in body or in limb, — what confidence he must have felt of relief, when he knew that the good neighbour who came to bathe his wounds, or assuage his inward torments, brought with him such things as "all-heal, break-stone, bruisewort, goutweed, fever-few" (*fugio*), and twenty other such comfortable mitigators of his afflictions; why their very names would almost charm away the sense of pain! The modern recipe contains no such terms of comfortable assurance: its meanings are all dark to the sufferer; its influence unknown. And then the good herbalist of old professed to have plants which were "all good:" they could assuage anger by their "loosestrife;" they had "honesty, truelove, and heartsease." The cayennes, the soys, the ketchups, and extratropical condiments of these days, were not required, when the next thicket would produce "poor man's pepper, sauce alone, and hedge-mustard;" and the woods and wilds around, when they yielded such delicate viands as "fat hen, lambs-quarters, way-bread, butter and eggs, with codlins and cream," afforded no despicable bill of fare. No one ever yet thought of accusing our old simplers of the vice of avarice, or love of lucre; yet their "thrift" is always to be seen: we have their humble "pennywort, herb twopence, moneywort, silverweed, and gold." We may smile, perhaps, at the cognomens, or the commemorations of friendships, or of worth, recorded by the old simplers, at their herbs, "Bennet, Robert, Christopher, Gerard, or Basil;" but do the names so bestowed by modern science read better, or sound better? it has "Lightfootia, Lapeyrousia, Hedwigia, Schkuhria, Scheuchzeria;" and surely we may admit, in common benevolence, such partialities as "good King Henry, sweet William, sweet Marjory, sweet Cicely, Lettuce, Mary Gold, and Rose." There are epithets, however, so very extraordinary, that we must consider them as mere perversions, or at least incapable of explanation at this period. The terms of modern science waver daily; names undergo an annual change, fade with the leaf, and give place to others; but the ancient terms, which some may ridicule, have remained for centuries, and will yet remain, till nature is swallowed up by art. No: let our ancient herbalists, "a grave and whiskered race," retain the honours due to their labours, which were most needful and important ones at those periods: by them were many of the casualties and sufferings of man and beast relieved; and, by aid of perseverance, better constitutions to act upon, and faith to operate, than we possess, they probably effected cures, which we moderns should fail to accomplish if attempted.

Upon an old bank, tangled with bushes and rubbish, we find in abundance that very early translated, and perfectly domesticated flower, the cottage snowdrop (*galanthus nivalis*): a plant that is undoubtedly a native of our island, for I have seen it in situations where nature only could introduce it, where it was never planted by the hand of man, or strayed from any neighbouring cultivation. Yet in most places where we find this flower, it is of manifest or suspicious origin; and with us it partakes of this latter character, though no remains of any ancient dwelling are observable near it. The damask rose, the daffodil, or the stock of an old bullace plum, will long remain, and point out where once a cottage existed; but all these, and most other tokens, in time waste away and decay; while the snowdrop will remain, increase, and become the only memorial of man and his labours. Many flowers present strong distinctive characters, or will, at least often do, excite in us variable feelings: the primrose, and the daisy, if not intrinsically gay, call forth cheerful and pleasing sensations; and the aspect or glance of some others will awaken different affections. The snowdrop is a melancholy flower. The season in which the "fair maids of February" come out, is the most dreary and desolate of our year: they peep through the snow that often surrounds them, shivering and cheerless: they convey no idea of reviving nature, and are scarcely the harbingers of milder days, but rather the emblem of sleety storms, and icy gales (snowdrop weather), and wrap their petals round the infant germ, fearing to admit the very air that blows; and, when found beyond the verge of cultivation, they most generally remind us of some deserted dwelling, a family gone, a hearth that smokes no more. A lover of cold, it maintains the beautiful ovate form of its flower only in a low temperature; warmth expanding the petals, vitiating its grace, and destroying its character. It seems to preserve its native purity free from every contamination; it will become double, but never wanders into varieties, is never streaked or tinged with the hues of other flowers.

Very many examples are known of the incredible length of time in which the vital principle of seeds has been retained, secured in the earth from atmospheric influence. Many further examples would be useless, but as one more instance of this duration of vitality has been very obvious with us this summer (1831), perhaps I may mention it. Labourers in quarrying stone on our down, opened to view several ancient contrivances for burning lime, unlike the kilns we now use, small, and probably sufficient for the purpose of the mason, and before the agriculturalist brought that substance much into use. So concealed were they by accumulation of turf and herbage, that there was no intimation of their existence. These walled pits had considerable quantities of sandy earth rammed around them to prevent the escape of the internal heat as much as might be. This arenaceous soil being thrown on the surface, produced the ensuing summer a dense crop of Henbane (*Hyoscyamus niger*). This soil one should suppose, must have been so heated by the lime-kiln fire, as to destroy every vital principle in so small a substance as Henbane seed. Yet it was preserved from a very distant period of time, and immediately produced its kind when exposed to the air. This plant was found locally and sparingly with us, diminishing annually, and latterly seemed to have disappeared. Yet by this extraordinary faculty, resistance to the powers of heat, moisture, and other agencies of time, excluded however from light, which so commonly vitiates the essential oil of seeds, thus destroying them, has the succession been maintained, when we thought it would appear no more with us. Some plants seem to derive the chief nutriment from moisture in the air or occasional rains, and little more than located or held in the soil. Plants on walls, sands, crevices of perpendicular buildings &c, others again require particular soils. But the propensities of plants, or the causes which influence vegetation to spring up on particular occasions, and in situations where no symptom of similar races existed before, upon the application of particular substances as lime, wood, or coal, ashes, bones, crushed or calcinated digested manures, is very inexplicable. That seeds may be preserved, excluded from atmospheric influence as above and vegetate when excited by the oxygen of the sun is comprehensible. But that a mineral or vegetable which has undergone combustion, should call into life, apparently confer the power of vegetation upon inert soil, is difficult to explain, for the combination that is accelerated, or decomposition effected, is performed in the dark. But possibly oxygen is bestowed by these accessories and brings about germination, or at times uniting oxygen with carbon, constituting carbonic acid, the rain water accelerating the root to imbibe the gas thus formed.

One of our pasture grasses is particularly affected by dry weather. Several are injured frequently by drought acting upon the stalk, not molesting the root, but withering the succulent base of the straw, which arises from the upper joint; in consequence of which, the panicle and connecting straw dry away, while the foliage and lower leaves remain uninjured. None are so obnoxious to this injury as the yellow oat-grass (*avena flavescens*), and in some seasons almost the whole of its panicles will be withered in a field of surrounding verdure. Pastures that are grazed must from circumstances be drier than those covered with herbage fit for the scythe; yet, from some unknown cause, this oatgrass seems less injured in this respect in grazing grounds, than in those where the herbage is reserved for mowing.

The plain, simple, unadorned vervain (*verbena officinalis*) is one of our most common, and decidedly waste-loving plants. Disinclined to all cultured places, it fixes its residence by way-sides, and old stone quarries, thriving under the feet of every passing creature. The celebrity that this plant obtained in very remote times, without its possessing one apparent quality, or presenting by its manner of growth, or form, any mysterious character to arrest the attention, or excite imagination, is very extraordinary, and perhaps unaccountable: most nations venerated, esteemed, and used it; the ancients had their Verbenalia, at which period the temples and frequented places were strewed and sanctified with vervain; the beasts for sacrifice, and the altars, were verbenated, the one filleted, the other strewed, with the sacred herb; no incantation or lustration was perfect without the aid of this plant. That mistletoe should have excited attention in days of darkness and ignorance is not a subject of surprise, from the extraordinary and obscure manner of its growth and propagation, and the season of the year in which it flourishes; for even the great Lord Bacon ridicules the idea of its being propagated by the operations of a bird as an "idle tradition," saying, that the sap which produces this plant is such as the "tree doth excerne and cannot assimilate." These circumstances, and its great dissimilarity from the plant on which it vegetates, all combine to render it a subject of superstitious wonder: but that a lowly, ineffective herb like our vervain should have stimulated the imaginations of the priests of Rome, of Gaul, and of Greece, the magi of India, and the Druids of Britain, is passing comprehension; and, as Pennant observes, "so general a consent proves that the custom arose before the different nations had lost all communication with each other." We might with some appearance of reason, perhaps, name the Druids of Gaul as the point, whence certain mysteries and observances were conveyed to the priesthood of various nations; but it would be difficult to assign a motive for their fixing upon such plants as vervain, and some others, to give efficacy to their ceremonies and rites. In some of the Welsh counties vervain is known by the name of "llyssiaur hudol," the enchanter's plant. It seems to have had ascribed to it the power of curing the bites of all rabid animals, arresting the progress of the venom of serpents, reconciling antipathies, conciliating friendships, &c. Gerard, after detailing some of its virtues from Pliny, observes, that "many odde old wives' fables are written of vervaine tending to witchcraft and sorcerie, which you may read elsewhere, for I am not willing to trouble you with reporting such trifles as honest ears abhorre to hear." To us moderns its real virtues are unknown; regular practice does not allow that it possesses any medicinal efficacy, and its fanciful peculiarities are in no repute; yet it seems to hanker after its lost fame, and lingers around the dwellings of man, for though not solely found about our habitations, as Miller thought, yet generally, when perceived, it is near some inhabited or ruined residence, not as a stray from cultivation, but from preference. Our village doctresses, an almost extinct race of useful, valuable women, the consolers, the comforters, and often mitigators of the ailments of the poor, still make use of vervain tea as a strengthener, and the dried powder of its leaves as a vermifuge; but probably in another generation all the venerated virtues of the vervain will be consigned to oblivion. This plant seems to be the native growth of many districts in Europe, Asia, and Africa.

One of our grazing grounds, bushy and rough, notable for an high embankment of a Roman encampment, and now serving as a common pasturage to several farms, has always possessed a bad name for annually disordering some of the young stock turned upon it, occasioning a complaint called the "red water," which frequently terminates fatally. Difficult as it may be to trace the cause of such injury amidst a great mass of luxuriant herbage, yet one year (1834) having been particularly marked by more than usual mortality (five head of horned cattle having died and fifteen others affected by this ailment) I was led to examine more carefully than general observation allowed, if any acrid plant could be detected, upon which we might fix the blame. *Heleborus viridus* (a scarce plant) is growing under some of the thickets, but yet too local to create suspicion, and indeed the cattle which suffered little besides to remain, left this perfectly untouched. Fox-glove is pretty plentiful there, yet probably its foliage is not particularly injurious. *Phellandrium aquaticum* (water Hemlock) and *Sium* (the water parsnip) abound in all the ditches and aquatic places and are sufficiently acrid and rank, but these plants are the common production of many other of our feeding grounds, are always eaten down by the cattle, and do not operate apparently prejudicially. But there is another plant *Pedicularis sylvaticus* (the louse wort) that abounds in such profusion as to cover in the spring months large patches of the dryer parts of the soil, overtopping the short herbage and giving a lively pink tint to all those parts of the common, nor do any other of our pasture grounds produce it by any means in equal abundance. I am inclined to suspect this to be the individual that occasions the mischief. Suspicions should not be attached without some reason, but Gerarde calls it "hurtful and an infirmitie of meadows". Witherings says it is an unwelcome guest there, others again say cattle will not touch it, but certainly it here is consumed from the soil, and beasts ranging the pastures crop many plants they decline when offered by the hand and unless this Pedicularis does occasion the evil, to no other can I attribute it. Our local dispensation for this malady which is at times efficacious, is "red shanks", (*Geranium Robertianum*) pounded to a pulp, and given night and morning, in a draught.

The dyers' weed, yellow weed, weld, or wold (*reseda luteola*), thrives in all our abandoned stone quarries, upon the rejected rubbish of the limekilns, and waste places of the roads, apparently a perfectly indigenous plant. Unmindful of frost, or of drought, it preserves a degree of verdure, when nearly all other vegetation is seared up by these extremes in exposed situations. It was, and is yet, I believe, cultivated in England for the use of the dyer. We import it, however, into Bristol from France, and it sells in that city for ten shillings per cwt. in a dry state. It gives a fine, permanent, yellow colour to cottons, silks, and woollens, in a variety of shades, by the aid of alum, &c. A blue tincture changes these to as fine a green. Injury has certainly been occasioned by writers on agricultural affairs recommending, without due inquiry, the culture of this or that crop; and I would not incur a censure that I blame in another; yet I cannot but suggest the possible profit that might arise from the culture of this plant. If foreigners derive sufficient encouragement to import it, notwithstanding the charges of freight, port duties, and various consequent expenses, why can it not be grown with us, and afford superior remuneration, not having such deductions to diminish the profits? The culture of it seems very simple, the manner of conducting the crop, and harvesting the product, attended with little trouble or risk. Marshal[14] prefers a good soil; others again say that it becomes stalky in a rich soil. With us it grows luxuriantly, three or four feet high, on a thin, stony, undressed soil, apparently the very station it prefers; and we have about us much land of this kind, not intrinsically worth ten shillings an acre. It might be rash to predict the amount of a crop in such soils, but a ton to an acre is said to be but a small allowance; yet the produce of only this quantity, which would procure in the market a return of ten pounds, without any expenditure for manure, no more manual labour after the seed is sown, for nine months, than three thinnings, and cleanings with the hoe, and the crop harvested within the year, would be no trifling profit, and may be deserving of some consideration[15]. The bark, the wood, the flower, the leaves of many of our native trees and plants afford a yellow dye; we have no colour so easily produced as this is; and it is equally remarkable that, amidst all the varied hues of spring, yellow is the most predominant in our wild and cultured plants. The primrose, cowslip, pilewort, globe-flower, butter-cup, cherlock, crocus, all the cabbage tribe, the dandelions, appear in this dress. The very first butterfly, that will

— aloft repair,
And sport and flutter in the fields of air,

is the sulphur butterfly (*gonepteryx rhamni*), which in the bright sunny mornings of March we so often see under the warm hedge, or by the side of

14 Rural Economy of Norfolk.
15 Article 'Reseda,' in Encyclopaedia Britannica.

some sheltered copse, undulating and vibrating like the petal of a primrose in the breeze. The blossoms of many of our plants afford for the decoration of the fair a vast variety of colours and intermediate tints, but they are all of them, or nearly so, inconstant or fugitive before the light of the sun, or mutable in the dampness of the air, except those obtained from yellow flowers: circumstances may vary the shade, but yet it is mostly permanent. Yellow is again the livery of autumn, in all the shades of ochre and of orange; the "sere and yellow leaf" becomes the general cast of the season, the sober brown comes next, and then decay.

Many impressions commonly fade away and become effaced as other objects create fresh sensations; but the love of nature, where the regard has been a settled principle, is more permanent, and influences the feelings as long as the occupations of life preserve any interest in our minds. As a child, I viewed the wild field flowers, and cropped them with delight; as a young botanist, culled with rapture the various species, returning often and again to my almost exhaustless treasure in the copse; and even now in the "sere and yellow leaf," when, in some mild vernal evening, I stroll through the grove, see the same floral splendour which year after year has been spread before me, I mark it with admiration and surprise, find it enchanting still, and fancy the present loveliness superior to all that has been before. There we see that beautiful little brilliant of the earth, like the name it bears (day's-eye), cheerful and pleasing to all. The exquisite chasteness of mien, and form of this flower, the contrast of its colours, and simplicity of attitude which it displays when springing from out its grassy tuft, can hardly be surpassed by any from another region. By its side peeps out the bright gleeful blue eyes of the little germander speedwell, in joyful gaiety — a lowly domestic plant that loves and seeks alliance with its kind, and in small family associations, by united splendour, decorates the foliage around; and there we find the stitch-wort, mingling her snowy bloom, immaculately pure, with pallid green: too delicate to vegetate alone, it seeks the shelter of the hedge or copse, trembles when the breeze goes by, and seems an emblem of innocence and grace. And there the bright-flowered lotus with its pea-like bloom, in social union glows as burnished gold, animating and gilding with its lustre all the tribes that spring near it; and fifty others, too, we note, which, though common and disregarded by reason of our familiarity with them, or expelled from favour by the novelty of far-fetched fair ones, deserve more attention than we are disposed to afford them. There are few plants which we look upon with more perfect contempt than that common product of every soil, the 'dandelion.' Every child knows it, and the little village groups which perambulate the hedges for the first offspring of the year, amuse themselves by hanging circlets of its stalks linked like a chain round their necks; yet if we examine this in all the stages of its growth, we shall pronounce it a beautiful production; and its blossom, though often a solitary one, is perhaps the very first that enlivens the sunny bank of the hedge in the opening year, peeping out from withered leaves, dry stalks, and desolation, as a herald, telling us that nature is not dead, but reposing, and will awaken to life again. And some of us, perhaps, can remember the pleasure it afforded us in early days, when we first noticed its golden blossoms under the southern shelter of the cottage hedge, thinking that the "winter was past," and that "the time of the singing of birds was come;" and yet, possibly, when seen, it may renew some of that childish delight, though the fervour of expectation is cooled by experience and time. The form of this flower, with its ligulate petals many times doubled, is elegant and perfect; the brightness and liveliness of the yellow, like the warm rays of an evening sun, are not exceeded in any blossom, native or foreign, that I know of: and this, having faded away, is succeeded by a head of down, which, loosened from its receptacle, and floating in the breeze, comes sailing calmly along before us, freighted with a seed at its base; but so accurately adjusted is its buoyant power to the burden it bears, that steadily passing on its way, it rests at last in some cleft or cranny in the earth, preparatory to its period of germination, appearing more like a flight of animated creatures than the seed of a plant. This is a very beautiful appointment! but so common an event as hardly to be noticed by us; yet it accomplishes effectually the designs of nature, and plants the species at distances and in places that no other contrivance could so easily and fitly effect. The seeds, it is true, might have fallen and germinated around the parent plant, but this was not the purport of nature; yet may seem to some a very unnecessary contrivance for the propagation of a common dandelion, whose benefits to mankind as a medicine, though retained in our pharmacopoeias, and occasionally resorted to, seem of no great importance. Nor are we sensible that its virtues are essential to any portion of the creation; but this very circumstance should abate our pride, our assumed pretensions of knowledge, as we may be assured that its existence, though hidden from us, is required in the great scheme of nature, or such elaborate and sufficient contrivances for its continuation and increase would never have been called into action by nature, who is so remarkably simple in all her actions, economical in her ways, and frugal of her means.

There are very obvious, but possibly not generally contemplated effects arising from cultivation, but more especially from horticulture, which remarkably instance the universal tendency of nature to supply and occupy, and from what simple and natural causes, life and being may originate. We may place ourselves on a wild and open heath, where little can be seen to invite the settlement of man, where nothing is heard but the whistling of the passing wind, the murmur of some solitary bee. We trench up the soil, introduce a vegetable, plant the rose, the bean, the apple, and they become immediately infested with numerous races of aphides, then come the *cochinellæ*, or ladybirds, to consume these creatures. Various tribes of little birds arrive to seek substance too. We set the current and the raspberry. The black-bird, throstle, red-breast, white throat appear. We place the cabbage there. The white butterfly and its larvæ soon abound. Troops of wasp, flies, spiders, ichneumons and myriads of others hover around, some to feed, some to convert the bodies of their companions into receptacles for their young. The grove springs up under our culture, warding off the winds from our nurselings, and where at first only crawled the beetle, and the worm, we now find numerous races feeding and nestling in the foliage. The woodland warblers come. The song of the nightingale is heard. There is the lichen, the moss, the fern, each with an inhabitant pasturing upon it. A voice seems uttered, and creation obeys the mandate - the simple culture of man calling from we know not where, the races of the earth. The heath becomes a paradise of flowers, fruits, and shades. Harmony and animation abound. We hear no more the icy whistle of the wind. The very atmosphere is changed; population follows. The entomologist, the botanist, the naturalist, human manners, human customs, all succeed and would have been unknown but for that wonder-working tool the spade! The labour of man manifesting that universal link, that binds all nature to union, that inaudible voice that calls to conjunction the separated energies of the earth. Whatever change may be accomplished by the powers of art, so will a spontaneous and ordained connection with other matters follow. The division of effects succeed, presenting to us the fact, that as there is a law which regulates the individual, there is a similar coordination providing for the succession, and provision of the whole - a power which

—through every age,
Through every moment up the track of time,
Adjusts – accommodates – and blesses all —

Some very extraordinary vegetable productions are now on the table before me. Though not gathered in this neighbourhood, I am induced to give them a place with our notables, because I believe that they have not been noticed, and afford a strong example of the persevering endeavours that plants exert at times to maintain existence. Plate 2. represents the tufted head and entire roots of a grass, gathered from a down fed by sheep from time immemorial. It is probably that of the hard fescue (*festuca duriuscula*), which, having been constantly eaten down by cattle, has never thrown up flowering stems, giving out only radicle leaves. These appear to have been cropped short, as soon as they have sprung up, the less succulent and strawy portions only being left, like a ball upon the surface, as a bush constantly clipped by the gardener's shears. The root appears to have annually increased, though the upper parts it was destined to nourish have been destroyed, until it became a lock of closely-compacted fibres, like a tuft of hair, six or eight inches in length. Furze bushes, growing upon many downs in Wales, Devon, and Cornwall, assume commonly the appearance of large, green, dense balls, every tender leaf being constantly shorn away by the sheep and rabbits that frequent those places, and present, upon a larger scale, the very appearance of these grass-balls. Our specimens are rather local than general, and were the produce of the Malvern hills.

The common brambles (*rubus cæsius* and *fruticosus*) may almost be considered as evergreens. Hedgers to be sure they are: but we have few, perhaps no other shrubby plant, naturally deciduous, excepting the privet, that will retain its verdure through the year, preserving, by a peculiar construction of its vessels, a portion of foliage unseared by frosts, and contending with gales that destroy and strip away all the honours of its neighbours. This circumstance enables us to observe a curious, strongly-defined line upon the leaves, like a glossy whitish film, meandering over the surface, becoming progressively larger, with a fine intestinal-like line running through the centre. (Plate 3. Fig. 3.) What occasioned this sinuous path long puzzled me satisfactorily to ascertain, considering it entirely of vegetable origin; and all the various polymorphous parasitics were successively thought of. At one time I deemed it like puccinia, which vegetates beneath the cuticle of leaves: but this was rejected; and probably I might long have wandered in error, had not the Rev. Mr. Kirby dissipated all my conjectures by informing me that it was the pathway of a small caterpillar. There are several species of them, which are placed by Reaumur in a tribe called "mineuses," all of which live upon the parenchyma, or pulpy substance found between the cuticles or skins of leaves gradually increasing in size until matured for transformation to the chrysalis, when they eat their way through the leaf, ultimately becoming moths, remarkable for the brilliant metallic lustre of their wings, the fine central line being the rejectments of the creature in the infant stages of its growth. Though several plants afford sustenance to these races, we have none on which this tortuous path is more strongly defined than the leaves of brambles, and the ever-blowing rose. But of these leaf mining insects we have one that most voraciously and effectually consume the parenchyma of a very common way-side dock (*Rumex obtusifolius*) like most of the race they are very uncertain visitors, but in June 1830, most of the leaves of this plant had the appearance of having been blistered by in hot sun, or seared by a morning frost as there were leaves as broad as one's hand, and near a foot long with the cuticle or outward skin perfectly separated, wrinkling to the touch and waving as a mere brown film only connected at the edge, and when held up to the light the larvae were in many instances to be seen near the termination of their work. The leaf of the hazel-nut is in some years blistered in numerous spots all over the surface in consequence of the feeding insects beneath; these larvae by the separation of the cuticle have a much less confined space to move in than those which mine their way in decayed wood, which are limited to the mere circumference of the feeding path. Very little of the discharged diet of the animal is found under the skin, by no means in proportion to the parenchyma consumed.

Notices of such incidents may perhaps be considered as too trifling to record; but the naturalist, from the habit of observing, sees many things not obvious to all persons: his province is to investigate all the operations of nature, and if he record them truly, he has done his duty; prolix and dull as his remarks will be to some, yet to another they may afford information, or tend to elucidate a conjecture. The bramble is a sadly reprobated plant, and I cannot say much in its favour as an independent individual, nor would I introduce it, to incommode by its society a thriving mound of white-thorn or of crab: but it generally introduces itself, and will flourish greatly, where other and better fences languish, and then by intertwining its long, flexile runners with the weakly products of the hedge-row, will compose a guard, where without it we could with difficulty have raised one. It will intrude, however, into many places where it is not required, originating probably from the rejectments of birds, and become a very unwelcome and tenacious inhabitant. Its long tendrils are much used by us as binders for thatching, being pegged down to prevent the straw coverings of ricks and such things being carried away by the winds, and we are satisfied with its performances. By the assistance of the bramble also, the new-placed turf is secured on the graves of our poorer neighbours, until it unites and forms a uniform sod; and during this service it will occasionally root itself and become an inhabitant not easily ejected from our churchyards. Badgers are said to feed much upon the fruit of the bramble. They are certainly very fat and fleshy about the time that the blackberry is ripe; but it is probable that the acorns and crabs, which it finds at the same season, contribute most to its nourishment.

THE MAPLE (*acer campestre*) is found growing in all our fences, generally reduced by the hedger's bill to serve the same humble purposes as the thorns and sloes associated with it. Sometimes, however, it is permitted to assume the rank of a tree, when, if not possessing dignity, it is certainly beautiful, and becomes an ornament in the hedgerow. It is the earliest sylvan beau that is weary of its summer suit; first shifting its dress to ochrey shades, then trying a deeper tint, and lastly assuming an orange vest; thus setting a fashion that ere long becomes the garb of all except the rustic oak, which looks regardlessly at the beau, and keeps its verdant robe unchanged. Soon tired of this, the maple takes a pattern from his sober neighbour ash, throws its gaudy trim away, and patiently awaits with all his peers the next new change. In spring the woodbine wreathes its knots of green around the rugged limbs of the maple; the rose beneath puts on its emerald gems, and then our gallant sir will wear such colours too, fluttering through all its summer's day. When first the maple begins to autumnize the grove, the extremities of the boughs alone change their colour, but all the internal and more sheltered parts still retain their verdure, which gives to the tree the effect of a great depth of shade, and displays advantageously the light, lively colouring of the sprays. We find the maple useful in our hedges, not from the opposition it affords, but by reason of its very quick growth from the stool after it has been cut, whence it makes a fence in a shorter time than most of its companions; and when firewood is an object, it soon becomes sufficiently large for this purpose. The singular ruggedness of the branches and shoots when they have attained a year's growth, and the depth of the furrows, give it a strongly marked character among our shrubs. The under side of the leaves in autumn, when they become yellow, and dashed here and there with a few specks of red and brown, appear, when magnified, like a very beautiful and perfect mosaic pavement, with all its tesserae arranged and fitted. If one of these rugged young shoots be cut through horizontally with a sharp knife, its corklike bark presents the figure of a star with five or more rays, sometimes irregularly, but generally exactly defined. A thin slice from this surface (see Plate 3. Fig. 1.) is a beautiful and curious object in the microscope, exhibiting the different channels, and variously-formed tubes, through which the sap flows, and the air circulates for the supply of all the diversified requirements of the plant; and it is good and delightful to contemplate the wonderful mechanism that has been devised by the Almighty Architect, for the sustenance and particular necessities of the simple maple, this "ditch trumpery," as Gilpin calls it; which naturally leads one to consider that, if he have so regarded such humble objects, how much more has he accounted worthy of his beneficence the more highly destined orders of his creation! As Evelyn says, on another occasion, "I beg no pardon for this application, but deplore my no better use of it." Modern practice records no medicinal virtues to be derived from the maple; but Pliny, in the quaint language of old Philemon Holland, tells us that a cataplasm made from the roots of this tree, is "singular to be applied for the griefs of the liver, and worketh mightily." In summer, the leaves of the hedgerow maple often assume a whitish mouldy look, which appears to be a mere exudation, as it neither presents any after character, nor have I observed that any thing results from it. The very damp autumn of 1830 occasioned some maple bushes in a shady situation to assume an unusual hoary appearance, and perfected what was before dubious, it now became evident that this whiteness was a species of mucor, or mould, the capsules of which were evident to the naked eye, and by aid of the microscope the little heads were observable in every stage of growth – green, brown, yellow, and some matured to ripeness, discharging their seed.

The young leaves, soon after their appearance in the spring, are beset with numerous fine spines of a bright red colour, most probably occasioned by the puncture of some insect, though I have never been able to discover any of the larvae inclosed in them. Some insects wound the leaves and sprays of plants for nutriment, though generally the object seems to be the formation of a nidus for their young, by the fluid that issues from the wound: but insects do something more than merely puncturing the parts to force a liquor to exude; a simple wound will not accomplish the desired object, as the sap not only hardens on the surface, but acquires a particular form and consistence, and even at times enlarges to a separate vegetable matter. The insect that wounds the leaf of the oak, and occasions the formation of the gall-nut, and those which are likewise the cause of the apple rising on the sprays of the same tree, and those flower-like leaves on the buds, have performed very different operations, either by the instrument that inflicted the wound, or by the injection of some fluid to influence the action of the parts. That extraordinary hairy excrescence on the wild rose (*cynips rosae*), likewise the result of an insect's wounds, resembles no other nidus required for such creatures that we know of; and these red spines on the leaf of the maple are different again from others. It is useless to inquire into causes of which we probably can obtain no certain result; but judging by the effects produced by different agents, we must conclude, that, as particular birds require and fabricate from age to age very different receptacles for their young, and make choice of dissimilar materials, though each species has the same instruments to effect it, where, generally speaking, no sufficient reason for such variety of forms and texture is obvious, so it is fitting that insects should be furnished with a variety of powers and means to accomplish their requirements, having wants more urgent, their nests being at times to be so constructed as to resist the influence of seasons, to contain the young for much longer periods, even occasionally to furnish a supply of food, or be a storehouse to afford it when wanted by the infant brood.

The wild clematis, or traveller's joy (*clematis vitalba*), thrives greatly in some of the dry stony parts of our parish, insinuating its roots into the clefts and passages of our limestone rocks, where those of many other plants could not find admission or support; and forms in our hedgerows a heavy shapeless mass of runners and branches, encumbering and overpowering its neighbours, many of which it often destroys; and we see the clematis clinging round a few stunted, half vegetating thorns, constituting the only fence, miserable as it is. The runners or branches are very strong and flexile, and are much used by our peasantry as a binding for hedge faggots. The tubes, lymph ducts, and air-vessels of this plant (Plate 3. Fig. 2.) appear in a common magnifier beautifully arranged, being large, and admitting the air freely to circulate through them. Our village boys avail themselves of this circumstance, cut off a long joint from a dry branch, light it, and running about, use it as their seniors do the tobacco-pipe. They call it "smoke wood," and the action of the breath constantly agitating the fire, it will long continue kindled. The pores are well seen by drawing some bright coloured liquor into them. I have often observed the long feathered part of the seed at the entrance of holes made by mice on the banks, and probably in hard seasons the seed may yield these creatures part of their supply. The diversity of form and arrangement in the pores of the roots, stems, and branches of plants, and the nerves, air-vessels, and fibres of the leaves, are extremely wonderful and beautiful; and it is possible that all the genera, species, and varieties, have more or less a different conformation of some of these parts. It is from the agency of these vessels, imbibing both from the air and the earth, compounding, decomposing, and discharging, in a way we know little about, that the sweetness of our fruits, the oil, the bread, and wine to glad the heart of man, proceed; and grateful should we be for them. From the vegetable world man derives his chief enjoyments: much of his fuel, most of his food, and the chief of his clothing, have once circulated in the tubes of a plant. The clematis plant possesses the power of preserving its verdure, and even thriving, in situations and seasons, when most other shrubby vegetation fails or languishes. With us its roots run amid loose stones, and in rocky places, far from any spring or apparent moisture; and yet, in those uncommonly dry summers of 1825 and 1826, it seemed to flourish with more than usual vigour, throwing out its long tendrils, of a fine healthy green colour, adorned with a profusion of blossoms, itself and the bramble being in some places the only thriving vegetation in a fence. It is marvellous how fibrous-rooted vegetables, the roots of which penetrate no depth into the soil, are enabled in some seasons to preserve any appearance of verdure, the earth they are fixed in seeming divested of all moisture by the power of the sun, and being heated like a sand-bath. The warmth of the earth in 1825 I omitted to record; but in the following year, which was more dry, and nearly as hot, the thermometer, buried in the earth to the depth of three inches, in a flower border where many plants were growing in that sort of languid state which they present in such exhausting seasons, indicated the heat of 110°.

Having said thus much of the clematis, the "withy-wind" of our peasantry, it must not be supposed that I advocate the advantages of this plant as a fence, but only tolerate it where we cannot induce much else to thrive, it making something of a boundary line; and perhaps that is all, for very frequently its numerous tendrils, and the downy clusters of its caudated seeds, are so interwoven, that the snow accumulates upon the bush, and presses the whole to the earth, so that in the spring we commonly find a gap to be repaired where the clematis has thriven. About February, or towards the end of winter, this plant becomes stripped of its feathery seeds, which is accomplished by mice, I believe the harvest and the long-tailed one (*mus sylvaticus*) principally; with these they form nestlike beds in the upper and thickest part of the hedge, resorting to them in the daytime, where they enjoy in tolerable safety the air and warmth of the season, in preference to their cold and damp apartments in the earth, and I have occasionally disturbed them in their dormitories; but at this time it is not observed that the seeds are much fed upon by them, and probably are only collected as shelter in a temporary dwelling.

T HE LITTLE EXCURSIONS of the naturalist, from habit and from acquirement, become a scene of constant observation and remark. The insect that crawls, the note of the bird, the plant that flowers, or the vernal green leaf that peeps out, engages his attention, is recognised as an intimate, or noted from some novelty that it presents in sound or aspect. Every season has its peculiar product, and is pleasing or admirable, from causes that variously affect our different temperaments or dispositions; but there are accompaniments in an autumnal morning's woodland walk, that call for all our notice and admiration: the peculiar feeling of the air, and the solemn grandeur of the scene around us, dispose the mind to contemplation and remark; there is a silence in which we hear every thing, a beauty that will be observed. The stump of an old oak is a very landscape, with rugged alpine steeps bursting through forests of verdant mosses, with some pale, denuded, branchless lichen, like a scathed oak, creeping up the sides or crowning the summit. Rambling with unfettered grace, the tendrils of the briony (*tamus communis*) festoon with its brilliant berries, green, yellow, red, the slender sprigs of the hazel, or the thorn; it ornaments their plainness, and receives a support its own feebleness denies. The agaric, with all its hues, its shades, its elegant variety of forms, expands its cone sprinkled with the freshness of the morning; a transient fair, a child of decay, that "sprang up in a night, and will perish in a night." The squirrel, agile with life and timidity, gambolling round the root of an ancient beech, its base overgrown with the dewberry (*rubus cæsius*) blue with unsullied fruit, impeded in his frolic sports, half angry, darts up the silvery bole again, to peep and wonder at the strange intruder on his haunts. The jay springs up, and screaming, tells of danger to her brood, the noisy tribe repeat the call, are hushed, and leave us; the loud laugh of the woodpecker, joyous and vacant; the hammering of the nuthatch (*sitta europæa*), cleaving its prize in the chink of some dry bough; the bumblebee, torpid on the disc of the purple thistle, just lifts a limb to pray forbearance of injury, to ask for peace, and bid us

Leave him, leave him to repose.

The cinquefoil, or the vetch, with one lingering bloom, yet appears, and we note it from its loneliness, spreading on the light foliage of the fern, dry and mature, the spider has fixed her toils, and motionless in the midst watches her expected prey, every thread and mesh beaded with dew, trembling with the zephyr's breath. Then falls the "sere and yellow leaf" parting from its spray without a breeze tinkling in the boughs, and rustling scarce audibly along, rests at our feet, and tells us that we part too. All these are distinctive symbols of the season, marked in the silence and sobriety of the hour; and form, perhaps, a deeper impression on the mind, than any afforded by the verdant promises, the vivacities of spring, or the gay, profuse luxuriance of summer. They seem so featured, of a character so entirely their own, as to fix upon the sober mind of men impressions of the deepest interest, awakened by, though unconnected with the things of this world. The common days of spring which so often assimilate those of autumn, make no impressions of this nature - in them everything is moving, life advancing in the green bud, the peeping blade from the soil, the insect that sports in the air, all is incipient, advancing existence. But in autumn all is universal repose, a gradual fading from being, its course seems finished, the conclusion of all at hand; and possibly few persons disposed to contemplations of an important nature, but would feel, without any pressure of dejection, or melancholy, an impulse upon the mind of calm meditation, perceptions very different from those excited by the business, the secularities of common life.

Such notes as these, such passing observations, are perhaps little fitted for, or deserving of, arrangement; yet, in a woodland autumnal ramble, we are naturally, almost irresistibly, led to contemplate that beautiful and varied race of vegetation included under the name of fungi, so particularly fostered by this season, and which so greatly delight to spring up in sylvan moisture and decay: nor is there, perhaps, any country better constituted for the production of the whole of this family than England is, particularly that portion of them denominated agarics. The various natures of our soil and pastures, the profusion of our woods and copses, the humidity of our climate, united with the general warmth of our autumn, accelerating rapid decay and putrescence of vegetable matter, all combine to give existence to this race. No county is, I believe, more favoured for the production of most of the kinds than Monmouth, with its deep dark woods and alpine downs. A residence in that portion of the kingdom for some years introduced to my notice a larger portion of this singular race than every botanist is acquainted with. A sportsman then, but I fear I shall be called a recreant brother of the craft, when I own having more than once let my woodcock escape, to secure and bear away some of these fair but perishable children of the groves. Travellers tell us of the splendour of this race in the jungles of Madagascar, but nothing surely can exceed the beauty of some old copse in Monmouthshire, deep in the valley, calm, serene, shaded by the pensile, elegant, autumnal-tinted sprays of the birch, the ground enamelled with every coloured agaric, from the deep scarlet to pallid white, the gentle gray, and sober brown, and all their intermediate shadings. Fungi must be considered as an appendage and ornament of autumn; they are not generally in healthy splendour until fostered by the evening damps and dews of September, and in this season no part of the vegetable world can exceed them in elegance of form, and gentleness of fabrication; but these fragile children of the earth are beauties of an hour:

> Transient as the morning dew,
> They glitter and exhale,

and must be viewed before advancing age changes all their features. There is a pale gray fungus (*agaricus fimiputris*) that may very commonly be observed in September on the edges of heaps of manure, and in pasture grounds, most beautifully delicate, almost like coloured water just congealed, trembling in the air from the slightness of its form, its sober tints softly blending with each other, lined and pencilled with an exactitude and lightness that defy imitation. The verdigris agaric (*agaricus æruginosus*) is found under tall hedgerows, and near shady banks, and few can exceed it in beauty when just risen from its mossy bed in all the freshness of morning and of youth, its pale green-blue head varnished with the moisture of an autumnal day; the veil irregularly festooned around its margin, glittering like a circlet of emeralds and topazes from the reflected colours of the pileus. But it is by examination alone that the beauties of this despised race can be perceived, not by a partial and inadequate description.

The certain appearance of many of the fungi can by no means be relied upon, they being as irregular in their visits as some of the lepidopterous class of insects. It is probable that decayed vegetable matter is in most cases the source whence this race of plants arises, while a certain degree of moisture and temperature, acting in concord with a precise state of decay, appears necessary to influence the sprouting of the seminal or radical matter. The beautiful floriform hydnum (*hydnum floriforme*) is very irregular in its appearance, whence it is a species seldom found by the botanist. The mitred helvella (*helvella mitra*) will abound, and then years may intervene and not a specimen be discovered. In 1825, a little, gray puff-ball (*lycoperdon cinereum*), about the size of a large pin's head, abounded, covering patches of grass in all our fields, looking like froth, and in decay, when discharging its seed, like a spongy curd; though it had not been observed, not having vegetated, or very sparingly, for upwards of ten years. Others, again, particularly the ligneous ones, remain permanently fixed for a long period. The fingered clavaria (*clavaria hypoxylon*) may be found vegetating on the stump of an old hazel in the orchard for twenty years in succession. That this elegant race has attracted so few votaries many reasons may be assigned. The agarics, in particular, are very versatile in their nature, and we frequently want an obvious, permanent character, to indicate the species, affording sufficient conviction of the individual. The rapid powers of vegetation in some will change the form and hues almost before a delineation can be made, or an examination take place, requiring nearly a residence with them to become acquainted with their various mutations; and we have no method of preserving them to answer the purpose of comparison. These are all serious impediments to the investigation of this class; yet, perhaps, I may with some confidence suggest, that any one, who is so circumstanced as to afford the time, so situated as to find a supply of these productions, and will bestow on them a patient examination, will find both pleasure and gratification in contemplating the beauty, the mechanism, the forms, the attitudes, of the whole order of fungi.

As far as we can observe, it appears to be an established ordinance of nature, that all created things must have a final period. This mandate is effected by various means, slow, and nearly imperceptible in some cases, but operative in all. As in the animal world, after disease or violence has extinguished life, the dispersion is accomplished by the agency principally of other animals, or animated creatures; so, in the vegetable world, vegetating substances usually effect the entire decomposition: for though, in the larger kinds, the high and lofty ones of the forest, insects are often the primary agents, yet other minute substances are commonly found to accelerate or complete the dissolution. Fungi in general, particularly those arranged as sphæria, trichia, peziza, and boletus, appear as the principal and most numerous agents, and we find them almost universally on substances in a certain state of decay, or approximation to it; though there are a few genera of this class which are attached to, and flourish on, living vegetation. The primary decline is possibly occasioned by putrescence of the sap, or defective circulation, and this unhealthy state of the plant affording the suitable soil for the germination of the parasitic fungus; for there must be an original though inert seed, till these circumstances vivify its principle. By what means the parasite finishes the dissolution is not quite obvious; but of that insidious race the byssi, of which family is the dry-rot (*byssus septica*), the radicles penetrate like the finest hairs into the substance, and thus destroy the cohesion of the fibres. So do the nidulariæ, many of the agarics, the boleti, and others; and it is not unlikely that this operation is the general principle of action of the whole race, though not so obvious in the minuter kinds. These terminators, many of which present but little character to the naked eye, under the microscope we find to be of various forms, though not always so distinguishable from each other as the flowers of our garden. Some of the genera of plants appear to have distinct agents assigned to them, and the detection and enumeration of them have been carried to considerable extent by some of the foreign naturalists; but, to point out the variety and curious organization of these substances, we will only instance four, to be found on the common plants of the garden or the copse: the laurel, the elm, the sycamore, and the beech.

The laurel (*prunus laurocerasus*) is not, properly speaking, a deciduous plant, though it casts its leaves in considerable numbers during the spring and summer seasons. These long resist the common agents of dissolution, like those of the holly, by means of the impenetrable varnish that is spread over them. This, however, wears off, and they decay; but their destruction is at times accelerated by a small excrescent substance, which fixes on the leaf, breaks the surface, and admits humidity. It appears in the form of a small black speck, and, when ripe, discharges a yellow powder from the centre; but as soon as one speck, which is the vessel containing the capsules, has fixed itself on one side of the leaf, a similar one will be found immediately opposite on the other; and hence it is well named by Lamarck the two-fronted uredo (*uredo bifrons*)[16]. This I believe to be peculiar to the laurel and the holly. (See Plate 4. Fig. 2.)

The leaf of the elm in autumn may commonly be observed marked with dark-coloured blotches, which are the "plague spot" of its destruction. These leaves remain in large proportions uninjured through the winter months; but when spring arrives, the spots become matured, the surface cracks, and the capsules discharge their seeds. (See Plate 4. Fig. I.) Lamarck names it *sphæria xylomoides*, but mentions another as a more early observer. At these spots the decay of the leaf generally commences.

Most persons must have observed that the upper surface of the leaves of the sycamore (*acer pseudoplatamis*) is blotched with dark-coloured spots (*xyloma acerinum*) in autumn. This leaf is detached by the earliest frosts, and falling to the ground, the spots commence their operations by corroding away the portions of the leaf that surrounds them, but continue attached themselves, appearing as raised, shining, vermicular lines. This has been mentioned by Lamarck and others, and is only now noticed to point out the variously constituted agents that accomplish the destruction of the foliage of plants.

The bark, the wood, have other deputed powers of destruction, many of which are very beautifully fabricated. To dwell on them would extend too much these remarks, designed rather as observations than details; yet I am tempted to introduce two. The sphæria coryli of Lamarck (*peziza coryli*) is occasionally to be found in the month of January, and through the winter until April, upon old hazel sticks, and engages our attention by the regularity of its tubercles. (See Plate 4. Fig. 3.) The seed, or first principle of production, whatever this may be, by means unknown to us, has been fixed upon the inner bark of the wood. Gently increasing, it bursts its way through the outer

---

16 Without close examination, this plant appears to be a uredo; but it is in fact a sphæria. Uredo differs from sphæria chiefly in the vessels not containing the capsules in cells, but loose. Hoffman observes, that both sphæria and uredo discharge pollen from an orifice; but, if the summit of this plant be cut off, the capsules are obvious.

bark, which now hangs as a fringe about it; the seed-vessels expand, and a dusty substance, being most probably the matter that continues the species, is dispersed around. A singular plant (*sphæria faginea?*) is found upon the decayed wood of the beech-tree[17], in the earlier part of the spring. It appears on the surface of it in little nodules, which, gradually uniting and increasing, form a regular black crust. Upon examination, we find, that little round bodies have forced a passage through the outer bark, and enlarged into small round tubes, which ultimately become the conductors of the seminal dust, discharged from round, beaked seed-vessels, imbedded beneath upon the inner bark. (See Plate 4. Fig. 4.) This plant presents us with a very remarkable instance of the attention of nature to the preservation of minute and little observed things; the protection of the seed-vessel, and the dissemination, being most particularly and carefully provided for.

These specimens are only individuals among hundreds which present us with a world of beauty, variety, and wonder. I would not wish it to be understood that it is maintained, by any thing here intimated, that the dissolution of vegetable matter is effected solely by the agency of insects or parasitic plants, nature having various ways of accomplishing her purposes; but only mean to contend, that, in numerous cases, these weak instruments are made use of to accelerate the decay and dispersion of it.

We are not favourably circumstanced for any great abundance of the race of fungi: the old fir grove, which produces such varieties, and the oak and birch copses, which have shed their leaves for ages, and given rise to many, are not found with us; yet we have a small scattering too, some of which are perhaps not undeserving of notice; and, though rather partial to a class which has afforded me many hours of gratification and delight, yet, sensible of the little interest they generally create, I must limit my mention to a very few.

The odorous agaric (*agaricus odorus*) may perhaps be locally found in plenty, but to me it has always been a plant of rare occurrence. Its colours are delicate and modest, rather than splendid, and a near acquaintance only makes us sensible of the justness of its name. We have another scented agaric (*agaricus fragrans*), much more commonly to be met with, which diffuses its fragrance to some distance: but the former species does not spread its fragrance, until brought into a temperate apartment, when it fills the room with an odour like that proceeding from the heliotrope, or from fresh bitter almonds, and communicates it to our gloves, or whatever it touches. I have found it sparingly here among dry beech leaves in Wolf-ridge copse.

17 I am uncertain whether this plant has been noticed. *Sphæria granulosa* of Sowerby, and *sp. tentaculata* of Batsch, may be it in a young stage of growth; *sp. faginea* of Lamarck does not accord well with it.

THE STINKING PHALLUS

There is a rare, local, and I believe unnoticed agaric, trailing its long roots in October among the small decayed fragments of some old hedge, elegant in itself, but more remarkable from the coloured fluid it contains, which upon being wounded it emits, not as a milky fluid, but like an orange-coloured, tasteless, spirituous extract, long retaining its colour upon paper, and tingeing the hand like the celandine, or bloodwort (*sanguinalis canadensis*); and hence I have called it a "stainer." Every part discharges this ichor, but it flows rather more copiously from the roots: in general appearance like *A. varius*. It may possibly be passed over as that species; but this is a race which, being local, precarious, mutable, or fugacious, is seen by the wandering naturalist alone, and we must leave these mysterious but beautiful productions of nature to their solitudes and woods[18]. — (See Plate 7.)

As weeds will grow with flowers, the unsightly with the beautiful, so do we meet with here much more abundantly that extraordinary and offensive production the stinking phallus (*phallus impudicus*). They do not dwell near each other, however; this being found in the month of June on many of our hedge banks. The smell it discharges has been thought to be like that arising from some decayed animal substance; but it is of a much more subtle kind, as if the animal fetor had been volatilized by carbonate of ammonia. Many persons, in their country walks at this period of the year, must have been occasionally surprised by a sudden disagreeable smell of this nature, and probably concluded that it proceeded from some dead animal, when most likely it was produced by this fungus: yet to find it is not always an easy matter; for the odour is so diffused on all sides, that it rather leads us astray from the object than aids our search, the plant being hidden frequently in the depth of the hedge. I have at times found it by watching the flight of the flies, which are attracted by its fetor. This strong smell is supposed to reside in the green gelatinous substance which is attached to the cell of the pileus; but the odour is at times discharged by this phallus, before the stem has arisen from the egg-like wrapper by which it is inclosed. This is a very unpleasant plant to delineate, as its odour, when in a room, is so very offensive, that few persons would willingly tolerate its presence; and its growth is so rapid in an increased temperature, that the form and appearance soon become changed. The seed is supposed to reside in the cells of the pileus and the gelatinous matter which we find on its summit;

18 Pileus — conical, one inch occasionally in diameter — pale gray, becoming ocherous, summit orange, flesh thin.
LAMELLAE — FIXED, WHITE, FOUR IN A SET, STAINED IN PLACES.
STIPES — FISTULAR, LONG, CHESTNUT AT THE BASE, UPWARDS PALE BROWN; ROOT LONG, TRAILING, WOOLLY.

and on this, and every part of the plant, slugs of various kinds are commonly found feeding, which, retiring to their holes in the earth, from the contents of their stomachs probably propagate this phallus. That many of our agarics, and those boleti which have central stems, are so diffused around by the agency of these creatures, it is reasonable to conclude; for it is a very usual thing to find the gills of these plants, in which the seed resides, so entirely eaten away by slugs as to have no remains perceptible, except a little of the flesh and the outer skin; and they prefer those plants which are somewhat advanced in age, and in which we suppose the seminal matter to be more perfected.

The various provisions which have been devised for the dispersion of the seeds of plants, and introducing them into proper situations for germination, are not the least admirable portion of the wonderful scheme of creation. Every class of beings appears appointed by collateral means to promote these designs; man, beasts, birds, and reptiles; and, for aught we know, the very fishes, by consuming, propagate the algae in the depths of the ocean. Even insects, by the fecundation of plants, perform an office equivalent to dissemination; and the multiplied contrivances of hooks, awns, wings, &c., and the elastic and hygrometric powers with which seeds are furnished, manifest what infinite provision has been made for the dispersion of seeds, and successive production of the whole race of vegetation.

The turreted puff (*lycoperdon fornicatum*) is one of our rare cryptogamous plants. I have had one specimen, in which the volvæ or wrappers of seven or eight individuals grew together, each throwing out a head or capitulum, forming a cluster the size of a doubled fist. It appears, from a close examination of this plant, that the upper part bearing the head was originally the inner skin or lining of the wrapper, which inclosed and shut it in. Upon the bursting of the wrapper, this inner skin peeled up, or loosened itself from the bottom, and rising, became finally detached from the wrapper in every part excepting at the points of the clefts, where it remained fixed; in the same manner as a man might be supposed able to pull up the skin from the hollow of the hand, and let it remain attached at the tips of the fingers. This puff dries remarkably well, and even shows the general form more distinctly than when recent.

The starry puff (*lycoperdon stellatum*) is rather difficult to find, but is a much more common plant, delighting to grow amidst the herbage of some dry bank, and so is hidden from common observation; — but the winds of autumn detach it from the banks, and it remains driving about the pastures, little altered until spring, when it decays.

We have the morell (*morchella esculenta*)[19], but to this I must subjoin "rarissimè." Bolton and Micheli represent the pileus as cellular, like a honeycomb. All that I have seen are mesenterically puckered. In what part of this morell the seeds reside is obscure: not in the hollows of the pileus, I think. That part of our morell, which in an agaric would be flesh, is found by the microscope to consist of fine woolly fibres united in a mass: and probably the seed is contained in this part; for when the plant is mature, and begins to dry, the outer coating cracks, and tears these filaments asunder, and gives the seminal matter, if contained in this part, a free passage for escape.

The bell-shaped nidularia (*nidularia campanulata*) is common with us, the smooth (*nidularia lævis*) is much less so. I do not mention them on account of their rarity, but to notice the singular size of the seeds of this genus. The principle, by which nearly the whole of the fungi are continued, is in most instances obscure. A dust, considered as seminal, is observable in some of the genera; in others, even this is imperceptible; but in the nidularia, the actual seeds, for they are not capsules, are visible at the bottom of the bell-shaped receptacle, of the size of a turnip seed, or of a large, flattened pin's head, loose, but attached by a filament, which in the striated species (*nidularia striata*) in moist weather, I have drawn out to nearly three inches in length. This thread appears designed to secure the vegetation of the seed, by affording it the power of deriving nutriment from the parent plant, during the period it is exerting its strength to vegetate in the earth. Heavy rains, I apprehend, fill the bells, and float out the seeds in the spring months, the filaments then stretching to their full extent. In severe weather we often find these bells emptied of their contents, and from observing the excrement of mice about the places of their growth, I conclude they are eaten by these creatures. The long mandibles of the little shrew are well fitted for this operation. I have never found the plant in such quantities as to yield them any considerable supply; yet it is remarkable, that the seeds of one genus only, out of such a numerous class, should be so visible, and of such a size, as to become an article of food to an animal like a mouse.

But we must dismiss the vegetable tribes, and enter upon the world of sensitive nature. The quadrupeds naturally present themselves first to our notice, but with us they are few in number; our population scares them,

19 This is the *phallus esculentus* of some; but Jussieu, Persoon, and others, have removed it from that genus, on account of its having no volva, but seeds in cells, not contained in a glareous mucus.

our gamekeepers kill them, and enclosures extirpate their haunts. Yet the marten (*mustela martes*) lingers with us still, and every winter's snow becomes instrumental to its capture, betraying its footsteps to those who are acquainted with the peculiar trace which it leaves. Its excursions generally terminate at some hollow tree, whence it is driven into a bag; and we are surprised that a predaceous animal, not protected by laws or arbitrary privileges, and of some value too, should still exist. Of all our animals called vermin, we have none more admirably fitted for a predatory life than the marten: it is endowed with strength of body; is remarkably quick and active in all its motions; has an eye so large, clear, perceptive, and moveable in its orbit, that nothing can stir without its observation; and it is supplied, apparently, with a sense of smelling as perfect as its other faculties. Its feet are well adapted to its habits, not treading upright on the balls alone, but with the joint bending, the fleshy parts being imbedded in a very soft and delicate hair, so that the tread of the animal, even upon decayed leaves is scarcely audible; by which means it can steal upon its prey without any noise betraying its approach. The fur is fine, and the skin so thin and flexible, as to impede none of its agile movements. Thus everything combines to render the marten a very destructive creature. It seems to have a great dislike to cold, residing in winter in the hollow of some tree, deeply imbedded in dry foliage, and when in confinement, covering and hiding itself with all the warm materials it can find. In genial seasons it will sleep by day in the abandoned nest of the crow or buzzard, and its dormitory is often discovered by the chattering and mobbing of different birds on the tree. It is certainly not numerous in England, our woods being too small, and too easily penetrated, to afford it adequate quiet and shelter. Its skin is still in some little request, being worth about two shillings and sixpence in the market; but it is used only for inferior purposes, as the furs of colder regions than ours are better, and more easily obtained.

But we have two little vivaceous creatures which haunt nearer the habitations of man plundering boldly his possessions even in day-light, which is never attempted by the marten; the egg in the nest becomes an unresisting prize, and they appear to be endowed with such a keen sense of smelling as to track almost undeviatingly such prey as avoids them by flight, and pursue with a persistency that generally succeeds. The weasel, and the stoat – where one of these creatures has discovered a brake, the abode of a rabbit or a leveret, it may be seen not unfrequently in active chase of the animal, following it through all its windings and deviations across the road, the path, through the briar mutely, but ardently; the rabbit may by exertion escape to a considerable distance, and rest, but the enemy never loses the trail, and is again upon it, the persecuted creature dares not seek its burrow, as from thence is no retreat. Thus harassed for hours, and worn out with fatigue, the vermin springs upon his neck, fixing with remarkable exactitude his teeth in the great vein behind the ear, and so fatal is this bite, that though I have seen the weasel fix, and disturbed him in action, yet never could rescue the victim, as death rapidly if not immediately ensues; very little injury is to be observed, a small moist space, and barely perceptible perforation, yet rapid extraction of arterial blood seems to occasion this speedy death, and this blood the object sought for, as the body is usually left unmutilated.

Notwithstanding all the persecutions from prejudice and wantonness to which the hedgehog (*erinaceus europæus*) is exposed, it is yet common with us; sleeping by day in a bed of leaves and moss, under the cover of a very thick bramble or furze-bush, and at times in some hollow stump of a tree. It creeps out in the summer evenings; and, running about with more agility than its dull appearance promises, feeds on dew-worms and beetles, which it finds among the herbage, but retires with trepidation at the approach of man. In the autumn, crabs, haws, and the common fruits of the hedge, constitute its diet. In the winter, covering itself deeply in moss and leaves, it sleeps during the severe weather; and, when drawn out from its bed, scarcely anything of the creature is to be observed, it exhibiting only a ball of leaves, which it seems to attach to its spines by repeatedly rolling itself round in its nest. Thus comfortably invested, it suffers little from the season. Some strong smell must proceed from this animal, as we find it frequently, with our sporting dogs, even in this state; and every village boy with his cur detects the haunts of the poor hedgehog, and as assuredly worries and kills him. Killing everything, and cruelty, are the common vices of the ignorant; and unresisting innocence becomes a ready victim to prejudice or power. The snake, the blind worm, and the toad, are all indiscriminately destroyed as venomous animals whenever found; and it is well for the last-mentioned poor animal, which, Boyle says, "lives on poison, and is all venom," if prolonged sufferings do not finish its being: but even we, who should know better, yet give rewards for the wretched urchin's head! that very ancient prejudice of its drawing milk from the udders of resting cows being still entertained, without any consideration of its impracticability from the smallness of the hedgehog's mouth; and so deeply is this character associated with its name, that we believe no argument would persuade to the contrary, or remonstrance avail with our idle boys, to spare the life of this most harmless and least obtrusive creature in existence.

If we were to detail the worst propensities of man, disgusting as they might be, yet the one most eminently offensive would be, cruelty —

a compound of tyranny, ingratitude, and pride; tyranny, because there is the power — ingratitude, for the most harmless and serviceable are usually the object — pride, to manifest a contempt of the weakness of humanity. There is no one creature, whose services Providence has assigned to man, that contributes more to his wants, is more conducive to his comforts, than the horse; nor is there one which is subjected to more afflictions than this his faithful servant. The ass, probably and happily, is not a very sensitive animal, but the poor horse no sooner becomes the property of man in the lower walks of life, than he commonly has his ears shorn off; his knees are broken, his wind is broken, his body is starved, and his eyes - ! ! I fear, in these grades of society, mercy is only known by the name of cowardice, and compassion designated simplicity and effeminacy; and so we become cruel, and consider it as valiance and manliness. Cruelty is a vice repeatedly marked in Scripture as repugnant to the primest attributes of our Maker, "because he delighteth in mercy." One of the three requisites necessary for man to obtain the favour of Heaven, and which was of more avail than sacrifice and oblation, was that of "showing mercy;" and He, who has left us so many examples in a life of compassion and pity, hath most strongly enforced this virtue, by assuring us, that the "merciful are blessed, for they will obtain mercy."

Hedgehogs were formerly an article of food; but this diet was pronounced to be dry, and not nutritive, "because he putteth forth so many prickles." All plants producing thorns, or tending to any roughness, were considered to be of a drying nature; and, upon this foundation, the ashes of the hedgehog were administered as a "great desiccative of fistulas."

The spines of the hedgehog are moveable, not fixed and resisting, but loose in the skin, and when dry, fall backward and forward upon being moved; yet, from the peculiar manner in which they are inserted, it requires more force to draw them out than may be at first sight expected. The hair of most creatures seems to arise from a bulbous root fixed in the skin; but the spines of the hedgehog have their lower ends fined down to a thin neck or thread, which, passing through a small orifice in the skin, is secured on the under side by a round head like that of a pin, or are riveted, as it were, by the termination being enlarged and rounded; and these heads are all visible when the skin becomes dry, as if studded by small pins thrust through. Hence they are moveable in all directions, and, resting upon the muscle of the creature, must be the medium of a very sensible perception to the animal, and more so than hair could be, which does not seem to penetrate so far as the muscular fibre. Now, this little quadruped, upon suspicion of harm, rolls itself up in a ball, hiding his nose and eyes in the hollow of his stomach, and thus the common organs of perception, hearing, seeing, smelling, are precluded from action; but by the sensibility of the spines, he seems fully acquainted with every danger that may threaten him; and upon any attempt to uncoil himself, if these spines be touched, he immediately retracts, assuming his globular form again, awaiting a more secure period for retreat.

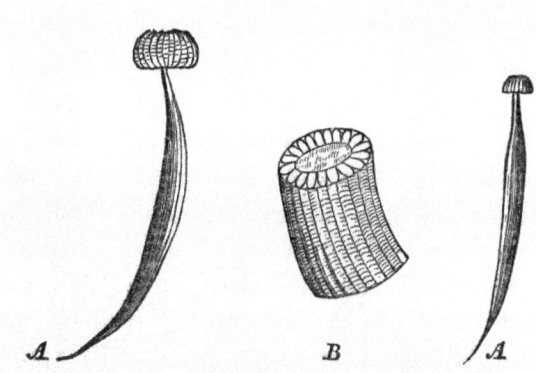

A, A, are spines of the hedgehog enlarged;
B, a segment, to shew the numerous tubes of communication.

Gipseys, and vagabonds, are constantly hunting the hedges for these creatures with their dogs, roasting them in the ashes of their fires, as a great delicacy. The privileges that favour the races, called game, in no way befriend the poor hedgehog, enclosures, and agricultural improvements so reducing their haunts and asylums. Persecution and ignorance tormenting them, I began to pity these poor harmless denizens of our thickets so obnoxious to destruction, and several were introduced into my kitchen garden; but they were very impatient of restraint, and though rather a large enclosure, they searched every means of escape, as a wild rabbit would do. They would feed from a pan placed for them, on scraps from the dinner table, even salt, or fresh meat, and fish were readily consumed, but bread and potatoes, seemed the favourites. In a natural condition worms, snails and coleopterous insects seem their principal diet, these last they collect beneath the droppings of cows. The spines seem very sensitive organs, for avoiding all wet places, they seek the driest banks the shelter of the briar or the furze. A sheltered nest in my garden the asylum of a kitten, was soon found out, but puss, the rightful proprietor, not pleased with these intruders and their spines, they were removed. One of the ladies produced me a brood of five young ones, but how protruded I know not, as at the age of two or three days their spines were as hard as their parents. These young ones were extraordinary little creatures, a good deal resembling the larvae of the Oniscus, their backs alone covered with prickles, all the lower parts hairy or naked. They were suckled by the old ones, as a hare would do, by night only, hiding herself by day, but she very soon became jealous, and dissatisfied by our knowledge of her treasures, and concealed them away from the nest. The various improvements in modern agriculture, scouring of ditches, and reducing fences, will ere long in many places extirpate this creature. Even now they are animals in such request in great towns, that seven and eight shillings are given for them alive, for the purpose of destroying the crickets which frequent in vast numbers bakehouses, and bakers' ovens for the sake of flour dropped in such places. I have heard of one thus usefully employed for eight years. In one severe spring, some who had the range of my garden, most perseveringly digged out the broad beans with their noses as pigs might do, as soon as the saccharine principle became developed by germination. A new planting became necessary.

The harvest mouse (*mus messorius*) in some seasons is common with us, but, like other species of mice, varies much in the numbers found. I have seen their nests as late as the middle of September, containing eight young ones, entirely filling the little interior cavity. Those nests vary in shape, being

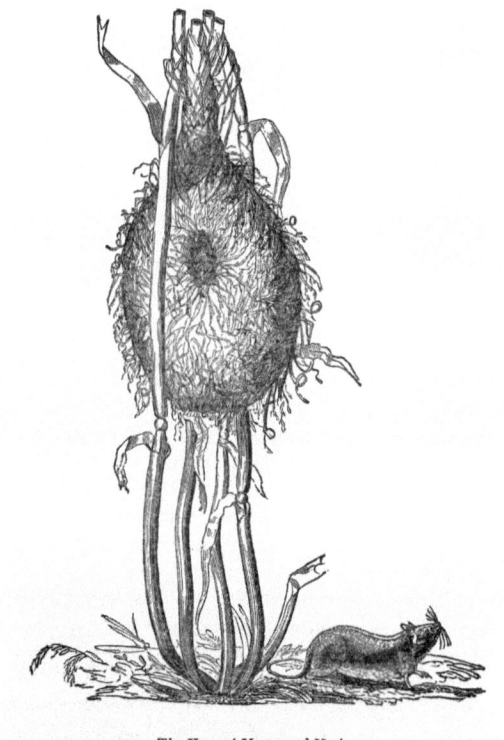

The Harvest Mouse and Nest.

round, oval, or pear-shaped, with a long neck, and are to be distinguished from those of any other mouse by being generally suspended on some growing vegetable, a thistle, a bean-stalk, or some adjoining stems of wheat, with which it rocks and waves in the wind; but to prevent the young from being dislodged by any violent agitation of the plant, the parent closes up the entrance so uniformly with the whole fabric, that the real opening is with difficulty found.

They are the most tame and harmless of little creatures; and, taking shelter in the sheaves when in the field, are often brought home with the crop, and found in little shallow burrows on the ground after the removal of a bean-rick. Those that remain in the field form stores for the winter season, and congregate in small societies in holes under some sheltered ditch-bank. An old one, which I weighed, was only one drachm and five grains in weight.

Our conceptions of the strength and power of animals are often founded upon the construction or apparent muscular means of the creatures but this very often is a most fallacious criterion as ability to accomplish their various purposes and requirements of condition, are afforded by other, and frequently by more efficacious agency than mere bodily strength. A feebler looking little creature than this harvest mouse, can hardly be presented, yet we are surprised to see how deeply and easily it can burrow in autumn into the hard banks, through all the impediments of roots, stones, to a safe and sheltered asylum. The quantity of earth which it removes to the surface, by means of its slender and delicate feet alone, is wonderful. The long- tailed mouse, is a stronger formed animal, yet with a very elegantly formed feet. When he mines his way into a loose, gravelly, or shattery stony soil, or bank, the accumulation of materials discharged at the entrance of his hole, all by the slender feet, have often been a subject of admiration; and affected too during a few hours in the night only, when watchfulness and precautions for safety against numerous and vigilant enemies must have consumed much of this interval.

Attending this day Novr. 30, the moving of some earth collected near the middle of a field, we discovered the association of three of these long tailed mice, and regret to say that a dash of the spade destroyed them, before it could be prevented. There were no more, nor could any have escaped, or remain concealed. We then found their hoarded provision for the winter, and a most extraordinary instance of industry and forethought it presented, these three creatures having provided above 400 acorns (for I counted them) in this magazine. As the nearest tree that could have finished this food was sixty eight yards distant, they must have travelled for each one hundred and thirty six yards, nearly thirty one miles in the whole.

The amazing burrows of the wild rabbit "a feeble folk which make their houses in the rock" is known to every one; the dilapidation of the house rat, when he works in earnest, is perhaps exceeded by no indigenous animal, but the mining of the insect, which forms her cavern in the wall, or winds a way amid the indurated fibre of the tree, or timber in the beam, is perhaps the most remarkable instance of persevering industry and gifted means, beyond apparent power of operation. We see that the forceps are sharp, and strong, that they exert considerable strength in the application of these engines, yet strength alone seems hardly adequate to accomplish what is affected by them: in some cases we know that secreted liquor softens the hard substances they abrade away, as in wasp-flies, and some others, but still action on the minutest portions, with long continued patient perseverance, seems to be the principal agent, and means whereby the vast achievements, of feeble bodied creatures are accomplished.

But though these mice rendered their labours useful in providing a store of provision for the winter, yet there is a disposition in all this race and in squirrels, for plunder, not always for the object of satisfying present wants, or as a precaution for the future. One season, my garden, which is always resorted to, was visited by more mice then chose to be entrapped. The plunder in my rows of beans and early peas, was proportionately great, but when the adjoining soil was turned up for other purposes, many of these seeds were found scattered in various places. Others germinated elsewhere, even the cavity where a branch had decayed in an old Apple tree received a portion, so that probably not one fourth of what was carried away was consumed. This habit is manifestly one of the great appointments of Nature for restoring the waste of time and replenishing the earth. The wind scatters the seeds, and propagates many species of plants. Some though are too weighty for dispersion by such means, but being carried away and secreted in the soil by the agency of animals, become in certain periods of time subservient for a new generation, when a former one has passed away. Numberless young oaks arise in our pastures far from the parent trees, most of which probably have been deposited in clefts of the soil, or hidden beneath the moss, and thus germinate in the warmth and moisture of spring. The nut, and the beach are extensively planted by the same little creatures, which may have perished themselves or forgotten their deposited food. Few things are more comprehensible than the increase of forests, and woodlands, in places where there is but little population to occupy the soil, or cattle to crop down the rising sapling. Mice and squirrels are perhaps the only animals that have the propensity to form stores, and yet we should imagine many creatures would have as great necessities in snows and severe seasons for the husbanded fruits of the summer; but yet it appears that all ruling nature has otherwise provided the means of subsistence, and supplied by sufficient means the requirements of all creation.

Mankind appears to be progressively increasing. It was an original command of his Creator, and the animals domesticated by him and fostered for his use, are probably multiplied in proportion to his requirements. But we have no reason to suppose that this annual augmentation proceeds in a proportionate degree with the wild creatures upon the surface of the globe; and we know that many of them are yearly decreasing, and very many that once existed have even become extinct. That there are years of increase and decrease ordained for all the inferior orders of creation, common observation makes manifest. In the years 1819 and 1820, all the country about us was overrun with mice; they harboured under the hassocks of our coarse grasses (*aira cæspitosa*), perforated the banks of ditches, occasioned much damage by burrowing into our potato heaps, and coursed in our gardens from bed to bed even during daylight. The species were the short-tailed meadow mouse, and the long-tailed garden mouse, and both kinds united in the spring to destroy our early-sown pease and beans. In the ensuing summer, however, they became so greatly reduced, that few were to be seen, and we have not had anything like such an increase since that period. It is probable that some disease afflicted them, and that they perished in their holes, for we never found their bodies, and any emigration of such large companies would certainly have been observed; yet the appearance and disappearance of creatures of this kind lead us to conclude that they do occasionally change their habitations.

A large stagnant piece of water in an inland county, with which I was intimately acquainted, and which I very frequently visited for many years of my life, was one summer suddenly infested with an astonishing number of the short-tailed water rat, none of which had previously existed there. Its vegetation was the common products of such places, excepting that the larger portion of it was densely covered with its usual crop, the smooth horsetail (*equisetum limosum*). This constituted the food of the creatures, and the noise made by their champing it we could distinctly hear in the evening at many yards' distance. They were shot by dozens daily; yet the survivors seemed quite regardless of the noise, the smoke and the deaths, around them. Before the winter, this great herd disappeared, and so entirely evacuated the place, that a few years after I could not obtain a single specimen. They did not disperse, for the animal is seldom found in the neighbourhood, and no dead bodies were observed. They had certainly made this place a temporary station in their progress from some other; but how such large companies can change their situations unobserved in their transits is astonishing. Birds can move in high regions and in obscurity, and are not commonly objects of notice; but quadrupeds can travel only on the ground, and would be regarded with wonder, when in great numbers, by the rudest peasant[20].

That little animal, the water shrew (*sorex fodiens*), appears to be but partially known, but is probably more generally diffused than we imagine. The common shrew in particular seasons gambols through our hedgerows, squeaking and rustling about the dry foliage, and is observed by every one; but the water shrew inhabits places that secrete it from general notice, and appears to move only in the evenings, which occasions it being so seldom observed. That this creature was an occasional resident in our neighbourhood, was manifest from the dead bodies of two or three having occurred in my walks; but it was some time before I discovered a little colony of them quietly settled in one of my ponds, overshadowed with bushes and foliage. It is very amusing to observe the actions of these creatures, all life and animation in an element they could not be thought any way calculated for enjoying; but they swim admirably, frolicking over the floating leaves of the pondweed, and up the foliage of the flags, which, bending with their weight, will at times souse them in the pool, and away they scramble to another, searching apparently for the insects that frequent such places, and feeding on drowned moths (*phalcæna potamogeta*) and similar insects. They run along the margin of the water, rooting amid the leaves and mud with their long noses for food, like little ducks, with great earnestness and perseverance. Their power of vision seems limited to a confined circumference. The smallness of their eyes, and the growth of the fur about them, are convenient for the habits of the animal, but impediments to extended vision; so that, with caution, we can approach them in their gambols, and observe all their actions. The general blackness of the body, and the triangular spot beneath the tail, as mentioned by Pennant,

20 As an event connected with the subject of temporary augmentation and diminution of creatures, I may be pardoned for noting the predominant increase of sex in some years. The most remarkable instance that I remember of late, was in 1825. How far it extended I do not know, but for many miles round us we had in that year scarcely any female calves born. Dairies of forty or fifty cows produced not more than five or six, those of inferior numbers in the same proportion, and the price of female calves for rearing was greatly augmented. In the wild state, an event like this would have considerable influence upon the usual product of some future herd. In the ensuing spring, we had in the village an extraordinary instance of fecundity in the sheep afforded us, one farmer having an increase of sixteen lambs from five ewes, four of which produced three each, and one brought forth four; however, only a small portion of these little creatures lived to maturity.

afford the best ready distinction of this mouse from the common shrew. Both our species of sorex seem to feed by preference on insects and worms; and thus, like the mole, their flesh is rank and offensive to most creatures, which reject them as food. The common shrew, in spring and summer, is ordinarily in motion even during the day, from sexual attachment, which occasions the destruction of numbers by cats, and other prowling animals; and thus we find them strewed in our paths, by gateways, and in our garden walks, dropped by these animals in their progress. It was once thought that some periodical disease occasioned this mortality of the species; but I think we may now conclude that violence alone is the cause of their destruction in these instances. The bite of this creature was considered by the ancients as peculiarly noxious, even to horses and large cattle; and variety of the most extraordinary remedies for the wound, and preventives against it, are mentioned by Pliny and others. The prejudices of antiquity, long as they usually are in keeping possession of the mind, have not been remembered by us; and we only know the hardy shrew now as a perfectly harmless animal, though we still retain a name for it expressive of something malignant and spiteful.

I think we have reason for suspecting that a shrew new to Britain exists in this neighbourhood. A pale blue shrew (*sorex Daubentonii* ? Cuvier) has been seen about the margins of our reenes, and the deep marsh ditches cut for draining the water from the low lands of the Severn; and something of the same kind, in a half-digested state, has been found in the stomach of the heron. If it exists with us, a similar tract of land in more fenny countries may contain it plentifully, though it has yet escaped detection.

The mole, want, mouldwarper, or mouldturner (*talpa europæa*), is common with us, as it appears to be in most places. No creature gives more certain indication of its presence, haunting, from preference, such places as its predecessors have done, though years may have intervened since they were frequented, and rains, and the treading of heavy cattle, have compressed to solid earth the ancient runs. However assiduously we may destroy them, should they appear again, it will probably be in the same places that have been formerly perforated by others. The earth that these animals eject from their runs, being obtained from very near the surface, and finely pulverised, has tempted me more than once to have it collected for my greenhouse plants, but not with the success that I had conjectured. Some persons have advocated the cause of moles, as being beneficial to vegetation, by loosening the soil about the roots of plants. Evelyn and others, again, censure them as injurious creatures; and there is a strange narration in Buffon, accusing them of eating all the acorns of a newly-set soil. I am not aware of any benefit occasioned by their presence; their warpings certainly give our pastures in the spring a very unsightly appearance, and in grounds designed to be mowed, occasion much trouble, by obliging us frequently to spread and remove them. In newly-sown corn lands, they disturb by their runnings the earth at the roots of the grain. But, perhaps, these trifling complaints, these almost imaginary grievances, are the only evils that can be attributed to them. In those wild creatures that are not immediately applicable to our use or amusement, we are more generally inclined to seek out their bad than their good qualities. Although I cannot produce any instance in which the utility of the mole is manifested, yet it is reasonable to conclude that they are eminently so, either directly or collaterally, Nature having provided, in an especial manner, for a constant supply[21], and their increase is prodigious when they are not molested. I have killed, for two years in succession, between forty and fifty each season, in a very few acres of ground; and, notwithstanding all our stratagems for their destruction, and the ease with which they are entrapped, still plenty always remain to recruit our annual waste of them. These creatures are supposed to have a very imperfect vision, and, like insects, have not any external ear, or manifest organ through which sounds can be received; yet we can in no way for a moment suppose that they have been created with any deficiency of power to accomplish all the objects of their being, but that every possible exigency has been provided for.[22] Perceptions may be conveyed, in very many instances, by intelligences unknown to us, and unquestionably are so. The defect of one power is frequently supplied by the increased activity of another; and the sense of smelling in the mole must be unusually acute, to enable it to pursue and capture its prey with the facility that it does. Its sole food, we believe, is worms; and these sensitive creatures retire immediately upon the smallest moving of the earth in which they reside. Now, as it follows them through all their meanderings, in which neither eyes nor ears would assist it, a fine sense of smelling seems necessary to enable it to catch them; and that its success is equal to its wants, and that it feeds plentifully, is manifest by the excellent condition in which the mole is at all seasons of the year. It will penetrate banks of earth after worms lodged in their interior,

---

21 See Ray's Synopsis.

22 Yet saying that the mole has no external organ of hearing is not correct, as this may be found near the usual seat of the ear in animals, yet is it so hidden in the fur, as in some manner to excuse my more than mole-like blindness in not noticing it at first; there is no elevated film, or membrane, no open space about it to lead to the discovery, but by carefully parting the fur a rather deeply seated orifice may be perceived, and thus the sensibility of the creature to very weak sounds becomes perfectly comprehensible.

hunt for them in the richest parts of the field, or on the edges of dung-heaps: in all which pursuits some unknown faculties may direct it; but no sense, that we are acquainted with, could promote its objects so effectually as that of smell. My talparius, a very skilful capturer of these animals, is so sensible of the power that moles are gifted with of readily discriminating smells, that his constant practice is, to draw the body of a captured animal through his traps, and the adjoining runs, and passages, to remove all suspicious odours which might arise from the touch of his fingers. Its feeling, too, must be acute; as, when casting up the earth, it is sensible of the pressure of a very gentle foot; and, unless our approaches are conducted with great caution, it ceases from its operation, and instantly retires. Should I be censured for needless prolixity in detailing these sensations of a common mole, and "telling of the mouldwarp and the ant," I trust forgiveness may be granted me, as endeavouring to remove all conceptions, should they exist, that anything, however vile and worthless it may seem to be, could be created with powers or means inadequate to supply its wants. Whoever will examine the structure of the body of a mole will, perhaps, find no creature more admirably adapted for all the purposes of its life. The very fur on the skin of this animal manifests what attention has been bestowed upon the creature, in providing for its necessities and comforts. This is singularly, almost impalpably, fine, yielding in every direction, and offering no resistance to the touch. By this construction the mole is in no degree impeded in its retreat from danger while retiring backwards, as it always does upon suspicion of peril, not turning round, which the size of its run does not permit, but tail foremost, until it arrives at some collateral gallery, when its flight is head foremost, as with other creatures.

If this fur had been strong, as in the rat or mouse, in these retreats for life it would have doubly retarded the progress of the creature; first by its resistance, and then acting as a brush, so as to choke up the galleries, by removing the loose earth from the sides and ceilings of the arched ways; thus impeding at least, if not absolutely preventing, retreat; but the softness of the fur obviates both these fatal effects.

The construction of the hair and fur of different creatures is very various and beautiful; and, if we believe in the beneficence of the All-wise Creator, we must conclude that such peculiar fabrications were resorted to for the purpose of being immediately useful, or as necessary to the condition of the animal. In a mere sketch like this, it would conduct me infinitely beyond my intentions to enumerate the many varieties of hair that are rendered manifest by the microscope; but three or four may be mentioned. The fur or clothing of the mole (Plate 3, Fig. 4, A) is internally composed of collateral bars. In man the hairs have at times a central tube, for the conveyance of medullary matter; as in bones, or some nutriment analogous to it; but in the mole there appears to be no communication with the body of the animal, unless the perspirable matter is conducted alternately from side to side along the bars. The fur of the bat (Fig. B) has knots like the rudiments of branches. The hairs of the hamster mouse (Fig. C) have a central perforation, apparently uninterrupted throughout their whole length. Some of the caterpillars (*callimorpha caja*) have spines proceeding from the hair that invests their bodies, (Fig. D[23].) All these, and the other various contrivances so manifest in the coverings of animals, are probably designed to convey off the perspirable fluids conducive to health in an appropriate manner; to discharge the superabundant heat, and keep the body temperate, in some cases; in others, again, to retard perspiration, and thus augment the warmth, by every possible gradation, or to increase the sensibility and perceptions of the animal. Many instances of these effects and modifications might be advanced, deserving a more extensive consideration.

The smell of the flesh of the mole is remarkably rank and offensive, as, from the nature of its food, might be expected; and it taints the fingers which have touched it with its peculiar odour, so that one washing does not remove it. It is reported of a late very eccentric nobleman, but with what truth I do not know, who essayed himself the flavour of every living thing, even to the eating of the large dew-worm, that the mole alone remained untasted by him, his stomach recoiling with disgust at the nauseous smell of the flesh of this creature. Foxes eat moles, and will at times dig out the traps containing them. The brown owl, too, feeds on them, when it can meet with them outside of their runs, hunting after dew-worms: and probably the smaller vermin do the same; but the cat and the dog turn from them with manifest aversion as food, though they will hunt and kill them as objects of the chase.

These animals, we might suppose, while in their subterranean dwellings, would be secure from all injury by such as generally pursue their prey upon the surface of the earth; but I have several times known the weasel caught in the mole-traps, making it manifest that it hunts after the mole for its food, and in doing so, according to our comprehensions, must encounter infinite danger

---

23   The organ which inflicts the pain, or sting, when we incautiously handle the nettle, is well known to be connected with a little vessel containing an acrid fluid, which, being compressed, rushes up the tube of the organ, and is thus conveyed into the wound; and it is rather singular, that the larvae of the peacock butterfly, which feeds upon the large hedge nettle, has the spines which arise from its body branched, and each collateral hair arises from a little bulb, similar to that of the plant on which it is chiefly found.

from suffocation; but it is more probable that so active a creature as the weasel is endowed with powers to accomplish its object with impunity, which we are not acquainted with.

During the course of a life passed much in the country, and perambulating the woods, the hedges, and the fields, I have contracted almost insensibly an acquaintance with the creatures that frequent them. Some have engaged my attention by their actions and manners; others have interested me by their innocency, and the harmlessness of their lives: and, perhaps, there is some little partial bearing toward others from long association, or from unknown, undefined causes. I tolerate, in despite of all their noise and all their litter, a colony of rooks, which have taken a liking to some tall elms near my dwelling. Not being ancient denizens there, they can claim no hereditary rights; but their contrivances, their regularity, and even their squabbles, are amusing; and, perhaps, there is mingled with this some little compassion for these dark, half-domesticated families of the grove, driven by the axe from an old abode, which may influence my forbearance.

The hedge sparrow, or shufflewing (*motacilla modularis*), is a prime favourite. Not influenced by season or caprice to desert us, it lives in our homesteads and our orchards through all the year, our most domestic bird. In the earliest spring, it intimates to us, by a low and plaintive chirp, and that peculiar shake of the wing which at all times marks this bird, but then is particularly observable, the approach of the breeding season, for it appears always to live in pairs, feeding and moving in company with each other. It is nearly the first bird that forms a nest; and this being placed in an almost leafless hedge, with little art displayed in its concealment, generally becomes the booty of every prying boy; and the blue eggs of the hedge sparrow are always found in such numbers on his string, that it is surprising how any of the race are remaining, especially when we consider the many casualties to which the old birds are obnoxious from their lameness, and the young that are hatched from their situation. The plumage of this motacilla is remarkably sober and grave, and all its actions are quiet and conformable to its appearance. Its song is short, sweet, and gentle. Sometimes it is prolonged, but generally the bird perches on the summit of some bush, utters its brief modulation, and seeks retirement again. Its chief habitation is some hedge in the rick-yard, some cottage-garden, or near society with man. Unobtrusive, it does not enter our dwellings like the redbreast, but picks minute insects from the edges of drains and ditches, or morsels from the door of the poorest dwelling in the village. As an example of a household or domestic bird, none can be found with better pretensions to such a character than the hedge sparrow.

The various manners and actions of the little creatures about us is very remarkable, though to our unseeing minds, no diversity of habit is required, it is marked and a fixed character, and consequently necessary to the requirements of their state of being: this hedge-sparrow will tend the brood when in the nest, with the same perseverance and anxiety, as is observable in all other hedge loving birds, but as soon as they leave that asylum we see no more attention to their wants, we never observe the parents feeding them, nor do they ever hover about the mother in pursuit of food. They are invisible to us except by stealth at times, but live scattered from each other in the deep foliage of the hedge, or the broad leaves of our garden plants with no family associations, and that the parents feed them is only made known to us by a little fondling squeak when she leaves their haunts. But they never follow or importune for support as other nestling birds; yet when they obtain maturity, and pair, no bird is more constant in its attachments, having mutually acquired a habit of social regard they never abandon their first formed associations.

I have often thought that this bird, the chaffinch, and some others, obtain much of their support in the winter and spring seasons, especially when the ground is covered with snow, by feeding upon the capsules or fertile heads of various mosses, having frequently noticed them pecking and masticating something upon the walls and in such places where these plants abound, and nothing besides, that could afford subsistence to any animated creature, particularly *bryum subulatum* (Dillineus), and these races perfect their capsules principally during those periods in which other matters which could afford them sustenance is sparingly found. The object of the existence of many of these lowly plants has been considered as obscure, and their profusion a general subject of admiration. Of these, *kœlreuteria* h*ygrometica* (Hedwig) presents a very prominent example of unvarying annual exuberance. If this conjecture is correct, that they afford nutriment to these poor little creatures in a season of destitution, it affords us another instance of the benevolence of their Creator, extending, as far as we can perceive, through every department of creation: we cannot trace this chain, because we are ignorant of consequences, nor perceive the termination, because it is of infinite duration; but to attest any perception of wisdom and of goodness is a laudable and just homage of the creature who observes it.

The abundance and variety of aliment that has been provided for all creatures, the peculiar formation of their digestive organs to reduce it to nourishment and their inclination for particular diet, is one of the many conclusive manifestations of the goodness and wisdom of that power which called them into being: many of the common creatures about us live we know not how, we see them in health, and in good condition, and acknowledge the hand that distributeth, but very often when we seek for this manna spread over the earth, it has melted in the morning sun. The simplest observer of nature, if he looks out upon the plot of turf before his door, will commonly perceive in the early days of spring, in the country at least, many little birds hopping about and pecking with great activity, hedge sparrows, common sparrows, chaffinches, wagtails &c – all obtaining subsistence from something found there; I have repeatedly endeavoured to find what this consisted of, the seeds of the old grasses consumed or driven away by the winter storms, are not found, the early *Poa* is not in sufficient abundance, *Cardamine hirsuta*, or *Draba*, occasionally appear but too sparingly to engage our little foragers to resort to them; from the actions of the birds, the number at one time and their absence at another, I conclude that minute insects arise at various periods, hatched in their subterraneous dwellings and constitute the banquet; the little feeders seem by their animation to find them in plenty, but so small are they, or our eyes so ill calculated for detecting them that nothing conclusive appears: but the eyes of birds appear formed according to their habits of life, those that prey upon animal substances and fish, with an extended vision, those which seek nutriment in seeds or upon insects with a short focus, by which near objects are discernible. A few animals or a few insects may instinctively make provision for an hour of want, but probably no bird does so, but seek their aliment as necessity prompts, one great beneficent power distributing to every one, the world being all before them and providence their guide.

  I always hear with delight the earliest chirpings of that pretty harbinger of spring, the willow wren (*motacilla trochilus*), trilling its wild and gleeful "chiff chaffs," as it chases the insects round the branches of the old oak in the copse, or on the yellow catkins of the sallow, itself almost like a coloured catkin too. But this elegant little bird is noticed only by the lovers and frequenters of the country; it animates the woods by its constant activity; the frequent repetition of its most cheerful modulation contributes essentially to the pleasing harmony of the grove; and its voice is most sprightly and frequent, when the morning is illumined with one of those mild, walk-enticing gleams, that render this short season the most delightful of our year. It builds its nest and rears its young with us; visits our gardens, but is no plunderer there, living almost entirely upon insect food; and its whole life is passed in harmlessness and innocence. As it is the earliest that arrives, so it is the last, I believe, of our feathered choir that leaves us, except a few lingering, irresolute swallows; and we hear it piping its final autumnal farewell even in October at times, and sporting with hilarity and joy, when all its congeners are departed.

The willow wrens, when they visit our gardens in autumn are accompanied by others of the same species, apparently by reason of the brightness of their plumage, the young of the season, the under parts of the body are beautifully coloured with yellow, the vent remaining silvery white, but below that the fine yellow colour is found again. From the gracefulness of their forms, sprightliness of action, and delicacy of colouring they become as elegant an ornament of our gardens, as the yellow wagtail is of our pastures – small as this *sylvia*, and its congener the 'chiff chaff' are, they appear at times to be jealous and pugnacious creatures chasing and assaulting any other little birds that approach their station. I lately saw a troop of fourteen or fifteen long-tailed-tits descend in their progress and settle upon a crab tree where a chiff chaff was busily feeding upon aphides constituting the American blight. The wren immediately attacked them unmindful of numbers, and successfully drave them from the tree. With the exception of hirundines, probably few species of our summer birds assemble in this Kingdom in such numbers, as these wrens of both species do, locally. Other species of birds may be more numerous, but these little creatures are scattered through every thicket, grove and village in the island, small, and when not in voice are but little noticed, but the attentive observer may generally find them. Large as the flocks must be, they steal unperceived into their stations, and we are ignorant of their presence, until some mild day, or summer hour give us notice by calling forth their song. The motive that urges the transit of these little birds across our narrow seas, may be reasonably conjectured, but that influencing mandate which assembles in great congregations wide spread creatures, and directs the arrangements of their motions in long and perilous travel, can be alone attributable to that intelligence which we can perceive every where, but define no where.

It is a difficult matter satisfactorily to comprehend the object of this bird in quitting another region, and passing into our island. The chief motives for migration seem to be food, a milder climate, and quiet during the period of incubation and rearing their young; but the willow wren, and some others of our insectivorous birds, appear to have other purposes to accomplish by their annual migrations. These little creatures, the food of which is solely insects, could assuredly find a sufficient supply of such diet during the summer months, in the woods and thickets of those mild regions where they passed the season of winter, and every bank and unfrequented wild would furnish a secure asylum for them and their offspring during the period of incubation. The passage to our shores is a long and dangerous one, and some imperative motive for it must exist; and, until facts manifest the reason, we may, perhaps, without injury to the cause of research, conjecture for what object these perilous transits are made. We know that all young creatures require particularly compounded nutriment during their infant state; and nature, as far as we are acquainted with it, has made in every instance provision for a supply of fitting aliment. In many instances, where the removal of station could not be conveniently accomplished, instinct has been given the parent to provide the fitting aliment for its new-born young. Thus insects, in some cases, store their cells with food ready for the animation of their progeny; in others, place their eggs in such situations as will afford it when they are hatched. The mammalia, at least the quadrupeds belonging to this class, which could least conveniently move their station, have supplies given them of a milky secretion for this purpose. Birds have nothing of this nature, and make no provision for their young; but they, of all creatures except fishes, can seek what may be required in distant stations with most facility. A sufficiency of food for the adult parent may be found in every climate, yet the aliment necessary for its offspring may not. Countries and even counties produce insects that differ, if not in species, at least in numbers; and many young birds we cannot succeed in rearing, or do it very partially, by reason of our ignorance of the requisite food. Every one, who has made the attempt, well knows the various expedients he has resorted to, of boiled meats, bruised seeds, hard eggs, boiled rice, and twenty other substances, that nature never presents, in order to find a diet that will nourish them; but Mr. Montague's failure in being able to raise the young of the cirl bunting[24], until he discovered that they required grasshoppers, is a sufficient instance of the manifest necessity there is for a peculiar food in one period of the life of birds; and renders it probable that, to obtain a certain aliment, this willow wren, and others of the insect and fruit-feeding birds, direct their flight to distant regions, and is the principal cause of their migrations.

It is some stimulus like this which urges that little creature, the golden-crested wren (*motacilla regulus*), that usually only flits from tree to tree, and never attempts upon common occasions a longer flight, to traverse the vast distance from the Orkneys to the Shetland Isles over stormy seas, that admit no possible rest during its long passage of above fifty miles! There it breeds its young; but, this one object accomplished, it leaves those isles, dares again this tedious flight, and seeks a milder clime. With us it never migrates, lives much in our fir groves during the winter, and breeds in our shrubberies in summer.

24 Linnæan Transactions, vol. vii.

Peculiar necessities, such as these, may incite the migration of many birds; but that certain species, which lead solitary lives, or associate only in very small parties, should at stated periods congregate from all parts to one spot, and there hold council on a removal, in which the very sexes occasionally separate, is one of the most extraordinary procedures that we meet with among animals.

If the sober, domestic attachments of the hedge sparrow please us, we are not less charmed with the innocent, blithesome gaiety of the linnet (*fringilla linota.*) But this songster is no solitary visitor of our dwellings: it delights and lives in society, frequenting open commons and gorsy fields, where several pairs, without the least rivalry or contention, will build their nests and rear their offspring in the same neighbourhood, twittering and warbling all the day long. This duty over, the families unite, and form large associations, feeding and moving in company, as one united household; and, resorting to the head of some sunny tree, they will pass hours in the enjoyment of the warmth, chattering with each other in a low and gentle note; and they will thus regularly assemble during any occasional bright gleam throughout all the winter season, —

And still their voice is song,

which, heard at some little distance, forms a very pleasing concert, innocent and joyous. The linnet is the cleanliest of birds, delighting to dabble in the water and dress its plumage in every little rill that runs by. The extent of voice in a single bird is not remarkable, being more pleasing than powerful; yet a large field of furze, in a mild sunny April morning, animated with the actions and cheering music of these harmless little creatures, united with the bright glow and odour of this early blossom, is not visited without gratification and pleasure.

The bull-finch (*loxia pyrrhula*) has no claims to our regard. It is gifted with no voice to charm us; it communicates no harmony to the grove: all we hear from it is a low and plaintive call to its fellows in the hedge. It has no familiarity or association with us, but lives in retirement in some lonely thicket ten months in the year. At length, as spring approaches, it will visit our gardens, an insidious plunderer. Its delight is in the embryo blossoms wrapped up at this season in the bud of a tree; and it is very dainty and curious in its choice of this food, seldom feeding upon two kinds at the same time. It generally commences with the germs of our larger and most early gooseberry; and the bright red breasts of four or five cock birds, quietly feeding on the leafless bush, are a very pretty sight, but the consequences are ruinous to the crop. When the cherry buds begin to come forward, they quit the gooseberry, and make tremendous havoc with these. I have an early wall cherry, a mayduke by reputation, that has for years been a great favourite with the bull-finch family, and its celebrity seems to be communicated to each successive generation. It buds profusely, but is annually so stripped of its promise by these feathered rogues, that its kind might almost be doubted. The Orleans and green-gage plums next form a treat, and draw their attention from what remains of the cherry. Having banqueted here awhile, they leave our gardens entirely, resorting to the fields and hedges, where the sloe bush in April furnishes them with food. May brings other dainties, and the labours and business of incubation, withdraw them from our observation.

The idea that has been occasionally entertained, that this bird selects only such buds as contain the embryo of an insect, to feed on it, and thus free us of a latent colony of caterpillars, is certainly not correct. It may confer this benefit accidentally, but not with intention. The mischief effected by bull-finches is greater than commonly imagined, and the ground beneath the bush or tree on which they have been feeding, is commonly strewed with the shattered buds, the rejectments of their banquet; and we are thus deprived of a large portion of our best fruits by this assiduous pillager, this "pick-a-bud," as the gardeners call it, without any redeeming virtues to compensate our loss. A snowy, severe winter makes great havoc with this bird. It feeds much in this season upon the fruit of the dog-rose, "hips," as we call them. When they are gone, it seems to pine for food, and is starved, or perhaps frozen on its roost, as few are observed to survive a long inclement winter. But it is not the buds of our fruit-bearing trees only that these destructive birds seek out; yet in all instances I think it will be observed, that such buds as produce leaves only are rejected, and those which contain the embryo of the future blossom selected: by this procedure, though the tree is prevented from producing fruit, yet the foliage is expanded as usual; but had the leaves, the lungs of the plant, been indiscriminately consumed, the tree would probably have died, or its summer growth been materially injured: we may thus lose our fruit this year, yet the tree survives, and hope lives too, that we may be more fortunate the next. The Tartarian honeysuckle (*lonicera Tart.*) and *corchorus Japonicus*, when growing in the shrubbery, are very commonly stripped of their bloom by bull-finches: the first incloses many separated blossoms in its calyx before expansion, and in that particular is analogous to the buds of icosandrious trees in the garden; and the full-petaled, swelling bloom of the latter affords a fine treat for their feasts; but we may permit these pretty birds to banquet here, though, if we expect a supply of summer fruit, we must unsparingly drive them away from the branches of our frugiferous trees. The blossoms of the peach, nectarine, and almond, I have never observed to be injured by these birds: the sparrow will pick away the buds of trees against walls when they frequent such places; but with this exception, I know none but the bull-finch which resort to that food as a regular supply.

The robin (*motacilla rubecola*), associated with malignants, is not, perhaps, in the place where it generally would be sought; but sad truths might be told of it too. It might be called pugnacious, jealous, selfish, quarrelsome, did I not respect ancient feelings, and long-established sentiments. A favourite by commiseration, it seeks an asylum with us; by supplication and importunity it becomes a partaker of our bounty in a season of severity and want; and its seeming humbleness and necessities obtain our pity; but it slights and forgets our kindnesses the moment it can provide for itself, and is away to its woods and its shades. Yet it has some little coaxing ways, and such fearless confidence, that it wins our regard; and its late autumnal song, in evening's dusky hour, as a monologue, is pleasing, and redeems much of its character. The universality of this bird in all places, and almost at all hours, is very remarkable; and perhaps there are few spots so lonely in which it would not appear, did we commence digging up the ground. I have often been surprised in the midst of woods, where no suspicion of its presence existed, when watching some other creature, to see the robin inquisitively perched upon some naked spray near me; or, when digging up a plant in some very retired place, to observe its immediate descent upon some poor worm that I had moved. The robin loses nearly all the characteristic colour from its breast in the summer, when it moults, and only recovers it on the approach of autumn, which, in some measure, accounts for the extraordinary assertion of Pliny, that the redbreast is only so in winter, but becomes a firetail in summer.

The object of the song of birds is not agreed upon by ornithologists, and we will not now think of it, but merely, in passing, note how singularly timid the song of the robin is. The blackbird and the thrush, in mild seasons, will sing occasionally throughout the winter; but the robin, after having been absent all the summer, returns to us late in autumn, and then commences its song, when most others of our feathered choristers are silent. An apparent contention in harmony ensues among them; at length the rivals approach, menace, and fight with a seeming vexation at each other's prowess. The song of no one bird is, perhaps, more observed and remembered than the autumnal and, at times, melancholy sounding farewell of the robin.

The chaffinch (*fringilla cœlebs*) appears to be universally spread throughout the English counties, and the male bird is remarkable for the cleanliness and trimness of his plumage, which, without having any great

THE CHAFFINCH

variety or splendour of colouring, is so composed and arranged, and the white on his wings so brilliant, as to render him a very beautiful little creature. The female is as remarkable for the quiet, unobtrusive tintings of her dress; and, when she lies crouching on her nest, elegantly formed of lichens from the bark of the apple tree, and faded mosses, she would hardly be perceptible, but for her little bright eyes, that peep with suspicious vigilance from her covert. With us the sexes do not separate at any period of the year, the flocks frequenting our barn doors and homesteads in winter being composed of both. In the northern parts of Europe, however, the females are said to migrate to milder regions, which induced Linnæus to bestow the name of "cœlebs" upon this species. In Gloucestershire and some of the neighbouring counties, they are little known by the name of chaffinches; but, from the constant repetition of one note, when alarmed or in danger, they have acquired the name of "twinks," and "pinks;" yet during incubation, the song of the male bird, though without any variation of tune, is very pleasing in the general concert, as most vernal notes, if not harsh and wearisome from monotony, are. These birds make sad havock with some of our spring flowers, and the polyanthus in March, in our sheltered borders, is very commonly stripped of all its blossoms by these little plunderers, I suppose to obtain the immature seeds at the base of their tubes. They will deflorate, too, the spikes or whorls of the little red archangel (*lamium purpureum*); and we see them feeding in the waste places where this plant is found in the spring, their little mouths being filled with the green seeds of this dead nettle. At this period, too, they are sad plunderers in our kitchen gardens, and most dexterously draw up our young turnips and radishes as soon as they appear upon the surface of the soil; but after this all depredation ceases, the rest of their days being past in sportive innocence. I have observed these birds in very hot seasons, to wet their eggs, by discharging moisture from their bills upon them, or at least perform an operation that appeared to be so. But this is an industrious little bird seeking his food with great assiduity, and as usual with so many of our little feathered associates, the labours of man supply the requirements, making his industry the chief object of his connection with him. I had often observed that our early crops of peas, just as they sprout from the earth, were cropped off, and the soil moved to obtain this shoot, but could not detect the depredator: at last I observed the chaffinch, hard at his work, searching for the tender plumate of the pea, the injury they do to the crop is at times extensive, but it is only in a very young state that it is agreeable to their taste, and as soon as the shoot obtains a vigorous growth, his step predations cease.

  We still continue here that very ancient custom of giving parish rewards for the destruction of various creatures included in the denomination of vermin. In former times it may have been found necessary to keep under or reduce the numbers of many predaceous animals, which, in a thickly-wooded country, with an inferior population, might have been productive of injury; and we even find parliamentary statutes enacted for this purpose: but now, however, our losses by such means have become a very petty grievance: our gamekeepers do their part in removing pests of this nature, and the plough and the axe leave little harbour for the few that escape; and thus we war on the smaller races of creation, and call them vermin. An item passed in one of our late churchwardens' accounts was, " for seventeen dozen of tomtits' heads!" In what evil hour, and for what crime, this poor little bird (*parus cœruleus*) could have incurred the anathema of a parish, it is difficult to conjecture. I know hardly any small animal that lives a more precarious life than the little blue tomtit. Indeed it is marvellous how any of the insectivorous birds, that pass their winter with us, are supplied with food during inclement seasons, unless they have greater powers of abstinence than we are aware of; but our small birds are generally much more active than those of a larger bulk: the common wren is all animation, its actions and movements bespeak hilarity and animal spirits; and that minute creature, too, the golden-crested wren, is always in motion, flitting from the yew hedge to the fir, or darting away to taller trees with a spring and a power we could not expect from its size. These muscular exertions must greatly counteract the effects of seasons, and enable these atoms of animals to support so cheerfully and gaily the winters of our climate. But in truth this tomtit perishes in severe winters in great numbers. It roosts under the eaves of our haystacks, and in little holes of the mows, where we often find it dead, perished by cold or hunger, or conjointly by both; yet the race survives, and this annual waste is recruited by the prolificacy of the creature, the nest of which will frequently contain from seven to nine young ones. Its chief subsistence is insects, which it hunts out with unwearied perseverance. It peeps into the nail-holes of our walls, which, though closed by the cobweb, will not secrete the spider within; and draws out the chrysalis of the cabbage butterfly from the chinks in the barn: but a supply of such food is precarious, and becomes exhausted. It then resorts to our yards, and picks diminutive morsels from some rejected bone, or scraps from the butcher's stall: yet this is the result of necessity, not choice; for no sooner is other food attainable, than it retires to its woods and thickets. In summer it certainly will regale itself with our garden pease, and shells a pod of marrowfats with great dexterity; but this, we believe, is the extent of its criminality. Yet for this venial indulgence do we proscribe it, rank it with vermin, and set a price upon its head, giving four-pence for the dozen, probably the ancient payment when the groat was a coin. However powerful the stimulus was then, we yet find it a sufficient inducement to our idle bat-fowling boys to bring baskets of poor toms' heads to our churchwarden's door.

THE WILES AND stratagems of every creature are deserving of attention, because they are, for the most part, the impulse of the weak and the feeble, instinctive efforts to preserve their own existence, or more generally to secure or defend that of their offspring. Few are able to effect these objects by bodily power; but all creatures probably exert a faculty of some kind to ward off injury from their young, though not observed by, or manifested to us. This poor little blue tomtit, which has neither beak, claws, nor any portion of strength to defend itself from the weakest assailant, will nevertheless make trial by menace to scare the intruder from its nest. It builds almost universally in the hole of a wall or a tree; and its size enables it to creep through so small a crevice, that it is pretty well secured from all annoyances but those of bird nesting boys: and these little plunderers the sitting bird endeavours to scare away, by hissing and puffing in a very extraordinary manner from the bottom of the hole, as soon as a finger is introduced, and so perfectly unlike the usual voice of a bird, that many a young intruder is deterred from prosecuting any farther search, lest he should rouse the vengeance of some lurking snake or adder. They who have seen much of birds, and attended to their actions, will in general be certain of the creature that flits past, by the manner of its flight, or that utters its note unseen by the peculiarity of voice; but the tribe of titmice (*parus*) especially in the spring of the year, emit such a variety of sounds, that they will occasionally surprise and disappoint us. Hearing an unusual voice, and creeping with caution to observe the stranger from which it proceeds, we perceive only our old acquaintance, the large tomtit (*parus major*), searching for food amid the lichens on the bough of an apple-tree. This bird, and that little dark species, the "coal," or "colemouse" (*parus ater*) in particular, will often acquire or compound a note, become delighted with it, and repeat it incessantly while sporting about the catkins of the alder, for an hour or so, then seem to forget or be weary of it, and we hear it no more.

One of our summer visiting birds the white throat, appears to be generally spread every where like the marten and the swallow. The male creature brings itself into notice by its particular restlessness and activity towards the commencement of the breeding season; the joyous little bird singing at intervals from early morning until the eve of a mild spring day, sometimes concealed in his leafy covert, or rising from his brake, will warble as he flutters in the air, distending his little snowy throat with all his energy. Sinking in the bush, we hear his 'cha, cha,' to warn his mate of suspicious appearances, for she is very timid and avoids appearing unless disturbed, nor can we well ascertain her arrival in our fields. Fearful that his warning should not be attended to, a second is given like the squeak of a mouse, and then the watchman carefully retires himself. The lesser white throat (*m. sylvia lin*^{i}). is very difficult to obtain a sight of by reason of its timidity, its voice is pretty, but not powerful, nor resembling that of the common one. It probably has often been mistaken for the female hay-chat, but it has none of the dark brown quill feathers of both the sexes of that species, and its fine silvery throat and stomach well indicate the lesser bird. The description of it by Bewick is good as far as it goes; the weight of the male bird is about 189 grains.

Our tall hedge-rows and copses are frequented by a very amusing little bird, the long-tailed titmouse (*parus caudatus*). Our boys call it the long-tailed tom-tit, long-tom, poke-pudding, and various other names. It seems the most restless of little creatures, and is all day long in a state of progression from tree to tree, from hedge to hedge, jerking through the air with its long tail like a ball of feathers, or threading the branches of a tree, several following each other in a little stream, the leading bird uttering a shrill cry of twit, twit, twit. Away they all scuttle to be first, stop for a second, and then are away again, observing the same order and precipitation the whole day long. The space travelled by these diminutive creatures in the course of their progresses from the first move till the evening roost must be considerable; yet, by their constant alacrity and animation, they appear fully equal to their daily task. We have no bird more remarkable for its family association than this parus. It is never seen alone, the young ones continuing to accompany each other from the period of their hatching until their pairing in spring. Its food is entirely insects, which it seeks among mosses and lichens, the very smallest being captured by the diminutive bill of this creature. Its nest is as singular in construction as the bird itself. Even, in years long passed away, when, a nesting boy, I strung my plunder on the benty grass, it was my admiration; and I never see it now without secretly lauding the industry of these tiny architects. It is shaped like a bag, and externally fabricated of moss and different herbaceous lichens, collected chiefly from the sloe (*lichen prunastri*), and the maple (*lichen farinaceus*); but the inside contains such a profusion of feathers, that it seems rather filled than lined with them: a perfect featherbed! I remember finding fourteen or sixteen pea-like eggs within this downy covert, and many more were reported to have been found. The excessive labour of the parent birds in the construction and collection of this mass of materials is exceeded by none that I know of. The exertions of two little creatures in providing for, and feeding, with all the incumbrances of feathers and tails, fourteen young ones, in such a situation, surpass in diligence and ingenuity the efforts of any other birds, persevering as they are, that I am acquainted with.

We might naturally suppose that, by the end of winter, all those little birds which are solely supported by insect food would find some difficulty in providing for their wants, having consumed by their numbers and exertions nearly all that store of provision which had been provided in the summer and deposited in safety. But I have found the stomachs of the tree-creeper, and this small titmouse, even in February, quite filled with parts of coleopterous creatures, which by their activity and perseverance they had been enabled to procure beneath the mosses on the branches, and from the chinks in the bark of trees, where they had retired in autumn. Such plenty being procurable after the supply of so many months, renders it apparent that there is no actual deficiency of food at any one period of the year. The small slugs, and some few insects, may perhaps be consumed by the severity of winter; but the larger portion of them are so constituted as to derive no injury from the inclemency of that season, but afford during many months, provender to other creatures, multitudes yet remaining to continue their races and animate the air, when the warm days of spring shall waken them to active life.

The construction and selected situations of the nests of birds are as remarkable as the variety of materials employed in them. The same forms, places, and articles, being rarely, perhaps never, found united by the different species, which we should suppose similar necessities would direct to a uniform provision. Birds that build early in the spring seem to require warmth and shelter for their young; and the blackbird and the thrush line their nests with a plaster of loam, perfectly excluding, by these cottage-like walls, the keen icy gales of our opening year. Yet should accident bereave the parents of their first hopes, they will construct another, even when summer is far advanced, upon the model of their first erection, and with the same precautions against severe weather, when all necessity for such provision has ceased, and the usual temperature of the season rather requires coolness and a free circulation of air. The house-sparrow will commonly build four or five times in the year, and in a variety of situations, under the warm eaves of our houses and our sheds, the branch of the clustered fir, or the thick tall hedge that bounds our garden, &c.; in all which places, and without the least consideration of site or season, it will collect a great mass of straw and hay, and gather a profusion of feathers from the poultry-yard to line its nest. This cradle for its young, whether under our tiles in March or in July, when the parent bird is panting in the common heat of the atmosphere, has the same provisions made to afford warmth to the brood; yet this is a bird that is little affected by any of the extremes of our climate. The wood pigeon and the jay, though they erect their fabrics on the tall underwood in the open air, will construct them so slightly, and with such a scanty provision of materials, that they seem scarcely adequate to support their broods, and even their eggs may almost be seen through the loosely-connected materials: but the goldfinch, that inimitable spinner, the Arachne of the grove, forms its cradle of fine mosses and lichens, collected from the apple or the pear-tree, compact as a felt, lining it with the down of thistles besides, till it is as warm as any texture of the kind can be, and it becomes a model for beautiful construction. The golden-crested wren, a minute creature, perfectly unmindful of any severity in our winter, and which hatches its young in June, the warmer portion of our year, yet builds its most beautiful nest with the utmost attention to warmth; and, interweaving small branches of moss with the web of the spider, forms a closely-compacted texture nearly an inch in thickness, lining it with such a profusion of feathers that, sinking deep into this downy accumulation, it seems almost lost itself when sitting, and the young, when hatched, appear stifled with the warmth of their bedding and the heat of their apartment; while the whitethroat, the blackcap, and others, which will hatch their young nearly at the same period, or in July, will require nothing of the kind. A few loose bents and goose-grass, rudely entwined, with perhaps the luxury of some scattered hairs, are perfectly sufficient for all the wants of these; yet they are birds that live only in genial temperatures, feel nothing of the icy gales that are natural to our pretty indigenous artists, but flit from sun to sun, and we might suppose would require much warmth in our climate during the season of incubation; but it is not so. The greenfinch places its nest in the hedge with little regard to concealment; its fabric is slovenly and rude, and the materials of the coarsest kinds: while the chaffinch, just above it in the elm, hides its nest with cautious care, and moulds it with the utmost attention to order, neatness, and form. One bird must have a hole in the ground; to another a crevice in a wall, or a chink in a tree, is indispensable. The bull-finch requires fine roots for its nest; the grey flycatcher will have cobwebs for the outworks of its shed. All the parus tribe, except the individual above mentioned, select some hollow in a tree or cranny in a wall; and, sheltered as such places must be, yet will they collect abundance of feathers and warm materials for their infants' beds. Endless examples might be found of the dissimilarity of requirements in these constructions among the several associates of our groves, our hedges, and our houses; and yet the supposition cannot be entertained for a moment that they are superfluous, or not essential for some purpose with which we are unacquainted.[25] By how many of the ordinations of Supreme Intelligence is our ignorance made manifest! Even the fabrication of the nests of these little animals exceeds our comprehension — we know none of the causes or motives of that unbodied mind that willed them thus.

One notice more of the parus tribe (the *parus cæruleus*), and these little creatures may retire to their leafy shades and be forgotten. I was lately exceedingly pleased in witnessing the maternal care and intelligence of this bird; for the poor thing had its young ones in the hole of a wall, and the nest had been nearly all drawn out of the crevice by the paw of a cat, and part of its brood devoured. In revisiting its family, the bird discovered a portion of it remaining, though wrapped up and hidden in the tangled moss and feathers of their bed, and it then drew the whole of the nest back into the place from whence it had been taken, unrolled and resettled the remaining little ones, fed them with the usual attentions, and finally succeeded in rearing them. The parents of even this reduced family laboured with great perseverance to supply its wants, one or the other of them bringing a grub, caterpillar, or some insect, at intervals of less than a minute through the day, and probably in the earlier part of the morning more frequently: but if we allow that they brought food to the hole every minute for fourteen hours, and provided for their own wants also, it will admit of perhaps a thousand grubs a day for the requirements of one, and that a diminished brood; and give us some comprehension of the infinite number requisite for the summer nutriment of our soft-billed birds, and the great distances gone over by such as have young ones, in their numerous trips from hedge to tree in the hours specified, when they have full broods to support. A climate of moisture and temperature like ours is

---

25 I remember no bird that seems to suffer so frequently from the peculiar construction of its nest, and by reason of our common observance of its sufferings obtains more of our pity, than the house marten. The rook will at times have its nest torn from its airy site, or have its eggs shaken from it by the gales of spring; but the poor marten, which places its earthy shed beneath the eave of the barn, the roof of the house, or in the corner of the window, is more generally injured. July and August are the months in which these birds usually bring out their young; but one rainy day at this period, attended with wind, will often moisten the earth that composes the nest, the cement then fails, and all the unfledged young ones are dashed upon the ground; and there are some places to which these poor birds are unfortunately partial, though their nests are annually washed down. The projecting thatch of the old farmhouse appears to be their safest asylum. The parent birds at times seem aware of the misfortune that awaits them; as, before the calamity is completed, we may observe them with great anxiety hovering about their nests.

peculiarly favourable for the production of insect food, which would in some seasons be particularly injurious, were we not visited by such numbers of active little friends to consume it.

The raven (*corvus corax*) does not build with us. A pair, indeed, attempted to raise a brood in our wych-elm; but they love retirement and quiet, and were soon scared away, and made no second trial. Ravens visit us, however, frequently, and always during the lambing season, watching for any weak and deserted creature, which, when perceived, is instantly deprived of its eyes; but they make no long stay in our pastures. They abide nowhere in fact, but move from place to place, where food may chance to be found. Should an animal die, or a limb of fresh carrion be on the hooks in the tree, the hoarse croak of the raven is sure immediately to be heard calling his congeners to the banquet. We see it daily in its progress of inspection, or high in the air on a transit to other regions, hastening, we conjecture, to some distant prey. With the exception of the snipe, no bird seems more universally spread over the surface of our globe than the raven, inhabiting every zone, the hot, the temperate, the severe — feeding upon, and removing noxious substances from the earth, of which it obtains intimation by means of a faculty we have little conception of. Sight it cannot be; and we know not of any fetor escaping from an animal previous to putrescence, so subtile as to call these scavengers of nature from the extremity of one county to that of another: for it is manifest, from the height which they preserve in their flight, and the haste they are making, that their departure has been from some far distant station, having a remote and urgent object in contemplation.

In England the raven does not seem to abound; but it is most common on the shores of harbours, or near great rivers, where animal substances are more frequently to be met with than in inland places. In Greenland and Iceland, where putrescent fishy substances abound, they appear to be almost domesticated. Horace calls the raven "*annosa cornix*;" and in a tame state it has attained a very long life. How long extended its existence may be, when roaming in an unrestricted state, we have no means of ascertaining. This liberty may be most favourable to longevity; yet, from the numerical contingencies attending the condition of these creatures, it is probable that few of them live out all their days, so as to become the "bird of ages." However, the supposed longevity they have attained, their frequent mention and agency in Holy Writ, the obscure knowledge we possess of their powers and motives, with the gravity of their deportment, like an "all-knowing bird," have acquired for them, from very remote periods, the veneration of mankind. The changes in our manners and ideas, in respect to many things, have certainly deprived them of much of this reverence; yet the almost supernatural information which they obtain of the decease, or approaching dissolution, of an animal, claims still some admiration for them. This supposed faculty of "smelling death" formerly rendered their presence, or even their voice, ominous to all, as

> The hateful messengers of heavy things,
> Of death and dolour telling;

and the unusual sound of their harsh croak, still, when illness is in the house, with some timid and affectionate persons, brings old fancies to remembrance, savouring of terror and alarm. I am no friend to the superstition of converting natural transactions, or occasional events, into signs and indications of coming things; superstitions are wearing out, and shortly will waste away, and be no more heard of; but, I fear, in their place, deism, infidelity, impiety, have started up, the offspring of intuitive wisdom: the first belief arises from weakness and ignorance; the latter disbelief is ingratitude, pride, wickedness.

Of the natural duration of animal life it is, from many circumstances, difficult to form an accurate statement, the wild creatures being in great measure removed from observation, and those in a condition of domestication being seldom permitted to live as long as their bodily strength would allow. It was formerly supposed that the length of animal life was in proportion to its duration in utero, or the space it remained in the parent from conception to birth, and the length of time it required to obtain maturity. This notion might have some support in reason and fact, occasionally, but, in many cases, was incorrect, and, in regard to birds, had no foundation. Herbivorous animals probably live longer than carnivorous ones, vegetable food being most easily obtainable in all seasons in a regular and requisite supply; whereas animals that subsist on flesh, or by the capture of prey, are necessitated at one period to pine without food, and at another are gorged with superfluity; and when the bodily powers of rapacious creatures become impaired, existence is difficult to support, and gradually ceases; but with herbivorous animals in the same condition, supply is not equally precarious, or wholly denied. Yet it is probable that few animals in a perfectly wild state live to a natural extinction of life. In a state of domestication, the small number of carnivorous creatures about us are sheltered and fed with care, seldom are in want of proper food, and at times are permitted to await a gradual decay, continuing as long as Nature permits, and, by such attentions many have attained to a great age; but this is rather an artificial than a natural existence. Our herbivorous animals, being kept mostly for profit, are seldom allowed to remain beyond approaching age; and when its advances trench upon our emoluments, by diminishing the supply of utility, we remove them. The uses of the horse, though time may reduce them, are often protracted; and our gratitude for past services, or interest in what remains, prompts us to support his life by prepared food of easy digestion, or requiring little mastication; and he certainly by such means attains to a longevity probably beyond the contingencies of Nature. I have still a favourite pony — for she has been a faithful and able performer of all the duties required of her in my service for upwards of two-and-twenty years — and, though now above five-and-twenty years of age, retains all her powers perfectly, without any diminution, or symptom of decrepitude: the fineness of limb, brilliancy of eye, and ardour of spirit, are those of the colt; and, though treated with no remarkable care, she has never been disabled by the illness of a day, or sickened by the drench of the farrier.[26] With birds it is probably the same as with other creatures, and the eagle, the raven, the parrot, &c., in a domestic state, attain great longevity; and though we suppose them naturally tenacious of life, yet, in a really wild state, they would probably expire before the period which they attain when under our attention and care. And this is much the case with man, who probably outlives most other creatures; for though excess may often shorten, and disease or misfortune terminate his days, yet naturally he is a long-lived animal. His "threescore years and ten" are often prolonged by constitutional strength, and by the cares, the loves, the charities of human nature. As the decay of his powers awaken solicitude, duty and affection increase their attentions, and the spark of life only expires when the material is exhausted.

Exemption from the fears of poverty and want, anxieties that wear both body and mind when contending with the world, are often favourable to the duration of human life. We have a beautiful instance at no great distance from hence of this, and of what care and kindness may accomplish. The inmates of two small almshouses in Bristol endowed by Alderman Stevens, the one for twelve the other for sixteen indigent women, have at this time (1834) attained a great age, they range variously from 93 to 71, giving an average of 81 years to each of these twenty-eight poor persons.

That rare bird, the crossbill (*loxia curvirostra*), occasionally visits the orchards in our neighbourhood, coming in little parties to feed upon the seeds of the apple; and, seldom as it appears, is always noticed by the mischief it does to the fruit, by cutting it asunder with its well-constructed mandibles, in order to obtain the kernels. A native of those extensive pine forests in the

---

26 Dec. 10, 1829. — But my faithful servant is gone: her life was an easy one — not shortened by violence, nor terminating in suffering.

neighbourhood of the Rhine, it makes excursions into various parts of Europe in search of change of food; and, though several instances are recorded of its visits to our island, I know but one mention of its having bred in England. A pair was brought to me very early in August, and the breast of the female being nearly bare of feathers, as is observable in sitting birds, it is very probable that she had a nest in the neighbourhood[27]

Gesner has called the common rook (*covis frugilegus*) a corn-eating bird. Linnæus has somewhat lightened this epithet by considering it only as a gatherer of corn; to neither of which names do I believe it entitled, as it appears to live solely upon grubs, various insects, and worms. It has at times great difficulty to support its life, for in a dry spring or summer most of these are hidden in the earth beyond its reach, except at those uncertain periods when the grub of the chaffer is to be found; and in a hot day we see the poor birds perambulating the fields, and wandering by the sides of the highways, seeking for and feeding upon grasshoppers, or any casual nourishment that may be found. At those times, was it not for its breakfast of dew-worms, which it catches in the gray of the morning, as it is appointed the earliest of risers, it would commonly be famished. In the hot summer of 1825, many of the young brood of the season perished from want; the mornings were without dew, and consequently few or no worms were to be obtained; and we found them dead under the trees, having expired on their roostings. It was particularly distressing, for no relief could be given, to hear the constant clamour and importunity of the young for food. The old birds seemed to suffer without complaint; but the wants of their offspring were expressed by the unceasing cry of hunger, and pursuit of their parents for supply, and our fields were scenes of daily restlessness and lament. Yet, amid all this distress, it was pleasing to observe the perseverance of the old birds in the endeavour to relieve their famishing families, as many of them remained out searching for food quite in the dusk, and returned to their roosts long after the usual period for retiring. In this extremity it becomes a plunderer, to which by inclination it is not much addicted, and resorts to our newly-set potato fields, digging out the cuttings. Ranks are seen sadly defective, the result of its labours, I fear; and the request of my neighbours now and then for a bird from my rookery, to hang up *in terrorem* in their fields, is confirmatory of its bad name. In autumn a ripe pear, or a walnut, becomes

---

27 In John Knapp's notebooks, the following extra remarks were written for insertion here, but then crossed out: "It appears that at times the Cross-bill migrates to this country in little flights, and this year 1835 in particular their numbers were such, that at Backwell in Somersetshire, one farmer in his apple orchard killed (apparently in mere wantonness) twenty five of these poor wanderers."

an irresistible temptation, and it will occasionally obtain a good share of these fruits. In hard frost it is pinched again, visits for food the banks of streams, and in conjunction with its congener, the "villain crow," becomes a wayfaring bird, and "seeks a dole from every passing steed."[28] Its life, however, is not always dark and sombre; it has its periods of festivity also. When the waters retire from meadows and low lands, where they have remained any time, a luxurious banquet is provided for this corvus, in the multitude of worms which it finds drowned on them. But its jubilee is the season of the cockchaffer (*melolantha vulgaris*), when every little copse, every oak, becomes animated with it and all its noisy, joyful family feeding and scrambling for the insect food. The power or faculty, be it by the scent, or by other means, that rooks possess of discovering their food, is very remarkable. I have often observed them alight on a pasture of uniform verdure, and exhibiting no sensible appearance of withering or decay, and immediately commence stocking up the ground. Upon investigating the object of their operations, I have found many heads of plantains, the little autumnal dandelions, and other plants, drawn out of the ground and scattered about, their roots having been eaten off by a grub, leaving only a crown of leaves upon the surface. This grub beneath, in the earth, the rooks had detected in their flight, and descended to feed on it, first pulling up the plant which concealed it, and then drawing the larvae from their holes. By what intimation this bird had discovered its hidden food we are at a loss to conjecture; but the rook has always been supposed to scent matters with great discrimination.

It is but simple justice to these often-censured birds, to mention the service that they at times perform for us in our pasture lands. There is no plant that I endeavour to root out with more persistency in these places than the turfy hair-grass (*aira cæspitosa*). It abounds in all the colder parts of our grass lands, increasing greatly when undisturbed, and, worthless itself, overpowers its more valuable neighbours. The larger turfs we pretty well get rid of; but multitudes of small roots are so interwoven with the pasture herbage, that we cannot separate them without injury; and these our persevering rooks stock up for us in such quantities, that in some seasons the fields are strewed with the eradicated plants. The whole so torn up does not exclusively prove to be

---

28 During the unusually severe winter of 1829-30, our rooks became certainly "corn-eaters:" the ground was bound down by the frost, and their favourite food hidden by the snow. They fixed themselves, by dozens, on the oat-ricks out in the fields; and the late sown, just germinating wheat was dug up from the soil, to a very injurious extent, by our half-famished birds; but they appeared to return to their common food upon the relenting of the frost.

the hair-grass, but infinitely the larger portion consists of this injurious plant. The object of the bird in performing this service for us, is to obtain the larvae of several species of insects, underground feeders, that prey on the roots, as Linnæus long ago observed upon the subject of the little nard grass (*nardus stricta*). This benefit is partly a joint operation: the grub eats the root, but not often so effectually as to destroy the plant, which easily roots itself anew; but the rook finishes the affair by pulling it up to get at the larvae, and thus prevents all vegetation; nor do I believe that the bird ever removes a specimen that has not already been eaten, or commenced upon, by the caterpillar.

The rook entices its young from the breeding trees, as soon as they can flutter to any other. These young, for a few evenings after their flight, will return with their parents, and roost where they were bred; but they soon quit their abode, and remain absent the whole of the summer months. As soon, however, as the heat of summer is subdued, and the air of autumn felt, they return and visit their forsaken habitations, and some few of them even commence the repair of their shattered nests; but this meeting is very differently conducted from that in the spring; their voices have now a mellowness approaching to musical, with little admixture of that harsh and noisy contention, so distracting at the former season, and seems more like a grave consultation upon future procedure; and as winter approaches they depart for some other place. The object of this meeting is unknown; nor are we aware that any other bird revisits the nest it has once forsaken. Domestic fowls, indeed, make use again of their old nests; but this is never, or only occasionally, done by birds in a wild state. The daw and rock pigeon will build in society with their separate kindred; and the former even revisits in autumn the places it had nestled in. But such situations as these birds require, the ruined castle, abbey, or church tower, ledge in the rock, &c., are not universally found, and are apparently occupied from necessity. The rooks appear to associate from preference to society, as trees are common every where; but what motive they can have in view in lingering thus for a few autumnal mornings, and counselling with each other around their abandoned and now useless nests, which before the return of spring are generally beaten from the trees, is by no means manifest to us.

The sense of smelling seems often to supply in animals the want of faculties they are not gifted with; and it is this power which directs them to their food with greater certainty, than the discernment of man could do. That we have every faculty given us necessary for the condition in which we are placed is manifest; yet the mechanical talents and intuition of the insect, the powers that birds and beasts possess, and the superior acuteness of some of their senses, of which, perhaps, we have little conception, makes it evident that all created things were equally the objects of their Maker's benevolence and care; the worm that creepeth, and the beast that perisheth, deserve our consideration, and claims from human reason mercy and compassion.

A short time since, circumstances induced me to save the entire offspring of my small rookery, and we calculated that about two hundred young birds escaped and joined the flights; but in the ensuing spring the numbers of nests in our trees were not sensibly increased, the old birds alone apparently building in them. The winter certainly may have consumed some, but probably not equal to the increase. All this seems in perfect unison with the tendency of nature in most of her departments, to spread and disperse rather than to accumulate, to fill up vacancies, and leave nothing unoccupied. I have repeatedly known districts, from which during the winter season every blackbird, thrush, gold, and bull finch had been killed, yet in the ensuing spring observed their places filled by others, and the song in the grove, and nesting in the brake, as harmonious and as plentiful as usual. Many sportsmen know that killing down their game does not universally prevent a supply in the ensuing season, nor the saving of whole covies assure in that part of the manor an unusual number of breeding pairs, but like mankind they colonise, and send out parties to inhabit unoccupied places: entirely domesticated, and half reclaimed creatures may remain where they were bred, but perfectly wild and free ones seem generally to obey this mandate of nature. With man, necessity, or some political cause usually incites this separation, but with animals it appears most often to be the mere fulfilment of an appointed ordination: the intrusions of man and his habits may drive timid creatures away from established abodes to another which is free from his annoyances, but the jealousy of some creatures, which incites them to drive others from the vicinity of their stations, and the bickerings and contentions of the elder birds with their offspring, forcing them to form new settlements, are only modes of accomplishing these designs of nature. But this must be understood in a limited sense, and not of universal, invariable occurrence, for there are various groups of animals and creatures dispersed over particular regions of the globe, indigenous to certain sites, and neither driven there by the increased multitudes of their fellows, nor the intrusions of man, but created for, and located to, a certain region, where alone they find congenial food, temperature, or some undefinable accordance fitting for their constitutional mechanism. Creatures habituated to a cold northern district, or the products and warmth of a solar clime, if driven southward or northward by any impulsive power, would pine, diminish, and ultimately perish: an alpine or a lowland creature would probably be very rarely able to support a mutual change of situation without deterioration, or gradual extinction; but where certain regions or countries similarly constituted become exhausted of their natural inhabitants by accident or by local acting causes, they seem universally to be resettled by the surplus of another, and uninhabited tracts become tenanted by families impelled to migrate by superior force or rival jealousy.

THE TALL TANGLED hedgerow, the fir grove, or the old, well-wooded inclosure, constitutes the delight of the magpie (*corvus pica*), as there alone its large and dark nest has any chance of escaping observation. We here annually deprive it of these asylums, and it leaves us; but it does not seem to be a bird that increases much any where. As it generally lays eight or ten eggs, and is a very wary and cunning creature, avoiding all appearance of danger, it might be supposed that it would yearly become more numerous. Upon particular occasions we see a few of them collect; but the general spread is diminished, and as population advances, the few that escape will retire from the haunts and persecutions of man. These birds will occasionally plunder the nests of some few others; and we find in early spring the eggs of our out-laying domestic fowls frequently dropped about, robbed of their contents. That the pie is a party concerned in these thefts we cannot deny, but to the superior audacity of the crow we attribute our principal injury. However the magpie may feed on the eggs of others, it is particularly careful to guard its own nest from similar injuries by covering it with an impenetrable canopy of thorns, and is our only bird that uses such a precaution, securing it from all common depredation, though not from the hand of the bird-nesting boy. When a hatch is effected, the number of young demand a larger quantity of food than is easily obtained, and whole broods of our ducklings, whenever they stray from the yard, are conveyed to the nest. But still the "maggot" is not an unuseful bird, as it frees our pastures of incredible numbers of grubs and slugs, which lodge themselves under the crusts formed by the dung of cattle. These the birds, with their strong beaks, turn over, and catch the lurking animals beneath, and then break them to search for more; by which means, during winter, they will spread the entire droppings in the fields; and by spring I have had, especially under the hedges, all this labour saved to me by these assiduous animals.

Natural affection, the love of offspring, is particularly manifested in birds; for in general they are timid and weak creatures, flying from apprehended dangers, and endowed with little or no power of defending themselves; but they will menace when injury is threatened to their brood, and incur dangers in order to obtain food for their young, that they will encounter in no other period of their lives.

The common jay (*corvus glandarius*) affords a good example of this temporary departure from general character. This bird is always extremely timid and cautious, when its own interest or safety is solely concerned; but no sooner does its hungry brood clamour for supply, than it loses all this wary character, and it becomes a bold and impudent thief. At this period it will visit our gardens, which it rarely approaches at other times, plunder them of every raspberry, cherry, or bean, that it can obtain, and will not cease from rapine as long as any of the brood or the crop remains. We see all the nestlings approach, and, settling near some meditated scene of plunder, quietly await a summons to commence. A parent bird from some tree surveys the ground, then descends upon the cherry, or into the rows, immediately announces a discovery by a low but particular call, and all the family flock in to the banquet, which having finished by repeated visits, the old birds return to the woods, with all their chattering children, and become the same wild, cautious creatures they were before. Some of our birds separate from their broods as soon as they are able to provide for themselves; but the jay and its family associate during all the autumn and winter months, taking great delight in each other's company, and only separate to become founders of new establishments. We see them in winter under the shelter of tall hedges, or on the sunny sides of woods and copses, seeking amid the dry leaves for acorns, or the crab, to pick out the seeds, or for the worms and grubs hidden under cowdung; feeding in perfect silence, yet so timid and watchful, that they seldom permit the sportsman to approach them. When disturbed, they take shelter in the depth of the thicket, calling to each other with a harsh and loud voice, that resounds through the covert. The Welsh call this creature "screch y coed," the screamer of the wood. The jay is a very heavy, inelegant bird. Its general plumage is sober and plain, though its fine browns harmoniously blend with each other: but the beautiful blue-barred feathers, that form the greater coverts of the wings, distinguish it from every other bird, and, in the days when feather-work was in favour with our fair countrywomen, were in such request, that every gamekeeper, and schoolboy brother with his Christmas gun, persecuted the poor jay through all his retirements, to obtain his wings.

The predaceous nature of this beautiful bird is but obscurely noticed by our writers upon ornithological subjects, not seeming assured of it from personal testimony. Unwilling as I am to supply any additional reason for injury and persecution to these our poor foresters, who hide from man as their bane, yet in truth I must say that the jay is well disposed to prey upon all the lesser birds, when his means enable him to do. For parts of the year these are not readily captured, and then he supports himself in other ways, but in summer having young ones themselves, they dash boldly into hedges after those of other birds, and plunder every nest they can find of its eggs or its young.

I have seen them even making off with an old bird in possession, and this summer one jay was seen to carry away from my orchard the nestlings of two pair of goldfinches, stealthily returning until the whole was accomplished. The misfortunes and depredations attending our small birds during the season of incubation are numerous, the capture of their eggs by the united boys of the village is great, those nests placed low in the bush, or in the tuft of grass at its base I have often found with broken eggs, or the contents carried off, in the first case probably mice are concerned, in the latter the snake; the jay, the shrike, cat, fitchet, assault them when their broods are hatched. The guns, the gins of winter assail them so that from all the casualties it is a matter of surprise that any of our little songsters, the animated companions of our summer hours, yet remain. Five pair of goldfinches one year hatched their eggs on my premises, and though defended with all care, not more than one matured their brood. More nests of the sky lark probably escape their plunderings than other birds, the careful mode of concealment, and open places in which they are situated, afford no clue for detection like the furze brake, the copse, the hedge, near which predacious animals harbour for prey and security, and the wandering boy looks into and has the chance only of observing. This accounts for the large flocks of larks which assemble in the winter.

The shrike, or butcher-bird (la*nius collurio*), is not uncommon with us, and breeds annually near my dwelling. It is one of our late birds of passage, but its arrival is soon made known to us by its croaking, unmusical voice from the summit of some tree. Its nest is large and ill concealed; and during the season of incubation the male bird is particularly vigilant and uneasy at any approach towards his sitting mate, though often by his clamorous anxiety he betrays it and her to every bird-nesting boy. The female, when the eggs are hatched, unites her vociferations with those of the male, and facilitates the detection of the brood. Both parents are very assiduous in their attentions to their offspring, feeding them long after they have left the nest, for the young appear to be heavy, inactive birds, and little able to capture the winged insects, that constitute their principal food. I could never observe that this bird destroyed others smaller than itself, or even fed upon flesh. I have hung up dead young birds, and even parts of them, near their nests, but never found that they were touched by the shrike. Yet it appears that it must be a butcher too, and that the name "*lanius*," bestowed on it by Gesner two hundred and fifty years ago, was not lightly given. My neighbour's gamekeeper kills it as a bird of prey, and tells me he has known it draw the weak young pheasants through the bars of the breeding-coops; and others have assured me that they have killed them when banqueting on the carcase of some little bird they had captured. All small birds have an antipathy to the shrike, betray anger, and utter the moan of danger, when it approaches their nests. I have often heard this signal of distress, and cautiously approaching to learn the cause, have frequently found that this butcher-bird occasioned it. They will mob, attack, and drive it away, as they do the owl, as if fully acquainted with its plundering propensities. As this bird seldom conceals itself in a bush, but sits perched upon some upper spray, or in an open situation, heedful of danger, or watching for its prey, it must be most mischievously inclined, if not a predatory bird. — May 23rd: — A pair of robins have young ones in a bank near my dwelling; the anxiety and vociferation of the poor things have three times this day called my attention to the cause of their distress, and each time have I seen this bird watching near the place, or stealing away upon my approach, and then the tumult of the parents subsided; but had they not experienced injury, or been aware that it was meditated, all this terror and outcry would not have been excited.

This evening July 12th. However, all conjectures as to the predacious habits of L. collurio, were completely satisfied; vociferous as these birds always are when we approach the nest or their young, I should not at this time have noticed that proceeding from a small party, had not their clamour been particularly violent, but we found a cock-bird sitting upon the branch of a white-thorn, earnestly feeding a young one, with something not well understood, which turned out to be a large Shrew Mouse, hanging from a stout thorn by its head, the creature had torn away the eyes and parts of the head, and was proceeding with the rest. Unwilling to disturb the supper, we walked on, but returning soon, we found the feast was ended and all consumed. It now appeared that this thorn was the regular shamble-hook of the party from having been much used, beneath were scattered the feathers of some small bird, the wing of a yellow hammer with the legs of dorrs. That this shrike should attempt to seize any creature excepting insects, and the weaker ones of its race, had not been supposed, but here we had proof of his boldness and powers in venturing upon and succeeding in the capture of a quadruped – having no talons sufficiently strong for holding his victims like birds of prey, an extraordinary instinct has been given to him to fix it upon a thorn, from whence he can pick it away in fragments, or reserve it in store, and we have seen the spiny sprays in the hedge near his haunts occasionally furnished with several coleopterous insects, which as they disappear we suppose are taken away when his wants require; we have thus three of our smaller birds which make use of artificial contrivance as assistance in obtaining their food; the thrush who seeks a stone to break the shell of his snail, sitta, who fixes the nut in the rough bark of the tree or in a cleft, as the larger tit will sometimes do a large seed, and this our butcher bird and his thorn.

Many birds are arranged in our British ornithology not known as permanent inhabitants, but which have occasionally visited our shores during inclement seasons, or been driven from their general stations by tempestuous weather. An event like this, the violent gale of All Hallows eve, in 1824, brought to us the stormy petrel (*procellaria pelagica*): a bird that resides far in the depths of the ocean, does not approach our shores, it is believed, except for the purposes of incubation, and we know only one place, the Isle of Sky, that it haunts even for this short period. It is a creature

— that roams on her sea-wing,
Unfatigued, and ever sleeps,
Calm, upon the toiling deeps.

It is a pretty good manifestation of the strength and extent of that hurricane, which could catch up a bird with a wing so powerful as to enable it to riot in the whirlwind and enjoy the storm, and bear it away irresistibly, perhaps, from the Atlantic waves, over such a space of land and ocean, and then dash it down on a rather elevated common in this parish, whence it was brought to me in a very perfect state. This little creature, scarcely as big again as a swallow, and the smallest of all our web-footed birds, has, like all the others of its genus, that extraordinary tube on its upper mandible, through which it spirts out an oily matter when irritated; but the real object of this singular provision seems unknown. Our seamen amuse themselves during the monotony of a voyage with the vagaries of "Mother Cary's chickens," as they have from very early times called this bird. The petrels seem to repose in a common breeze, but upon the approach, or during the continuation of a gale, they surround a ship, and catch up the small animals which the agitated ocean brings near the surface, or any food that may be dropped from the vessel. Whisking with the celerity of an arrow through the deep valleys of the abyss, and darting away over the foaming crest of some mountain wave, they attend the labouring bark in all her perilous course. When the storm subsides they retire to rest, and are no more seen. The presence of this petrel was thought in times past to predict a storm, and it was consequently looked upon as an unwelcome visitant.

The notice that seamen now take of the petrel seems almost universal, and yet there does not appear to be any manifest reason for it. Particular regard from any motive of reverence is in these times seldom bestowed by any Christian people upon the inferior orders of creation. The stork in some fortunate hour obtained this respect, and yet preserves it, not only with some of our continental neighbours, but with a few asiatic nations. But with ourselves I can name none that we reverence; the robin, now poor fellow! can expect but little of the regard that he seems to have experienced from former generations, though he still approaches our dwellings in the winter of the year with an indirect and silent supplication for relief, with a fearless confidence in our humanity; and some few may not have forgotten the impressions of childhood, nor the leaves strewed by his care over injured innocence. But the petrel is by no means the most familiar, or only bird of the ocean, or in such association with the mariner as to obtain a fair title to his regard; even the existence of so small a bird in the immensity of the abyss, amidst elementary turmoil, when those which were larger and stronger sought their safety upon the rocks or the land; no race of men perhaps think more deeply, when they do think, upon sacred things, than those who occupy their business in deep waters. Any superstitious feeling to become general, should have a motive, which in this case is not obvious, and if this bird of the storm is yet held in reverence by the seamen of all the European nations, the influence seems of a modern date. It was possibly not noticed by any of the elder navigators. Not even Dampier, I believe, who enters much into descriptions of natural things, makes mention of it. The voyagers for discovery in the Pacific both English and French, by their frequent mention of "Mother Cary's chickens," if not the first noticers, have made the name, and regard familiar to us.

The wryneck (*jynx torquilla*) visits us annually, but in very uncertain numbers, and from some unknown cause, or local changes, in yearly diminishing quantities. In one short season after its arrival we hear its singular monotonous note at intervals through half the day. This ceases, and we think no more about it, as it continues perfectly mute; not a twit or a chirp escapes to remind us of its presence during all the remainder of its sojourn with us, except the maternal note or hush of danger, which is a faint, low, protracted hissing, as the female sits clinging by the side or on the stump of a tree. Shy and unusually timid, as if all its life were spent in the deepest retirement away from man, it remains through the day on some ditch bank, or basks with seeming enjoyment, in any sunny hour, on the ant-hills nearest to its retreat;

and these it depopulates for food, by means of its long glutinous tongue, which with the insects collects much of the soil of the heaps, as we find a much larger portion of grit in its stomach than is usually met with in that of other birds. When disturbed it escapes by a flight precipitate and awkward, hides itself from our sight, and, were not its haunts and habits known, we should never conjecture that this bustling fugitive was our long-forgotten spring visitant, the wryneck. The winter or spring of 1818 was, from some unknown cause, singularly unfavourable for this bird. It generally arrives before the middle of April; and its vernal note, so unlike that of any of its companions, announces its presence throughout all the mild mornings of this month, and part of the following; but during the spring of that year it was perfectly silent, or absent from us. The season, it is true, was unusually cheerless and ungenial.

Some of our birds are annually diminishing in numbers; others have been entirely destroyed, or no longer visit the shores of Britain. The increase of our population, inclosure, and clearage of rude and open places, and the drainage of marshy lands, added to the noise of our fire-arms, have driven them away, or rendered their former breeding and feeding stations no longer eligible to many, especially to the waders and aquatic birds. The great Swan Pool, near the city of Lincoln, on which I have seen at one time forty of these majestic creatures sailing in all their dignity, is, I am told, no longer a pool; the extensive marshes of Glastonbury, which have afforded me the finest snipe-shooting, are now luxuriant corn-farms; and multitudes of other cases of such subversions of harbour for birds are within memory. An ornithological list, made no longer ago than the days of Elizabeth, would present the names of multitudes now aliens to our shores. The nightingale was common with us here a few years past; the rival songs of many were heard every evening during the season, and in most of our shady lanes we were saluted by the harsh warning note of the parent to its young; but from the assiduity of bird-catchers, or some local change that we are not sensible of, a solitary vocalist or so now only delights our evening walk. The egg of this bird (Plate 3, Fig. 5) is rather singularly coloured, and not commonly to be obtained. Our migrating small birds incur from natural causes great loss in their transits; birds of prey, adverse winds, and fatigue, probably reduce their numbers nearly as much as want, and the severity of the winter season, does those that remain; and in some summers the paucity of such birds is strikingly manifest. Even the hardy rook is probably not found in such numbers as formerly, its haunts having been destroyed or disturbed by the felling of trees, in consequence of the increased value of timber, and the changes in our manners and ideas. Rooks love to build near the habitation of man; but their delight, the long avenue, to caw as it were in perspective from end to end, is no longer the fashion; and the poor birds have been dispersed to settle on single distant trees, or in the copse, and are captured and persecuted.

Old-fashion'd halls, dull aunts, and croaking rooks,

a modern Zephalinda would scarcely find now to anticipate with dread. In many counties very few rookeries remain, where once they were considered as a necessary appendage, and regularly pointed out the abbey, the hall, the court-house, and the grange.

The starling (*sturnus vulgaris*) breeds with us, as in most villages in England. Towards autumn the broods unite, and form large flocks; but those prodigious flights with which, in some particular years, we are visited, especially in parts of those districts formerly called the "fen counties," are probably an accumulation from foreign countries, We have seldom more than a pair, or two, which nestle under the tiling of an old house, in the tower of the church, the deserted hole of the woodpecker, or some such inaccessible place. The flights probably migrate to this country alone, as few birds could travel long, and continue such a rapid motion as the starling. The Royston crow, the only migrating bird with which it forms an intimate association, is infinitely too heavy of wing to have journeyed with the stare. The delight of these birds in society is a predominant character; and to feed, they will associate with the rook, the pigeon, or the daw; and sometimes, but not cordially, with the fieldfare: but they chiefly roost with their own families, preferring some reedy, marshy situation. These social birds are rarely seen alone; and should any accident separate an individual from the companions of its flight, it will sit disconsolate on an eminence, piping and plaining, till some one of its congeners join it. Even in small parties they keep continually calling and inviting associates to them, with a fine clear note, that, in particular states of the air, may be heard at a considerable distance. This love of society seems to be innate; for I remember one poor bird, that had escaped from domestication, in which it had entirely lost, or probably never knew, the language or manners of its race, and acquired only the name of its mistress; disliked and avoided by its congeners, it would sit by the hour together, sunning on some tall elm, calling in a most plaintive strain, Nanny, Nanny, but no Nanny came; and our poor solitary either pined itself to death, or was killed, as its note ceased. They vastly delight, in a bright autumnal morning, to sit basking and preening themselves on the summit of a tree, chattering altogether in a low song-like note. There is something singularly curious and mysterious in the conduct of these birds previous to their nightly retirement, by the variety and intricacy of the evolutions they execute at that time. They will form themselves, perhaps, into a triangle, then shoot into a long, pear-shaped figure, expand like a sheet, wheel into a ball, as Pliny observes, each individual striving to get into the centre, &c., with a promptitude more like parade movements than the actions of birds. As the breeding season advances, these prodigious flights divide, and finally separate into pairs, and form their summer settlements; but probably the vast body of them leaves the kingdom.

Travellers tell us, that starlings abound in Persia and the regions of Caucasus. No birds, except sparrows, congregate more densely than stares. They seem continually to be running into clusters, if ever so little scattered; and the stopping of one, to peck at a worm, immediately sets all its companions hastening to partake. This habit in the winter season brings on them death, and protracted sufferings, as every village popper notices these flocks, and fires at the poor starlings. Their flesh is bitter and rank, and thus useless when obtained; but the thickness of the flights, the possibility of killing numbers, and manifesting his skill, encourages the trial. The flight of these birds, whether from feeding to roost, or on their return to feed, is so rapid, that none with any impediment can keep company; and in consequence we see many, which have received slight wing or body wounds, lingering about the pastures long into spring, and pining after companions they cannot associate with.

These birds are very assiduous in their attentions to their young, and in continual progress to collect worms and insects for them. However strong parental affection may be in all creatures, yet the care which birds manifest in providing for their nestlings is more obvious than that of other animals. The young of beasts sleep much; some are hidden in lairs and thickets nearly all the day, others take food only at intervals or stated periods, the parent ruminating, feeding, or reposing too: but birds, the young of which remain in their nests, as most of them do, excepting the gallinaceous and aquatic tribes, have no cessation of labour from early morning till the close of eve, till the brood can provide for themselves. What unceasing toil and perseverance are manifest in the rooks, and what distances do they travel to obtain nourishment for their clamorous brood! It is a very amusing occupation, for a short time, to attend to the actions of a pair of swallows or martens, the family of which have left the nest, and settled upon some naked spray, or low bush in the field, the parents cruising around, and then returning with their captures to their young: the constant supply which they bring, the celerity with which it is given and received, and the activity and evolutions of the elder birds, present a pleasing example of industry and affection. I have observed a pair of starlings for several days in constant progress before me, having young ones in the hole of a neighbouring poplar tree, and they have been probably this way in action from the opening of the morning — thus persisting in this labour of love for twelve or thirteen hours in the day! The space they pass over in their various transits and returns must be very great, and the calculation vague; yet, from some rude observations, it appears probable that this pair in conjunction do not travel less than fifty miles in the day, visiting and feeding their young about a hundred and forty times, which, consisting of five in number, and admitting only one to be fed each time, every bird must receive in this period eight and twenty portions of food or water! This excessive labour seems entailed upon most of the land birds, except the gallinaceous tribes, and some of the marine birds, which toil with infinite perseverance in fishing for their broods; but the very precarious supply of food to be obtained in dry seasons by the terrestrial birds, renders theirs a labour of more unremitting hardship than that experienced by the piscivorous tribes, the food of which is probably little influenced by season, while our poor land birds find theirs to be nearly annihilated in some cases. The gallinaceous birds have nests on the ground; the young leave them as soon as they escape from the shell, are led immediately from the hatch to fitting situations for food and water, and all their wants are most admirably attended to; but the constant journeyings of those parent birds that have nestlings unable to move away, the speed with which they accomplish their trips, the anxiety they manifest, and the long labour in which they so gaily persevere, is most remarkable and pleasing, and a duty consigned but to a few.

We have no bird more assiduous in attentions to their young, than the red start (*steort*, Saxon, a tail,) one or other of the parents being in perpetual action, conveying food to the nest, or retiring in search of it; but as they are active, quick-sighted creatures, they seem to have constant success in their transits. They are the most restless and suspicious of birds during this season of hatching and rearing their young; for when the female is sitting, her mate attentively watches over her safety, giving immediate notice of the approach of any seemingly hostile thing, by a constant repetition of one or two querulous notes, monitory to her or menacing to the intruder: but when the young are hatched, the very appearance of any suspicious creature sets the parents into an agony of agitation, and, perching upon some dead branch or a post, they persevere in one unceasing clamour till the object of their fears is removed: a magpie near their haunts, with some reason, excites their terror greatly, which is expressed with unremitting vociferation. All this parental anxiety, however, is no longer in operation than during the helpless state of their offspring, which, being enabled to provide their own requirements, gradually cease to be the objects of solicitude and care; they retire to some distant hedge, become shy and timid things, feeding in unobtrusive silence.

Soon after the broods are matured, we lose sight of the male red start. As we generally see him in the spring for a few days unaccompanied by his mate, we might suppose that he has again unsocially separated from his family; but whatever imperative motive may actuate him in the former period, he does not now seem to be influenced by it, but he moults so completely as to part with all the characteristic featherings of the sex; and we may observe many of these birds flitting about the sunny retired hedgerows in July without a single male attendant, yet, before their final departure from us, his fine plumage is renewed, and they all disappear for a season[29].

A pair of Fire-tails have taken a very particular liking to a comfortable hole in my garden wall, and for several years have built their nest in it; but not more than once have they succeeded in raising the brood, for by reason of some undefined cause, the sparrows have plundered it of the eggs and we find them broken and dropped near the nest, nor do they occupy the place afterwards themselves; the fire-tails are not usurpers of any prior rights, nor are the nests sufficiently near those of their destroyers to create a reasonable jealousy, but the poor birds are forced in the end to seek a more quiet asylum.

The great diversity in character in birds, the short movements, and almost local sphere of action in some, and the ceaseless transition of others; the lonely retirement of a part, and the fearless domestication of others with man, must commonly be remarked by many; and the chief motive which probably influences much of this variety of deportment may arise from the nature of the food on which they subsist. There is one of these creatures with which we are all acquainted in his season, whose rambling nature and restless flight are always subjects of admiration, — the swift (*hirundo apus*), which visits us about the beginning of May, and soon after, as, having no time to lose, commences preparation for its brood; that operation over, they again depart, remaining with us about fourteen or fifteen weeks: and all this period seems passed in hurry and precipitation, in unceasing action on the wing, from early morning, till hidden by the shades of night, not calmly sailing on his way, but rushing and struggling through the air with unremitting haste; and such is their muscular power, violent as these exertions are, they present no diminution of celerity or lassitude of body — they seem the destined couriers

---

29 Editor's note: Knapp noted that he wanted the entire section on brown starlings, which had appeared at this point in the previous four editions, to be deleted. This has been done in this edition and a paragraph which he had written on Fire-tails has been inserted in its place.

of the sky, coursing for life. But these laborious progressions of the swift do not proceed from the mere pleasure resulting from a life of action; they are rendered necessary by the vagrant habits of his prey, as we generally observe him occupied in pursuit or capturing his victim: at one moment, perhaps, circling round us like the head of a barbed arrow, snatching up some hatch of creatures just risen from the soil; that effected, away he goes like the passing breeze, now skimming with noiseless wing the surface of the pool, then ranging high in the azure of the air.

But the rivers and waters near large towns, where there is a constant decomposition of animal matters, giving life to a profusion of insects, are the great haunts of this black marten, and in some seasons their numbers are incredible. I have known the river about Chelsea, or Fulham, swarm with such myriads as to distress the eye that watched them, by the rapidity and intricacy of their motions; nor is the harmless manner in which these crossings and traversings are performed the least admirable part of the scene: apparently intent alone upon the capture of their prey, they dash along in every adverse direction, yet each conducts its progress with such observance and precision, that they never strike against, or impede the progress of another.

The swallow tribe live almost perpetually upon the wing, even the high soaring hawks, rest upon their feet a part of the day, but this race circles nearly from hour to hour in the sun beam finding repose in calmly floating upon the air; their lives here appear to be passed in ease and enjoyment, & when the temperature of our climate becomes ungenial they depart with their families to other regions unrestrained by impediments or doubts. The young broods towards autumn collect upon some roof or ridge, and then as by a signal dart into the atmosphere, a momentary race, return after a short excursion and commence their frolick again. These birds present us with an instance of that carelessness and unsuspicion of harm which a life of innocence is thought to create, arriving here after a long journey, they resort to our houses, near their families, sporting about us in a fearless familiarity, neither injuring men nor his possessions. She becomes one of the few birds man suffers to associate about him, without requiring at some period a tribute for his sufferance.

Wearied as every one must be with the conflicting opinions regarding the migration of swallows, I must add my testimony to that of others, in maintaining that a portion of these birds under some circumstances, do pass their winter in England. Near the bridge over the Dart at Totness, is, or was, a high sand stone cliff, perforated with the holes of the marten. Many years past upon the morning preceding Christmas day, returning from snipe shooting, the ground sprinkled with snow and frozen, and the sun shining brightly, there were a dozen or more birds, apparently the common house swallow, cruising along this cliff, in the sunny gleam, nor did they venture beyond this shelter; the weather changed, and they were not again visible during my stay: I have often regretted that the species was not fully ascertained at the time and a specimen preserved. But no man, as the proverb says, is wise at all times

It was a very ancient observation (and modern investigation seems fully to confirm it), that many of the serpent race captured their prey by infatuation or intimidation; and there can be no doubt of the fact, that instinctive terror will subdue the powers of some creatures, rendering them stupified and motionless at the sudden approach of danger. We have two kinds of petty hawks, the sparrow-hawk (*falco nisus*) and the kestrel (*falco tinnunculus*), that seem fully to impress upon their destined prey this species of intimidation. A beautiful male bullfinch, that sat harmlessly pecking the buds from a blackthorn by my side, when overlooking the work of a labourer, suddenly uttered the instinctive moan of danger, but made no attempt to escape into the bush, seemingly deprived of the power of exertion. On looking round, a sparrow-hawk was observed on motionless wing gliding rapidly along the hedge, and, passing me, rushed on its prey with undeviating certainty. There was fully sufficient time from the moment of perception for the bullfinch to escape; but he sat still, waiting the approach of death, an unresisting victim. We have frequently observed these birds, when perched on an eminence, insidiously attentive to a flock of finches and yellowhammers basking in a hedge, and after due consideration apparently single out an individual. Upon its moving for its prey, some wary bird has given the alarm, and most of the little troop scuttle immediately into the hedge; but the hawk holds on its course, and darts upon a selected object. If baffled, it seldom succeeds upon another; and so fixed are its eyes upon this one individual, that, as if unobservant of its own danger, it snatches up its morsel at our very sides. A pigeon on the roof of the dovecot seems selected from its fellows, the hawk rarely snatching at more than one terror-stricken bird. The larger species of hawks appear to employ no powers excepting those of wing, but pursue and capture by celerity and strength.

We converse annually upon early and late seasons; and such things there are. A mild winter, a warm February and March, will influence greatly the growth of vegetation: not that a primrose under that bank, or a violet under the shelter of this hedge, affords us any criterion of earliness; but a general shading of green, an expansion of buds, an incipient unfolding of leaves, gives notice of the spring's advance. The principal blossoming of plants usually takes place at nearly stated periods; but particular mildness in the atmosphere, and additional warmth in the soil, accelerate this season; and of all the evils which threaten the horticulturist, an early spring is most to be deprecated. An April breathing odours, wreathed in verdure and flowers, the willow-wren sporting in the copse, the swallow skimming over the pool, lambs racing in the daisied mead, may be a beautiful sight to comtemplate, —

Fair laughs the morn, and soft the zephyrs blow;

but it is like the laugh of irony, the smile that lures to ruin,

Which, hush'd in grim repose, awaits his certain prey.

Then comes a ruthless May, with Winter in her train, who, with his frosty edge, unpitying shears away all the expectancies, the beautiful promise of the year; and we have to await returning seasons, and patient hope for better things. A garden, pining and prostrate from the effects of a churlish, frosty May, leaves crisp and blackened, flowers withered, torn, and scattered around, is a melancholy sight — the vernal hectic that consumes the fairest offspring of the nursery. There is a plant, however, the white-thorn (*mespilus oxycanthus*), the May of our rustics, common in all places and situations, that affords a good example of general steadiness to time, uninfluenced by partial effects. An observation of above twenty years upon this plant has proved how little it deviates in its blossoming in one season from another; and, under all the importunities and blandishments of the most seductive Aprils, I have in all that period never but twice seen more than a partial blossom by the first of May. We hail our first-seen swallow as a harbinger of milder days and summer enjoyments; but the appearance of our birds of passage is not greatly to be depended upon, as I have reason to apprehend from much observation. They will be accelerated or retarded in the time of their departure by the state of the wind in the country whence they take their flight; they travel much by night, requiring in many instances the light of the moon to direct them; and the actual time of their arrival is difficult to ascertain, as they steal into our hedges and copses unperceived. If the weather be bright or warm, their voices are heard; if gloomy and cold, they will lie secreted till the call of hunger or of love intimates their presence. Though we rarely see these birds in their transits, yet I have at times, on a calm bright evening in November, heard high in the air the redwing and the fieldfare, on progress to a destined settlement, manifested by the signal notes of some leading birds to their scattered followers. These conductors of their flocks are certainly birds acquainted with the country over which they travel, their settlements here being no promiscuous dispersion; it being obvious that many pairs of birds return to their ancient haunts, either old ones which had bred there, or their offspring. The butcher-bird successively returns to a hedge in one of my fields, influenced by some advantage it derives from that situation, or from a preference to the spot where hatched; but we have perhaps no bird more attached to peculiar situations than the grey flycatcher (*muscicapa grisola*); one pair, or their descendants, frequenting year after year the same hole in the wall, or the same branch on the vine or the plum. Being perfectly harmless, and hence never molested, they become

> Enamour'd with their ancient haunts,
> — and hover round.

 I once knew a pair of these birds bring off two broods in one season from the same nest. This flycatcher delights in eminences. The naked spray of a tree, or projecting stone in a building, or even a tall stick in the very middle of the grass-plot, is sure to attract its attention, as affording an uninterrupted view of its winged prey; and from this it will be in constant activity a whole summer's day, capturing its food, and returning to swallow it.

 July 28th a very pleasing instance of parental anxiety and contrivance was presented this week, by a grey flycatcher, whose nest had been constructed upon the branch of an honeysuckle, trained against an eastern wall in my garden. The previous weather having been damp and chilly, the bird hatched her young without experiencing any inconvenience from her situation. But on the 26th the sun became so powerful for nearly one week as to raise the thermometer in that situation to 80 degrees, and greatly distressed the young by its warmth. To obviate this as much as possible, for the last two days the poor mother has been sitting close above her brood, with expanded wings canopying the nest, her bill open painting in the gleam; and so assiduously continuing this shade, as not to remove, though she saw the thermometer placed within a yard of her station. By 12 o'clock the sun moved off, and then the suffering parent relinquished her painful attentions, commencing her animated pursuit of insects for her young. The male bird, though he left the domestic contrivances to his mate, was not indifferent to his duty or comforts, but brought his captures to the nest with such perseverance as apparently to exclude his obtaining any supply himself, but indeed birds having young ones to feed, by the small portion which they consume, seem to have lost the necessity or the inclination for food. This bird seems to prefer an eastern aspect, as in very many instances I have seen it chosen; and for three successive years honey-suckle has been in estimation. I am very well aware that an anecdote relating to the same bird similar to what is stated above, is in White's "Selborne," but I have not suppressed my statement, it being a full corroboration of that account given by my great and amiable predecessor.

The digestion of some birds must be remarkably rapid, to enable them to receive such constant replenishments of food. The swift and the swallow are feeding from the earliest light in the morning till the obscurity of evening; the quantities of cherries and raspberries that the blackcap and pettichaps will eat are surprising, as they are unremittingly consuming from morning till night; and this flycatcher seems to require a proportion of food equal to any bird, being in constant progress, capturing one moment, and resting the next. A pair of large tomtits (*parus major*) will attach themselves to a crop of peas in our gardens, and unremittingly persevere in the business of consuming them from morning until night, without any abatement of appetite or lassitude from employ: but all wild creatures require quantities, and delight in a variety of food; and we see them, even when abundance of one sort is to be obtained, frequently partaking of several: cattle, in particular, will resort to hedges, brakes, and pools, consuming the promiscuous herbage found in such places. But most probably much of this diet acts as a corrective to what has been eaten, and, by certain medicinal properties, is assistant to digestion; and, as wild creatures have greater facility of obtaining these alternatives than domesticated ones have, the superior health they commonly enjoy may, in part, be attributable to this preventive diet. But fruit and insects are with us, only for a short season; and their privations, when these no longer afford a supply, indicate that they possess the power of abstinence, as well as that of consumption.

This year (1833) was more deficient than almost any other which I remember, in the number of migrating birds. Hirundines, even the strong winged swift, were very scarce, white-throats, black-caps, red-starts, and all our little hedge minstrels which make our spring months so joyous with their harmony, and marking the seasons by variety of voice, were scarcely heard. The solitary pe-pe-pe- of the wryneck, or chatter of some wary babillard, could alone be noticed where dozens in other years intimated their presence; a single pair of butcher birds, in a circle commonly visited by many. It is probable that some adverse gale of wind, or season of storms drew the little wanderers from their course to other stations, or overwhelmed them in their passage to our island. Even the common yellow wagtail was only scattered about our fields, where numbers are generally to be found glancing in little flights in the evening sunshine. The grey wagtail (*M. boarula*) always the scarcest species, was now a rare bird. Each of these species was probably influenced by the same contingency that annihilated others. Yet one of our summer visitants, the nightingale, escaped all accidents whatever they might be, and after having almost abandoned our copses, appeared now in numbers equal to any former period, or perhaps more generally diffused; and we could not but remark how little the produce of any coming year could be predicted by the circumstance of preceding seasons. The winter was remarkably mild, the spring dry, and month of May unusually warm, and as we considered, favourable to all insect produce, yet scarcely any of these tribes except the commonest and apparently least influenced by seasons and their change, were hovering about us.

We observed this summer two common thrushes frequenting the shrubs on the green in our garden. From the slenderness of their forms, and the freshness of their plumage, we pronounced them to be birds of the preceding summer. There was an association and friendship between them, that called our attention to their actions; one of them seemed ailing, or feeble from some bodily accident; for though it hopped about, yet it appeared unable to obtain sufficiency of food: its companion, an active sprightly bird, would frequently bring it worms, or bruised snails, when they mutually partook of the banquet; and the ailing bird would wait patiently, understand the actions, expect the assistance of the other, and advance from his asylum upon its approach. This procedure was continued for some days, but after a time we missed the fostered bird, which probably died, or by reason of its weakness met with some fatal accident. We have many relations of the natural affection of animals; and whoever has attended to the actions of the various creatures we are accustomed to domesticate about us can probably add many other instances from their own observation. Actions which are in any way analogous to the above, when they are performed by mankind, arise most commonly from duty, affection, pity, interest, pride; but we are not generally disposed to allow the inferior orders of creation the possession of any of these feelings, except perhaps the last: yet when we have so many instances of attachment existing between creatures similar and dissimilar in their natures, which are obvious to all, and where no interest can possibly arise as a motive; when we mark the varieties of disposition which they manifest under uniform treatment, their various aptitudes and comprehensions, sensibility or inattention to sounds, &c., it seems but reasonable to consider them as gifted with latent passions; though, being devoid of mind to stimulate or call them into action by any principle of volition or virtue, how excited to performance we know no more than we do the motives of many of their bodily actions! The kindnesses and attentions which the maternal creature manifests in rearing its young, and the assistance occasionally afforded by the paternal animal during the same period, appear to be a natural inherent principle, universally diffused throughout creation;

but when we see a sick or maimed animal supplied and attended by another, which we suppose gifted with none of the stimuli to exertion that actuate our conduct, we endow them by this denial with motives with which we ourselves are unacquainted; and at last we can only relate the fact, without defining the cause.

The throstle is a bird of great utility in a garden where wall-fruit is grown, by reason of the peculiar inclination which it has for feeding upon snails, and very many of them he does dislodge in the course of the day. When the female is sitting, the male bird seems to be particularly assiduous in searching them out, and I believe he feeds his mate during that period, having frequently seen him flying to the nest with food, long before the eggs were hatched: after this time the united labours of the pair destroy numbers of these injurious creatures. That he will regale himself frequently with a tempting gooseberry or bunch of currants, is well known, but his services entitle him to a very ample reward. The blackbird associates with these thrushes in our gardens, but makes no compensation for our indulgences after his song ceases, as he does not feed upon the snail; but the thrush benefits us through the year by his propensities for this particular food, and every grove resounds with his harmony in the season; and probably if this race suffered less from the gun of the Christmas popper, the gardener might find much benefit, in his ensuing crop of fruit, from the forbearance.

We have no bird, I believe, more generally known, thought of, or mentioned with greater indifference, perhaps contempt, than the common sparrow (*fringilla domestica*), "that sitteth alone on the house-top;" yet it is an animal that Nature seems to have endowed with peculiar characteristics, having ordained for it a very marked provision, manifested in its increase and maintenance, notwithstanding the hostile attacks to which it is exposed. A dispensation that exists throughout creation is brought more immediately to our notice by the domestic habits of this bird. The natural tendency that the sparrow has to increase, will often enable one pair of birds to bring up fourteen or more young ones in the season. They build in places of perfect security from the plunder of larger birds and vermin. Their art and ingenuity in commonly attaching their nests beneath that of the rook, high in the elm, a bird whose habits are perfectly dissimilar, and with which they have no association whatever, making use of their structure only for a defence to which no other bird resorts, manifest their anxiety and contrivance for the safety of their broods. With peculiar perseverance and boldness, they forage and provide for themselves and their offspring; will filch grain from the trough of the pig, or contend for its food with the gigantic turkey; and, if scared away, their fears are those of a moment, as they quickly return to their plunder; and they roost protected from all the injuries of weather. These circumstances tend greatly to increase the race, and in some seasons their numbers in our corn-fields towards autumn are prodigious; and did not events counteract the increase of this army of plunderers, the larger portion of our bread corn would be consumed by them. But their reduction is as rapidly accomplished as their increase, their love of association bringing upon them a destruction, which a contrary habit would not tempt. They roost in troops in our ricks, in the ivy on the wall, &c., and are captured by the net: they cluster on the bush, or crowd on the chaff by the barn-door, and are shot by dozens at a time, or will rush in numbers, one following another, into the trap. These and various other engines of destruction so reduce them in the winter season, that the swarms of autumn gradually diminish, till their numbers in spring are in no way remarkable. I have called them plunderers, and they are so; they are benefactors likewise, seeming to be appointed by Nature as one of the agents for keeping from undue increase another race of creatures, and by their prolificacy they accomplish it. In spring and the early part of the summer, before the corn becomes ripe, they are insectivorous, and their constantly-increasing families require an unceasing supply of food. We see them every minute of the day in continual progress, flying from the nest for a supply, and returning on rapid wing with a grub, a caterpillar, or some reptile; and the numbers captured by them in the course of these travels are incredibly numerous, keeping under the increase of these races, and making ample restitution for their plunderings and thefts. When the insect race becomes scarce, the corn and seeds of various kinds are ready; their appetite changes, and they feed on these with undiminished enjoyment.

We have scarcely another bird, the appetite of which is so accommodating in all respects as that of the house sparrow. It is, I believe, the only bird that is a voluntary inhabitant with man — lives in his society, and is his constant attendant, following him wherever he fixes his residence. It becomes immediately an inhabitant of the new farm-house, in a lonely place or recent enclosure, or even in an island; will accompany him into the crowded city, and build and feed there in content, unmindful of the noise, the smoke of the furnace, or the steam-engine, where even the swallow and the marten, that flock around him in the country, are scared by the tumult, and leave him: but the sparrow, though begrimed with soot, does not forsake him; feeds on his food, rice, potatoes, or almost any other extraneous substance he may find in the street; looks to him for his support, and is maintained almost entirely by the industry and providence of man. It is not known in a solitary and independent state.

Most of our small birds in retreating to their rest, seek the thatch, the hole, or the shelter of the leafy bush in quiet, retiring in little parties or singly to their nightly abode; the black-bird when mounting the ivyed pollard will at times indeed give notice of his object, and the Shuffle-wing tells his mate in a low and plaintive voice that bed time is come, this agreed upon, they frisk away and we hear of them no more: but the house sparrow will prepare for roosting perhaps an hour before he is settled, and then all the occurrences of the day seem to be detailed, as they will chatter in the depth of the yew bush, or the holly, after they retire as long as a slant of sunshine, or a ray of light remains, in apparent altercation for some wished for place or spray the possessor is unwilling to give up, but when it becomes dusky they gradually settle down into quiet rest. The undisturbed lodging of birds in their nightly roost seems an object of great importance to them, feeling probably how perfectly indefensible their condition is when asleep: to what distances from their feeding grounds will rooks traverse, as their great height in the air upon their advance and return sufficiently manifest, for this purpose; one might suppose that any tall sheltered tree would answer the object, but this is by no means the case, as places have a preference from a cause not obvious; how careful will swallows, and wagtails, seek the long straggling sallows or low bushes that stretch over the pool; as the evening becomes dusky we may observe the former in little parties rush by us in a low flight hasting to the sheltered pond, take a few turns round it, then dart into the station they occupy for the night, at times forming considerable numbers. Starlings, for years from autumn until spring have passed by us in very large bodies to their nightly rest with great timely regularity, at first in a low flight, but persons having awaited their passage to shoot them, they ultimately took a higher range, but always eastward. But this last year they have reversed their practice flying westward, taking up their abode in Kings Weston park near the mouth of the Avon, some favourite reeds, bushes, or trees, in their original station having probably been removed, or annoyance arisen. The starling is one of those birds which from some unperceived cause, some favourable circumstances in the hatching and rearing of their young, that periodically increase amazingly: our last great year of abundance was 1826, after that, or the following year, no particular numbers of them have been observable, but this year 1835 their numbers are prodigious, so much so as to occasion the notice of very indifferent persons. In the months of June and July we could see small parties running about most of our pasture lands, but in their favourite feeding haunts, the marsh lands of the lower level of the Severn, their flights in the evening when assembling for their rest were very remarkable. They did not at this time associate in the vast masses which we are accustomed to notice in autumn or winter sweeping away to some distant roost, but now sought the shelter of some tall tree in parties only, yet in some cases so plentiful were they that they might be seen upon every spray that could support them, and when the periodical union of these separate bodies arrive they will occasion much surprise. To these generally silent retirements however we must admit an exception in this bird (the starling) whose evening social chatter fully equals that of the sparrow, and as their numbers are usually very great so is the prattle proportionately excessive, and few who were not convinced of it, would believe that the sound which proceeds from the head of the tall tree upon which they settle could be that of birds, it being much like the falling waters of a cascade heard at a little distance, a low soothing monotone, a soft lulling murmur. In 1839 the number of these birds exceeded our general supply; but the singular circumstance was that very many hundreds of them never paired, never obeyed the usual impulse of nature but continued in these associated myriads. In May and the early part of June, when others were performing the duties of incubation, they resorted inattentive to their roost: male birds generally abide near the retirements of their mates, nor do I know any other instance of their congregating in the breeding season away from their nesting station, or manifesting a general negligence to the appointments of nature.

THOUGH I REMEMBER no bird so peculiarly associated with the human race as this is, yet there are other animals that seem dependent on man for support, or at least that find his means subservient to their comforts, and domesticate themselves with him. The meadow and the long-tailed mouse occasionally become foragers in our gardens and domains, when a natural supply of food becomes difficult of attainment, yet they are not wholly settlers with us; but the common mouse (*mus domesticus*) resorts entirely to our premises, and seems to exist wholly on food of our providing. In towns it accommodates its appetite to the variety of sustenance it finds there; and will enjoy the preserve in the pot, the cheese in the rack, or the pie in the pantry. In the country it will ransack the cupboard, live in the barn, or colonize in our ricks. Still, in all these cases, the store and provision of man are its delight, and its only resource; and it will even quit a residence which is abandoned by its provider. It is true, it maintains the same love of liberty as its celebrated ancestor is reported to have done; but the simplicity of manners and taste of the sage, the "hollow tree, the oaten straw," have been abandoned; it has become pleased with household comforts, and a luxurious citizen in its appetite.

The rat (*mus rattus*), too, perhaps, may be united with these companions of mankind. Not knowing it in an independent state, we cannot say what its resources might be, but so sagacious and powerfully-endowed an animal could always provide for its own necessities; yet it prefers our provision to any precarious supply from its own industry. In summer it partially quits our dwellings, the heat and dry ness of our buildings becoming irksome to it, and the occasional difficulty of obtaining water, in which it delights, prompts it to resort to hedges and banks for a certain period; but it always returns when our barns are filled, and ready for it.

The house fly (*musca carnaria*) is another creature that appears domesticated with us; in some seasons a very numerous, and always a very dirty inmate. It associates in our windows at times with a similar insect (*stomoxys calcitrans*) that loves to bask on stones and posts, and which is now biting my legs with the most teasing perseverance. But this phlebotomist has not the same attachment to our habitations, is a more solitary insect, and does not unite in those little social parties that circle for hours in a sober uniformity of flight below the ceilings of our chambers. Wherever man appears, this house fly is generally to be seen too: and instances are known, when islands have been taken possession of very far removed from the main land, that for a time no flies were visible, yet ere long these little domestic insects have made their appearance; neither natives of the isle, nor can we reasonably suppose them to have taken flight from a distant shore; but probably the offspring of parents that came with the stores in the vessel of the party.

We may have some few other instances of these apparent dependences of animals on man; yet, if we consider the relative situations of both, we shall find them existing, with very few exceptions, independent of him, and that he is more indebted to them for their services than they are for his protection and support. Man from the earliest periods began to subject the animal world to his dominion, and avail himself of its properties and powers to improve his own condition. As his wants or propensities occurred, he compelled to his aid such animals as he could subdue, or were adapted to his purposes. The chief objects for which we require the aid of animals, are for food, clothing, vigilance, and strength. Though the two former are highly essential to our comforts, they are not indispensable; the vegetable world supplies them in abundance to large portions of the inhabitants of the globe, and the companionable qualities, watchfulness, and swiftness of the dog might be dispensed with. It is the strength of animals that makes us sensible of our own weakness. By their power we build our dwellings, effect an intercourse with distant places, obtain much of our food, and the fuel of our hearths: a state of civilization requires, as an indispensable requisite, these things and others, rendering most manifest our obligations to the animal world. Animals were created before man; but some of them were apparently endowed with their useful and valuable properties for his comfort and assistance; for he had the dominion of them consigned to him, and was commissioned to subdue them. Having used their products for food and clothing, conjointly with the fruits and seeds of the vegetable world, and their bodies for the carriage of his burdens, after a long age of abstinence he began to feed on their flesh; and they have continued his faithful and assiduous servants, contented with their destiny, and submissive to his desires. He gives them food and shelter in payment of service, attending them with diligence and care: all this may be for his own emolument and pleasure, yet the well-being of the creature, had it continued wild, would not have required it: most of them live longer, and have more enjoyment, in a wild and unreclaimed state, than when domesticated with him. By art, and for profit, he has in many instances altered the very nature of the animal, and created ailments, rendering his cares and attentions necessary, which in a state of nature are not required. The lives of many of them, even when subjected to the best of treatment, are consumed with labour and fatigue; and when their unhappy destiny consigns them to the power of poverty and evil passions, what an accumulation of misery and suffering do these wretched creatures undergo! If these arguments have any foundation in truth, it will appear, that animals are not necessarily dependent on man, and generally derive no benefit from their intercourse and association with him; but that, in conformity with original appointment, they aid him to acquire the enjoyments and accomplish the necessities of civilized life. Yet there is one creature, that seems designed by its natural habits to be the servant and dependant of man; and of all that fall under his dominion, not one receives an equal portion of his care, or is more exempt from a life of exhaustion in his service. The dog is fed with him, housed, and caressed; associates with him in his pleasures, is identified with and enjoys them with his master; living with him, he acquires the high bearing and freedom of his lord; feels he is the companion and the friend; deports himself as a partaker of the importance and superiority, we might almost say of the sorrows and pleasures, of the man; is elated with praise, and abased by rebuke; submissive when corrected, and grateful when caressed: his anxiety and tremor when he has lost his master, and, with him, himself, is pitiable; when deserted by his lord, he becomes the most forlorn of animals, a never-failing victim to misery, famine, disease, and death. His ardour may excite him at times, until overpowered by fatigue; but he is not generally stimulated by pain or menace to attempts beyond his natural powers: view him in all his progress, his life will be found to be an easy, and frequently an enjoyable one; and though not exempt from the afflictions of age, yet his death, if anticipated, becomes a momentary evil. When in a native state, he is a wretched creature, a common beast of the wild, with no innate magnanimity, no acquired virtues; has no elevation, no character to maintain, but passes his days in contention and want, is base in disposition, meagre in body, a fugitive, and a coward.

The wheatear (*sylvia œnanthe*) frequents annually our open commons and stone quarries, and breeds there. I have seen it with nesting materials in its bill, and have had its eggs, though rarely, brought me. This bird visits England early in the spring, and continues with us till nearly the end of September, that is, during the entire breeding-season. Yet it is remarkable, notwithstanding its numbers, and the little concealment which its haunts afford, how rarely its nests are found. Its principal place of resort is the South Downs in Sussex; and it appears, from the accounts of the most experienced and credible persons of that county, from whom I have my information, that the females are performing their duties of incubation during the month of March; as at that time scarcely any but male birds are visible, of which hundreds are then flying about; while the females with their families appear early in May, and are captured afterwards in great numbers: yet the oldest shepherds have seldom seen their nest! But in fact no bird conceals its nest with more artifice than the wheatear; and in consequence of this circumstance, and the retired places in which it fixes its summer residence, very many of the young ones are produced. This summer (June 15, 1828) I appointed a boy to watch two hen birds to their retreat, and after some hours of vigilance he succeeded and gave me notice; one had made her nest deep in the crevice of a stone-quarry, so carefully hidden by projecting fragments as not to be observed from without until part of the rock was removed: her fabric was large and rudely constructed with dried bents, scraps of shreds, feathers, and rubbish, collected about the huts on the down, and contained four pale blue eggs, about the size of those of the sky-lark. The other bird had descended through the interstices of some rather large loose stones, as a mouse would have done, and then proceeded laterally to a hollow space in a bank against which the stones were laid; and so deep had she penetrated that many of the stones had to be removed before we could discover her treasure: as no appearances led to any suspicion of a nest, it would never have been detected but for our watchfulness. With us the wheatear stays only to hatch her brood. When this is effected, and the young sufficiently matured, it leaves us entirely, and by the middle of September not a bird is found on their summer stations. They probably retire to the uplands on the sea-coasts, as we hear of them as late as November in these places, where it is supposed they find some peculiar insect food, required by them in an adult state, and not found, or only sparingly, in their breeding-stations, in which the appropriate food of their young is probably more abundant. Thus united on the coasts, they can take their flight, when the wind or other circumstances favour their passage — all of them departing upon the approach of winter.

Partial as I am to the habits and all the concerns of the country, I regret to say that rural amusements, connected as they commonly are with the creatures about us, are frequently cruel; and that we often most inconsiderately, in our sports, are the cause of misery and suffering to such as nestle around our dwellings, or frequent our fields, which, from some particular cause or motive, become the object of pursuit. I say nothing of the birds known as game, as perhaps we cannot obtain them by less painful means than we are accustomed to inflict, and the pursuit is frequently conducive to recreation and health; but the sportsman's essaying his skill on the swallow race, that "skim the dimpled pool," or harmless glide along the flowery mead, when, if successful, he consigns whole nests of infant broods to famine and to death, is pitiable indeed! No injury, no meditated crime, was ever imputed to these birds; they free our dwellings from multitudes of insects; their unsuspicious confidence and familiarity with man merit protection, not punishment, from him. The sufferings of their broods, when the parents are destroyed, should excite humanity, and demand our forbearance. But the wheatear, in an unfortunate hour, has been called the English ortolan, and is pursued as a delicate morsel through all its inland haunts, when hatching and feeding its young — the only period in which it frequents our heaths. I execrate the practice as most cruel: their death evinces no skill in the gunner; their wretched bodies, when obtained, are useless, being embittered by the bruises of the shot, and unskilful operations of the picker and dresser. No, let the parental duties cease; and when the bird retires to its maritime downs, if doomed to suffer, the individual dies alone, and no starving broods perish with it. I supplicate from the youthful sportsman his consideration for these most innocent creatures, the summer wheatear and the swallow.

The eggs produced by the wheatear are uniform in colour and similar in shape; but the eggs of birds in general vary much, and are occasionally very puzzling to identify, when detached from their nests, as the colourings and markings differ greatly in the same species, and even nest. Those of one colour, like this wheatear's, retain it, with only shades of variation; but when there are blotchings or spots, these are at times very dissimilar, occasioned, in great measure, probably by the age of the bird; though this cannot account for the difference of those in an individual nest. None vary more than the eggs of the common sparrow. Those of marine birds, especially the guillemot (*colymbus troile*) are often so unlike each other, that it requires considerable practice to arrange them. The plumage of birds has probably never varied, but remains at this hour what it originally was: but whether these markings on the eggs have any connexion with the shadings on the feathers, it is difficult to determine; as we know that eggs entirely white will produce birds with a variety of plumage. The shell of the egg appears to be designed consists of carbonate and phosphate of lime, is to unite with the white of the egg, and form, during incubation, the feathers and bone of the future young ones; but as a large portion of this covering remains after the young are produced, its other object is to guard from injury the parts within. As far as I have observed, in eggs of one hue, the colouring matter resides in the calcareous part; but where there are markings, these are rather extraneous to it than mixed with it. The elegant blue that distinguishes the eggs of the fire-tail and the hedge-sparrow, though corroded away, is not destroyed by the muriatic acid. The blue calcareous coating of the thrush's egg is consumed; but the dark spots, like the markings upon the eggs of the yellowhammer, house-sparrow, magpie, &c., still preserve their stations on the film, though loosened and rendered mucilaginous by this rough process. Though this calcareous matter is partly taken up during incubation, the markings upon these eggs remain little injured, even to the last, and are almost as strongly defined as when the eggs are first laid. These circumstances seem to imply, that the colouring matter on the shells of eggs does not contribute to the various hues of the plumage, but, it is reasonable to conclude, are designed to answer some particular object, not obvious to us: for though the marks are so variable, yet the shadings and spottings of one species never wander so as to become exactly figured like those of another family, but preserve, year after year, a certain characteristic figuring. Few animal substances, in a recent state, contain more hepatic gas than an eggshell, as is manifest from the very offensive smell that proceeds from it when burned. A little of this is caused by the gluten that cements the calcareous matter, but the overpowering fetor comes from the inner membrane that lines the shell.

The amazing variety, aptitude, and efficiency that all the works of Providence present, though the keenest observation of man can discover but the utmost skirts of wisdom and divinity that invests them, yet in everything is this remarkable: let us take up the most common production before us, tossed about, used and forgotten, the egg of a bird, and without entering into the elaborate conformation of its parts, give it a passing thought. It is probable, and as far as observation goes, is true, that birds in a natural state lay a certain number of eggs, incubate and produce their species. Should an accident destroy the first store, a second portion may be produced, and in a few cases should circumstances injure these, a third, though usually decreasing in number each time, possibly by reason of the difficulty of bringing up full broods when immediate and most favourable season has passed; still the number is limited by the particular economy of nature. Yet it is a curious fact that the arts of domestication can in many instances subvert this limitation of nature, seeming to bring about a strength of constitution, an animal efficacy that is not usually bestowed upon the creature. The common hen as far as we can learn, is like others of the gallinaceous birds, laying from ten to twenty eggs, and cease providing more, but in a domestic state this bird will furnish eggs in many instances for the larger portion of the year. A neighbour has one which with the omission of fourteen days, has produced an egg daily for three years, another has furnished two hundred and thirty in the year, and two hundred in the same period is no uncommon number. It then becomes a subject of some wonder, how the animal material can furnish without exhaustion this excess of substance! The hard external coating consists of several substances, but having extracted these (phosphate, water, gluten), about 78 grains of pure carbonate of lime remains. A creature then laying two hundred eggs, must have elaborated nearly 2 ¾ lbs of lime from its substance annually to cover its eggs, from a resource, and by a process, and with a rapidity of which we have little intelligence. For this calcareous covering is obtained in the ovi-duct and probably by the agency of the albumen; yet how this glareous fluid secretes it is not obvious, and possibly an hour before its exclusion, it possessed only a filmy envelope, though now surrounded with a beautiful incrustation of Limestone, cemented by animal gluten. Though the creature's bones are in part composed of like calcareous matter, it is not probable that they furnished this supply, which would reduce the firmness requisite for support of the frame. From whence then is this reservoir of limestone, the machinery that elaborates a grain of barley into a stone? The abundance of substance that animals can produce is instanced in several of the spiders, silkworms, and others, having magazines supplied or augmented by the agency of food with requisite matter for the emergency; some of the lactiferous animals as the cow, furnish remarkable portions of substance from the system, food in many cases bringing on that state of repletion called fat, but in others supplying milk; yet if we consider the quantity of food a cow consumes as daily supply, its produce is not in proportion to that of a hen. The contents of a common sized egg, albumen and yoke only, may weigh about eleven and a half drams, so that an animal as above, elaborates from its substance an annual supply equal to more than nine pounds, to say nothing of the lime, yet apparently without consequent debility, whereas in a natural state the produce would have been about nine ounces! And yet in many cases the proportion or allowance of food obtained in a domestic state is but little if anything but what would have been acquired in that of a wild one. The rapid manner in which nature can effect a purpose when required, is obvious to every observer in endless instances; but we had a common circumstance to notice this summer connected with the incidents related above, that was a striking illustration of it. A water-hen (*Gallinula chloropus*) had been sitting on nine eggs in one of my ponds, and when near hatching the nest was destroyed, she immediately prepared another, and in ten days, as nearly as we could ascertain, she furnished this with seven eggs, and eventually brought out the brood. It seems to be considered that whatever number of rudiments of eggs may be in the ovarium, that a certain number only are detached into the ovi-duct, where they undergo the requisite preparations fitting for exclusion. This egg tube being cleared of its contents, the ovarium remains in a quiescent state until the ensuing season; but it appears that upon this necessity, an exciting action took place, the ovarium commenced again to deposit its stone, perfecting in an unusual rapid manner all the requirements, which under ordinary circumstances would not have taken place, or have been delayed to a distant period.

This gallinule, though without webs to its feet, is particularly fitted for the confined waters in which its station is assigned. It swims in the open parts sufficiently well for its habits of life, feeding as it proceeds, and it can assume a buoyancy when required to run by aid of its wings along the surface to concealment. Even the young ones, whose wings are too short to afford much assistance, manage this running upon the water well enough to escape to their harbour. But when the pool is furnished with floating aquatic plants, which is commonly the case, particularly the broad leaved plantain, the mother when alarmed with all her little woolly sooty tribe can course along the surface with a facility equal to what they could do upon the turf. By their fleetness, they escape from injury when from the narrow bounds of their station without this rapidity it would be difficult to accomplish, and manifests among multitudes of examples the peculiar fitness of creatures for their assigned stations, not the result of habit or necessity, but given as an endowment to preserve life and fulfil the ends of their being. Place the water-hen and her brood upon the open lake, the broad river with the duck or the teal, the chances of escape from injury would be greatly narrowed, probably not effected.

The superstitions and fancies of persons, though we may often contemn them, are yet at times deserving of notice, being occasionally to be traced to some former-received belief or national custom, and perhaps, when charactered by emblems or ceremonies, may be considered as certainly originating from the tenets of some sect or popular observance; the partiality manifested by the English in general for flowers and horticultural pursuits is recently, from a sentence in Pliny (Nat. Hist. XIV., chap, iv.), supposed to have been acquired from their Roman conquerors; and probably many other attachments and practices, though obscured and perverted by time, have been retained from the example of some of the various nations who have ruled in our island. Bird-nesting boys, I suppose, are yet to be met with in many a rural village, being a habit from immemorial antiquity, pursued with eagerness in contention with their fellows for numbers and rarity; but that accomplished, like so many of our pursuits in after life, the pleasure ceases when rivalry is no more: but, regarding these birds' eggs, we have a very foolish superstition here; the boys may take them unrestrained, but their mothers so dislike their being kept in the house, that they usually break them; their presence may be tolerated for a few days, but by the ensuing Sunday are frequently destroyed, under the idea that they bring bad luck, or prevent the coming of good fortune, as if in some way offensive to the domestic deity of the hearth: having occasionally enquired for these plunders of our small birds at the cottages, to supply some deficiencies in a collection, I have found so general a prepossession against retaining them, as in most cases to fail of success.

The kite (*falco milvus*) is one of our rarest birds. We see it occasionally, in its progress to other parts, sailing along sedately on its way; but it never visits us. Our copses present it with no enticing harbourage, and our culture scares it. In former years I was intimately acquainted with this bird; but its numbers seem greatly on the decline, having been destroyed, or driven away to lonely places, or to the most extensive woodlands. In the breeding season it will at times approach near the outskirts of villages, seeking materials for its nest; but in general it avoids the haunts of man. It is the finest native bird that we possess, and all its deportment partakes of a dignity peculiar to itself, well becoming a denizen of the forest or the park; for though we see it sometimes in company with the buzzard, it is never to be mistaken for this clumsy bird, which will escape from the limb of some tree, with a confused and hurried flight, indicative of fear; while the kite moves steadily from the summit of the loftiest oak, the scathed crest of the highest poplar, or the most elevated ash — circles round and round, sedate and calm, and then leaves us. I can confusedly remember a very extraordinary capture of these birds when I was a boy. Roosting one winter evening on some very lofty elms, a fog came on during the night, which froze early in the morning, and fastened the feet of the poor kites so firmly to the boughs, that some adventurous youths brought down, I think, fifteen of them, so secured! Singular as the capture was, the assemblage of so large a number was not less so, it being in general a solitary bird, or associating only in pairs.

The blackcap (*motacilla atracapilla*) is our constant visitor, but very uncertain in its numbers, as it fully participates in all the casualties of our migratory tribes; not by any great diminution probably in its winter residence, but by loss in its transits of autumn or spring. We have years when every little copse resounds with harmony; at other periods only a few solitary songsters are to be heard; and the blackcap is the principal performer in the band of our domestic vocalists. In the scale of music it is the third for mellowness, and the third perhaps, too, for execution and compass. As this melody, however, continues only during the period of incubation, we hear it but for a short time; for this bird wastes no time in amusements, appearing to be in great haste to accomplish the object of its visit, and to depart. Thus, immediately upon its arrival, we observe it surveying and inspecting places fitting for nidification, and commencing a nest; but so careful and suspicious is it, that several are often abandoned before finished, from some apprehension or caprice: any intrusion is jealously noticed; and during the whole period of sitting and rearing its young, it is timid and restless. I have observed that both birds will occasionally perform the office of incubation.

It seems to live entirely by choice on fruits; and as soon as the brood can remove, it visits our gardens, feeding with delight and almost insatiable appetite on the currant and the raspberry; and so much is it engaged when at this banquet, that it suffers itself to be looked at, and forgets for the moment its usual timidity: but its natural shyness never leaves it entirely; and though it remains in our gardens or orchards as long as any of its favourite fruits continue, it avoids observation as much as possible, and hides itself in the foliage from all familiarity or confidence. This exceeding dislike of man is very extraordinary. Larger or more important birds might have an instinctive fear of violence, but this creature is too small and insignificant to have ever experienced or to apprehend injuries from him. It may arise from a long residence in wilds and solitary places, seldom visited by human beings during those eight or nine months when it is absent from us, so that man becomes an unknown creature, and injury is suspected. Our native small birds, that reside all the year with us, and see us often, though they may retire at our near approach, do not exhibit such shyness and avoidance as several of our migrating birds. The grey flycatcher, and the swallow tribe, which seek their food, we conclude, all the year near the dwellings of man, where most abundantly found, manifest familiarity with us rather than dislike, are accustomed to the sight of human beings, and do not fear them; but whatever may be the cause that influences the precipitate retreat of certain birds, we note the original mandate, and see that the "fear of us, and the dread of us," are still in operation with many of these little "fowls of the air," that would never receive harm from our hands. The blackcap finishes its feast here with the jargonel pear, when it can meet with it, then leaves us for other fruits and milder climes.

"And the fear of you, and the dread of you, shall be upon every beast of the earth, and upon every fowl of the air, and upon all that moveth upon the earth." This vesture of universal dread, which was to envelop man, though appointed from the beginning of time, has never been removed, but most signally and remarkably attaches to him still. It was ordained to be so; and so it is. In some few instances only does this awe of man subside: in extreme cases of want, for individual preservation, or when protection is required. In such cases, the fear or sensibility of pain, love of life, or a paramount duty, becomes the stronger principle, annihilating the weaker; and the dread of man's supremacy is no more. The weakest, the very insect, then assails him, and at times becomes the victor. Does any conceivable or visible cause exist from which this awe can proceed? Does "his sublime countenance, contemplative of the heavens," the image that he bears, or his deportment, afford any ascendant influence productive of this impression? In bodily power he is more weak and obnoxious to injury than many that shrink from a contest with him; his natural arms and means of protection are inferior often to those of the beings which he subdues; yet, from an undefinable cause, he is omnipotent over all. Terror in man most commonly arises from a knowledge of power, apprehension of ills from accident, or fear of the evil inclinations of another. What the fowls of the air or the beasts of the field perceive, or are impressed with, we know not; but none of these causes can exist in a brute mind without intelligence or experience. These are the reflections of a thoughtful hour. The cause, "though a man labour to seek out, yet shall he not find it; and though a man think to know it, yet shall he not be able." But the contemplation is not wholly an unworthy occupation of time. All ages, all people, must have perceived the admitted power and universal dread occasioned by the presence of man; but no reason, no motive, could have been assigned for it; but in these days, by revelation, we know the cause, have impressed upon our minds the immutable truth of that Being which ordained, and of that volume which has proclaimed his mandate to us. But man has the power assigned him of calling to his aid a visible object of dread, confided to him from the earliest periods; and he alone, of all created beings, has the agency of this terror. All the inferior orders have a fear of it, and flee from it, even when its effects could never have been known or experienced, but which appears to be innate and inseparable from all. Man alone has the knowledge, the means, of calling heat into action; and though warmth is the delight, and essential to the being of most, yet, rouse it into active operation, producing fire, and terror and flight succeed enjoyment and rest: it deters the approach of the most ferocious, and man and his charge abide unharmed when surrounded by the terror he has raised. In addition to the many characters given as a definition of man, we might call him a fire-producing creature.

The end of our summer months, and the autumnal season, afford us frequently the best periods for observing some of our occasional visiting birds. Upon their first arrival, and for a time afterwards, their notes announce their presence; but they are not always to be seen with satisfaction, and scattered in retired places, or occupied in the business of incubation, when they are particularly wary and suspicious, they are but casually noticed: but in the times above stated, our gardens, shrubberies, and orchards, become their resort, seeking for the fruits usually produced in those places. And, first, the pettychaps, with all her matured brood, is certain to be found, feeding voraciously upon our cultivated berries, or mining a hole in the fig or jargonel pear; and so intent are they upon this occupation, that they will permit a reasonable examination of their form and actions, but at other periods it is difficult to approach them. The blackcap discontentedly flits about our inclosures and thickets all the summer through, building her nest or tending her young; the fine clear harmony of the male bird resounding in the morning from the brake, yet, timid and alarmed, he ceases and hides himself if we approach: but now he introduces all his progeny to our banquet; cautious still, we can yet observe his actions, and easily distinguish the black or brown heads of the sexes, as they are occupied beneath the foliage of an Antwerp raspberry. The white-throats now, too, leave their hedges, and all their insect food, which for months had been their only supply, and in the thick covert of the gooseberry extract with great dexterity the pulp of the fruit, or strip the currant of its berry. The elegant, slender form of the female, her snowy throat and silvery stomach, render her very conspicuous as she scuttles away to hide herself in the bush: her plain, brown-backed mate seems rather less timid, but yet carefully avoids all symptoms of familiarity. Other doubtful little birds likewise appear, and are gone, several of which, however, are probably the young of ascertained species. And here the little willow-wren is often to be seen: he comes in company with his travelling friends, not as a partaker of their plunder, appearing never to abandon his appetite for insect food: the species may change with the season, but still it is animal: he glides about our rows of peas, peeps under the leaves

ANIMAL INSTINCT — EARLY AWAKENING OF BIRDS

of fruit-trees for aphides and moths, continuing this harmless pursuit until the cold mornings of autumn drive him to milder regions. All these fruit-eating birds seem to have a very discriminating taste and a decided preference for the richest sorts — the sweetest variety of the gooseberry or the currant always being selected; and when they are consumed, less saccharine dainties are submitted to: but the hedge blackberry of the season our little foreign connoisseurs disdain to feed on, leaving it for the humbler-appetited natives — they are away to sunnier regions and more grateful food.

June 14. — I was much pleased this day by detecting the stratagems of a common wren to conceal its nest from observation. It had formed a hollow space in the thatch, on the inside of my cow-shed, in which it had placed its nest by the side of a rafter, and finished it with its usual neatness; but lest the orifice of its cell should engage attention, it had negligently hung a ragged piece of moss on the straw-work, concealing the entrance, and apparently proceeding from the rafter; and so perfect was the deception, that I should not have noticed it, though tolerably observant of such things, had not the bird betrayed her secret, and darted out. Now from what operative cause did this stratagem proceed? Habit it was not; — it seemed like an after-thought; — danger was perceived, and the contrivance which a contemplative being would have provided was resorted to. The limits of instinct we cannot define[30]: it appeared the reflection of reason. This procedure may be judged, perhaps, a trifling event to notice; but the ways and motives of creatures are so little understood, that any evidence which may assist our research should not be rejected. Call their actions as we may, they have the effect of reason; and loving all the manners and operations of these directed beings, I have noted this, simple as it may be.

At one period of my life, being an early waker and riser, my attention was frequently drawn "to songs of earliest birds;" and I always observed that these creatures appeared abroad at very different periods as the light advanced. The rook is, perhaps, the first to salute the opening morn; but this bird seems rather to rest than to sleep. Always vigilant, the least alarm after retirement

30 I know not any definition of what we term "animal instinct" more comprehensive and accordant with truth than the following, given in the "Elements of Entomology" by Messrs. Kirby and Spence. "Without pretending to give a logical definition of it (instinct), which, while we are ignorant of the essence of reason, is impossible, we may call the instincts of animals those unknown faculties implanted in their constitutions by the Creator, by which, independent of instruction, observation, or experience, and without a knowledge of the end in view, they are impelled to the performance of certain actions tending to the well-being of the individual, and preservation of the species."

rouses instantly the whole assemblage, not successively, but collectively. It is appointed to be a ready mover. Its principal food is worms, which feed and crawl upon the humid surface of the ground in the dusk, and retire before the light of day; and, roosting higher than other birds, the first rays of the sun, as they peep from the horizon, become visible to it. The restless, inquisitive robin now is seen too. This is the last bird that retires in the evening, being frequently flitting about when the owl and bat are visible, and awakes so soon in the morning, that little rest seems required by it. Its fine large eyes are fitted to receive all, even the weakest rays of light that appear. The worm is its food too, and few that move upon the surface escape its notice. The cheerful melody of the wren is the next we hear, as it bustles from its ivied roost; and we note its gratulation to the young-eyed day, when twilight almost hides the little minstrel from our sight. The sparrow roosts in holes, and under the eaves of the rick or shed, where the light does not so soon enter, and hence is rather a tardy mover; but it is always ready for food, and seems to listen to what is going forward. We see it now peeping from its penthouse, inquisitively surveying the land; and, should provision be obtainable, it immediately descends upon it without any scruple, and makes itself a welcome guest with all. It retires early to rest. The blackbird quits its leafy roost in the ivied ash; its "chink, chink," is heard in the hedge; and, mounting on some neighbouring oak, with mellow, sober voice it gratulates the coming day. "The plainsong cuckoo grey" from some tall tree now tells its tale. The lark is in the air, the marten "twitters from her straw-built shed," all the choristers are tuning in the grove; and amid such tokens of awakening pleasure it becomes difficult to note priority of voice. These are the matin voices of the summer season; in winter a cheerless chirp, or a hungry twit, is all we hear; the families of voice are away, or silent; we have little to note, and perhaps as little inclination to observe.

During no portion of the day can the general operations of nature be more satisfactorily observed than in the early morning. Rosy June — the very thought of an early summer's morning in the country, like enchantment, gives action to the current of our blood, and seems to breathe through our veins a stream of health and enjoyment! All things appear fresh and unsoiled; the little birds, animated and gratulous, are frisking about the sprays; others, proceeding to their morning's meal, or occupied in the callings of their nature, give utterance by every variety of voice to the pleasures that they feel: the world has not yet called us; and with faculties unworn, we unite with them, partake of this general hilarity and joy, feel disposed to be happy, and enjoy the blessings around us: the very air itself, as yet uninhaled by any, circulates about us replete with vitality, conveying more than its usual portion of sustenance and health, "and man goeth forth unto his labour." Night-feeding creatures, feeling the freshness of light, and the coming day, are all upon the move, retiring from danger and observation; and we can note them now unhidden in their lairs, unconcealed beneath the foliage in the hedge: the very vegetation, bathed in dew and moisture, full fed, partakes of this early morning joy and health, and every creeping thing is refreshed and satisfied. As day advances, it changes all; and of these happy beings of the early hour, part are away, and we must seek them; others are oppressed, silent, listless; the vegetable, no longer lucid with dew, and despoiled of all the little gems that glittered from every serrature of its leaf, seems pensive at the loss. When blessed with health, having peace, innocence, and content, as inmates of the mind, perhaps the most enjoyable hours of life may be found in an early summer's morning.

Oct. 9. — A brilliant morning! warm, without oppression; exhilarating, without chilling. Imagination cannot surely conceive, or caprice wish for an atmospheric temperature more delightful than what this day affords; having mingled with it just that portion of vital air which brisks up animality, without consuming the sustenance of life; satisfying the body with health, and filling the heart with gratitude. Fine threads of gossamer float lazily along the air, marking by this peculiar feature the autumn of our year. On our commons, and about our thistly hedge-rows, flocks of goldfinches (*fringilla carduelis*), the united produce of the summer months, are sporting and glistening in the sunny beam, scattering all over the turf the down of the thistle, as they pick out the seed for their food. But this beautiful native has only a few short weeks in which it will have liberty to enjoy society and life. Our bird-catchers will soon entrap it; and of those that escape their toils, few will survive to the spring, should our winter prove a severe one. Long as I have noticed this bird, it has appeared to me that it never makes any plants generally its food, except those of the syngenesia[31] class, and on these it diets nearly the whole year. In the spring season it picks out the seeds from the fir cones. During the winter months it very frequently visits our gardens, feeding on the seeds of the groundsel (*senecio vulgaris*), which chiefly abounds in cultivated places, and vegetates there throughout the coldest seasons. This, however, is a humble plant; and when covered by the snow, the poor birds are half famished for want. We then see them striving to satisfy their hunger by picking some solitary green head of the plant remaining above the frozen snow; and so tame, that they will suffer a very near approach before they take flight. As the frost continues, our little

---

31 Goldfinches occasionally pick out the seeds of the field scabious, the florets of which, united in one head, present an appearance not unlike the syngenesious plants.

garden visitors diminish daily, and by spring only a few pairs remain of all the flocks of autumn. Yet it is very remarkable, notwithstanding this natural predilection, how readily this bird conforms to a perfect change in its diet, and in all the habits of its life. Most of our little songsters, when captured as old birds, become in confinement sullen and dispirited; want of exercise, and of particular kinds of food, and their changes alter the quality of the fluids: they become fattened, and indisposed to action by repletion; fits and ailments ensue, and they mope and die. But I have known our goldfinch, immediately after its capture, commence feeding on its canary or hempseed — food it could never have tasted before; nibble his sugar in the wires like an enjoyment it had been accustomed to, frisk round its cage, and dress its plumage, without manifesting the least apparent regret for the loss of companions or of liberty. Harmless to the labours or the prospects of us lords of the creation, as so many of our small birds are, we have none less chargeable with the commission of injury than the goldfinch; yet its blameless, innocent life does not exempt it from harm. Its beauty, its melody, and its early reconciliation to confinement, rendering it a desirable companion, it is captured to cheer us with its manners and its voice, in airs and regions very different from its native thistly downs, and apple-blossom bowers.

The goldfinch is by no means to be considered as a regular domestic bird, there are five months in the year, unless driven by snow to cultivated lands, that he is only occasionally seen near dwellings, keeping about thistly commons and downs, and places in the fields where the groundsel vegetates, feeding in little animated flocks near each other. Very early in April, as the season of incubation advances, they pair, immediately abandoning their distant abodes, seeking the neighbourhood of houses and villages, to secure themselves probably from injuries they seem at other seasons unmindful of; these situations, the smallness of the fabric, and its very cautious concealment, affording apparently the desired safeguard. Trees by the sides of public roads are a favourite station for these nests, undeterred by noise or passing objects, and when the foliage has been stripped away by autumnal frosts, these elegant little constructions, still firmly united, will frequently be seen hanging over the much frequented path, or the dusty turnpike.

Though goldfinches in general feeds far away from our habitations, yet in cold seasons as the evening advances, we may often perceive them assembled in our flocks in gardens, preparing to pass the night upon the sprays of some fir-tree, or any other tall evergreen, that will afford them shelter from the chilling atmosphere; but when morning comes, they return to feed on their thistly downs.

THE TREE CREEPER – THE YELLOW WAGTAIL

The tree creeper (*certhia familiaris*) is as little observed as any common bird we possess. A retired inhabitant of woods and groves, and not in any manner conspicuous for voice or plumage, it passes its days with us, creating scarcely any notice or attention. Its small size, and the manner in which it procures its food, both tend to secrete him from sight. It feeds entirely on small insects, which it seeks between the crevices in the bark of trees, or under the mosses and lichens that invest their limbs. In these pursuits its actions are more like those of a mouse than of a bird, darting like a great moth from tree to tree, uttering a faint trilling sound as it fixes on their boles, running round them in a spiral direction, when with repeated wriggles having gained the summit, it darts to another, and commences again; and so intent is it on the object of pursuit, and unsuspicious of harm, that I have seen it swept from the tree with a stick. Mr. Pennant thinks that it retires into milder regions upon the advance of winter; but many certainly remain with us. In the early part of the spring, when food is comparatively scarce in the woods, it will frequent the mossy trees in our orchards and gardens; but after a very short examination of them, is away to its usual retirements, seeking no familiarity with us, notwithstanding the social epithet it has obtained. This little creature is observed in no great numbers; yet its actions and manners seem to be such as would tend to its increase. The female lays eight or nine eggs: it roosts securely in the holes of large trees; and from its manner of feeding, and the places it inhabits, it can scarcely be destroyed by birds of prey; yet, from some counteracting cause, our little certhia, instead of increasing, apparently becomes a scarcer bird. The limits that are appointed to the increase of all the inferior orders of creation are very worthy of remark. There may be periods when a great augmentation of individual species takes place; but this circumstance is local, or temporary, and future numbers do not result from it. Some motive for the increase, no doubt, existed; but, the object being accomplished, it ceases; and apparent events, or imperceptible causes, reduce the profusion of the race, so that certain numbers only continue. This little tree-creeper, though always active, seems to possess most animation and restlessness in the autumnal months.

The yellow wagtail (*motacilla flava*) is so regularly seen with us in his season, as to be quite a common bird, breeding in our fields; yet, generally observed as he is, he always invites our attention, by his graceful form and brilliant plumage, either actively running in our path, or sporting in the pastures with that animation and ease so remarkable in all this family, that we may justly distinguish them as the gentles of our fields. With manners and habits similar to the common grey ones, yet there seems to be but little intimate association between the species; and though they are occasionally intermixed, we most commonly observe them feeding by themselves, and frolicking with their own particular race. In autumn, when their broods are united with them, they assemble in large parties towards the evening preparatory to their nightly roost, selecting low-spreading bushes hanging over the pool, or as near the water as they can, and thus become secured from capture by nocturnal vermin. Being in full beauty at this time, the fine yellow breasts of the male birds render them very conspicuous as they glance about the dry bents of the pasture. Autumn advancing, we lose these flights; but now and then a single bird will appear in one of those occasional bright sunny days that even winter will produce, looking like some deserted straggler who has lost its passage, or from some other cause remaining with us, chasing the gnat on the margin of the sheltered pool, and then, when the sunny ray passes away, he departs with it, is hidden we know not where, supported by means we are not acquainted with, till another partial gleam allures him from retirement. In April, the flights once more appear with all the fine feather and freshness of autumnal birds, running about the furrows in arable fields, and catching the insects disturbed by the plough in its progress. Soon building their nest, and attending their families, they become bleached by the sun and rain of the season, and remain shabby for weeks. Though they may follow the course of the swallow and other migrating birds, yet their peculiar manner of flight seems to preclude long-continued exertion; not sailing and poising in air like the hirundines and others, but proceeding by jerks, by risings and sinkings, which at every pause require muscular action to set them in progress anew, which, for any length of time, could hardly be continued. It is probable that their migrations are not very remote. The mode of life assigned to these creatures requires great activity of body; for living solely upon insects and winged animals, they are constantly capturing or pursuing; and their length of tail, which is perpetually in motion, seems to aid and balance the operations of the body. In the evening, when the winged creatures are at rest, or, from the state of the atmosphere, in repose, the wagtail resorts to the pastures, feeding under the very bodies and noses of the cattle, who now become the starters of his game, which, moving from the animal, are captured by the bird. Being drowsy, and settling almost as soon as disturbed, their prey would escape, was the wagtail less nimble in his actions— for he does not appear to perceive the insect, except when it moves. How differently formed is this bird and the grey fly-catcher! Though both are solely insectivorous, yet they secure their prey by very distinct means, the latter seldom capturing on the ground or using his legs in pursuit; the other uses actively his slender legs and extended wings to aid him. The swallow race, again, feed unlike them both, and haunting the pool, the stream, the mead, or the higher regions of the air, which his fraternity possess as a peculiar domain, satisfy their wants in peace, without collision or contention for the object. The pied Wagtail will occasionally capture its prey in a peculiar manner, for rising above the bents in the pasture, it will hover backwards and forwards spreading its tail to the entire display of the white side feathers, possibly for the purpose of beating up the couching insects, as eagles are said to act by game, which it then gets sight of, settles and obtains as a reward for its ingenuity.

Admirably adapted to the requirements of each creature as their dispositions and institutions are known to be, yet their peculiar modes of dieting, or inclination for particular food, and formation of the organs that digest it, should not be utterly unheeded, because by these appointments of Omniscience, abundance is produced for every race of created things in all places, without variance or unfitting exertions to procure it. Could we unite into one district a human being from every square mile upon the surface of the globe, unshackled by bigotry or the tenets of any faith, they probably, without reluctance, having the means, might feed upon and be nourished by one natural diet — we will say the flesh of the ox, with potatoes or rice — but this is by no means the case with the inferior animals. Most of them, having different conformations and inclinations, are supported by variety of diet; by which means every station and place is made an abode, and maintains its inhabitants, for the "Creator hath opened his hand, and filled all things living with plenteousness."

As a brief note, not a disquisition, upon the subject is designed, we will pass over the habits and dispositions of beasts and insects, strongly charactered as they are, and only instance a few of our land-birds, as affording the most familiar instances; and we shall find that it is not the genera only, but the individuals which compose them in many instances, that are supported by different aliment. And first, those birds which we denominate as Rapacious, such as falcons, hawks, owls, live upon animal food which they capture, kill, and devour; abstaining, unless stimulated by necessity, from creatures they may find dead. Then come the pies: of these, the raven and crow likewise eat animal food, but it is generally such as has been killed by violence or ceased to exist, only in cases of want[32] killing for themselves. The rook, the daw, the

32 The crow in the spring, when food is difficult of attainment, will kill young pigeons; and the magpie having young ones, captures the new hatches of our domestic poultry: but these are cases of necessity rather than habit. The raven has

magpie, consume worms, grubs, and are not addicted, except from hunger, to eating other animal matters. The two first feed at times in society; the latter associates with neither, but feeds in places remote from such as are frequented by them. The jay too eats grubs and such things, but seeks them out under hedges, in coverts and places which others of his kind abandon to him. The cuckoo seems principally to live upon the eggs of birds, with a few insects and larvae occasionally; the wryneck upon emmets, from heaps under hedges near concealment — the woodpeckers upon insects found upon trees; and when they seek for the emmet, they prefer the ant-heaps of commons and open places; — the halcyon upon small fishes: — thus all these creatures, even when they require similar aliment, diet at their separate boards. Of the Gallinaceous birds, the wood-grouse is supported by the young shoots of the pine in his forests; but the black and red grouse live upon berries found on the moor, the seeds and tops of the heath; the partridge upon seeds in the field, blades of grass or of corn; the pheasant upon mast, acorns, berries from the hedge or the brake. The bustard is content to live upon worms alone, found in early morning upon downs and wide extended plains, where none dispute his right or compete with him, but one species of plover. The doves make their principal meals in open fields, upon green herbage and seeds. The stare again feeds upon worms and insects, but in places remote from the bustard, nor does he contend with the rook, or the daw, but takes his meat and is away.

The Passarine birds, indeed, are remarkably dissimilar in their manner of feeding. The missel-thrush will have berries from the misseltoe, or seeks for insects and slugs in wild and open places, the heath or the down. The song-thrush makes his meal from the snail on the bank, or worm from the paddock; but the blackbird, though associating with him, leaves the snails, contenting himself with worms from the hedgeside, or berries from the briar or the bush. The fieldfare consumes worms in the mead or haws from the hedge. The crossbill will have seeds from the apple, or cone of the fir — the greenfinch, seeds from the uplands, or door of barn, or rickyard. The bunting is peculiarly gifted with a bony knob in the roof of his bill, upon which he breaks down the hard seeds he is destined to feed upon. The bullfinch selects buds from trees and bushes. The goldfinch is nurtured by thistle seeds, or those of other syngenesious plants. Sparrows feed promiscuously. Linnets shell out seeds from the cherlock, or the rape, or the furze on the common.

One lark will feed in the corn-field, another in the mead, another in the

> a decided inclination for the eyes of creatures, and finding lambs in a weak state, immediately plucks them out, and when the animal is recently dead, commences his depredations upon these parts.

THE COMMON BUNTING 153

woodlands — one titmouse upon insects frequenting the alder and willow; some upon those which are hidden under mosses, and lichens on large trees; a third upon coleopterous creatures, secreted in the hedge-row and the coppice. The grey wagtail finds food with us all the year; but the yellow one must seek it in other regions. The nightingale diets upon a peculiar grub, and when that is not found in the state he prefers, he departs. The domestic swallow feeds round our houses or in the meadow; but the bank swallow never comes near us, chases his food beneath the crag, and along the stream. The swift prefers the higher ranges of the air, dieting upon the flies that mount into those regions. The goatsucker does not notice the creatures of the day, capturing the moths and dors of the night. The wheatear feeds only upon such insects as he finds upon fallow lands, the down or the heath; and thus almost every individual might be characterized by some propensity of appetite, by some mode or place of feeding; and hence individuals are found as tenants of the homestead, the wild, the stream, the air, rock, down, and grove — in every place finding plenty, and fulfilling their destination without rivalry or contention: nor, perhaps, is there any race of creatures that associates more innocently, or passes their lives more free from bickering and strife, than these our land-birds do, persevering, from period to period, with undeviating habits and propensities, manifesting an original appointment and fixed design of Providence, whose bounteous table, wherever we look round, is spread for all, and good things meted out to each by justice, weight, and measure.

I am neither inclined to seek after, nor desirous of detailing, the little annoyances that these wildings of nature, in their hard struggles for existence, may occasionally produce; being fully persuaded that the petty injuries we sometimes sustain from birds are at others fully compensated by their services. We too often, perhaps, notice the former, while the latter are remote, or not obtrusive. I was this day (Jan. 25) led to reflect upon the extensive injury that might be produced by the agency of a very insignificant instrument, in observing the operations of the common bunting (*emberiza miliaris*); a bird that seems to live principally, if not entirely, upon seeds, and has its mandibles constructed in a very peculiar manner, to aid this established appointment of its life. In the winter season it will frequent the stacks in the farmyard, in company with others, to feed upon any corn that may be found scattered about; but, little inclined to any association with man, it prefers those situations which are most lonely and distant from the village. It could hardly be supposed that this bird, not larger than a lark, is capable of doing serious injury; yet I this morning witnessed a rick of barley, standing in a detached field, entirely stripped of its thatching, which this bunting effected by seizing the end of the straw, and deliberately drawing it out, to search for any grain the ear might yet contain; the base of the rick being entirely surrounded by the straw, one end resting on the ground, the other against the mow, as it slid down from the summit, and regularly placed as if by the hand; and so completely was the thatching pulled off, that the immediate removal of the corn became necessary. The sparrow and other birds burrow into the stack, and pilfer the corn; but the deliberate operation of unroofing the edifice appears to be the habit of this bunting alone.

Every one probably has observed the vagaries of several of our male birds in the spring of the year, as extraordinary actions exhibited to their mates. Of these few can be more conspicuous than those of the green finch (*loxia chloris*). He will rise from the spray or bush, without any obvious motive, and like a wounded creature in pain, stretch is wings to the utmost, waving and balancing himself here and there, in a kind of forlorn flight apparently ready to drop and expire; then having gone through these devious motions for perhaps a minute, resume his natural mode of flight, and return to his station.

Old simplicities, tokens of wind and weather, and the plain observances of rural life, are every where waning fast to decay. Some of them may have been fond conceits; but they accorded with the ordinary manners of the common people, and marked times, seasons, and things, with sufficient truth for those who had faith in them. Little as we retain of these obsolete fancies, we have not quite abandoned them all; and there are yet found among our peasants, a few who mark the blooming of the large white lily (*lilium candidum*), and think that the number of its blossoms on a stem will indicate the price of wheat by the bushel for the ensuing year, each blossom equivalent to a shilling. We expect a sunny day, too, when the pimpernel (*anagallis arvensis*) fully expands its blossoms; a dubious, or a moist one when they are closed. In this belief, however, we have the sanction of some antiquity to support us; Sir F. Bacon records it; Gerarde notes it as a common opinion entertained by country people above two centuries ago; and I must not withhold my own faith in its veracity, but say that I believe this pretty little flower to afford more certain indication of dryness or moisture in the air, than any of our hygrometers do. But if these be fallible criterions, we will notice another, that seldom deceives us. The approach of a sleety snow-storm, following a deceitful gleam in spring, is always announced to us by the loud untuneful voice of the missel-thrush (*turdus viscivorus*), as it takes its stand on some tall tree, like an enchanter calling up the gale. It seems to have no song, no voice, but this harsh predictive note; and it in great measure ceases with the storms of spring. We hear it occasionally in autumn, but its voice is not then prognostic of any change of weather. The missel-thrush is a wild and wary bird, keeping generally in open fields and commons, heaths, and unfrequented places, feeding upon worms and insects. In severe weather it approaches our plantations and shrubberies, to feed on the berry of the misseltoe, the ivy, or the scarlet fruit of the holly or the yew; and should the redwing or the fieldfare presume to partake of these with it, we are sure to hear its voice in clattering and contention with the intruders, until it drives them from the place, though it watches and attends, notwithstanding, to its own safety. In April it begins to prepare its nest. This is large and so openly placed, as would, if built in the copse, infallibly expose it to the plunder of the magpie and the crow, which at this season prey upon the eggs of every nest they can find. To avoid this evil, it resorts to our gardens and our orchards, seeking protection from man, near whose haunts those rapacious plunderers are careful of approaching: yet they will at times attempt to seize upon its eggs even there, when the thrush attacks them and drives them away with a hawk-like fury; and the noisy warfare of the contending parties occasionally draws our attention to them. The call of the young birds to their parents for food is unusually disagreeable, and reminds us of the croak of a frog. The brood being reared, it becomes again a shy and wild creature, abandons our homesteads, and returns to its solitudes and heaths.

The extraordinary change of character which many creatures exhibit, from timidity to boldness and rage, from stupidity to art and stratagem, for the preservation of a helpless offspring, seems to be an established ordination of Providence, actuating in various degrees most of the races of animated beings; and we have few examples of this influencing principle more obvious than this of the missel bird, in which a creature addicted to solitude and shyness will abandon its haunts, and associate with those it fears, to preserve its offspring from an enemy more merciless and predaceous still. The love of offspring, one of the strongest impressions given to created beings, and inseparable from their nature, is ordained by the Almighty as the means of preservation under helplessness and want. Dependant, totally dependant, as is the creature for every thing that can contribute to existence and support, upon the great Creator of all things, so are new-born feebleness and blindness dependant upon the parent that produced them; and to the latter is given intensity of love, to overbalance the privations and sufferings required from it. This love, that changes the nature of the timid and gentle to boldness and fury, exposes the parent to injury and death, from which its wiles and cautions do not always secure it; and in man the avarice of possession will at times subdue his merciful and better feelings. Beautifully imbued with celestial justice and humanity as all the ordinations which the Israelites received in the wilderness were, there is nothing more impressive, nothing more accordant with the divinity of our nature, than the particular injunctions which were given in respect to showing mercy to the maternal creature cherishing its young, when, by reason of its parental regard, it might be placed in danger. The eggs, the offspring, were allowed to be taken; but "thou shalt in anywise let the dam go;" "thou shalt not, in one day, kill both an ewe and her young." The ardent affection, the tenderness, with which I have filled the parent, is in no way to lead to its injury or destruction: and this is enforced, not by command only, not by the threat of punishment and privation, but by the assurance of temporal reward, by promise of the greatest blessings that can be found on earth, length of days and prosperity.

The jack snipe (*scolopax gallinula*) is with us here, as I have always known it, a transitory visiter in the winter only — a solitary, unsocial bird — an anchorite from choice. With the exception of our birds of prey, the manner of whose existing requires it, and a few others, all the feathered tribe seem to have a general tendency toward association, either in flocks, family parties, or pairs; but the individuals of this species pass a large portion of their lives retired and alone, two of them being rarely, or, perhaps never, found in company, except in the breeding season. They are supposed to pair and raise their young in the deep marshy tracts or reedy districts of the fen-counties, which afford concealment from every prying eye, and safety from all common injuries. Driven by the frosts of winter from these watery tracts, their summer's covert, they separate, and seek for food in more favoured situations, preferring a little, lonely, open spring, trickling from the side of a hill, tangled with grass and foliage, or some shallow, rushy streamlet in a retired valley. Having fixed on such a place, they seldom abandon it long, or quit it for another, and though roused from it, and fired at repeatedly through the day, neither the noise, nor any sense of danger seems to alarm them; and, if we should seek for the little judcock on an ensuing morning, we find it at its spring again. The indifference with which it endures this daily persecution is amazing. It will afford amusement or vexation to the young sportsman throughout the whole Christmas vacation; and, from the smallness of its body, will finally often escape from all its diurnal dangers. The rail, and several other birds, confide for safety more in their legs than their wings, when disturbed; but this snipe makes little use of its feet, and takes to its wings with such reluctance, from an apparent indolence of disposition, that, could it be seen in the rushes, or tufts of herbage, where it hides, it might be captured by the hand. It leaves us early in the spring. Fond of concealment as this little bird usually is, yet there are times when it is infinitely less so than at others; and, I think, upon the relenting of a frost, or when there is a tendency to a thaw, it shows unusual alacrity, springs from its rushy drain almost as readily as the common snipe, and occasions, for the moment, a doubt of the species. The mandible of this species is of a weak and spongy nature.

The causes that influence this snipe to lead so solitary a life are particularly obscure, as well as those which stimulate some others to congregate, as we comprehend no individual benefit to arise from such habits. Wild fowl, the rook, and some other birds, derive security, perhaps, from feeding in society, as a sentinel appears to be placed by them at such times to give notice of danger; but our congregating small birds take no such precaution: security or mutual protection does not seem to be obtained by it, as the largeness of the flocks invites danger; and warmth in the winter season it does not afford. For the purposes of migration, such associations are in many respects serviceable and consistent; but in our resident species, considered in its various results, it becomes rather a subject of conjecture, than of explanation. Timid creatures associate commonly upon the apprehension of danger, and, without yielding any mutual support, become only the more obnoxious to evil; and this snipe, though its habits are the very reverse of connexion with its species, yet affords no clue to direct us to the causes of its unusual habits. These associations of some, and retirement of others, are not the capricious actions of an hour in a few individuals, but so regularly and annually observed in the several species, that they are manifestly appointed provisions of nature, though the object is unknown. This half-snipe, as our sportsmen call it, has rather generally been considered by our young shooters as the male of the larger species, or common snipe (*scolopax gallinago*); yet it is difficult to assign any reason for the prevalence of such an idea, with those who have had many opportunities of observing the dissimilarity in the mode of life, the manners, and plumage of the birds. I know not any bird that lays so large an egg, in proportion to its size, as the snipe.

There is another little bird which frequently visits us, with habits as decidedly solitary as the jack snipe, but it is not a winter but a summer visitant only: the green legged sand piper (*tringa ochropus*: Pennant). It frequents the margins of ponds upon one farm, running after the flies and water insects, found at the muddy mouths of those places, but is never observed with any associate. It is a very delicate, pretty little creature, active and shy, but I know it only as one of our local visitants; the beautiful marking upon the feathers under the wings, and its long green legs, readily distinguish it from others of the species. It appears to be by no means a common bird, and from the season at which it comes to our ponds, is probably the male bird, separated from its mate and the cares of incubation.

A few pairs of the peewit (*tringa vanellus*) visit annually some of our larger ploughed fields to breed; but they are so frequently disturbed by those necessary processes of husbandry, hoeing and weeding, that they seldom succeed in the object of their visit. On our adjoining heath they escape better, and bring off many of their young: but the larger portion of them keep their station on the banks and dikes of the great drains and sewers in the marsh lands; and the traveller, who happens, in the spring of the year, to pass along any of the roads bordering upon these haunts, where many pairs are settled, will long remember the wearying and incessant clamour of these birds, which, rising as he approaches, wheel about him in an awkward, tumbling flight, accompanied by the unremitting, querulous cry of "peewit, peewit," continued by the perseverance of successive pairs, as long as he remains near their habitation; which generally being a flat, aguish, uninteresting country, where little is heard but the whispering of the wind in the reeds and sedges, the teazing monotony of this bird gives a very peculiarly dreary and melancholy character to parts of our lowland roads. In some counties these cold, wet districts go by the name of "peewit or pewety lands" At this period of the year, the bird is bold and fearless, and menaces the intruder with all its vociferous powers, when he approaches its haunts; but the broods being fledged, the families unite, form large flocks, and retire to open meadows, unenclosed commons and downs, feeding on slugs and worms, and become wild and vigilant creatures. It is well known that the glareous liquor, or white of the egg of this bird, upon being boiled, becomes gelatinous and translucent, not a thick opake substance like that of the hen; a circumstance that is likewise observable in the eggs of the rook, and of many of our small birds. The latter are not sufferers by it; but the eggs of the poor rook, though bearing little resemblance to those of this plover, are in some places not uncommonly taken and sold conjointly with them in the London market; and probably the habitual eater of them only can distinguish a sensible difference.

July 3d. A little bird was obtained this day, which upon examination proved to be a common white throat (*sylvia cincerea*) and was remarkable by the plumage being entirely white, without any shading of the natural brown hues, the legs, and beak straw coloured, the eyes red, and luminous like those of many white creatures, it was a bird of the year but full grown and perfectly feathered. When first observed it was living, and apparently uninjured, but it died shortly after its capture, having been probably by its dissimilarity obnoxious to its congeners, and so pecked and buffeted to death, as mule birds, and those escaped from confinement frequently are, but a native terrestrial bird with unusual plumage, has no chance of escape itself; and a white lark, rook, partridge, bring destruction on their companions, the flocks are persecuted from day-to-day, until the unfortunate object of curiosity has been obtained.

The roots of the hair in many animals become deranged or the fibre of the flesh changed from its natural state by a scar, the proper nutriment seems to be by such accidents diverted or sparingly supplied; some birds, and animals are subject to annual mutations in their clothing at the approach of the cold period of the year, a beautiful ordination of Providence, occasioned it appears, from withdrawing at those seasons the black matter deposited between the skins of animals, which supply the external covering with its new treatment. But this alteration which takes place at times in the colour of animals in severe seasons, from dark hues to white, as a provision, in one case, to mitigate the radiation of heat from their bodies and thus render the change of temperature less sensible to them, has no connection with these casual circumstances of variation of natural hues, the one being a provision, the other arising from accidental causes: but colour, its objects, its cause, is one of the great secrets of nature, be it observable in the flower, or in the animal. Nor can we satisfactorily explain how that vast diversity in the clothing is brought about, in the many animals domesticated around us, which only very casually occur in an unreclaimed state. The cat, rabbit, ox, horse, pigeon and others, hourly as the occurrence is, the operating cause seems obscure - and if we say that the refraction of the light becomes changed by alteration of form, or texture - we notice the effect, but do not make manifest the reason of this change. Some of our old writers call "white" a "penurious colour", and when accidental, is probably occasioned by a vitiated or scanty supply. The robin, swallow blackbird, with others found thus deviating from their congeners, and a milk-white linnet, has been with the flocks all this summer in our fields, but this specimen of the "white throat", was the only one of that race I had seen varying so remarkably from general appearances. I have before me now a rat, mole, house mouse, and shrew, each of this "penurious colour."

April 28, 1829 — A violent windy day, beating down and bruizing all young vegetation, and scattering the petals of our flowers round the garden; but in my rookery the consequences are disastrous indeed, bending the trees, and tossing out the nests, young and all, to the ground; many of the old birds seem to have forsaken the trees, and are away from this scene of desolation. In some of the neighbouring colonies I hear the destruction has been more extensive still: many birds having lost their first nests construct others, and succeed in rearing up later broods occasionally; but this the poor rook cannot hope to accomplish: building early, they hatch their young in April, and all take flight in May; for at no period of the year is there a greater probability of obtaining food than now: the general moisture of the season, and universal business of turning up the ground, preparing for the barley crop, &c., on every farm, exposes multitudes of worms and grubs, so that supply is abundant everywhere. Should the young be produced later, when the plough would be more at rest, and the heats of June and July dry up the ground, sufficiency of food would with difficulty be obtained, and the broods be in want. That universal diet, too, the worm, in mild weather, seeks the surface of the earth, and is obtained by all who require it; when the soil becomes dry, heated, or chilled, it retires. Yet, occasionally, some unfortunate pair of these birds will reconstruct their edifice, and when their congeners have left the trees, this solitary couple have all the parental duties to perform; and the constant clamour of the young for food, so unusual in nestling birds, renders it manifest that the labour and exertion of the parents cannot supply a sufficiency for their requirements. This habit of birds in forming a second establishment, after the first has been destroyed, is in conformity with that universally diffused ordination of the great Creator, to continue the species: it is an effort, but commonly an unsuccessful one, the numbers produced being frequently fewer than in the first case, and the advance of the season rendering it difficult to bring up even those brought forth. It is very pleasing to observe how timed and regulated all these hatchings are: some very early, as noted in the blackbird and thrush, and this rook, that their offspring may have all the benefit of the nutriment suited to their condition; others again progressively later, as the seeds of plants ripen, or the peculiar insects on which they feed are to be obtained: this order may be varied by the inclinations of creatures, but, like most other deviations from the courses of nature, with very equivocal effect, or a total failure of success.

After the perseverance of years to favour this my dark colony of rooks, which had attached themselves to me, the attempt has been abandoned. Like so many earthly things, however to be desired, possession was found undeserving the contention required. Nightly plunder and clandestine persecution have nearly driven my poor birds from the trees. But amidst all their troubles, that love of offspring, which at periods becomes the predominating principal of animal life, was strongly manifested, and the timidity which arises for self-preservation, when the former feeling ceases, seemed, though in existence, subdued by a more powerful passion. A nest here and there with young ones lately hatched, remained. When the others were forsaken, or the birds flown away, yet did the poor parents of then wailing creatures attend them by stealth and feed them at a daily risk of the own lives, which they appeared fully sensible of, by the vigilance they manifested, and the briefness of the visits. But the urgings of parental love, subdued the terrors, and they encountered death, to obey more positive dictates of nature. And in the main, success attended their efforts

Prognostications and signs, a great amusement, and the groundwork of belief to our forefathers, have, in general, pretty much declined with us; the repeated falsity of most of them having destroyed their reputation. We know so little, if any thing, of the actuating causes of seasons and their change, or the combinations effecting results, that no safe conclusion can be formed of any present events influencing the future. Whatever our almanacs may do, few persons of credit will venture now to predict, from what we call natural causes, a hot summer, or a severe winter; yet that very ancient idea, "amongst country people, that years of store of haws and heps do commonly portend cold winters," still lingers with us. However warmly we assent to the fundamental truth, the merciful consideration of Providence, in providing food for the necessities of the little fowls of the air, which, perhaps, piously gave rise to the observation, almost every year proves, that any conclusions drawn from these "stores of haws and heps" are perfectly fallacious. The birds that feed chiefly upon the fruit of the white thorn, and the wild rose, are the fieldfare (*turdus pilaris*), and the redwing (*turdus iliacus*); and that they do so, every sportsman has had the most manifest conviction: yet it has been said recently, that these creatures do not eat these fruits; and said, too, by an eminent and amiable[33] man, with whom I have frequently had the honour of conversing, and always with profit[34]. Were he living, his love of science would encourage my observations, though not in unison with his opinion: my breath shall not agitate his ashes, nor will his spirit, I am certain, frown in anger at my lines. It must be premised, that these birds, generally speaking, give the preference to insect food and worms; and when flights of them have taken their station near the banks of large rivers, margined by lowlands, we shall find that the bulk of them will remain there, and feed in those places; and, in the uplands, we shall observe small restless parties only. But in the midland and some other counties, the flocks that are resident have not always these meadows to resort to, and they then feed on the haws as long as they remain. In this county, the extensive lowlands of the river Severn in open weather are visited by prodigious flocks of these birds; but as soon as snow falls, or hard weather comes on, they leave these marshy lands, because their insect food is covered or become scarce,

---

33 Editor's note: the "eminent and amiable man" was the celebrated Dr Edward Jenner of Berkeley (1749-1823).

34 Substance of a paper read before the Royal Society, Nov. 27, 1824. See Zoological Magazine, vol, i.

visit the uplands to feed on the produce of the hedges, and we see them all day long passing over our heads in large flights on some distant progress, in the same manner as our larks, at the commencement of a snowy season, repair to the turnip fields of Somerset and Wiltshire. They remain absent during the continuance of those causes which incited their migration; but, as the frost breaks up, and even before the thaw has actually commenced, we see a large portion of these passengers returning to their worm and insect food in the meadows, attended probably by many that did not take flight with them — though a great number remain in the upland pastures, feeding promiscuously as they can. In my younger days, a keen, unwearied sportsman, it was always observable, that in hard weather these birds increased prodigiously in number in the counties far distant from the meadow lands, though we knew not the reason; and we usually against this time provided tempting bushes of haws, preserved in a barn, to place in frequented hedges, near our secret standings. When the fieldfare first arrives, its flesh is dark, thin, and scurfy; but, having fed a little time in the hedges, its rump and side veins are covered with fat. This is, in part, attributable to suppression of perspiration by the cold, and partly to a nutritive farinaceous food; its flesh at the time becoming bluish and clean. The upland birds are in this state, from perhaps the end of November till the end of January, according as the hedge fruit has held out; and at this period they are comparatively tame: afterward, though the flights may be large, they become wild; and the flesh, assuming its darkness, manifests that their food has not been farinaceous. The distant foreign migrations, which have been stated to take place from the meadows of the Severn, I believe to be only these inland trips; and that the supposed migrators returned to those stations fat and in good condition, owing to their having fed during their absence on the nutricious berry of the white thorn. I have several times seen the fruit on our hedges refused by these birds, and this too in no very temperate season; but in all these cases, the summer had been ungenial — the berries had not ripened well, they were nipped by the frosts of October, and hung on the sprays dark in colour, small, and juiceless in substance. The summer of 1825 produced the finest and largest haws I ever remember. They were in general of a bright red hue, and filled with farinaceous pulp; and in consequence, though the season was uncommonly mild and open, long before Christmas, little wandering parties of these birds consumed the whole of them.

Perfectly gregarious as the fieldfare is, yet we observe every year, in some tall hedge-row, or little, quiet pasture, two or three of them, that have withdrawn from the main flocks, and there associate with the blackbird and the thrush. They do not appear to be wounded birds, which from necessity have sought concealment and quiet, but to have retired from inclination; and I have reason to apprehend that these retreats are occasionally made for the purpose of forming nests, though they are afterward abandoned without incubation; as I have now before me the egg of a bird, which I believe to be that of a fieldfare, taken from a nest somewhat like that formed by the song-thrush, in 1824. Its colour is uniform — a rather pale blue; it is larger than that of the thrush, obtuse at both ends, and unlike any egg produced by our known British birds. These retiring birds linger with us late in the season, after all the main flights are departed, as if reluctant to leave us; but towards the middle or end of April these stragglers unite, form a small company, and take their flight.

Rural sounds, the voices, the language of the wild creatures, as heard by the naturalist, belong to, and are in concord with the country only. Our sight, our smell, may perhaps be deceived for an interval by conservatories, horticultural arts, and bowers of sweets; but our hearing can in no way be beguiled by any semblance of what is heard in the grove or the field. The hum, the murmur, the medley of the mead, is peculiarly its own, admits of no imitation, and the voices of our birds convey particular intimation, and distinctly notify the various periods of the year, with an accuracy as certain as they are detailed in our calendars. The season of spring is always announced as approaching by the notes of the rookery, by the jangle or wooing accents of the dark frequenters of its trees; and that time having passed away, these contentions and cadences are no longer heard. The cuckoo then comes, and informs us that spring has arrived; that he has journeyed to us, borne by gentle gales in sunny days; that fragrant flowers are in the copse and the mead, and all things telling of gratulation and of joy: the children mark this well-known sound, spring out, and cuckoo! cuckoo! as they gambol down the lane: the very plough-boy bids him welcome in the early morn. It is hardly spring without the cuckoo's song; and having told his tale, he has voice for no more — is silent or away. Then comes the dark, swift-winged marten, glancing through the air, that seems afraid to visit our uncertain clime: he comes, though late, and hurries through his business here, eager again to depart, all day long in agitation and precipitate flight. The bland zephyrs of the spring have no charms with this race; but basking and careering in the sultry gleams of June and July, they associate in throngs, and, screaming, dash round the steeple or the ruined tower, to serenade their nesting mates; and glare and heat are in their train. When the fervour of summer ceases, these lovers of the sun depart. The evening robin, from the summit of some leafless bough, or projecting point, tells us that autumn is come, and brings matured fruits, chilly airs, and sober hours, and he, the lonely minstrel now that sings, is understood by all.

These four birds thus indicate a separate season, have no interference with the intelligence of the other, nor could they be transposed without the loss of all the meaning they convey, which no contrivance of art could supply; and, by long association, they have become identified with the period, and in peculiar accordance with the time.

We have a rural sound here daily to be heard, and perhaps it is a local one; it may be noticed by many passing or residing, and yet the larger portion be ignorant of its object, which must be my apology for introducing it. Servants on the farm are usually called home, or to their meals, by the sounding of a horn, a loud shout, a bell, and other expedients; but when females only are left in the house, and a horn cannot be sounded, no sufficient shout given, and the wind obscure the ring of the bell, and so no notice be conveyed, we have here a practice that is fully answerable to the end: — the dairy-maid lifts up her milk-pail, and, by beating it with a stick, obtains so loud a sound, that it has often surprised me by the distances at which I have heard it — equal, perhaps, to the noise of a drum. The keen ears of a hungry man readily distinguish this domestic tattoo, — always a practicable, and a competent mode of indicating the dinner hour.

The Woodpeckers too are conducive to various rural sounds, the peculiar cachinnation, or resemblance to a vacant laugh of the common green one, as he clings to the bole of the tree, is unlike any other voice of the season, and has a wildness attendant on it, indicative of the absence of man. Then comes the trill of the thrush — the harsh scream, the shake of the pheasant; the coo-coo of the cushat — the low murmur of the dove, and other utterances of the sylvan race, producing a succession of sounds, the concert of the wild, harmonising with the gravity of mind, the temperament of the woodlands. It is very pleasing in some warm calm evening to listen to the various sounds and voices that become audible in that silent hour. Far over the fields the baying of some farmers dog, the mournful hooting of the owl, the jingle of the bell on the sheep, the steady measured tread of the steed upon the hard road pacing homeward, the moan of some beast in his lair, the night call of the bird, the crickets chirrup in the break, the crook of the frog; all these heard in the various distances, come softened on the ear with peculiar influence. In the day they would be mingled with variety of impressions, but in the evening have from the temperature of mind come to contemplation, or more distinctly heard as Humboldt thinks, from the uniform undisturbed density of the air, give pleasing sensations, which in sun-light would be almost annoyance.

We note birds in general more from their voices than their plumage; for the carols of spring may be heard involuntarily, but to observe the form and decoration of these creatures requires an attention not always given. Yet we have some native birds beautifully and conspicuously feathered; the goldfinch, the chaffinch, the wagtails, are all eminently adorned, and the fine gradations of sober browns in several others are very pleasing. Those sweet sounds, called the song of birds, proceed only from the male; and, with a few exceptions, only during the season of incubation. Hence the comparative quietness of our summer months, when this care is over, except from accidental causes, where a second nest is formed; few of our birds bringing up more than one brood in the season. The redbreast, blackbird, and thrush, in mild winters, may continually be heard, and form exceptions to the general procedure of our British birds; and we have one little bird, the woodlark (*alauda arborea*), that, in the early parts of the autumnal months, delights us with its harmony, and its carols may be heard in the air commonly during the calm sunny mornings of this season. They have a softness and quietness, perfectly in unison with the sober, almost melancholy, stillness of the hour. The skylark also sings now, and its song is very sweet, full of harmony, cheerful as the blue sky and gladdening beam in which it circles and sports, and known and admired by all; but the voice of the woodlark is local— not so generally heard — from its softness, must almost be listened for, to be distinguished, and has not any pretensions to the hilarity of the former. This little bird sings likewise in the spring; but, at that season, the contending songsters of the grove, and the variety of sound proceeding from every thing that has utterance, confuse and almost render inaudible the placid voice of the woodlark. It delights to fix its residence near little groves and copses, or quiet pastures, and is a very unobtrusive bird, not uniting in companies, but associating in its own little family parties only, feeding in the woodlands on seeds and insects. Upon the approach of man, it crouches close to the ground, then suddenly darts away, as if for a distant flight, but settles again almost immediately. This lark will often continue its song, circle in the air, a scarcely visible speck, by the hour together; and the vast distance from which its voice reaches us in a calm day is almost incredible. In the scale of comparison, it stands immediately below the nightingale in melody and plaintiveness; but compass of voice is given to the linnet, a bird of very inferior powers. The strength of the larynx and of the muscles of the throat in birds is infinitely greater than in the human race. The loudest shout of the peasant is but a feeble cry, compared with that of the golden-eyed duck, the wild goose, or even this lark. The sweet song of this poor little bird, with a fate like that of the nightingale, renders it an object of capture and confinement, which few of them comparatively survive. I have known our country bird-catchers take them by a very simple but effectual method. Watching them to the ground, the wings of a hawk, or of the brown owl, stretched out, are drawn against the current of air by a string, as a paper kite, and made to flutter and librate like a kestrel over the place where the woodlark has lodged; which so intimidates the bird, that it remains crouching and motionless as a stone on the ground; a hand-net is brought over it, and it is caught.

We must assuredly conclude that all permanent confirmation general in animate or inanimate nature, is ordained for wise and requisite purposes, and it is perhaps a laudable exercise in the creature to search out with humble heart the object of those appointments, which must lead to the exaltation of the Creators wisdom and benevolence in devising such contrivances for the accomplishment of his designs and the well being of his creation: yet it is humiliating to find how little of this machinery we can comprehend, though it may be most simple and unincumbered. There is a race of birds commonly to be observed consisting of not more than six or seven species arranged by Linnaeus under the name of larks, (*Alauda*) all of which are notable by reason of the length of their hinder claw, but it does not appear certain for what purpose this projecting nail was given, yet perhaps a conjecture may be formed: these birds including the tit-lark (*anthus pratⁱ:*) we presume always roost upon the ground, because in the trenches or little hollows in the stubbles, beneath the low bushes of gorse, or tufts of grass in the pastures frequented by them, we constantly observe the accumulated rejectments of the night: and this race with the wheat-ear, are I think the only small birds that do so[35] — we may apprehend then that this peculiar construction accords with the particular habits of the creature, and it may possibly act as a prop to sustain the upright position of the bird when at rest, at which time the head being placed under the wings removes the balance of the body from the feet, throwing it behind: birds that perch clasp their claws and feet three of which are in front, round the supporting branch and are thus held in their upright position, but those that remain upon the ground have nothing to grasp and seem to require aid when the balance of the head is lost to keep them upright. This long claw to one species indeed becomes at times a great impediment, for in moist seasons of winter, or after thaws, it almost prevents the Skylark feeding in his favourite situation on the green wheat lands, by collecting such masses of dirt as to incommode his progress, and necessitate resorting to the grass-lands, picking up in those places what he can, until the moisture of the herbage clears away the accumulated soil from the claw, when he returns to his usual haunts.

35 Field-fares, and Redwings, when feeding in large meadows removed from trees or hedges repose at times in those places, but such are not the general habits of the birds.

From various little scraps of intelligence scattered through the sacred and ancient writings, it appears certain, as it was reasonable to conclude, that the notes now used by birds, and the voices of animals, are the same as uttered by their earliest progenitors. The language of man, without any reference to the confusion accomplished at Babel, has been broken into innumerable dialects, created or compounded as his wants occurred, or his ideas prompted; or obtained by intercourse with others, as mental enlargement or novelty necessitated new words to express new sentiments. Could we find a people from Japan or the Pole, whose progress in mind has been stationary, without increase of idea, from national prejudice or impossibility of communication with others, we probably should find little or no alteration in the original language of that people; so, by analogy of reasoning, the animal having no idea to prompt, no new want to express, no converse with others, (for a note caught and uttered merely is like a boy mocking the cuckoo,) so no new language is acquired. With civilized man, every thing is progressive; with animals, where there is no mind, all is stationary. Even the voice of one species of birds, except in particular cases, seems not to be attended to by another species. That peculiar call of the female cuckoo, which assembles so many contending lovers, and all the various amatorial and caressing language of others, excites no influence generally, that I am aware of; with all but the individual species it is a dialect unknown. I know but one note which animals make use of, that seems of universal comprehension, and this is the signal of danger. The instant that it is uttered, we hear the whole flock, though composed of various species, repeat a separate moan, and away they all scuttle into the bushes for safety. The reiterated ("twink twink" of the chaffinch is known by every little bird as information of some prowling cat or weasel. Some give the maternal hush to their young, and mount to inquire into the jeopardy announced. The wren, that tells of perils from the hedge, soon collects about her all the various inquisitive species within hearing, to survey and ascertain the object, and add their separate fears. The swallow, that shrieking darts in devious flight through the air when a hawk appears, not only calls up all the hirundines of the village, but is instantly understood by every finch and sparrow, and its warning attended to. As nature, in all her ordinations, had a fixed design and foreknowledge, it may be that each species had a separate voice assigned it, that each might continue as created, distinct and unmixed: and the very few deviations and admixtures that have taken place, considering the lapse of time, association, and opportunity, united with the prohibition of continuing accidental deviations are very remarkable, and indicate a cause and original motive. That some of the notes of birds are as language designed to convey a meaning, is obvious from the very different sounds uttered by these creatures at particular periods: the spring voices become changed as summer advances, and the requirements of the early season have ceased; the summer excitements, monitions, informations, are not needed in autumn, and the notes conveying such intelligences are no longer heard. The periodical calls of animals, croaking of frogs, &c., afford the same reasons for concluding that the sound of their voices by elevation, depression, or modulation, conveys intelligence equivalent to an uttered sentence. The voices of birds seem applicable, in most instances, to the immediate necessities of their condition; such as the sexual call, the invitation to unite when dispersed, the moan of danger, the shriek of alarm, the notice of food. But there are other notes, the designs and motives of which are not so obvious. One sex only is gifted with the power of singing, for the purpose, as Buffon supposed, of cheering his mate during the period of incubation; but this idea, gallant as it is, has such slight foundation in probability, that it needs no confutation: and after all, perhaps, we must conclude, that listened to, admired, and pleasing, as the voices of many birds are, either for their intrinsic melody, or from association, we are uncertain what they express, or the object of their song. The singing of most birds seems entirely a spontaneous effusion produced by no exertion, or occasioning no lassitude in muscle, or relaxation of the parts of action. In certain seasons and weather, the nightingale sings all day, and most part of the night; and we never observe that the powers of song are weaker, or that the notes become harsh and untunable, after all these hours of practice. The song-thrush, in a mild, moist April, will commence his tune early in the morning, pipe unceasingly through the day, yet, at the close of eve, when he retires to rest, there is no obvious decay of his musical powers, or any sensible effort required to continue his harmony to the last. Birds of one species sing in general very like each other, with different degrees of execution. Some counties may produce finer songsters, but without great variation in the notes. In the thrush, however, it is remarkable, that there seems to be no regular notes, each individual piping a voluntary of his own. Their voices may always be distinguished amid the choristers of the copse, yet some one performer will more particularly engage attention by a peculiar modulation or tune; and should several stations of these birds be visited in the same morning, few or none probably will be found to preserve the same round of notes; whatever is uttered seeming the effusion of the moment. At times a strain will break out perfectly unlike any preceding utterance, and we may wait a long time without noticing any repetition of it. During one spring, an individual song-thrush frequenting a favourite copse, after a certain round of tune, trilled out most regularly some notes that conveyed so clearly the words, lady-bird! lady-bird! that every one remarked the resemblance. He survived the winter, and in the ensuing season, the lady-bird! lady-bird! was still the burden of our evening song; it then ceased, and we never heard this pretty modulation more. Though merely an occasional strain, yet I have noticed it elsewhere— it thus appearing to be a favourite utterance. Harsh, strained, and tense, as the notes of this bird are, yet they are pleasing from their variety. The voice of the blackbird is infinitely more mellow, but has much less variety, compass, or execution; and he too commences his carols with the morning light, persevering from hour to hour without effort, or any sensible faltering of voice. The cuckoo wearies us throughout some long May morning with the unceasing monotony of its song, and though there are others as vociferous, yet it is the only bird I know that seems to suffer from the use of the organs of voice. Little exertion as the few notes it makes use of seem to require, yet, by the middle or end of June, it loses its utterance, becomes hoarse, and ceases from any further essay of it. The croaking of the nightingale in June, or the end of May, is not apparently occasioned by the loss of voice, but a change of note, a change of object; his song ceases when his mate has hatched her brood; vigilance, anxiety, caution, now succeed to harmony, and his croak is the hush, the warning of danger or suspicion to the infant charge and the mother bird.

Of the several intimations relating to the voice of animals as preserved to us in Scripture, we have none more deserving of attention than the "crowing of the cock" throughout the night, there being a first crowing about midnight, and a second again as day began to dawn; and this so regularly proceeded in, as to be made use of to mark the progress of time from a very early period, it being pointed out as a well-known and established occurrence above eighteen centuries gone by. Though this vociferation of the bird is yet persevered in, it seems to be without any regularity, except, perhaps, the general clamour of the early morning, as in particular nights this crowing may be heard at various intervals during the darkness. Night-travelling birds sound a signal for the guidance of their followers; but these creatures, usually when at rest, or feeding in the gloom, observe a profound silence, and perhaps the cock is the only creature that notifies to any enemy within hearing his asylum on the roost. If such are the habits of these creatures in an unreclaimed state, it must very frequently be productive of injury to them and their families around them. But in this his domesticated state, it is a voice which, heard during some sleepless hour, in the deep quiet of the night, becomes most impressive and solemn, brings past events to our recollection, and has, perhaps, often produced holy thoughts and meditations.

But here I must close my notes of birds, lest their actions and their ways, so various and so pleasing, should lure me on to protract

> My tedious tale through many a page;

for I have always been an admirer of these elegant creatures, their notes, their nests, their eggs, and all the economy of their lives; nor have we, throughout the orders of creation, any beings that so continually engage our attention as these our feathered companions. Winter takes from us all the gay world of the meads, the sylphs that hover over our flowers, that steal our sweets, that creep, or gently wing their way in glittering splendour around us; and of all the miraculous creatures that sported their hour in the sunny beam, the winter gnat (*tipula hiemalis*) alone remains to frolic in some rare and partial gleam. The myriads of the pool are dormant, or hidden from our sight; the quadrupeds, few and wary, veil their actions in the glooms of night, and we see little of them; but birds are with us always: they give a character to spring, and are identified with it; they enchant and amuse us all summer long with their sports, animation, hilarity, and glee; they cluster round us, suppliant in the winter of our year, and, unrepining through cold and want, seek their scanty meal amidst the refuse of the barn, the stalls of the cattle, or at the doors of our house; or, flitting hungry from one denuded and bare spray to another, excite our pity and regard: their lives are patterns of gaiety, cleanliness, alacrity, and joy.

THERE ARE VERY many subjects and employments of mankind, which, if we would obtain a competent knowledge of them, will require an almost undivided attention; yet, after all our "rising early and late taking rest," we shall know too little to be weighed in competition with what is beyond our attainment or comprehension. As in ascending mountainous regions we may reach the summit of one hill with comparative ease, that of a higher with more laborious efforts, and a still higher is attained by a gifted few, beyond which our breath fails us, our natural powers become inadequate; — so a small number may ascend the Alps of science, but pant, unable to attain the Himmalya ranges of their wishes. If proficiency be the object, all the branches of natural history require undivided attention; but amusement, admiration, and intelligence, may be obtained by even superficial observation; and of all these departments, perhaps entomology, or the investigation of the insect world, from the variety it embraces, the season, the subjects, and the vigilance necessary to catch every momentary action, requires from its followers a homage more absolute, an attention more devoted, than most others. Amid those few branches of science on which I have sought for blossoms, that of entomology I have least investigated: yet, perhaps, it may be said, that such slight notices as the foregoing need not have usurped the time that the study of this department required, To this truth I cannot but assent, and say, with the eminent man whose "Centuries of Experiments" I have often quoted, that they are indeed more the suggestions of "light than of fruit;" proficiency was beyond my powers; I have sought for amusement, and gratefully record the many peaceful hours and oblivion of pain, which the perusal of Nature's volume gave me, superficial as that perusal was.

On whatever side we turn our attention in this world of wonders by which we are surrounded, we constantly find some subject that calls forth our admiration; and, as far as our very imperfect vision is permitted to penetrate, we observe the same unremitting order and provision for a seemingly mean and worthless purpose, as is bestowed upon a higher and apparently more worthy object. We consider insects as one of the lower orders of creation, but are as perfectly unacquainted, generally speaking, with the objects of their being, though they have for ages crawled and winged their way around us, as the first man, Adam, was; yet is there a care manifested for the preservation and accommodation of these, which we often designate as contemptible creatures, that is most elaborate and wonderful. The forethought with which many of them have been furnished to deposit their eggs in safety from the contingencies of seasons and hostile incidents, and precisely in the situation most fitting, must call forth the admiration of all who have observed it. Some of these are lodged in summer and autumn deep in the earth, on that part of a plant which in due time is to be raised up, constituting a stalk or blade, bearing with it by gentle steps these eggs, to be vivified by the summer's air and warmth.

Others fix them on some portion of an herb hidden beneath the mud in the pool; and this being elevated by the warmth of spring, conveys them with its growth above the element that protected them; and they hatch, the infants feeding on the substance that has borne them to the air. In their chrysalis state, a cradle of preparation for a final change, the same wisdom and care are more particularly obvious from their size and frequent occurrence. There is that marvellous fabrication in the insect world which tends more to exalt our conceptions of the truly Almighty architect, than perhaps any other portion of creation; the construction of the vegetable, the beauty of the flower and foliage, gracefulness of growth, utility or pleasantness of its fruit, or other essential inherents, must ever be a constant theme of admiration, though the actual mode, or cause of its growth and construction is perfectly mysterious to us. But the insect in its mechanical functions is more comprehensible, because more nearly assimilating to those of ourselves. The common gnat which rests on our window-frames, or the scarcely visible atom of a fly, that creeps up the glass, with limbs finer than the slenderest hair, has integuments to actuate its motions, valves, nerves, joints, sockets, tubes, with circulating fluids, brushes to cleanse its body, and hooks to adhere by, placed in a space so small as to be hardly visible! And what else may be contained in those slender limbs unrequired by our motions, what perceptions and faculties unknown to human beings, must be constantly present to our imagination, but beyond our comprehension. But to enlarge sufficiently upon the contrivances and manifestations of regard brought to our observance in all the stages of an insect's life, would almost require a detail of the race.

A particularly curious covering for a moth, or butterfly, *(phalæna pavonia?)*, fell into my hands, which might be well known to a more experienced entomologist, but was new to me. The species I do not know, as it never arrived at perfection. This case was formed of the fine silky substance that wraps up so many of the race. The summit, for some cause, was less closed than usual; but to obviate any injury to the creature from this circumstance, a conical hood of similar materials was placed over the exposed part of the aurelia, through which it received air in perfect security. This veil being formed of elastic threads, and opening upon pressure, would constitute no impediment to the escape of the fly when perfected. A (Plate 5, Fig. 1), the case and its hood; B, the hood detached. More care and forethought than these contrivances manifest, we are not acquainted with for any order of beings. I conjecture it would have produced the emperor moth.

June 16. — I this day captured in a neighbouring meadow a fine specimen of the four-spotted dragonfly (*libellula quadrimaculata*), and note this for my entomological friends; being the first certain instance I am acquainted with of its being taken in England of late years, for Ray mentions it. Another, I believe, escaped by its shyness. It is a handsome creature, about three inches in breadth between the extremities of its wings. The two dark linear marks on the upper margin of each wing, and tapering downy body, distinguish this fly from any other. I can add nothing regarding its history or manners.

The ghost moth (*hepialus humuli*) is commonly seen here, as I believe it to be in most other places, but is mentioned to point out to any young person unacquainted with this insect its singular habit when on the wing, which at once distinguishes it from any other moth. The larva which produces this creature is hidden in the ground during the season of winter; the fly being formed in the month of May, and soon rising from the soil, then commences its short career. At this time one or more of them may frequently be observed under some hedge in a mead, or some low place in a damp pasture, only a few feet from the ground, persevering for a length of time together in a very irregular flight, rising, and falling, and balancing about in a space not exceeding a few yards in circumference — an action not observable in any other, and fully indicating this moth. This procedure is not the meaningless vagary of the hour, but a frolicsome dance, the wooing of its mate, which lies concealed in the herbage over which it sports. The two insects are something similar in their general form, but very differently marked. The male exhibitor is known by its four glossy, satiny, white wings, bordered with buff; the lady reposer has her upper wings of a tawny yellow, spotted and banded with deep brown. They are very inert creatures, easily captured; and their existence appears to be of very short duration, as we soon cease to observe them, either in action or at rest. The male probably becomes the prey of every bird that feeds by night; his colour and his actions rendering him particularly obnoxious to dangers of this nature; and the frequency with which we find his wings scattered about, points out the cause of death to most of them. The bat pursues with great avidity all those creatures that fly in the evening; and by its actions it seems to meet with constant employment, and has greater probability of success, than some insectivorous birds that feed by day, as all the myriads which abound at this time are the sole prey of itself and a few nocturnal ramblers. From this singular flight in the twilight hour, haunting as it were one particular spot, the fancy of some collector, considering it as a spectrelike action, named it the "ghost moth."

The fern owl, but chiefly, I conjecture, the larger bats, are the creatures that have caused me to experience at times both envy and regret, when I have observed scattered in some woodland path, amidst the fragments of their nightly banquet, the relics of such beautiful insects as the emperor of the woods, the verdigris moth, and twenty other rare insects, to be obtained only after the patience of years, or fortune of the hour; and yet our merciless birds devour these choice dainties without compunction or regard. This ghost moth discharges her eggs in a very singular manner, and frequently immediately upon capture, not deliberately protruding them, but dismissing them from the oviduct in rapid succession, until it is exhausted, with a slight elastic force, that conveys them clear from the abdomen. They are perfectly dry and unadhesive. As the life of the male of this species is sportive and brief, yet that of the female is of much longer duration, her greater concealment during the flight of the other sex is possibly one cause for preservation, and when all the white flutterers have disappeared she is frequently to be observed slowly moving a few feet from the ground about the ditch bank, or among the lowly herbage under the hedge; and many a young collector has been disappointed in finding the capture of the evening which seemed a prize, to be only this common insect.

It requires more than usual delicate management to preserve an uninjured specimen of the male of this species, as the slightest touch robs the wings of the fine scaly plumage which is affixed to their film or substance by an extreme point, as is the case with most others of our moths, but in this instance so loosely that a very gentle friction rubs it off. The plumage which covers the wings and bodies of many of our lepidopterous insects is variously coloured, and, like the feathers of birds, gives them their splendour: in the butterflies I have not observed it to vary greatly in form, but in the moths the same uniformity does not appear to be maintained, as a few specimens will manifest: —

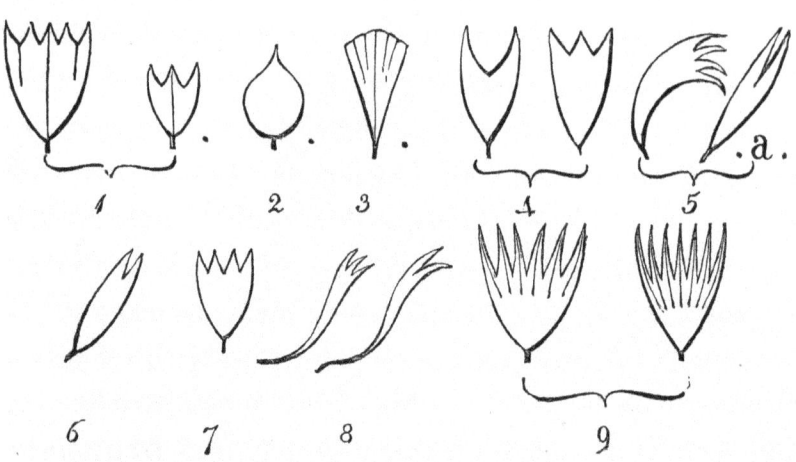

No. 1. Scales from wings of phalæna pronuba — yellow underwing. 2. Ghost moth. 3. Phalaena bucephala — buff tip. 4. Ph. vinula — puss moth. 5. Ph. potatoria, a, the female — drinker moth. 6. Papilio brassicæ — great white. 7. Pap. Napi — green-veined butterflies. 8. Large brown moth (name omitted). 9. Acherontia atropos — death's head.

But the variety of clothing with which insects are decorated is most admirable and curious! The upper and the under vestiture of the wings, their fringes, that which covers the body in different parts, varies greatly; the bird, splendidly habited as he sometimes is, frequently will be found draped with less variety of form and colour than the insect which escapes our notice by his actions, and the power of our eyes by the smallness of its parts. Our lepidopterous creatures seem to be most characteristically framed and constituted for the different hours and places in which they delight to move; so much so, that I think if we were to invert the order of their appearance, the singular unfitness of many of them for their stations would be immediately manifest to us. The butterfly, light, airy, joyous, replete with life, sports in the sunshine, wantons on the flower, and trips from bloom to bloom, gay as the brilliant morn, and cheerful as the splendour of heaven: heat and light appear to be the very principle of his being; in a cloudy or a chilly atmosphere his energies become suspended, and, closing his wings, he reposes like a sickly thing upon some drooping flower: but let the cloud disperse, the sun break out, he springs again to active life; associating with the birds of day, and denizen of the same scenes, he only seems of a less elevated order. But the moth, though possessing at times sufficient activity for self-preservation, is less buoyant, less sprightly on the wing, avoids the heat and light, the higher ranges of the air, and seeks his mate or his food in the shelter of the hedge or the ditch, amidst foliage and shade, where we may see him hovering sedately around some flower, or passing on his way with quiet steady flight, accordant with the silence and twilight of the hour: companion of the owl and the bat, his grave actions are quite unsuitable to the gaiety, the flutter of a summer's sun; the former is emblematic of levity and display, the latter of retirement and shade. And thus each, though but slightly seen, is in admirable harmony with the season in which it moves, manifesting the peculiar fitness of things to their several stations in this vast world of wisdom — an observation obvious to all, and a truth everywhere admitted; yet, as a Christian moralizer, I could not pass by unheeded any evidence of foresight and of power.

The gay clothing of creatures, the down, the pile of the insect, the feather of the bird, the hair of the beast are constantly before us, becoming often the subject of remark. When duly investigated, their office and fabrication seen, our admiration of them must increase as with every other production of nature, whenever we can obtain even a superficial knowledge of it: but there are other animals whose vesture is brought less before our notice, yet it is finished with peculiar care and beautifully decorated. In all the works of creation there appears to be a great economy in the expenditure of material, nothing as far as our knowledge extends, but what is required for the object, leading us to conclude that whenever superior excellence of execution in external decoration is observable, it is essential to some purpose of the creatures life or comfort, or necessary to the perfecting of some organ, the action of some function in inanimate being, not worthily appreciated possibly, and even hidden in obscurity from the only creature who can respect or meditate upon such things; intimating the vast mine of wisdom and contemplation that is in store for the future generations of mankind. In the outer covering of fishes the iridescent hues and gleams of lustrous colouring are often beautiful, much of this splendour is occasioned by the refraction of the viscous substance which invests their bodies when recently captured, but still a portion of it is produced by their scales, these little integuments have not been so generally noticed I think as their great variety and curious fabrication may deserve; The scale of the sole has been represented by Hook, in that beautiful manner so very observable in all his plates; Leeuwenhoek, Adams and a few others have noticed them;[36] from their size and transparency they are good microscopic objects and different parts of the creatures present some variation of markings; a few only are now noticed as sufficient to indicate the variety and manner of ornament bestowed upon them.

Fish do not appear to renew their scales yearly as birds and beasts do their clothing: of the little which we know of these creatures, their sensibility to atmospheric pressure or temperature is apparent by their swimming in different depths, resting at the bottom of their stations, or shifting their situations in changes of weather; their avidity for food in one state of air, the cessation of inclination for it upon the slightest change, indicating a remarkable perception of feeling; as such influences are transmitted from one element to another in which they reside, notable for its resistance to compression, slowness in parting with, or receiving warmth. In viewing the careful manner in which these external coverings of fish are ornamented, one cannot but apprehend that they are designed in some way unknown to us, to be instrumental to this manifested consciousness of the creatures, or to accelerate other impulses of which we have no information.

Basking in the glare of an August or a July sun, in our pastures we see the little elegant blue argus butterfly (*papilio argus*), noted and admired by all, now warmed into active life. A few of our lepidopterous creatures, especially the common white butterflies of our gardens, are contentious

---

36 Since writing the above lines, Mr. Yarrells admirable work upon the Brit: Fishes (1836) have appeared, in which are delineated several scales of these creatures, but as the object of this gentleman seems to be chiefly to display the extraordinary mechanism of the mucous tubes in the lateral line of fishes, my trifling suggestions will not interfere with his superior observations.

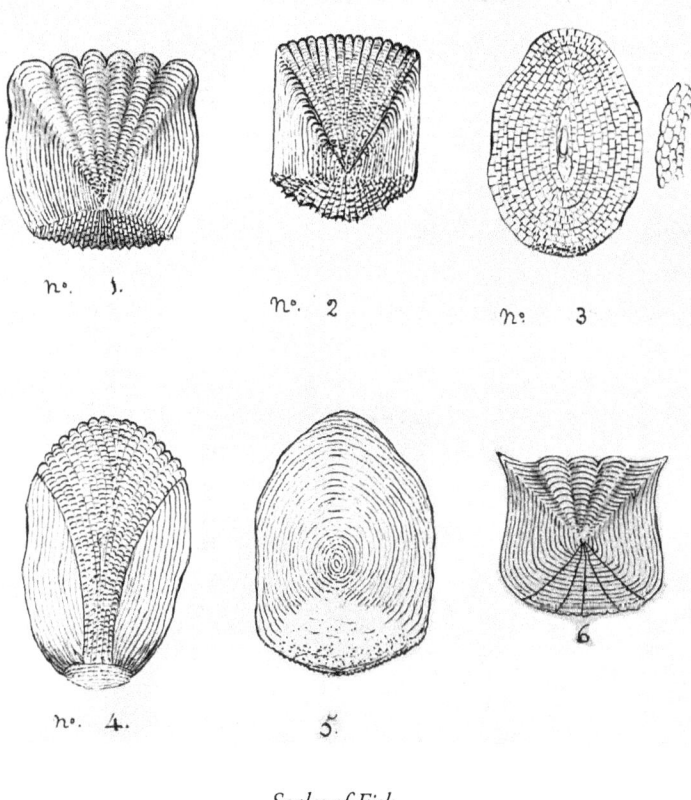

Scales of Fish

N°. 1 scale of the common Perch (Perca fluviatilis) N°. 2 the Basse (P. labrax) – N°. 3 the Cod (Gadus morhua) N°. 4 the Brill, or Kite (Pleu^s. Rhombus) No. 5 the Salmon (Salmo Salar) N°. 6 the Roach (Cyprinus Rutilus)

animals, and drive away a rival from their haunts. We see them progressively ascending into the air, in ardent, unheeding contest; and thus they are observed, captured, and consumed in a moment by some watchful bird: but we have few more jealous and pugnacious than this little argus. When fully animated, it will not suffer any of its tribe to cross its path, or approach the flower on which it sits, with impunity; even the large admiral (*vanessa atalanta*) at these times it will assail and drive away. There is another small butterfly (*papilio phlæas*) however, as handsome, and perhaps still more quarrelsome, frequenting, too, the same station and flowers; and a constant warfare exists between them. We shall see these diminutive creatures, whenever they come near each other, dart into action, and continue buffeting one another about till one retires from the contest; when the victor returns in triumph to the station he had left. Should the enemy again advance, the combat is again renewed; but should a cloud obscure the sun, or a breeze chill the air, their ardour becomes abated, and contention ceases.

The papilio phlaeas enjoys a combat even with its kindred. Two of them are seldom disturbed, when basking on a knot of asters in September, without mutual strife ensuing. Being less affected by cold and moisture than the argus, they remain with us longer, and these contentions are protracted till late in the autumn. The pugnacious disposition of the argus butterfly soon deprives it of much of its beauty; and, unless captured soon after its birth, we find the margins of its wings torn and jagged, the elegant blue plumage rubbed from the wings, and the creature become dark and shabby.

This spring, 1827, fostered into active life an insect unknown in our district, or at least unnoticed before by me; a pretty little blue butterfly, for which I know no common appellation, and so have named it the "spring azure," (*papilio argiolus*). It appeared quite at the end of April, and in some numbers, but was yet a transient visitor with us, as after the first week in May only a lingering specimen or so was visible. Few wild flowers are then in bloom; but, leaving all herbaceous plants, it frequented chiefly the holly, the laurel, and the black currant, feeding on the honey secreted by the nectaries in their blossoms. If this butterfly be anywhere common, it may be mistaken by indifferent observers for the little blue argus of our pastures; but it appears some months earlier than that insect is accustomed to do; does not flit from blossom to blossom, and bask upon the disks of the lowly herbs; and, though a feebler creature on the wing, takes a much higher range in flight, and sports in altitudes which the argus, with all its animation, is very rarely inclined to attempt. When in captivity, the dark margins of the upper wings, the black specks, not eyes, and the pale blue of the reverse, without any other character, render it perfectly distinguishable from the papilio argus, corydon, or any other butterfly found with us. A small hatch again takes place about the end of July, arid this pretty insect haunts anew our currant bushes; but, enlivened by the warmth of the season, it becomes more wild and wary, and avoids our approach.

The hummingbird hawkmoth (*sphinx stellatarum*) visits us annually, and occasionally in some numbers, frisking about all the summer long, and in very fine seasons continues with us as late as the second week in October. The vigilance and animation of this creature are surprising, and seem to equal those of its namesake, that splendid meteoric bird of the tropics, "that winged thought," as some one has called it; though our plain and dusky insect can boast none of its glorious hues. Our little sphinx appears chiefly in the mornings and evenings of the day, rather avoiding the heat of the mid-day sun, possibly roused from its rest by the scent, that "aromatic soul of flowers" which is principally exhaled at these periods; delighting in the jasmine, marvel of Peru, phlox, and such tubular flowers; and it will even insert its long, flexible tube into every petal of the carnation, to extract the honeylike liquor it contains. It will visit our geraniums and greenhouse plants, and, whisking over part of them with contemptuous celerity, select some composite flower that takes its fancy, and examine every tube with rapidity, hovering over its disk with quivering wings, while its fine hawklike eyes survey all surrounding dangers. The least movement alarms it, and it darts away with the speed of an arrow; yet returns, and with suspicious vigilance continues its employ, feeding always on the wing. Nature seems to have given this creature some essential requisites for its safety; its activity, when on the wing, renders its capture difficult; and when it rests, it is on a wall, the bark of a tree, or some dusky body, that assimilates so nearly to its own colour, as to render it almost invisible, though watched to its settlement: the larva is seldom found. We sometimes see it enter our rooms, attracted by flowers in the open windows; but it seems to be immediately aware of its danger, disappears in an instant, and is safe from capture. Wild and fearful as this creature is by nature, yet continued gentle treatment will remove much of its timidity, and render it familiar to our presence. Perfectly free from any annoyance as they are when ranging from sweet to sweet on my borders, and accustomed to a close inspection of all their operations, I have frequently touched their wings with my fingers, while hovering over a flower, and dipping their long tubes into the corolla of a geranium: they would retire a little, confused with such freedoms and interruptions, but, experiencing no harm, they would return and finish their meal, unmindful of such petty annoyances. I have known this creature, like some other insects, counterfeit death when apprehensive of danger, fall on its back, and appear in all respects devoid of life when in a box; and, as soon as a fit opportunity arrived, dart away with its usual celerity.

On the blue heads of the pasture scabious (*scabiosa succisa*) we occasionally see, toward the end of the summer, the painted lady butterfly (*papilio cardui*); but this is a creature that visits us at very uncertain periods, and is vivified by causes infinitely beyond the comprehension of the entomologist, seeming to require a succession and variety of seasons and their change, and then springing into life we know not how. This was particularly obvious in the summer of 1815, and the two following, which were almost unceasingly cold and rainy; scarcely a moth or butterfly appeared. And in the early part of 1818, the season was not less ungenial; a few half-animated creatures alone struggled into being; yet this "painted lady" was fostered into life, and became the commonest butterfly of the year: it has, however, but very partially visited us since that period. The keenest entomologist, perhaps, would not much lament the absence of this beauty, if such cheerless seasons were always requisite to bring it to perfection. Some years ago a quantity of earth was raised in cutting a canal in this county; and, in the ensuing summer, on the herbage that sprang up from this new soil on the bank, this butterfly was found in abundance, where it had not been observed for many years before.

In some particular seasons, we have acres of this scabious in bloom, during the months of September and October, giving a tender shade of lavender colour to the whole field, affording now great pleasure to the entomologist, by reason of the multitude of insects that resort to it for the honey in the tubular florets of the plant. Late as this period is, I have seen, in some bright morning, besides multitudes of bees, flies, and such creatures, eleven, different species of lepidopterous insects, feeding and balancing on the blue heads, and glancing their gay wings in the sunny beam.

The marble butterfly (*papilio galathea*) is an equally capricious visitant of our fields. I have known intervals of ten or twelve years when none could be found, and in some following seasons it would be a prevailing species.

The common wasp (*vespa vulgaris*) is infinitely uncertain in its numbers. A mild winter, and a dry spring or summer, we might conclude to be favourable circumstances for the increase of this creature; yet such is not always the case. Years productive of the plum are said to be congenial likewise to the wasp. A local rhyme will have it, that

> When the plum hangs on the tree,
> Then the wasp you're sure to see.

The wasps which appear in spring are those of the preceding summer which have passed the winter in a dormant state secured from the cold and damps of the season in crevices or the thatch of outhouses, behind the paper in bed rooms, and a favourite place is under the dry bark of some decayed tree. Before retiring to those places they appear from the state we find them in, very carefully to secure their wings, for any injury to these membranes would under revival be of no avail for the purposes of life: she bends them from the socket in which they are fixed to the body, and from which they now appear loosened, bringing them entirely under her stomach, folds them over each other, closes her legs above them, and bending her body a little, appears like a wingless creature, and thus secures them as effectually as possible, from common contingencies during her winter sleep – the house fly, in its winter dormitory has its wings carefully lapped over each other to save them, but not so cautiously secured as those of the wasp.

Amid the tribes of insects so particularly influenced by seasons, there are a few which appear little affected by common events; the brown meadow butterfly (*papilio janira*), so well known to every one, I have never missed in any year; and in those damp and cheerless summers when even the white cabbage butterfly is scarcely to be found, this creature may be seen in every transient gleam, drying its wings, and tripping from flower to flower with animation and life, nearly the sole possessor of the field and its sweets. Dry and exhausting as the summer may be, yet this dusky butterfly is uninjured by it, and we see it in profusion hovering about the sapless foliage. In that arid summer of 1826, the abundance of these creatures, and of the ladybird (*coccinella septem punctata*), was so obvious, as to be remarked by very indifferent persons. We could scarcely select a less conspicuous butterfly than this meadow brown, its dark umber colour, and ochrey yellow, constitute a vest of the plainest hues, little calculated to excite attention, and its actions are in perfect uniformity with this matronly garb, not dashing away or flaunting over the field, but humbly flitting in short trips from one little elevated herb to another, threading a passage often between the benty grasses to rest upon some lowly clover or dwarfish betony, dotting along attentive to its own simple pursuits unmindful of other objects or concerns.

There is a large yellow underwing moth (*phalæna pronuba*), too, which is generally abundant. It hides itself during the day in the thickest foliage, and screens itself from the light in the moist grass crops of the mead, where it is perpetually disturbed, and roused from its rest, by the scythe of the mower. That elegant little bird, the yellow wagtail tail, is a great destroyer of this insect. It is very soon apprised of these movements, and will often attend the steps of the mower, fearless of harm, to watch for its prey. As soon as the moth rises, it is chased; and its exertions and shiftings to escape, and the activity and perseverance of the bird to capture it, are very amusing.

Our lepidopterous insects feed upon various substances in their several states; and most of our butterflies, when perfected, appear to extract the sweet liquor from the tubes or nectaries of plants, and many of our moths obtain their nourishment by similar means: but one butterfly alone, the admirable (*v. Atalanta*), and at times the peacock (*v. Io*), feed upon the juices of our autumnal fruits; and in the months of September and October we may frequently see these beautiful creatures basking and regaling themselves upon the rejected fragments of our wall fruit. They seldom prey upon the growing produce, like the hornet, wasp, and hive bee; but when it has fallen and advanced to a state of fermentation, it becomes the most grateful to them. Nothing can be less injurious than this propensity; and it seems that fruit in such a state is requisite for them by some constitutional formation, as they appear only at the termination of a season when the product of our trees is in a state of great ripeness and decay. The life of this creature appears to be remarkably short, and we have more certain means of ascertaining its duration than are afforded us for others of the race. It very rarely appears until late in September, and then so perfect and fresh in its plumage as to manifest its recent production from the chrysalis. In some years they abound, and we may see twenty of these beautiful creatures expanding and closing their brilliant wings under the fruit trees on our walls, or basking upon the disc of some autumnal flower; and at another, perhaps, hardly a specimen is to be obtained: nor do they seem, like the wasp, to be scarce or abundant according to the deficiency or plenty of the season, but influenced by other causes. Many of our butterflies are produced by successive hatches, supplying the places of those which have been destroyed, and hence it is difficult to mark the duration of an individual; and others, as the nettle, peacock, and wood tortoise, in many instances, survive the winter, hidden in some recess or sheltered apartment, appearing in the spring time — worn and shabby. But van. Atalanta appears only in the autumn, not as a preserved creature, but a recent production; and hence we can ascertain the period of its life to be comprised only between those few days that intervene from the end of September to the end of October, by which time its food in our gardens has pretty well disappeared. Some sheltered wall, garnished with the bloom of the ivy, may prolong its being a little longer, but the cold and dampness of the season soon destroy it; rendering the life of this creature, the most beautiful of our lepidopterous tribes, of very brief duration.

In the autumn of one year I had the decayed fruit and windfalls of my garden, placed in a sheltered nook beneath a southern wall, with the design of attracting all the wandering individuals of this splendid species, that I might observe their actions and their beauty without molesting them, and in the earlier part of a sunny morning a little flock, ten or a dozen, attended to partake of my banquet in full enjoyment, basking, flapping, and displaying their wings in the warm beam, and running about the ground and fruit, more freely and easily that I have observed any other butterfly do, and so tame as to permit their wings to be touched, circling only around and then returning; but the Admiral alone came to feed at my table, *hesp^a. aegeria*, now and then inspected the contents of my dishes, but found them unpalatable and was away: as the year advanced, my guests, fickle beauties, all left me and became wanderers and associates with bees and flies upon the ivied wall.

Of the few faculties that we perceive in creatures, the intelligence they all manifest of obtaining access to their food is very observable. Insects in particular will pass over a district unmindful of its product, if their peculiar nutriment or fancy, is not contained in it, or they will visit a situation in which they are unusually found, when a favourite plant is growing in it. How this knowledge is obtained we know not, but probably other faculties to guide them besides sight or smell, or these greatly exalted beyond human power, are possessed by them: I have often admired the sagacity of the hive bee, and the wasp, when attempting in my garden to procure the honey secreted on the nectaries of the scarlet salvia's (*splendens,* & *rubra*) unable from the closing of the mouth, flatness, and smallness of the tube of the blossom to penetrate to the liquor, they watch it falling from the calyx, which their weight often promotes, and immediately descending to the ground fix upon the white extremity and collect the honey pretty copiously found in that part of the tube: an incurious observer might notice numbers of these insects buzzing over the plant, and crawling upon the ground, but without watching their actions might be puzzled to guess what their objects were. The odours proceeding from the flowers are probably as various as the individuals themselves, and it seems rather singular that amidst the vast variety of foreign plants introduced into our gardens, how few of them are noticed by our indigenous insects, generally speaking, the extensive collections and varieties found even in nursery gardens do not seem to attract them in the manner so much gaiety might lead us to suppose would be the case, and all the splendour of our borders becomes often slighted for a thistle, a scabious, a charlock, or a tuft of thyme; plants long naturalized as mignonette, the phloxes, asters, lavender &c, seem preferred to more recent introductions: farina, and the requirements for the construction of the cells of the bees, may be sought for in some cases, but food is the chief demand of insects, and possibly the faculties of these creatures enable them to detect in the odour, the presence of nectareous liquors, and fix, or avoid such as may be strange or dubious.

The gamma moth (*phalæna gamma*) is also another creature that seems in no way affected by moist seasons, which retard the appearance, or apparently destroy so many others of its kind. This creature has imprinted on its dark wings a white character, something like the letter Y, but more like the small Greek gamma, and hence has received a pertinent name. Like Cain, it bears with it, in all its wanderings, a mark that distinguishes it from others of its race. Its habits, also, are quite unlike those of other moths, as it feeds principally in the day-time; and we see it late in the summer whisking about with all the activity and action of the hummingbird sphinx. Like the latter it keeps its wings, while feeding, in a constant state of vibration; haunts clover fields, and the yellow blossoms of the wild mustard, and the heads of the pasture scabious. It seems little mindful of the common frosts of October, retiring from us with such reluctance, that, should the autumn be fine, we not uncommonly find it in some piece of aftergrass, enjoying there the few flowers which linger out the approaches of November.

But I remember no instance more notable than that which occurred in the spring of 1834. We observed as early as March 20th innumerable patches of some lepidopterous creature, feeding in a cold and clayey field upon the young foliage of *Scabiosa suc*[a]: or basking upon a decayed leaf in the feeble sun-shine of the season, though in general the severity of the period seemed unheeded by them: these patches were so numerous and the colour of the larvae so dark that they appeared like little heaps of soot, and when the creatures dispersed very early in April the field seemed almost as if sowed with caterpillars; the small birds passed them by, probably deterred by their spiny coating: upon assuming the chrysalis state a few were suspended but many retired beneath the soil. About May 20th, the fly appeared as *Melitaea artemis*, an insect so scarce with us that in the long space of above twenty years, two or three only were annually observed, but were now swarming in almost incredible profusion, some suspended cause at this time operating, or it may be that many years are required to perfect the eggs deposited in the soil, as there were no visible parents to give being to such myriads of offspring. Many of the larvae never attained chrysalis state, having been punctured by a predaceous insect and become surrounded with numerous little white cocoons, which in the space of a week produced a fly, apparently *Microgaster glomeratus* – the existence of the Melitaea was brief, the plumage soon became shabby and after about a fortnight of frolick, they all disappeared.

In the following August in the same places which had been the scenes of their parents festivity, we observed numbers of these little sooty larvae again appear congregated in a web upon the foliage of the Scabious, from this web they did not wander, adhering to that plant it appeared which was to support them in spring, but remained in a lethargic state. Those we confined died, but those in the fields retired beneath the soil, to remain we apprehend until the spring, as the larvae of a few of our Lepidopterous creatures do, *Arctia vil*[a], and others – and appear as a fly in May – and thus the eggs being hatched towards the end of the summer, they must remain in the larva state for nine months, seven of which would be in that of rest, a condition one would suppose of peculiar hazard in such a climate of moisture and caprice as ours. Yet as such is the appointed order, it must be ordained from a reason the wisdom and expedience of which we are ignorant of.

We have no moth more strikingly handsome, or which engages more attention when seen, than the common tiger (*Arctia caja*): it is by no means unusual in many places, at certain periods, but, like so many others of these races, makes its appearance at very uncertain intervals. The larvae may usually be found among the foliage of the strawberry in June, covered with long grey hair, dark on the back, and rufous on the undersides; and they delight to feed upon the leaves of the lettuce, appearing as a moth in July or August. Some of the very heavy-bodied moths seem only rarely to make use of their wings, not flitting about from sweet to sweet, or traversing the hedge in search of their mates, but depositing their eggs near the spot where they were perfected, soon after expire. Annually viewing this creature, yet I have never seen it in flight, but stationary, or crawling into retirement; still it appears, that during some hour of the night it has used its wings, from the frequency with which we find its handsome relics scattered in glades, or places frequented by night-birds of the larger kind, most of which creatures capture their insect food when on the wing, not seeking it in a dormant state, like many of our day-feeding birds. The hair with which so many of our caterpillars are supplied has uses not quite obvious to us; it is eminently useful at times to preserve some of the heavy-bodied ones from being bruised by falling on the ground by accident, or from the influence of winds or storms; but some equally heavy, and subject to similar casualties, are not furnished with these elastic hairs, and others of the smaller and light-bodied species are yet very hairy. This matter, again, contributes to the perceptions of the creature, as is evident from their shrinking, or avoidance, when touched; but those that are destitute of it have no apparent equivalent for perception, rendering this circumstance only a partial incident. Those that are supplied with hairy coating make great use of it for fabricating their cases: drawing a web, like that supplied by the spider, from an interior receptacle, they weave it up into a rough mantle, wrapped in which they undergo their change, and finally assume a perfect state. The hairless ones abrade particles of wood and other matters, and, by aid of a peculiar gluten, compose a case impenetrable to moisture, and not injured by the severities of the winter season.

It is very pleasing to observe the provisions that are made by creatures for the security, and, in many cases, comfort of their young: we see the land-birds collect a variety of materials, and, where requisite, of warm substances, to shelter their broods. Others require no such provision. The water birds provide down from their bodies to line their nests: vegetable matters would soon become damp, but this plumage contracts little moisture, and hence the eggs are kept dry in humid situations. The stop, or nest of a rabbit, is a very conspicuous instance of maternal care: the mother, plucking off nearly all the hair and fur from her stomach, and mingling it with short dry grass, forms a mass of materials for the comfort of her young ones, securing them in it with great art, and visiting them with the utmost vigilance and caution. The insect weaves up the hair of its skin to form a covering, a web of cotton or of silk impervious to the contingencies of the weather — all tending to the security and well-being of its young; manifesting the deep and settled affection for its offspring impressed by the Creator upon the parent, in whatever grade or state it may be placed; "for even the sea monsters draw out the breast." We have one example given us of indifference, and it seems advanced as a solitary circumstance, a deviation from the general analogy of nature beyond the comprehension of man — that of the ostrich, which is "hardened against her young;" and why? because "God hath deprived her of understanding," hath especially withheld from her, from motives not assigned, that ascendant tenderness with which he has endowed all others. We may not understand a motive — we may misconceive an object; but to render it adverse to an apparent universal appointment, cannot safely be attempted.

In the autumn of the year 1827, the larvae of the goat moth (*phalæna cossus*) abounded beyond any customary proportion, and we could commonly see the traces made by these creatures in the dust. They had apparently fed during the summer in the earth, and were now proceeding in search of a retreat during winter to some old hedge-row tree — a part to repose, and those which approached maturity to abrade the softer wood, and form their cases, preparatory to changing to a final perfect state in the spring. At times we observed them coursing along our paths with great strength and activity; and when not seen, that peculiar subtile smell, which proceeds from them, and has been thought to resemble that of the goat, was perceptible in all our walks. The object and seat of this odour seem not well understood. Some have conjectured it to proceed from a fluid evacuated from the mouth, and discharged to soften the wood in which they burrow. But it seems inconsistent with any probability, that this creature, which is furnished with such very powerful mandibles, should be gifted with an auxiliary aid to accomplish its object; while, of the many insects that perforate timber, most of them with inferior means, no other possesses an equivalent agent to facilitate its labours; for not one of them, as far as we know, is so supplied. Besides, if such were the purpose, the discharge would be made only when required, and thus this unpleasant odour not always be perceptible. The microscope, too, does not manifest the exudation of any fluid. The larva is furnished with eight curious retractile processes on its under side, in the manner of what entomologists call the "prolegs." These are encircled with little hooks, made use of probably to remove the fragments of the wood, when broken off by the mandibles above, and clear the passages. The strength of these jaws is so great that they will very soon destroy any common chip box in which the animal may be placed, by abrading the edges, to effect its escape. With us they chiefly inhabit the ash; and we very commonly see at the roots of our aged trees the fragments removed by them in forming their passages. In breaking up the decayed pollards, we not unusually find the grub in all the stages of its growth; but more generally observe them without inhabitants, yet perforated with paths large enough to admit the finger. I suspect that these "augerworms" are the primary cause of the decay of the tree; having often observed their perforations, and found them, both large and small, in the solid spur or root of the tree, when the upper portion, having been bored, and in a state of decline, is abandoned by them. Those that are full fed appear to form their cases in that part which has lost coherency, while the younger and imperfected creatures mine their way, and obtain nutriment in the solid timber, thus killing the tree by inches; when rain and moisture find lodgement, and complete the dissolution. One year's preparation is the period usually assigned to the larvæ of most insects, before they arrive at their perfect state; but by the goat moth three years are required before it attains its winged state from the egg. Consequently, for the larger portion of its life it is occupied in these destructive operations; and thus this creature becomes a very powerful agent in reducing these Titans of the vegetable world, crumbling them away to their original dust: for what was decreed to be the termination and punishment of man is found in active operation throughout the whole chain of Nature's works, which are but dust, and unto dust return, continuing an endless series of production and decay, of restoration and of change. All these larvæ which I have observed in the colder portions of our year, were hard, stiff, and torpid, but soon became relaxed and animated by the warmth of the hand: thus they probably remain quiet during the winter months, but revive in spring, and recommence their ravage in the tree. The caterpillar of this moth I believe to be the largest of any of those of the British lepidopteræ; and, when full fed, exceeds in size that of the death's-head sphinx. To those who dislike the appearance of things of this nature, it is particularly disgusting; not only from its magnitude and smell, but from its colour, which is a lurid red, so compounded with a dingy yellow, as to give it a lividness of look, conveying the idea of something raw. Common as the grub is in some years, I have seldom been able to obtain the moth without the often tedious process of feeding the larvæ, and waiting for its change.

The goat moth seems remarkably tenacious of life, and possessed with a strength of vitality that is unsubduable by the common agents which effect the destruction of others; even Prussic acid does not act with decision in this case, and at times appears to have no influence.

Of those caterpillars which feed upon the foliage of vegetation, a considerable portion are picked off and consumed by the numerous little birds which are constantly hunting after them, as food for themselves or their young ones; and many of those which are supported by the roots of plants, and remain covered in the soil, are detected by the perception of rooks, and birds of that order: but those which feed upon the internal parts of trees seem exempted from any of these causes or destruction. This is possibly a reason that the larvae of *phalæna cossus* is so plentifully found; but yet it is pretty certain that some other and equally fatal visitation assails them, and reduces their numbers during the long period which is required to perfect their state: for though, by feeding and care (for they are very impatient of confinement), we can obtain the moth in numbers, yet few seem to survive and become perfected by the common processes of Nature — at least I have seldom found them in this state, though the larvæ is so plentifully seen.

THE DESIGNS OF Supreme Intelligence in the creation and preservation of the insect world, and the regulations and appointments whereby their increase or decrease is maintained, and periodical appearance prescribed, are among the most perplexing considerations of natural history. That insects are kept in reserve for stated seasons of action we know, being commonly made the agents of Providence in his visitations of mankind. The locust, the caterpillar, the palmer worm, the various family of blights that poison in the spring all the promise of the year, are insects. Mildew, indeed, is a vegetable; but the wireworm destroys the root, the thrips the germe of the wheat, and hunger and famine ensue. Many of the coleopteræ remove nuisances, others again incumbrances, and worms manure the soil; but these are trite and isolated cases in the profusion of the animal world; and left alone, as we are, in the desert of mere reason and conjecture, there is no probability that much satisfactory elucidation will be obtained. They are not, perhaps, important objects of inquiry; but when we see the extraordinary care and attention that have been bestowed upon this part of creation, our astonishment is excited, and forces into action that inherent desire in our minds to seek into hidden things. In some calm summer's evening ramble, we see the air filled with sportive animated beings: the leaf, the branch, the bark of the tree, every mossy bank, the pool, the ditch, all teeming with animated life, with a profusion, an endless variety of existence; each creature pursuing its own separate purpose in a settled course of action, admitting of no deviation or substitution, to accomplish or promote some ordained object. Some appear occupied in seeking for the most appropriate stations for their own necessities, and exerting stratagems and wiles to secure the lives of themselves or their offspring against natural or possible injuries, with a forethought equivalent or superior to reason; the aim in some others we can little perceive, or, should some flash of light spring up, and give us a momentary glimpse of Nature's hidden ways, immediate darkness closes round, and renders our ignorance more manifest. We see a wonderfully-fabricated creature struggling from the cradle of its being, just perfected by the elaboration of months or years, and decorated with a vest of glorious splendour; it spreads its wings to the light of heaven, and becomes the next moment, perhaps, with all its marvellous construction, instinct, and splendour, the prey of some wandering bird! and human wisdom and conjecture are humbled to the dust. That these events are ordinations of Supreme Intelligence, for wise and good purposes, we are convinced; but are blind, beyond thought, as to secondary causes; and admiration, that pure source of intellectual pleasure, is almost alone permitted to us. If we attempt to proceed beyond this, we are generally lost in the mystery with which the divine Architect has thought fit to surround his works; and, perhaps, our very aspirations after knowledge increase in us a sense of our ignorance: every deep investigator into the works of Nature can scarcely possess other than an humble mind.

In all our pursuits we shall find in Nature, wheresoever we can penetrate, a formation, a faculty adapted to all the wants and comforts of the creature; yet the objects of infinite wisdom in the creation of this world of matter, animate and inanimate, will probably never be made known to mankind; for though knowledge is in a constant progressive state, and the attainments of science in latter years have been comparatively prodigious, yet these acquirements are, in fact, but entanglements: they lead us deeper into surprise and perplexity, and the little perceptions of light which we obtain serve to show how hopeless any attempt must be to penetrate the secrets of infinity — a conviction, if we "dwell deep in the valley of humility," that will in no manner discourage our pursuits, but rather incite our ardour to investigate so exhaustless a store, which will lead us, from contemplation, to admiration, to devotion.

That pretty sparkler of our summer evenings, so often made the ploughboy's prize, the only brilliant that glitters in the rustic's hat, the glowworm (*lampyris noctiluca*), is not found in such numbers with us as in many other places, where these signal tapers glimmer upon every grassy bank; yet, in some seasons, we have a reasonable sprinkling of them. Every body probably knows, that the male glowworm is a winged, erratic animal, yet may not have seen him. He has ever been a scarce creature to me, meeting, perhaps, with one or two in a year; and, when found, always a subject of admiration. Most creatures have their eyes so placed as to be enabled to see about them; or, as Hook says of the house-fly, to be "circumspect animals;" but this male glowworm has a contrivance by which any upward or side vision is prevented. Viewed when at rest, no portion of his eye is visible, but the head is margined with a horny band, or plate, being a character of one of the genera of the order Coleoptera, under which the eyes are situate. This prevents all upward vision; and blinds, or winkers, are so fixed at the sides of his eyes as greatly to impede the view of all lateral objects. (See Plate 5, Fig. 2.) The chief end of this creature in his nightly peregrinations is to seek his mate, always beneath him on the earth; and hence this apparatus appears designed to facilitate his search, confining his view entirely to what is before or below him. The first serves to direct his flight, the other presents the object of his pursuit; and as we commonly, and with advantage, place our hand over the brow, to obstruct the rays of light falling from above, which enables us to see clearer an object on the ground, so must the projecting hood of this creature converge the visual rays to a point beneath. This is a very curious provision for the purposes of the insect, if my conception of its design be reasonable. Possibly the same ideas may have been brought forward by others; but as I have not seen them, I am not guilty of any undue appropriation; and no injury can be done to the cause I wish to promote, by detailing again such beautiful and admirable contrivances.

Glowworms emit light only for a short period in the year; and I have but partially observed it after the middle of July. I have collected many of these pretty creatures on a bank before my house, into which they retire during the winter, to shine out again when revived by the summer's warmth; but in this latter season I have frequently missed certain of my little proteges, and have reason to apprehend that they formed the banquet of a toad, that frequented the same situation.

Observing, above, that the glowworm does not emit light after the 14th of July, I mean thereby that clear, steady light which has rendered this creature so remarkable to all persons; for I have repeatedly noticed, deep in the herbage, a faint particularly manifested September the 28th, 1826. The evening was warm and dewy, and we observed on the house-bank multitudes of these small evanescent sparks in the grass. The light displayed was very different from that which they exhibit in the warm summer months. Instead of the permanent green glow that illumines all the blades of the surrounding herbage, it was a pale transient spot, visible for a moment or two, and then so speedily hidden, that we were obliged, in order to capture the creature, to employ the light of a candle. The number of them and their actions, creeping away from our sight, contrary to that half-lifeless dulness observed in summer, suggested the idea that the whole body had availed themselves of this warm, moist evening, to migrate to their winter station. A single spark or so was to be seen some evenings after this, but no such large moving parties were discovered again. If we conclude that the summer light of the glowworm is displayed as a signal taper, the appearance of this autumnal light can have no such object in view, nor can we rationally assign any use of it to the creature itself, unless, indeed, it serves as a point of union in these supposed migrations, like the leading call in the flight of night-moving birds. The activity and numbers of these insects, in the above-mentioned evening, enabled me to observe the frequent presence and disappearance of the light of an individual, which did not seem to be the result of will, but produced by situation. During the time the insect crawled along the ground, or upon the fine grass, the glow was hidden; but on its mounting any little blade, or sprig of moss, it turned round and presented the luminous caudal spot, which, on its falling or regaining its level, was hidden again.

Few animals seem feebler and less disposed to action than the female glowworm, yet scarcely any one is more uneasy in confinement, nor more successful in effecting its escape. Inclosed under a cup, or in any common box, if there is the least crevice, it will force its way through and depart before the ensuing morning, especially if confined late in the season; and this it appears to effect by its strength and the ability it possesses of contraction. As the faculties or powers bestowed upon creatures are not given them for any useless display, but to accomplish some immediate necessity, or remote requirement, we must consider the strength possessed by some insects as subservient to their peculiar habits. Many insects are endowed with remarkable muscular powers, the common *Melol. vulg.* (chafer), the *Geotrupes ster.* (dorr), held in the hand or placed under a weight, manifest prodigious strength. The first, a creature perfected beneath and inclosed in the soil, requires great power to liberate itself from confinement, which, probably, had it been more feeble, it would not have been enabled to do; and the perforations required by the female when she deposits her eggs, are the efforts of much strength. The employ of the dorr, again, is to mine holes in the soil, remove the earth, and secrete the nuisances and incumbrances that may be found upon the surface; and this no weak animal could accomplish: but the strength of this beetle almost exceeds credibility. It has little power as a draught animal, but his business is to heave up the earth, entombing matters, and his muscular means appear to be situated in his legs, the upper joint of which is very large and firm. Having repeatedly placed one of these creatures, weighing 15 grains, under a weight equal to 4796 grains, sufficient, it would be considered, to crush its body, 319 times its own weight! it heaved it up and withdrew, and the same pressure, being placed on its leg, was immediately disengaged by the powers of the other. Man effects his objects by the reasonings of his mind, mechanical agencies, or the strength of others: had he depended upon, mere animal power to accomplish his wishes, in order to equal the means of a common beetle, he must have raised his body from an incumbent pressure of perhaps 20 tons! — Our glowworm requires all its faculties, retiring in autumn into the crevices of a stony or earthy soil, where it passes its inanimate hours: before the spring arrives, all these passages by which it entered would probably be closed by the decomposition of the one or mouldering of the other, through the agency of frosts and rains; and it is thus probably endowed with strength, and the faculty of contraction and flattening its body, in order that it may remove the weights, or squeeze through the impediments that check the return to light and warmth, and the accomplishment of the purport of its being. These are but lowly things to converse upon, creatures fashioned beneath in the earth; yet, hallowed by their Creator's hand, they manifest his omniscience, and we cannot but revere his wisdom and goodness.

My labourer this day (July the 18th), in turning over some manure, laid open a mass of snake's eggs (*coluber natrix*), fifteen only; and they must have been recently deposited, the manure having very lately been placed where they were found. (Plate 5, Fig. 4.) They were larger than the eggs of a sparrow, obtuse at each end, of a very pale yellow colour, feeling tough and soft, like little bags of some gelatinous substance. The interior part consisted of a glareous matter like that of the hen, enveloping the young snake — imperfect, yet the eyes and form sufficiently defined. Snakes must protrude their eggs singly, but probably all at one time, as they preserve no regular disposition of them, but place them in a promiscuous heap. At the time of protrusion they appear to be surrounded with a clammy substance, which, drying in the air, leaves the mass of eggs united wherever they touch each other. I have heard of forty eggs being found in these deposits; yet, notwithstanding such provision for multitudes, the snake, generally speaking, is not a very common animal.

The kite, the buzzard, and the raven, which prey on it occasionally, are too seldom found greatly to reduce the race; and its deep retirement in the winter seems to secure it from fatal injuries by the severity of the weather: yet, in the warm days of spring, when it awakens from its torpidity and basks upon our sunny banks, the numbers that appear are not proportionate to what might be expected from the number of eggs produced, Few creatures can assail it in its dormitory, yet its paucity proves that it is not exempt from mortality and loss. The mole may follow it in its retirement, but would hardly attempt to seize so large an animal. The polecat and the weazel, too, can enter its runs; are sufficiently bold and strong to attempt the conquest; and not improbably in the winter season resort to such food, the poor snake having no power of defending itself, or of avoiding the assault. The common snake of this country is a very harmless, unobtrusive creature; so timid, as to avoid the presence of man whenever he appears, hiding itself as much as possible in bushes and rugged places from his sight. At times a strong fetor proceeds from it; but this appears to be sexual, or made use of as the means of annoying its enemies. It possesses no power to commit injury, and has apparently no inclination to molest any thing beyond its requirements for food, as frogs and mice. When a young man, I have repeatedly handled it with impunity; and though often bitten, a temporary swelling, with slight inflammation, was the only result; but in these experiments the viper must not be mistaken for the common snake. Yet this poor creature, under the curse of ignorance and cruelty, never escapes unscathed from power and opportunity. All the snake tribe, innocuous and pernicious, seem to be viewed with horror and aversion by mankind. This horror, from the knowledge of their power of inflicting harm in countries where such kinds are found, is natural, and often preservative of life; but the aversion generally felt, and that shuddering occasionally noticed, at the sight of our harmless snake, are like a deep-rooted principle. We imbibe in infancy, and long retain in remembrance, the impression of injuries from the wiles of the serpent; and the "enmity between it and the seed of the woman" appears still in full operation, and is possibly more extensively and insensibly diffused among mankind than we are aware of. The harmless nature of our snake seems to be fully known to the little birds of the hedge, as they in no way give intimation of its presence by any warning of avoidance to their young, or that insulting vociferation so observable when any really injurious creature is perceived, but hop and sport about the basking snake without fear or notice.

All the human race seem to have inherited the original anathema against this creature; for though the capricious cruelty of man is very frequently exerted to the injury of many that his power enables him to tyrannize over, yet the serpent appears to be a peculiar object of his enmity, as if it was understood to be an absolute duty to "bruise his head" whenever the opportunity should be afforded. It is very remarkable how few noxious creatures, animals which annoy man, inhabit with us; beasts and birds we have none, for the petty depredations occasionally made on his property are undeserving of attention. The gnat, and perhaps a few insects, may at times puncture our skin; but the period of action is brief, the injury only temporary. The wasp and the hornet, I believe, very rarely use their weapons wantonly, only in self-defence, and when persecuted: thus leaving the balance incalculably in favour of innocency and harmlessness. But of all the guiltless beings which are met with, we have none less chargeable with criminality than the poor slow worm (*anguis fragilis*), yet none are more frequently destroyed than it — included as it is in the general and deep-rooted prejudice attached to the serpent race. The viper and the snake, though they experience no mercy, escape often by activity of action; but this creature, from the slowness of his movements, falls a more frequent victim. We call it a "blind worm," possibly from the supposition that, as it makes little effort to escape, it sees badly; but its eyes, though rather small, are clear and lively, with no apparent defect of vision. The natural habits of the slowworm are obscure; and, living in the deepest foliage, and the roughest banks, he is generally secreted from observation; but loving warmth, like all his race, he creeps half torpid from his hole, to bask in spring-time in the rays of the sun, and is, if seen, inevitably destroyed. Exquisitely formed as all these gliding creatures are, for rapid and uninterrupted transit through herbage and such impediments, it is yet impossible to examine a slow worm without admiration at the peculiar neatness and fineness of the scales with which it is covered. All separate as they are, yet they lap over, and close upon each other with such exquisite exactitude, as to appear only as faint markings upon the skin, requiring a magnifier to ascertain their separations; and, to give him additional facility of proceeding through rough places, these are all highly polished, appearing lustrous in the sun, the animal looking like a thick piece of tarnished copper wire. When surprised in his transit from the hedge, contrary to the custom of the snake or viper, which writhe themselves away into the grass in the ditch, he stops, as if fearful of proceeding, or to escape observation by remaining motionless, but if touched he makes some effort to escape: this habit of the poor slowworm becomes frequently the cause of his destruction.

Of all the active, vigilant creatures that animate our paths, we have none superior to the little, beelike bombylius (*bombylius medius*); but this creature is to be seen only in the mornings of a few bright days in spring, seeming to delight in the hot, windy gleams of that season, presenting an emblem of that

THE GREAT BLACK ANT

portion of our year, fugitive and violent. It is, I believe, plentiful nowhere. Particularly solicitous of warmth, it seeks the dry, sunny reflection of some sheltered gravel-walk, or ditch-bank in a warm lane; and here it darts and whisks about, in seeming continual suspicion of danger; starting away with angry haste, yet returning immediately to the spot it had left; buffeting and contending with every winged fly that approaches, with a jealous, pugnacious fury, that keeps it in constant agitation. This action, its long projecting proboscis, and its pretty, spotted wings, placed at right angles with its body, distinguish our bombylius from every other creature. It appears singularly cautious of settling on the ground. After long hovering over and surveying some open spot, with due deliberation and the utmost gentleness it commits its long, delicate feet to the earth; but on the approach of any winged insect, or on the least alarm, is away again to combat or escape. Associates it has none: the approach even of its own race excites its ire, and, darting at them with the celerity of thought, it drives them from its haunts. When a captive, it becomes tame and subdued, and loses all its characteristic bustling and activity, the inspiration of freedom. The contention of a pair of these creatures at times when sufficiently animated, affords an example of ardour not often witnessed, the length of time it continues, the hostile murmurs excited indicate peculiar irritation, but the rapidity of their circlings, and when the eye can follow them, the precise distances they observe in their gyrations, as if they were attached by some rigid body that kept them from retreat or approach is very remarkable, and how such regularity can be preserved under this celerity of movement, and variety of evolution, is not easy to conjecture, or find a counterpart.

The great black ant (*formica fuliginosa*) is commonly found in all little copses, animating by its numbers those large heaps of vegetable fragments which it collects and is constantly increasing with unwearied industry and perseverance as a receptacle for its eggs. The game-fowl, the woodpecker, the wryneck, and all the birds that feed upon the little red ant, and soon depopulate the hillocks which they select, do not seem equally to annoy this larger species. These systematic creatures appear always to travel from and return to their nests in direct lines, from which no trifling obstacle will divert them; and any interruption on this public highway they resent, menacing the intruder with their vengeance. A neighbour related to me an instance of this unyielding disposition, which he witnessed in one of our lanes. Two parties of these black ants were proceeding from different nests upon a foraging expedition, when the separate bodies happened to meet each other. Neither would give way; and a violent contest for the passage ensued. After a time the combat ceased, and all animosity subsided, each party retiring to its nest, carrying with it its dead and maimed companions. This encounter seemed quite accidental; and the disposition to move in a uniform line, which their meeting prevented, the sole cause of their hostility, combat, and mutual injury. The strength of some creatures, especially insects, considering the smallness of their size, is in several instances prodigious. Man, by his reason and power, calls to his aid mechanical means, and other agents, to effect his objects; but unreasoning beings accomplish their purposes by contrivance and bodily powers. The strength of these black ants is manifested by the quantity and magnitude of the materials which they collect for their heaps; but the common little red ant (*formica rubea*) a much smaller creature, gives daily proofs of its abilities to remove heavy substances, equal to any that we meet with. One of these little creatures, thirty-six of which only weigh a single grain, I have seen bear away the great black fly as its prize, equal to a grain in weight, with considerable ease; and even the wasp, which exceeds forty times its own weight, will be dragged away by the labour and perseverance of an individual emmet. These little ants are occasionally and profusely deprived of their lives by some unknown visitation. In the year 1826, in particular, and again in the following year, I observed, in the month of August, a lane strewed with their bodies They had bred during the summer in an adjoining bank; but some fatality had overwhelmed them when absent from their nests, and nearly annihilated the fraternity, as only a few scattered survivors were to be seen, feebly inspecting the bodies of their associates. The task of removal, however, with all their industry, appeared beyond their powers to accomplish, as on the ensuing day few had been taken away. Had these creatures been destroyed in combat by rival contention, the animosity must have been excessive; but it is more probable that they met their death by some other infliction.

One year, on the 3d of March, my labourer being employed in cutting up anthills, or tumps, as we call them, exposed to view multitudes of the yellow species (*formica flava*) in their winter's retirement. They were collected in numbers in little cells and compartments, communicating with others by means of narrow passages. In many of the cells they had deposited their larvæ, which they were surrounding and attending, but not brooding over or covering. Being disturbed by our rude operations, they removed them from our sight to more hidden compartments. The larvæ were small. Some of these anthills contained multitudes of the young of the woodlouse (*oniscus armadillo*), inhabiting with perfect familiarity the same compartments as the ants, crawling about with great activity with them, and perfectly domesticated with each other. They were small and white; but the constant vibration of their antennae, and the alacrity of their motions, manifested a healthy vigour, The ants were in a somewhat torpid state; but on being removed into a temperate room, they assumed much of their summer animation. How these creatures are supported during the winter season it is difficult to comprehend, as in no one instance could we perceive any store or provision made for the supply of their wants. The minute size of the larvæ manifested that they had been recently deposited; and consequently that their parents had not remained during winter in a dormant state, and thus free from the calls of hunger. The preceding month of February, and part of January, had been remarkably severe; the frost had penetrated deep into the earth, and long held it frozen; the ants were in many cases not more than four inches beneath the surface, and must have been enclosed in a mass of frozen soil for a long period; yet they, their young, and the onisci, were perfectly uninjured by it; affording another proof of the fallacy of the commonly-received opinion, that cold is universally destructive to insect life. Some creatures may be injured or destroyed by frost; but the larger portion of them Nature has provided with constitutions to which it is innocuous, or furnished with instinct to prevent its harming them. These emmets had probably received no sustenance, or required any, from the time of their retirement in the autumn — a period of full six months; were inclosed during the space of thirty days in a mass of frozen earth, and yet remained perfectly uninjured by this long abstinence and frost.

I suspect from some observations, that the common hill ants are heliotropes, and that they perform in the chillier portions of our seasons, the daily circuit of their little mounts, obtaining the first rays of the morning at the eastern edges, following the course of that luminary until it declines in the west. The power of induring cold in insects, or as it seems the salubrity of it for some species, is strongly manifested in a moderately sized moth, a night flyer, which we frequently find drowned in pools or broad ditches, where it remains frozen on the surface, embalmed in ice. But in all cases its death was probably occasioned by drowning, not by cold, as even in the genial days of our winterly months we yet find many of these moths floating upon the surface, and with the exception of the winter gnat, is the only winged insect visible in this season. It is difficult to conceive what that food must be, which sustains life in these creatures in January and February, as the honey secreted by flowers is as far as we know, the only sustenance. These grey moths vary so much, and though so numerous, are not satisfactorily described by any authority I have consulted. But one of the *tineidæ, phalæna colonella,* of Linnæus is the most accordant with them that I have met with, or the *ilythia collonella* of Latreille.

In contemplating that wonderful creation which is spread before us upon the surface of the earth, which most men at times either partially or generally must in some manner do, we cannot avoid resting for a moment upon that immense though often minute part of it, the innumerable tribes of animated beings called insects, which abound in every region and space of it. We see that they are in existence from age to age, and that ample provision for this mighty host, which no man can number, is provided, likewise, leading us to wonder from what source is this profusion and variety obtained. The herbs, the fruits of the soil, juices and secretions of plants supply a part, and the substance of other beings become the only meant of some, but there is a very large portion of creation which not apparently supported by these means, but are continued in being by establishments or contrivances not manifested to us.

When we observe in a fine summer evening the entire atmosphere filled as far as vision extends with prodigious myriads of winged insects, we conclude that every portion beyond our eyesight is equally occupied with the same profusion of life, and enquire what food can supply the fragile and minute bodies of these multitudes. We do not observe that they alight upon any vegetable matter to obtain their requirements, in any numbers, and they would obscure it if they did so, most of them seeming to preserve a certain range, and perform all the gyrations of the hour within it. During any mild gleam even in the severity of our winter, large troops of *tipulæ* are enticed into action. The gleam disappears and they vanish. Now all these creatures of the wood, the marsh, the mead, the wintry hour, which we see in constant action, "sport", as we call it, upon the sunny beam. Why then is the insect world thus abandoned to liberty like this, if no object is to be obtained? Young creatures at times are full of frolic, and a momentary prank or gaiety may activate the more advanced in age, but the rest, feeding, or seeking supply, are the occupations of the life of creatures. The fly that cruises almost the day long beneath our ceilings, the gnat that dances for hours in those little social troupes, and all the varieties that float upon the slanting beam of the evening, have apparently no concert with their fellows, but move in individual spaces occupied with themselves. If we must consider this a recreation, we become involved in perplexing considerations. We must admit something analogous to intellect, an incitement to gratifications not observable in any other portion of creation. Have we not then mis-called these actions of insects, and are they not reducible to the same common objects and attainments as influence the whole visible creation?

As proof in such cases may be with difficulty obtained, perhaps conjecture when not a violation of reason, may be admitted – this conjecture is that these creatures imbibe oxygen from the atmosphere, which when so imbibed, unites with the carbon, and hydrogen of the animal, existing in the oil, the fat, the gelatine, the albumen, the ammonia and the water of the creature. They may even separate the necessary gases from other sources, and these united will form sugar, a saccharine matter for nutriment, and the action which we have called the "sporting of the insect", is perhaps a process accelerating the combination of the gases, to effect the formation of the saccharine nutriment, or the insect's mode of feeding: pasturing in the atmosphere. The common gnat may frequently be observed fixed firmly upon some support, with inclined head, and in that state continue rapidly agitating its body up and down for a long space. Has not this motion an appearance of being a means of producing the like results, a feeding without volant action?

To all created beings a certain sphere of operation must be assigned, a destination of usefulness. And yet how little do we know of the agencies of insects, of these infinite millions which animate the earth and fill the atmosphere! To the vegetable world, actions of great importance are allowed, and may we not arrange insects in the rank of utility with these often very lowly products of the earth, considering them as a portion of that vast machinery for neutralizing the carbonic and hydrogen gases which are constantly floating about, to mix with and contaminate the air. Thus, by an act which manufactures their own subsistence, they purify the element for the respiration of other races of beings. A provision of this nature renders the subsistence of the insect world, even those which continue in activity during the most destitute periods of our year, a matter of facility, when other means of supply are not apparent, and is fully accordant with that simplicity, mercy and wisdom which the Almighty power has in every thing manifested to us.

Water, in a state of rest over decayed and putrescent vegetable matter, is peculiarly favourable for the residence of many of the insect world. The eggs that are lodged there remain undisturbed by the agitation of the element, and the young produced from them, or deposited there by viviparous creatures, remain in quiet, tolerably secure from accidental injuries; but there are natural causes which render these apparent asylums the fields of ravenousness and of death. To these places resort many of those voracious insects and other creatures which prey upon the smaller and helpless; for all created things seem subordinate to some more powerful or irresistible agent, from the hardly visible atom that floats in the pool, to man, who claims and commands the earth as his own. But we have no animal that seems to commit greater destruction in these places than the common newt (*lacertus aquaticus*). In some of these well-stored magazines this reptile will grow to a large size, and become unusually warty, and bloated with repletion; feeding and fattening upon the unresisting beings that abound in those dark waters wherein it loves to reside. It will take a worm from the hook of those that angle in ponds; and in some places I have seen the boys in the spring of the year draw it up by their fishing lines, a very extraordinary figure, having a small shell-fish (*tellina cornea*) attached to one or all of its feet; the toes of the newt having been accidentally introduced into the gaping shell, in its progress on the mud at the bottom of the pool, or designedly put in for the purpose of seizure, when the animal inhabitant closed the valves and entrapped the toes. But from whatever cause these shells became fixed, when the animal is drawn up hanging and wriggling with its toes fettered all round, it affords a very unusual and strange appearance.

This animal like so many other aquatic creatures deposits it spawn in the spring upon the under side of some floating foliage, the broad leaf of the common Pond weed (*Potamogeton natans*) is a very favourite situation: there are probably several reasons why such situations are eligible, but the most obvious are to secure the spawn from the undue influence of light, the sudden alternations of temperature, as water parts slowly with its heat, & to preserve it is a constant state of moisture, otherwise it would shrivel up and be destroyed. By examining these leaves about the month of June, a small gelatinous tubercle may be found, so very translucent that the young may be perceived coiled up within, the eyes, head, and the dark markings of the adult animal being visible. In a cup of water it will soon break from its confinement, exhibit great vivacity, and sport about like a small fish.

Water, quiet, still water, affords a place of action to a very amusing little fellow (*gyrinus natator*), which, about the month of April, if the weather be tolerably mild, we see gamboling upon the surface of the sheltered pool; and every schoolboy, who has angled for a minnow in the brook, is well acquainted with this merry swimmer in his shining black jacket. Retiring in the autumn, and reposing all the winter in the mud at the bottom of the pond, it awakens in the spring, rises to the surface, and commences its summer sports. They associate in small parties of ten or a dozen, near the bank, where some little projection forms a bay, or renders the water particularly tranquil; and here they will circle round each other without contention, each in his sphere, and with no apparent object, from morning until night, with great sprightliness and animation; and so lightly do they move on the fluid, as to form only some faint and transient circles on its surface. Very fond of society, we seldom see them alone, or, if parted by accident, they soon rejoin their busy companions. One pool commonly affords space for the amusement of several parties; yet they do not unite, or contend, but perform their cheerful circlings in separate family associations. If we interfere with their merriment they seem greatly alarmed, disperse, or dive to the bottom, where their fears shortly subside, as we soon again see our little merry friends gamboling as before.

THE DORR BEETLE

This lively little animal, arising from its winter retreat shortly after the frog, at times in March, continues its gambols all the summer long, remaining visible generally until the middle of October, thus enjoying a full seven months of being — a long period of existence for insects, which are creatures subject to so many contingencies, that their lives appear to be commonly but brief, and the race continued by successive productions. All these water creatures must be endowed with much perception. Cold as this element is in early spring, when the ice of winter is hardly dissolved, and the fluid only six or seven degrees above freezing, yet they become immediately sensible of this temperature, and are excited to animation and the vocations of their being.

I have never observed the larvae of this creature in any state. When they retire in the autumn, these insects appear of an uniform size, and emerging in the spring they are all apparently full grown, and during the summer none of smaller dimensions associate with the family parties. This plain, tiny, gliding water-flea seems a very unlikely creature to arrest our young attentions; but the boy with his angle has not often much to engage his notice; and the social, active parties of this nimble swimmer, presenting themselves at these periods of vacancy, become insensibly familiar to his sight, and by many of us are not observed in after life without recalling former hours — scenes of, perhaps, less anxious days: for trifles like these, by reason of some association, are often remembered, when things of greater moment pass off, and leave no trace upon our mind.

July 29. — We frequently notice in our evening walks the murmuring passage, and are often stricken by the heedless flight, of the great dorr beetle (*scarabæus stercorarius*), clocks[37], as the boys call them. But this evening my attention was called to them in particular by the constant passing of such a number as to constitute something like a little stream; and I was led to search into the object of their direct flight, as in general it is irregular and seemingly inquisitive. I soon found that they dropped on some recent nuisance: but what

---

37 Multitudes of words are retained in our language derived from very ancient dialects; and possibly the name "clock," as given to this beetle, conveying no meaning to our present comprehensions, is a corruption of some syllable in former use. Its subterranean residence might have been signified by the old word "cloax," a vault, a creature from below. Or, burrowing in filth and ordure, as it does, the epithet "clocca," the offspring of a common shore, or jakes, would not have been insignificant of its origin and habits. Fancy, too, playing with trifles, amuses itself in bandying about even its more general appellative, dorr. In old times a "dorr" was a stupid, blundering fellow; and "to dorr," was to din, or trouble with noise; both meanings applicable to the heedless flight, and loud noise, made in all the transits of this dung beetle.

powers of perception must these creatures possess, drawn from all distances and directions by the very little fetor which in such a calm evening could be diffused around! and by what inconceivable means could odours reach this beetle in such a manner as to rouse so inert an insect into action! But it is appointed one of the great scavengers of the earth, and marvellously endowed with powers of sensation, and means of effecting this purpose of its being. Exquisitely fabricated as it is to receive impressions, yet probably it is not more highly gifted than any of the other innumerable creatures, that wing their way around us, or creep about our path, though by this perceptible faculty, thus "dimly seen," it excites our wonder and surprise. "How wondrous, then, the whole!"

This creature affords us a good example of that extraordinary artifice to which some insects have recourse upon the apprehension of danger — the counterfeiting of death. The dorr, with a violent and noisy flight, proceeds on its way, or circles around with an apparent fearlessness of harm; yet the instant it is touched, or interrupted in its progress, though in no way injured, it will immediately fall to the ground, generally prostrate on its back, its limbs extended, stiff, and seemingly devoid of life, and suffering itself to be handled without manifesting any signs of animation. In time, finding no harm ensues, it resumes its former state. If our conjectures be correct, that the object of this stratagem is to preserve its life, it is difficult to comprehend how far it can be successful. Several birds feed on it, as we observed; and that others do so likewise, is evident from their castings. Of these, the owl and the nightjar catch it when on the wing; and the crows, rooks, magpies, &c., seem to have no hesitation in picking it to pieces, as well as all the other beetles that put on the semblance of death, in whatever state they find them. One or two beasts, it is said, when captured, feign death. With these exceptions, we remember none of the other orders of creation that have recourse to such an expedient upon any emergency; but with insects it is by no means an uncommon procedure, most probably resorted to by them for a motive we are not fully acquainted with, and which is in all likelihood attended with the success it was designed to effect.

The perfect cleanliness of these creatures is a very notable circumstance, when we consider that nearly their whole lives are passed in burrowing in the earth, and removing nuisances; yet such is the admirable polish of their coating and limbs, that we very seldom find any soil adhering to them. The meloe, and some of the scarabæi, upon first emerging from their winter's retreat, are commonly found with earth clinging to them; but the removal of this is one of the first operations of the creature; and all the beetle race, the chief occupation of which is crawling about the soil, and such dirty employs, are, notwithstanding, remarkable for the glossiness of their covering, and freedom from defilements of any kind. But purity of vesture seems to be a principal precept of Nature, and observable throughout creation. Fishes, from the nature of the element in which they reside, can contract but little impurity. Birds are unceasingly attentive to neatness and lustration of their plumage. All the slug race, though covered with slimy matter calculated to collect extraneous things, and reptiles, are perfectly free from soil. The fur and hair of beasts in a state of liberty and health are never filthy, or sullied with dirt. Some birds roll themselves in dust, and occasionally cover themselves with mire; beasts have the same habit; but this is not from any liking or inclination for such things, but to free themselves from annoyances, or to prevent the bites of insects. Whether birds in preening, and beasts in dressing themselves, be directed by any instinctive faculty, we know not; but they evidently derive pleasure from the operation; and thus this feeling of enjoyment, even if the sole motive, becomes to them an essential source of comfort and of health.

IT MAY BE noted probably by some, how frequently I recur to the causes and objects of the faculties, manners, and tendencies of animate and inanimate things. This recurrence springs from no cavil at the wisdom, no suspicion of the fitness, of the appointment, nor, I trust, from any excitement to presumptuous pryings into paths which are in the great deep, and not to be searched out; but are humbly indulged, from the pleasure which the contemplation of perfect wisdom, even in a state of ignorance, affords; and if by any consideration we can advance one point nearer to the comprehension of what is hidden, we infinitely increase our satisfaction and delight.

Surrounded as we are by wonders of every kind, and existing only by a miraculous concurrence of events, admiration seems the natural avocation of our being; nor is it easy to pronounce, amidst such a creation, what is most wonderful. But few things appear more incomprehensible than the constant production and re-absorption of matter, impressed upon us even by these very dorrs. An animal falls to the ground and dies; myriads of creatures are now summoned by a call, by an impulse of which we have no perception, to remove it, and prepare it for a new combination; chemical agencies, fermentation, and solution, immediately commence their actions to separate the parts; and in a short time, of all this great body, nothing remains but the framework or bones, perhaps a little hair or some wool — all the rest is departed we know not whither! Worms and insects have done their parts; the earth has received a portion, and the rest, converted into gases and exhalable matters, has dispersed all over the region, which, received into vegetable circulation, is again separated and changed, becomes modified anew, and nourishes that which is to continue the future generations of life. The petal of the rose; the pulp of the peach; the azure and the gold on the wing of the insect; all the various productions of the animal and vegetable world; the very salts and compounds of the soil, are but the changes some other matters have undergone, which have circulated through innumerable channels since the first production of all things, and no particle been lost; bearing in mind this assured truth, that all these combinations have not been effected by chance or peculiarity of circumstances, but by the predetermination of an Almighty Intelligence, who sees the station, progress, and final destination of an atom. What an infinity of power and intellective spirit does this point out! an omnipotence, which the bodied minds of us poor creatures cannot conceive! Truly may we say, "who can find out the Almighty to perfection ?"

May 24, 1827. — Abundance of cock chaffers (*melolantha vulgaris*) are flying about, yet by no means in the profusion of some years. How much at times the interests of man and the wild creatures about him are at variance! Those that are domesticated and precluded from obtaining food but by his permission, have their welfare in part identified with his — they may share in his abundance, or pine from his parsimony; but the independents of the field are differently circumstanced. The appearance of these chaffers, in any numbers, is very uncertain and partial; but in those summers when they abound, very extensive injuries frequently ensue. In the grub state, they will entirely destroy the pastures where they inhabit, by consuming the roots of the grasses; acres and fields are deprived of their produce, becoming brown as stubbles, with only a sprig or tuft of green useless vegetation observable in them; the grain crop likewise totally fails when the larvæ of this chaffer feed in the field. Upon assuming their winged state, they devour the foliage of the oak and other trees so effectually, that entire copses may be seen early in June defoliated by their depredations. So much for their injury to man: but now the feast of the wilding commences — the plough in April dislodges multitudes of these long white grubs. Dogs then seek them eagerly to eat, but they seem to be surfeited by the food; for, though fattened at first, they afterwards become diseased, and lose their hair. Rooks and crows are running over the ridges, busily seeking for this larvæ; the swine find it out, and come in for their share, and, having finished here, they commence grubbing in the grass lands. The insect now soon takes wing, and then every tree in the wood or the brake becomes a scene of plunder and delight to all the train from the rookery — the cats will eat him — every sparrow that flies by has a chaffer in its mouth, captured on the wing or snatched from the spray, and now to be pecked to pieces on the ground — the thrush feasts, too, and all the poultry in the yard are running after chaffers, or chasing each other for the prize; and thus this insect supplies, in one state or another, a general feast to many.

Our extensive cultivation of the potato furnishes us annually with several specimens of that fine animal, the death's-head moth (*acherontia atropos*), and in some years I have had as many as eight brought me in the larva or chrysalis state. Their changes are very uncertain. I have had the larva change to a chrysalis in July, and produce the moth in October; but generally the aurelia remains unchanged till the ensuing summer. The larvæ or caterpillars, "strange ungainly beasts," as some of our peasantry call them, excite constant attention when seen, by their extraordinary size and uncommon mien, with horns and tail, being not unusually five inches in length, and as thick as a finger. This creature was formerly considered as one of our rarest insects, and doubtful if truly indigenous; but for the last twenty years, from the profuse cultivation of the potato, is become not very uncommon in divers places. Many insects are now certainly found in England, which former collectors, indefatigable as they were, did not know that we possessed; while others, again, have been lost to us moderns. Some probably might be introduced with the numerous exotic plants recently imported, or this particular food may have tended to favour the increase of rarely existing natives; but how such a creature as this could have been brought with any plant is quite beyond comprehension. We may import continental varieties of potatoes, but the death's-head moth we have never observed to have any connexion with the tuber itself, or inclination for it. As certain soils will produce plants by exposure to the sun's rays, or by aid of peculiar manures, when no pre-existent root or germ could rationally be supposed to exist, so will peculiar and long-intervening seasons give birth to insects from causes not to be divined. We may, perhaps, conclude, that some concurrence produced this sphinx, and then its favourite food, the potato plant, nourished it, to the augmentation of its species.

Superstition has been particularly active in suggesting causes of alarm from the insect world; and where man should have seen only beauty and wisdom, he has often found terror and dismay. The yellow and brown-tailed moths, the deathwatch, our snails (as mentioned in p. 216), and many others, have all been the subjects of his fears; but the dread excited in England by the appearance, noises, or increase of insects, are petty apprehensions, when compared with the horror that the presence of this acherontia occasions to some of the more fanciful and superstitious natives of northern Europe, maintainers of the wildest conceptions. A letter is now before me from a correspondent in German Poland, where this insect is a common creature, and so abounded in 1824, that my informant collected fifty of them in the potato-fields of his village, where they call them the "deaths-head phantom," the "wandering death-bird" &c. The markings on its back represent to these fertile imaginations the head of a perfect skeleton, with the limb bones crossed beneath; its cry becomes the voice of anguish, the moaning of a child, the signal of grief; it is regarded not as the creation of a benevolent being, but the device of evil spirits — spirits enemies to man, conceived and fabricated in the dark; and the very shining of its eyes is thought to represent the fiery element whence it is supposed to have proceeded. Flying into their apartments in the evening, it at times extinguishes the light, foretelling war, pestilence, hunger, death, to man and beast. We pity, rather than ridicule, these fears; their consequences being painful anxiety of mind and suffering of body. However, it seems these vain imaginations are flitting away before the light of reason and experience. In Germany, as in England, they were first observed on the jasmine, but now exclusively upon the potato, though they will enter the beehives, to feed on the honey found in them. This insect has been thought to be peculiarly gifted in having a voice, and squeaking like a mouse, when handled or disturbed; but, in truth, no insect that we know of has the requisite organs to produce a genuine voice. They emit sounds by other means, probably all external. The grasshopper and the cricket race effect their well-known and often wearisome chirpings by grating their spiny thighs against their rigid wings; and this acherontia atropos appears to produce the noise it at times makes, which reminds us of the spring call of the rail or corncrake, by scratching its mandible, or the instrument that it perforates with, against its horny chest. The object of this noise is, apparently, a mere sexual call. Heavy and unwieldy creatures, they travel badly, and from the same cause fly badly and with labour; and as they commonly hide themselves deep in the foliage and obscurity, without some such signal of their presence a meeting of the parties would seldom be accomplished.

Another of the ravenous creatures that infest our pools is the great water-beetle (*ditiscus marginalis*); and perhaps it is the most ferocious of any of them, being adapted by every provision for a life of rapine, endued with great muscular power, armed with a thick and horny case over its body, and having its eyes large to observe all the creatures about it, and powerful mandibles to seize and reduce them to fragments. It riots, the Polyphemus of the pool; and having thinned its herd in one place, is supplied with wings to effect a removal to a fold better furnished. It even eats the young of the frog; and its bite is so powerful, as to be painfully felt by the hand that holds it a captive, though defended by a glove. In the larva state it is almost equally destructive; it swims admirably; its hinder legs are long and brawny, beside being aided by a fringe of hairs, so that they are powerful oars to propel its body with celerity and ease. Nor must we omit a peculiarity attending the constitution of this beetle, which marks it as a creature especially endowed for the station in which it is placed. Multitudes of insects exist in the larva state, for a certain space of time, in water; and, having accomplished a given period in this state, perfecting their forms, they take wing, and become aerial creatures, after which a return to the element whence they sprang would be death to them. But this beetle, when it has passed from the larva state and obtained its wings, still lives in that water which nourished it to this state of perfection, without any inconvenience, as long as it suits its inclination; when weary of this place, or its food becoming scarce, it wings its way to another pool, into which it immediately plunges, and recommences its life of rapine. Having deposited its eggs in autumn, we suppose it to die in the winter; yet many may survive this season, and, arising from the mud in the spring, be undistinguished from the recently perfected larvæ. Such little notices and indications of the habits of these obscure creatures, though certainly unimportant, are not, perhaps, wholly unprofitable; for we so darkly see our way, and proceed so slowly in acquiring intelligence of the paths of Nature, that nothing should be considered as beneath regard that we meet with in them; and every advancing step must elevate the mind, as it affords us additional knowledge of the solicitude and provision of the great Architect of creation in the appointment and endowment of his creatures; since, though we are very rarely able to comprehend even the object of existence, we see sufficient to convince us, that such care and such powers were not bestowed except for some wise and good purpose. It seems hardly possible that mankind can ever obtain anything approaching to the comprehension of the motives of Providence, because they have not, as far as is apparent to us, individual and separate bearings, but are connecting and in concordance with a series of influences, and consequently the whole should be seen, fitly to understand a part; and this mighty mechanism what human mind can embrace? — Heaven metes out to man by degrees something of its laws and ordinances; but no life, no period, can exhaust that store of hidden wisdom by which these mandates have been decreed: every little transitory view that we obtain should be received with gratitude as an advance in knowledge, a progress in the wisdom of Him who hath ordained all things in truth.

A very elegant nest of an insect has this day occurred pendant beneath a flat stone in an ancient wall, the young had been hatched and escaped, but the remains of the eggs manifested that it had been well stored, there is every reason to suppose that it had been fabricated by some undetected species of spider - it was formed of the finest cotton, so white and prettily shaped that it immediately gave one the idea of a pearl drop - such as might have ornamented Cleopatra's ear at that celebrated banquet with Anthony.

Nest of an unknown spider

The eye of the naturalist, prying about in places where those of indifferent persons are rarely fixed, sees many things that others do not notice, or observe without interest, from forming no connexion with any previous subject of pursuit. Few, perhaps, would stay to inspect the clay hairworm (*gordius argillaceus*), yet it is a very curious creature. We find it at the bottoms of drains and ditches, chiefly in the spring of the year. Its colour is a pale yellow; and it appears like some long vegetable fibre, or root, coiled up and twisted together (Plate 6, Fig. 2). The whole body of the animal consists of numerous annulations, or rings, by means of which it has the power of contracting its substance, as it has likewise of extending it, until it becomes nearly a foot in length, and smooth as a wire. The extreme points are transparent and tapering, formed of apparently harder materials than the body. The designation of most of our small land and water creatures, in the economy of creation, is very obscure; and owing to the places they frequent, and the secrecy of their actions, amidst mud and vegetation, we have little opportunity of becoming acquainted with their habits. This hairworm, however, is rather less mysterious in its movements than some others; and there is cause to suppose that its chief occupation is that of forming perforations and openings in clayey soils, admitting, by this means, water to pervade the mass, and open it; the finer roots of vegetables then find entrance, and part it yet more, or decay in it, and meliorate and fertilize the substance.

Wonderful as all the appointments and endowments of insects are, there is no part of their economy more extraordinary than the infinite variety of forms and materials to which they have recourse in the fabrication of their nests; and, as far as we can comprehend, their expediency for the various purposes required. Among those with which I am acquainted, none pleases me more than that of a solitary wasp (*vespa campanaria*) which occasionally visits us here. It is not a common insect, but I have met with their nests. One was fixed beneath a piece of oak bark, placed in a pile; another was pendent in the hollow of a bank of earth. The materials which composed these abodes seemed to be particles scraped or torn from the dry parts of the willow, sallow, or some such soft wood, and cemented again by animal glue, very similar in texture to that provided by the common wasp, which makes great use of the half-decayed wood of the ash, and will penetrate through crevices in the bark, to abrade away the dry wood beneath. They seem to have but small families, ten or twelve cells only being provided. These are situate at the bottom of an egg-shaped cup, contracted at the lower end, where an orifice is left for the entrance. This again is covered, in the part where the cells are placed, by a loose hood, or shed, extending about half way down the inner one. The pendent situation of the whole, and this external hood, round which the air has a free circulation, are admirably contrived for securing the cells from injury by water. The nest, when hanging in its proper situation, is like the commencement of some paper-work flower, and can never be observed but with admiration at the elegance of its structure; and the unusual appearance of the whole must excite the attention of the most incurious observer of such things (see plate 6, Fig 1).

But this wasp does not appear always to fabricate her nest upon an uniform principle, but varies it as circumstances necessitate; for this summer (1829) another of these constructions, of a similar nature, was obtained from a very different situation. A weak swarm of the common bee had possession of a hive, but whether they abandoned it from dislike, or were dispossessed, is uncertain: however the hive, feeling light, was examined, and a nest of the species of wasp above described was found fixed upon a common waxen comb, commenced by the hive bee.

It is rather singular that so small a community as the constructors of this beautiful flower-like nest must have been — not having formed above eighteen cells — should have had the means of expelling the original proprietors, or of so offending them as to occasion the voluntary abandonment of their labour and home. But the forcible seizure of the rights of another seldom prospers: these usurpers, having constructed their edifice, deserted it in their turn, and never made use of their cells. It differed from that above noticed, by being of a more globular form; and instead of one hood covering the inner envelope of the cells, it had five, not attached, but surrounding each other, having spaces between, equal in the whole to a diameter of two inches. The inner egg-shaped cup was shorter — the first hood covering it entirely, and forming the entrance; the others gradually becoming less extended, as the petal of a rose appears when in its perfect beauty. This nest is not, I apprehend, the same as that described by Reaumur — his having two ranges of many small cells — whereas ours, though perfectly formed, had only one, and those in the centre nearly as large and deep as the compartments of the comb on which it was fixed. I think Messrs. Kirby and Spence mention having once seen a similar production, but it probably is of unusual occurrence in England.

BEES' NESTS

August 2. — Cut out this morning, from an old oaken rail, the nest of the carpenter bee (*Megach. centuncularis*) — a curious receptacle, well known to many persons; but yet it merits attention, because it seems to be a construction upon which more than usual foresight has been manifested — it not being merely an asylum in which the young may be matured in quiet, but secured against external annoyance from its most destructive enemy by a contrivance so unusual and effective, that it appears formed from the result of a reasoning upon probabilities. This creature— a short, stout, plain bee — mines a tubular channel into some decayed woody substance — a post, pale, or some such thing — boring in the direction of the fibre, making her repository at the bottom, being sufficiently deep for her purpose.

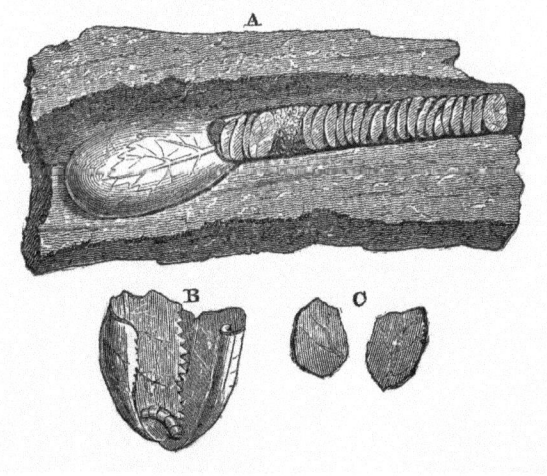

A, the rail cut open, showing the channel and nest. B, part of the nest, with the larva. C, patches filling up the channel.

She cuts from the leaf of a rose several large pieces, often an half, conveying them to the bottom of her cave, and rolling them up so as to form a case: in this she deposits an egg. The mouth of the cave is then covered with five or six circular patches of leaves; fragments of wood like saw-dust are then lodged over them, and the remainder of this channel filled up with other patches of the leaf, requiring, perhaps, twenty or more to accomplish it. This requires no great effort or time to effect, the active workman labouring with great assiduity; but the conveyance of the fragments which form the nest seems to be attended with much fatigue, for I have often seen this bee, with its burden drawn between its legs, retire to rest against the bole of a tree, or upon a wall, the abdomen contracting and dilating with rapidity, like the sides of a creature after exertion. Having acquired strength, her labours are resumed. All these layers and sawdust are designed to secure the larva from punctures by the ichneumon fly, which is constantly seeking for caterpillars of various insects, in which to deposit her egg: this hatches and grows with their growth, feeding on their vitals, consuming them and perfecting itself. But all these guards effectually prevent the purport of the fly, the larva remaining safe in its asylum until sufficiently matured; it then penetrates through all these layers, assuming the form of the parent bee. I believe the rose and the cytisus laburnum only are made use of for these circular layers. In this specimen the larva was perfected; but at an earlier period, an egg would have been found, with a reservoir of glutinous matter near it, to constitute the food of the hatched creature.

This action is a very remarkable example of a superior directing Power; — the bee, never having experienced the effect of the ichneumon's mischief, could have no perception of danger — no consciousness of a possible injury; but she was given a precept to evade, by a peculiar contrivance, the wiles of an enemy: and we have here one of the few instances of the provisionary wisdom of creatures that we can comprehend. It affords us infinite pleasure to witness the labour and success of the insect, and it directs our minds to the overruling cause of all action and effect.

There are few things less explainable than the appearance at peculiar periods, and disappearance after a long residence of insects – the universal prevalence at one season, and absence at another, from some momentary prejudice, or unknown influence – we were this year (1842) visited by a bee, in considerable numbers, which was not observable before – It was a rose-leaf-cutter, probably that known as the "upholsterer" – they selected for their nest a dry bank in my garden having a southern aspect. In this they excavated their holes in such numbers that a foot square might contain from 16 to 20 nests – though similar to the carpenter bee (megachile centuncularis) its habits were very different. The carpenter with us always selects wooden work of some sort for her operations, and cuts her segments from the foliage of several species of roses, and other plants, but this bee passed over a variety of sorts, selecting at last a Noisette, and with such perseverance did she cut her portions, that from a large bush, scarcely a leaf remained unmutilated. With these she formed cylindrical cases as thick as a child's finger, in some instances two inches in length, fixing them in the earth – each consisting of several cells, in each of which she deposited a larva, each cell separated by a few circular patches – in the bottom of each where the mouth of the larva was placed, was deposited a considerable portion of pollen of some plant – with peculiar art she continued to bend the flat segments cut from the leaf so as to wrap firmly over each other without any cement, and so fixed were they as to remain closely adhering , after the hatching of the lava. The celerity with which she cut her segments, and the unremitting perseverance of her labours was most admirable; but the exertions of insects to fabricate their nests is wonderful! – the period for work is short, and seems an imperative duty to accomplish, as in most cases the architect speedily afterwards disappears and in most cases probably perish.

Every-day events manifest to very superficial observation, that no created being, from the monster of the ocean, "that makes the deep boil like a pot of ointment," to the insect that feebly creeps on the ground, exists free from the persecutions or annoyance of another. Some may be subject to fewer injuries than others, but none are wholly exempt: the strong assail by power, and become assaulted themselves by the minute or weak. This year (1826) the hornet (*vespa crabro*) abounded with us in unusual numbers, and afforded constant evidence of its power and voracity that could not have been exceeded by any ravenous beast. In our gardens the imperious murmur of four or five of them at a time might be frequently heard about our fruit-trees. They would occasionally extract the sweet liquor from the gage, or other rich plums; but the prime object of their visit was to seize the wasps that frequented the same places. This they not only did when the creature was feeding on the fruit, but would hawk after them when on the wing; capture them with a facility to which their heavy flight seemed unequal; bear them to some neighbouring plant, and there feed on the insect, which seemed perfectly overpowered by the might of the hornet. The first operation was to snip off the head, then to cut away the lower part by the waist; and, when near, we could hear them shearing away the outer coat from the body, and crushing it with their strong mandibles; sometimes devouring it, but generally only sucking the juices it contained. Their avidity for this sort of food is very manifest, when the grape ripens on the walls being commonly the only remaining fruit, the wasp abounds there; the hornets flock to the prey, and we may see them in constant progress, bearing their victims from the bunches. The wasp itself seizes the house fly; but this seems rather the display of wanton power than for food, as it bears the fly about with it for a length of time, and drops it unconsumed. The fly, in its turn, is conducive, after its manner, to the death of many an animal. We know not any insect that destroys the hornet; but its power and being are terminated by some very effective agent, as in particular years it is almost unknown[38]. Though we may not often perceive the means by which certain

38 The hornet is a very pugnacious animal. They will fight desperately with each other at times, when they meet in pursuit of prey, biting each other's body, and trying to get their mandibles under the head of their opponents, to snip it off. I one

races are reduced in number, more than their multiplication effected, yet we are frequently sensible that it is accomplished.

I do not recollect any creature less obnoxious to harm than the common snail (*helix aspersa*). of our gardens. A sad persevering depredator and mangler it is; and when we catch it at its banquet on our walls, it can expect no reprieve from our hands. But our captures are partial and temporary; and, secured in its strong shell, it seems safe from external dangers; yet its time comes, and one weak bird destroys it in great numbers. In the winter season, the common song-thrush feeds sparingly upon the berries of the whitethorn, and the hedge fruits, but passes a great portion of its time at the bottoms of ditches, seeking for the smaller species of snails (*helix hortensis* and *helix nemoralis*), which it draws out from the old stumps of the fence with unwearied perseverance, dashing their shells to pieces on a stone; and we frequently see it escaping from the hedge bank with its prize, which no little intimidation induces it to relinquish. The larger kind at this season are beyond its power readily to obtain; for, as the cold weather advances, they congregate in clusters behind some old tree, or against a sheltered wall, fixing the openings of their shells against each other, or on the substance beneath, and adhering so firmly in a mass, that the thrush cannot by any means draw them wholly, or singly, from their asylum. In the warmer portion of the year they rest separate, and adhere but slightly; and should the summer be a dry one, the bird makes ample amends for the disappointment in winter, intrudes its bill under the margin of the opening, detaches them from their hold, and destroys them in great numbers. In the summers of 1825 and 1826, both hot and dry ones, necessity rendered the thrush unusually assiduous in its pursuits; and every large stone in the lane, or under the old hedge, was strewed with the fragments of its banquet. This has more than once reminded me of the fable of the "Four Bulls;" united invincible, when separated an easy prey; but, with the exception of this season, and this bird, I know no casualty to which the garden snail is exposed.

Yet periods to arrive when this creature is visited by the visitation that awaits all created things and extensive mortality ensues. In the long and severe winter of 1829 and 1830, they were frozen in their retirements in masses as they clustered, and the entire substance of the animal limax became as brittle as ice, nor did many of them revive when the milder season of spring arrived. The small coloured snails seem to have been better secured,

day confined, under a glass, two of these creatures which had been fighting. One had evidently the mastery; but both had been so injured in the contest, that they soon died; and it is most probable that they fall victims to each other's voracity, in the cold, damp season that usually terminates the autumn of our year.

and fewer of them perished, but the garden snail (*H. aspersa*), was in great measure destroyed, and remained in masses behind the stumps, and in the crevices where they had withdrawn themselves; and yet those little shelless limaces, yellow, white, and black, which we call slugs, remained perfectly uninjured, probably defended by the slimy gluten that invests their bodies, and by being removed in crevices, and depths beyond the penetration of the frost. They abounded, crawling about the withered herbage in every temperate interval, secreting themselves under the decayed foliage in less propitious weather, smearing and devouring in the moisture of April, the tenderlings of our gardens and flower beds.

Snails when they attached themselves to stones &c to repose during winter, throw off an operculum, or horny covering, between their bodies and the substance on which they fix. But this year we observed that four of these filmy veils were formed, not united, but with intervals between, possibly upon the principle that we construct double window frames for our houses, to preserve an uniformity of temperature in our apartments. For as the atmospheric air is a non conductor, the warmth of our rooms is prevented from flying off to augment the temperature of a colder medium. The animal having formed its usual film, seems as the cold increased, to have contracted its substance, and there formed another, and so successfully continued to shrink in its shell, until the space it had voided was occupied by these four coverings. But animal energies at length ceased, and the means attempted to preserve life failed, by the great severity of the winter. They seem in these seasons to attach themselves by adhesive matter which indurates, whereas limpets &c. exhaust the air, and become fixed by incumbent pressure.

That these filmy veils were formed at intervals, as necessity prompted, appears probable from the circumstance that in the ensuing year on November 15th, only one operculum was formed, but a large space remained between it and the animal. By December 12th, in general two were formed, and in a few instances three - as some frosty mornings had lowered the temperature of the air, but none occurred with four, as in the preceding more inclement season. But the winter of 1837 and 1838, so fatal to a large portion of our little birds and the long existing plants in our gardens, proved most destructive to our garden snails. A few of the smaller species survived, probably by creeping into deeper retirements, but *H. hortensis* with very few exceptions, perished. In several which I examined, as many as two of these films were formed, but either from the severity, or suddenness of the cold, all energy then ceased, and the shells contained only the animal remains. Where hundreds were usually seen, scarce a specimen was to be obtained.

Ignorant as we are of the scope, limitation, and even existence, of certain faculties in animals, we can frequently do little more than conjecture the means whereby they perform many of the functions of life. This ignorance leads us naturally at times to refer these powers to the agency of senses, like our own; but, in most instances, probably without any foundation in truth. No creature seems less qualified to commit the depredations which it does, than the garden snail. We grieve to see our fruit mangled and disfigured by these creatures, but cannot readily comprehend by what means they obtain the knowledge that its maturity is approaching— though we find that they must be endued with some faculty capable of accomplishing the purpose; for no sooner does a plum, a fig, a nectarine, or other fruit, begin to ripen on the wall, and long before any sensible odour can be diffused from it, even before an experienced eye can detect the approach to maturity, than those creatures, the slug and the snail, will advance from their asylums, though remotely situate, and proceed by very direct paths to the object. This cannot probably be by the guidance of any known faculty. Eyesight was once considered to be situate on the summit of their horns; but this is now known to be erroneous, and we do not know that they have any vision. The acoustic organ of worms and insects is unknown; and it is not by any means ascertained that these creatures ever hear.[39] If they possess the faculty of smelling, in them it must be a very exquisite sense, beyond any delicacy we can comprehend. Thus, excluding human means of comprehension, which appear inadequate, we more reasonably conclude them to be endowed with intelligences for effecting intentions of which we have no perception, and which we have no capacity for defining. The contemplative man finds pleasure in viewing the ways and artifices of creatures to accomplish a purpose, though he knows not the directing means; and it fortifies the convictions of the believer, by giving him fresh evidences of the universal superintendence of his Maker, that even the slug and the snail, which are arranged so low in the scale of creation, are yet, equally with all, the object of his benevolence and care.

Connected with this subject of snails, a circumstance that took place in this neighbourhood is brought to my remembrance, which discovered yet latent in a few of us, notwithstanding our boasted enlightenment, some leaven of the superstition of darker ages; and that any occurrence, not the event of every coming day, may be made a subject of wonder by the ignorant, and a means for the artful to deceive the credulous. A little banded snail (*helix virgata*) is a very common species on most of our arid, maritime pastures, and

---

39 That bees are attracted by the hiving-pan is generally considered as fallacious; and the practice useless.

the sheep-downs of many inland places. It happened, from some unknown cause, that those inhabiting a dry field in an adjoining parish were in one season, a few years ago, greatly increased, so as to become an object of notice to a few, then to more, till at length this accumulation was noised about as a super-natural event. The field was visited by hundreds daily from neighbouring villages and distant towns. People who could not attend purchased the snails at a halfpenny each; and there were persons who made five shillings a day by the sale of them. As this increase of the creature was not certainly to be accounted for, some had the impudence to assert that they had witnessed their fall from the clouds; and many declared their belief that some great public or private misfortune was indicated by it. The proprietor of the field being supposed not to maintain the same sentiments as the commonalty upon a political circumstance which at that moment greatly agitated the country, it was considered as a manifestation of heavenly displeasure, precursive of malady, misfortune, death. However, autumn came, these snails retired to their holes in the banks, and the worthy man lived on, — and long may he live, esteemed and respected by all, unscathed by snails or misfortunes!

Little obnoxious to injury as this garden snail appears to be, there is another creature (and that a very important one in the operations of Nature) that is surrounded by dangers, harassed, pursued unceasingly, and becomes the prey of all: the common earthworm (*lumbricus terrestris*). This animal, destined to be the natural manurer of the soil, and the ready indicator of an improved staple, consumes on the surface of the ground, where they soon would be injurious, the softer parts of decayed vegetable matters, and conveys into the soil the more woody fibres, where they moulder, and become reduced to a simple nutriment, fitting for living vegetation. The parts consumed by them are soon returned to the surface, whence, dissolved by frosts and scattered by rains, they circulate again in the plants of the soil —

Death still producing life.

Thus eminently serviceable as the worm is, it yet becomes the prey of various orders of the animal creation, and perhaps is a solitary example of an individual race being subjected to universal destruction. The very emmet seizes it when disabled, and bears it away as its prize: it constitutes throughout the year the food of many birds; fishes devour it greedily; the hedgehog eats it; the mole pursues it unceasingly in the pastures, along the moist bottoms of ditches, and burrows after it through the banks of hedges, to which it retires in dry seasons. Secured as the worm appears to be by its residence in the earth from the capture of creatures inhabiting a different element, yet many aquatic animals seem well acquainted with it, and prey on it as a natural food, whenever it falls in their way; frogs eat it; and even the great water-beetle (*ditiscus marginalis*) I have known to seize it when the bait of the angler, and it has been drawn up by the hook. Yet, notwithstanding this prodigious destruction of the animal, its increase is fully commensurate to its consumption, as if ordained the appointed food of all; and Reaumur computes (though from what data it is difficult to conjecture), that the number of worms lodged in the bosom of the earth exceeds that of the grains of all kinds of corn collected by man.

Worms, generally speaking, are tender creatures, and water remaining over their haunts for a few days drowns them; they easily become frozen, when a mortification commences at some part, which gradually consumes the whole substance, and we find them on the surface a mucilaginous mass; and their retiring deeper in the soil is no bad indication of approaching cold weather: but no sooner is the frost out of the earth, than they approach the surface to feed on decayed vegetable matter. Greatly beneficial as these creatures are, by drawing leaves and decayed matters into the earth, where their dissolution is accomplished, yet they are sad tormentors to us gardeners, and occasion the loss of more young plants than even the slug, by drawing in the leaf, which throws out the root; so that in the morning we find our nursling inverted. It is the same propensity, or ordination, for removing decayed matters that influences them in these actions; as they are the faded leaves that are seized by them, such as newly-removed plants present before the root draws nutriment from the earth. Even stones of some magnitude are at times drawn over their holes. The horticulturist, perhaps, encounters more mortification and disappointment than any other labourer upon the earth from insects, elementary severity, the slug, and the worm; yet, if the depredations of this last creature do at times excite a little of our irascibility, we must still remember the nightly labours, and extensive services, that are performed for the agriculturist by this scavenger of the earth, and manurer of the soil.

Besides, worms are essentially useful in draining our lands from superfluous moisture, which in many cases, without their agency, would be detained upon or near the surface of the earth, chilling and deteriorating our pastures. A few inches of soil, resting upon a substratum of clay, would commonly, without some natural or artificial drainage, be soaked with water after heavy rains, and thus become a bog, or produce coarse water herbage, rather than good grasses; but these worms greatly facilitate the passage of the water by draining horizontally along the bed of clay, and aid the emission of the water by this means; as I have often observed in the trenches, which we cut in our retentive soils, numerous worm-casts on their sides a few days after they had been made, being the exits of the horizontal runs; and through these the water drains into the trenches, and runs off. I do not assert that water would not in any case be discharged without the agency of worms, but that the passages which they make expedite it; which, in situations where the operation would be subjected to delay from the position of the ground, or the under stratum, is of infinite advantage. Thus the soil is not only rendered firm, allowing the admission of cattle, but the good herbage, which the long residence of water would vitiate or destroy, is saved from injury, and the aquatic and useless plants starved or checked in their growth; but after great gluts of rain, when the supply of water is greater than can be speedily carried off, it becomes stagnant; and those worms which cannot burrow beyond its influence, soon perish, and we lose the services of these very beneficial creatures. Drainage is therefore one of the most important operations in our agricultural concerns. As by irrigation we turn a quantity of nutritive water over our lands, or by reason of its higher temperature foster the growth of grasses; so, by draining cold and superfluous moisture off, we promote the growth of valuable vegetation. I would advocate the cause of all creatures, had I the privilege of knowing the excellency of them; not willingly assigning vague and fanciful claims to excite wonder, or manifest a base pride by any vaunt of superior observation; but when we see, blind as we are, that all things are formed in justice, mercy, truth, I would tell my tale as a man, glory as a Christian, and bless the gracious Power that permitted me to obtain this knowledge. Every age has been the dupe of empiricism, and the greater its darkness, the more impudent appear to have been the pretensions of knavery.

RESIDING, AS I constantly do, in the country, and having been long observant of rural things, and the operations of Infinite Wisdom, through the very feeble organs with which I have been endowed, I have often thought that we, who are daily made sensible of so many manifestations of creative power and mercy, should be more seriously disposed, more grateful for the beneficences of Providence, than those who live in societies removed from these evidences; but yet I neither know nor believe that we in any respect give greater proof of this disposition, or are more sensible of the benevolence of an overruling power, than others. The manufacturer by the combination of artful contrivances effects his purposes, and by aid of man's wisdom brings his work to perfection; the artisan may eat his bread with all thankfulness and humility of heart, solace his labours and mitigate his fatigue by the grateful flavour and juices of fruits purchased at the stall, but he sees nothing of the machinery, the gradual elaborations of Nature, nor can he be conversant with the multiplicity of influences and events which are requisite to bring them to his hand. He who lives in the country knows that an omnipotent impulse must be constantly in action; he may till his land, and scatter his corn, but the early and latter rain must soften his furrows; the snow, as wool, must cover the soil; the hoarfrost, like ashes, lighten his glebe; the sunshine animate the sprouting shoot; and winds evaporate noxious moisture; insects and blights, that hover around, or circulate through the air, must be guided away, or our labours become abortive, or are consumed; we see the bud, the blossom, leaf, and germ, all progressively advance, to afford plenty or yield us enjoyment, we see these things accomplished by the influencing interpositions of a beneficent Providence, and in no way effected by the machinery or artifices of our own hands; and it should operate more powerfully in disposing those who witness them to particular resignation and gratitude, than others who cannot behold them, but view the ingenuity of man as the agent and means of his prosperity; yet how it happens that this principle is not in more active operation within us, I cannot perceive.

A simple glass of water, cold and brilliant as the crystal that enshrines it, may often be a subject of contemplation and gratitude more than the costly vintages of the Rhine, all other beverage is a fabrication, a preparation of man for his sole comfort and advantage – milk is an animal or in a few cases vegetable produce – but water springs prepared from the source of benevolence and power for the happiness of all creation; the wild hart of the desert smells it from afar, and leaps with joy; the wandering bird sips of the stream shakes his wing with delight and passes on; and how often has the way-faring man in a weary land drank of the brook in his path, and thanked the universal Father for renewed life and strength, whispering departing praise – "there is no natural blessing that the bounty of our creator has bestowed upon us but is either fostered for our use or safely dispensed to us by this medium."

We may even now, perhaps, swallow a few matters, the arcana of the needy or the daring, in the various compositions of powders, draughts, and pills, which are not quite agreeable to our palates or our stomachs; but our forefathers had more to encounter, as they had more faith to support them, when they were subjected, for the cure of their maladies, to such medicines as *album græcum*, or the white bony excrement of dogs, bleached on the bank, for their heartburns and acidities; the powder produced from burnt mice, as a dentifrice; millepedes, or woodlice, for nephritic and other complaints; and the ashes of earthworms, administered in nervous and epileptic cases.

Our apple-trees here are greatly injured, and some annually destroyed, by the agency of what seems to be a very feeble insect. We call it, from habit, or from some unassigned cause, the "American blight" (*aphis lanata*); this noxious creature being known in some orchards by the more significant name of "white blight." In the spring of the year a slight hoariness is observed upon the branches of certain species of our orchard fruit. As the season advances this hoariness increases, it becomes cottony; and toward the middle or the end of summer the under sides of some of the branches are invested with a thick, downy substance, so long as at times to be sensibly agitated by the air. Upon examining this substance, we find that it conceals a multitude of small, wingless creatures, which are busily employed in preying upon the limb of the tree beneath. This they are well enabled to do, by means of a beak terminating in a fine bristle (Plate 5, Fig. 3); this being insinuated through the bark, and the sappy part of the wood, enables the creature to extract, as with a syringe, the sweet, vital liquor that circulates in the plant. This terminating bristle is not observed in every individual: in those that possess it, it is of different lengths, and is usually, when not in use, so closely concealed under the breast of the animal as to be invisible. In the younger insects it is often manifested by protruding like a fine termination to the anus; but as their bodies become lengthened, the bristle is not in this way observable. The alburnum, or sap wood, being thus wounded, rises up in excrescences and nodes all over the branch, and deforms it; the limb, deprived of its nutriment, grows sickly; the leaves turn yellow, and the part perishes. Branch after branch is thus assailed, until they all become leafless, and the tree dies.

Aphides, in general, attack the young and softer parts of plants; but this insect seems easily to wound the harder bark of the apple, and by no means makes choice of the most tender part of the branch. They give a preference to certain sorts, but not always the most rich fruits; as cider apples, and wildings, are greatly infested by them, and from some unknown cause other varieties seem to be exempted from their depredations. The Wheeler's russet, and Crofton pippin, I have never observed to be injured by them; and the insect is so fastidious in its selections, that it will frequently attack the stock or the graft, leaving the one or the other untouched, should it consist of a kind not to its liking. This insect is viviparous, or produces its young alive, forming a cradle for them by discharging from the extremities of its body a quantity of long, cottony matter, which, becoming interwoven and entangled, prevents the young from falling to the earth, and completely envelops the parent and offspring. In this cottony substance we observe, as soon as the creature becomes animated in the spring, and as long as it remains in vigour, many round pellucid bodies, which, at the first sight, look like eggs, only that they are larger than we might suppose to be ejected by the animal. They consist of a sweet, glutinous fluid, and are probably the discharges of the aphis, and the first food of its young. That it is thus consumed, I conjecture from its diminution, and its by no means increasing so fast as fæcal matter would do, from such perpetually-feeding creatures. I have not, in any instance, observed the young to proceed from these globular bodies, though they are found of various ages at all times during the season. This lanuginous vestiture seems to serve likewise as a vehicle for dispersing the animal; for though most of our species of aphis are furnished with wings, I have never seen any individual of this American blight so provided; but the winds, wafting about small tufts of this downy matter, convey the creature with it from tree to tree throughout the whole orchard. In the autumn, when this substance is generally long, the winds and rains of the season effectually disperse these insects, and we observe them endeavouring to secrete themselves in the crannies of any neighbouring substance. Should the savoy cabbage be near the trees whence they have been dislodged, the cavities of the under sides of its leaves are commonly favourite asylums for them. Multitudes perish by these rough removals, but numbers yet remain; and we may find them in the nodes and crevices, on the under sides of the branches, at any period of the year, the long, cottony vesture being removed, but still they are enveloped in a fine, short, downy clothing, to be seen by a magnifier, proceeding apparently from every suture or pore of their bodies, and protecting them in their dormant state from the moisture and frosts of our climate. This aphis, in a natural state, usually awakens and commences its labours very early in the month of March; and the hoariness on its body may be observed increasing daily: but if an affected branch be cut in the winter, and kept in water in a warm room, these aphides will awaken speedily, spin their cottony vests, and feed and discharge as accustomed to do in a genial season.

The long tailed titmouse occasionally visits these infected apple trees, to feed on the aphides; but the golden crested wren, may very frequently be observed in the autumnal and spring months, busily occupied about the branches, apparently deriving a considerable portion of its supply in these seasons, by means of these insects. This little creature flits about our shrubberies and fir groves so unobtrusively as to escape general notice, very far from being an uncommon bird, and it feeds so silently too, not moving a leaf, or breaking a seed, that it is often passed without observance. We may stand beneath a tree on which several are occupied in capturing the minute insects which constitute their food, hearing not even the snapping of their bills, and be only sensible of their presence by their actions, as noiseless and as fearless as an evening moth - though at times a gentle and a pleasing voice announces their presence-

It is often very difficult to ascertain the first appearance of many creatures not natives of our climate, though, from the progress of science, and more general observation, many things will be recorded. The first visit of the death's-head moth is very obscure; an extraordinary snail (*testacellus halotideus*[40]) is now spreading by transplantation in many places, and may hereafter occasion inquiry. The first visit of this aphis to us is by no means clear. The epithet of American blight may be correctly applied; but we have no sufficient authority to conclude that we derived this pest from that country. Normandy and the Netherlands, too, have each been supposed to have conferred this evil upon us; but extensively as this insect is spread around, and favourable as our climate appears to be to its increase, it bids fair to destroy in progression most of our oldest and long-esteemed fruit from our orchards. The same unknown decree, which regulates the increase and decrease of all created beings, influences this insect; yet wet seasons, upon the whole, seem genial to its constitution. In the hot dry summer of 1825, it was abundant everywhere; in the spring of 1826, which was unusually fine and dry, it abounded in such incredible luxuriance, that many trees seemed at a short distance as if they had been whitewashed; in the ensuing summer, which was a very dry and hot one, this cottony matter so entirely disappeared, that to superficial observation the malady was not in existence; and it did not become manifest again until September, when, after the rains of that season, it reissued in fine cottony patches from the old nodes on the trees. Many remedies have been proposed for removing this evil, efficacious perhaps in some cases upon a small scale; but when the injury has existed for some time, and extended its influence over the parts of a large tree, I apprehend it will take its course, and the tree die. Upon young plants, and in places where a brush can be applied, any substance that can be used in a liquid state, to harden into a coat, insoluble by rain, will assuredly confine the ravages of the creature, and smother it. Hard rubbing with a dry brush crushes many, but there are crevices into which the bristle cannot enter: thus some escape, and

---

40   This creature was first observed, I am told, about the year 1819, in the nursery-garden of Messrs. Miller and Sweet, near Bristol — introduced, as is supposed, on some imported plant. It increases readily in our climate. The white moss-rose (*rosa muscosa, var. alba*): this beautiful variety was first produced about the year 1808, in the garden of Gabriel Goldney, Esq, at Clifton, near Bristol; a branch of the common red moss-rose, becoming diseased, produced its flowers white. A neighbouring nurseryman, being employed by that gentleman's gardener to lay down the branch, from cuttings propagated the variety, and shortly after dispersed many plants.

the propagation continues. I have very successfully removed this blight from young trees, and from recently-attacked places in those more advanced, by an easy application. Melt about three ounces of resin in an earthen pipkin, take it from the fire, and pour into it three ounces of fish-oil; the ingredients perfectly unite, and, when cold, acquire the consistence of honey. A slight degree of heat will liquefy it, and in this state paint over every node or infected part in your tree, using a common painter's brush. This I prefer doing in spring, or as soon as the hoariness appears. The substance soon sufficiently hardens, and forms a varnish, which prevents any escape, and stifles the individuals. After this first dressing, should any cottony matter appear round the margin of the varnish, a second application to these parts will, I thinly be found to effect a perfect cure.

The prevalence of this insect gives some of our orchards here the appearance of numerous white posts in an extensive drying-ground, being washed with lime from root to branch — a practice, I apprehend, attended with little benefit. A few creatures may be destroyed by accident; but as the animal does not retire to the earth, but winters in the clefts of the boughs, far beyond the influence of this wash, it remains uninjured, to commence its ravages again when spring returns.

All the natural fruit of the hedge and the copse appear at one period of their growth, to serve as food for some creature or another, ripening at various seasons, and by that means many sorts become propagated over the earth, which if otherwise disposed of might be lost or of local occurrence. The crab, the acorn, and the nut, fall to the earth and are eaten by squirrels, badgers, and mice, numbers of which being hidden in the soil by these creatures as reserved food, but from the transitory nature of their lives are often not reclaimed, thus germinate and become trees. The garden feeding birds disseminate many seeds. The sloe, the hip, the ivy, furnish food in rotation and are consumed as necessity requires, affording the principal supply of these creatures during the inclement periods of year: yet the privet (*ligustrum vulgare*) seems to be an exception to this observation; few plants produce more berries, yet the racemes remain untouched even until the end of March frequently, and then become detached by the natural operation of the plant, producing few additional shrubs in the uncultured state, being smothered by surrounding herbage, the plant increasing by stoles from the roots. Our antipathies and likes, are no criterions of the inclinations of animals differently endowed, but to our smell and taste the berries are nauseous, and rank, and their remaining so long without being consumed renders it probable that poor hungry wildings have no inclination for them and long abstain from using them, though in severe and deficient seasons they will be partially fed upon.

July 1829. — All the early part of this month has been hot and damp, tending apparently to the production of the insect race; yet in general the vegetation is remarkably free from their depredations, and, excepting the commonest insects, very few indeed are to be observed, and our migrating small birds, which feed upon them, are as scarce as their food: but for some days there has been a profusion of aphides upon everything, and the under sides of many leaves are almost hidden by the numbers which have taken shelter there. Even plants removed from cover are soon spotted with the winged aphis, not apparently feeding, but resting, seemingly an immense migrating body: but, slightly formed as they are, few creatures seem less calculated for a distant removal; and from whence do they come? Only two plants are in any way remarkably infected with this insect; and it seems singular that these two, so different in their characters and nature, should commonly be equally sufferers at the same time, which I believe they are — the bean-plant and the hop; the former we do not cultivate to an extent sufficient to give being to such numbers, and the hop-grounds of Worcester and Hereford are too distant (having an adverse wind) to inundate us with these moving myriads. It is more probable that the late rains have given such a vigour to the bean-plant as to render its circulating juices unpalatable to the creatures abounding on them; and hence I conjecture they have left the districts where this legumen is extensively cultivated in Somersetshire, and journeyed from a distance of, perhaps fifteen miles, with the wind, in search of more suitable provender. But the appearance and generation of insects, and this race in particular, are involved in much mystery: plants growing in houses removed from all connexion with those that are infected, will yet regularly produce them— the cinerarias universally, and one species, *c. cruenta*, most remarkably so. This is a plant whose herbage annually dies down to the root, and hence has no vegetation remaining to conceal eggs, if there were any; for these creatures bring forth their young alive, in all the cases in which I have noticed them. We may fresh pot the root, renewing it with separate earth, or keep it in the windows of our habitation, where no insect of the kind is observed to reside; yet in due time it will become infested with the aphis, whose depredations, as far as I have observed, are the cause of the dying down of the herbage. Feeding upon the sap of plants, the foliage, deprived of its nutriment, languishes; and if there is not sufficient vigour to renew this loss of vitality, it perishes. Voracious beasts might ravage our flocks and our herds, but could scarcely by their powers accomplish greater injuries to the labours and possessions of man than the seeming despicable creatures, weevils, wireworms, thrips, aphides, or those atoms which we denominate the blight of some seasons. We should accustom ourselves to view no portion of creation with contempt: the particle which we may brush away to-day with contumely, may, to-morrow, become an instrument of our punishment or ruin.

Where all is wonderful, it is difficult to pronounce what is chiefly marvellous; yet the insect world exhibits most astonishing construction, viewed as to its splendour or fabrication. This feeble aphis, now crawling over my paper, with limbs indescribably slender, seems yet endowed with every requisite given to a larger body — joints, integuments, circulation of fluids, and every mechanical action requisite for its being; and yet the whole is so fragile as to be overturned by a puff of my breath. But smallness of bulk is no criterion of inferiority of power: an apple-tree, several feet, perhaps, in its circumference, spreading its branches over a rood of land, sickens and dies from the puncture of the *aphis lanata*, a creature so small as to be imperceptible on its limbs. How many ages may be required to find out or comprehend the mechanism, objects, agreements, and gradations of Nature; and the expansion which the human intellect must undergo to receive such an accession of knowledge, is beyond the reach of thought. We frequently expatiate upon the increased wisdom and acquirements of mankind, considerable as they may be; yet how trifling has our annual accession of information been, when we consider how little we in fact know even of the perceptible wonders of creation, and what inconceivable stores are hidden from us, and five thousand years of man's life have been expended in obtaining almost nothing! Hence, perhaps, we may conjecture, that any infinite increase in worldly wisdom is not necessary to our being here, or we should have been given faculties and intelligence commensurate to our necessities. This consideration must not check our investigation, or the workings of such powers as we are endowed with: so much remains to be done in an intermediate region that is great for "little man," suitable to his reason, and required by his endowments, that no want of adequate materials for action can ever be feared, and fitting employment that may make us wiser and happier here, and expand and calm the mind to contemplations of superior and better wisdom. I seldom take my daily ramble be it in the frosts of winter, or amidst the gay profusion of summer hours, or as chance the season may be, but I find some enjoyment in the natural scene about me, some manifestation of creative benevolence and wisdom. The interest, the comfort I have found in lowly converse with all my friends around, they only who delight in Nature's works can experience; be it with the stately tree canopying the earth with verdure, feeding myriads of animated beings with its foliage, the humble moss, the lichen on its bark, or the fungi at its foot, the gay scarlet cup of the peziza peeping from a tuft of hypnum, the fern on the bank unrolling its hairy leaf,

& breaking its way to light & life, the floating weed on the ditch sheltering the wreathed snail, or the worm, or the little bending tribes of the earth – all in their turns have stayed my steps, awakened wonder and consideration; little that I have seen, and little indeed that I could comprehend, still the very occupation of mind, the perception of such variety, such vital energy, such unceasing evidence of goodness, wisdom, and of power, became a blessing, and my return to my dwelling, though grieving at my ignorance, was in tranquil peace with all.

Seasons arrive and pass away, the general features alone remaining impressed upon our minds; but they often produce consequences not commonly expected; and a departed summer or winter has frequently been the cause of some event which we consider as exclusively occasioned by atmospheric changes or present temperature. A warm, dry summer generally occasions a healthy spring-blossom the ensuing year, the bearing wood being ripened and matured to produce in its most perfect state. A wet, damp one usually effects the reverse, by occasioning an abundant flow of sap, producing wood and foliage rather than blossom; and the following spring, in such cases, from the floral vigour being diverted, has generally its blossom weak, and, though perhaps not defective, incompetent to mature the germen. This is mere reasoning upon general consequences; but so imperfect are our theories, and so many circumstances counteract the calculations, the predictions of human wisdom, which can rarely even "discern the face of the sky," that results must more often be looked for than known. The recording of events is the province of the naturalist; and, perhaps, occasionally by comparing existing circumstances with past events, something approximating to probability may be obtained. The two burning summers of 1825 and 1826 are remembered by all; but it was in the succeeding year only that the result of this heat and drought was manifested to us, by effects upon our pasture lands which we did not expect.

Not only in those on the limestone substratum, but in many that were sandy, and in the clayey which were chapped by the heat, the roots of the grasses, which we have generally considered as not being subject to such injuries, were destroyed in some cases, and greatly injured in others; and in their places frequently sprang up crowfeet (*ranunculus acris*, and *bulbosus*), and dandelions, a mere useless vegetation, which, as long as the grasses flourished, were kept in subordination and obscurity by their superior growth; while bare patches in other places told us of aridity and failure: the meadow grass (*poa*) and ray grass (*lolium perenne*) were great sufferers; the dog's tail (*cynosurus*) supported itself better; the cockfoot (*dactylis*) though not killed, was so much hurt that its ensuing vegetation, instead of the coarse luxuriance it generally manifests, was dry, hard, and deficient in succulency, or, as our labourers emphatically say, was "stunned;" and bent-grass (*agrostis vulgaris*), that certain indicator of a dry soil, appeared more than it commonly does. But this destruction of the roots, in very many places, was not obvious — the turf, as it was, remaining; yet some injury was apparent in the succeeding summer and autumn. The crop cut for hay was unusually abundant, and seemed to have exhausted the roots by its growth, as no after-grass sprang up; nor did the pastures which were fed afford more than a dry, hard, yellow provender, looking tanned, as if seared by severe frost: and in September, when, in general; we expect our fields to yield an abundance of grass, as food for months, they presented commonly the aspect of hard-fed lands in March, though so much rain had fallen, both in July and August, as to lead us to expect profusion. It did not appear that the roots had actually perished, which could not have been the case, by producing the mowing crops that they did; but this was a single effort: the injury was manifested by the deficiency of the autumnal vigour; this was the actual result, difficult as it is to assign a satisfactory reason. Perhaps these effects upon our pasture lands were unprecedented; but these things pass away, unless recorded; and though we may resort to the oldest memory for evidence, yet memory is oblivious, often exaggerative, and cannot safely be trusted.

Our premises this year, 1830, was visited by a little animal which is in general a stranger in our fields, not finding probably the concealment he delights in, and our population scares him – a squirrel, finding himself in the power of obtaining certain agreeable things, without molestation, he soon introduced a companion. We presently found the traces of their visitation, attacking first a standard cherry-tree, that was soon cleared, even the not half ripe Morello, did not escape a trial. Indeed, we cultivate no fruit more universally a favourite with all wildings of the air, than the cherry. Jays, daws, thrushes, blackbirds, sparrows, and divers besides banquet upon this fruit, and are enticed by its presence, when others are unheeded; nor do our stratagems perfectly preserve it, pecking through the meshes of the net with avidity for this tempting fruit. The large fruit of the Wilmot strawberry next attracted attention. Roseberrys and Keens, were not in such request, but along the margin of a bed of Caledonias, we could see the little rogues, even at mid-day seated with erect tail, secured from defilement, with the berry in their paws, consuming it most comfortably, dropping the calyx, and commencing another. Early morning was one of the principal periods for dieting, seeming to be no nocturnal feeders, the scattered fragments being those of the day.

Coursing about in the sun, and running with great facility up walls, they seemed enabled to accomplish any kind of action inclination prompted: the little bird fearful of his predaceous looks, never fails to mob him as they would a cat or other dangerous vermin. The range of the garden was perfectly free, but the common fruits were left untouched through the summer. They remained with us long enough to try the filbeards, but they probably were not sufficiently matured, or this fickle almost untameable creature became annoyed by the notice of human life, and left us for more congenial solitude and freedom from observance.

Most of our garden fruits became in turn, the banquet of some of domestic visitors, a sort of tribute for their music, and services in consuming many insect depredators; squirrels are known, where plentiful, to be very fond of wall fruit, detaching a Peach with great dexterity - the short tailed mouse too, I have repeatedly seen, running from the beds with a Carolina strawberry in his mouth, ending nearly an inch to his length, and when not sufficiently near to be identified, presents the appearance of a very extraordinary animal: but the strawberry[41] seems less liable to plunder then the generality of our garden produce, the wasp, a great pillager, is not in activity when the fruit ripens - the thrushes and the blackbirds, only sparingly taste them, the young of the centipede (*scolopendra forficate?*) in wet seasons mine away the substance and nestle in the hollows, but with these exceptions I know no plunderer that banquets in our strawberry beds. Squirrels are accused of occasioning great injury in young plantations of firs by gnawing off in the winter and spring months the leading shoots of these trees, which probably afford them a portion of their food in those seasons. That many young shoots are frequently found broken off, or detached by animals on the ground in fir groves is certain and this little animal is apparently the offending creature.

July 10th. — Hay-making now general, rendering, during this fine weather, our fields a scene of animation, and affording us pleasure in the prospect of merited success to the cultivator, and abundant wholesome food to the poor animals in a season of particular need. It is a sad sight to observe the hungry beast tossing about his dusty, mouldy provender, and wandering from crib to crib in frost and snow, discontented and in want, seeking for some more grateful mouthful than a rainy hay-season has afforded. Yet all this inflorescence around and odours in the air are more than some of us find agreeable; and I have known persons painfully affected, during the

---

41 Our common people call this fruit the "Stray-berry" and perhaps this is the real origin of our name. Few plants ramble more. That it is occasionally collected, and conveyed on straw, is hardly of sufficient notoriety, to give it a name.

whole of this period, by reiterated sneezings, occasioned probably by the invisible farina of the various herbage floating in the air: but the emission of scents from plants is in no way similar to this dispersion of the pollen, but an imperceptible body that has uninterruptedly flowed from certain races, when in particular states. Viewing the faculties of man in all their bearings, each present a character, by the sensations or perceptions excited, most extraordinary and inexplicable; nor is this sense of smelling attended with less remarkable effects than others: we may have before us innumerable fruits or flowers that emit odours, each exciting a different sensation by the gas or vapour that proceeds from them, distinguishing them as readily as the sight of the individual would do. Of what does this vapour consist? Is it compounded of solid particles, each emission differently formed, and acting in a way peculiar to itself upon the nasal organ? The odour may be pleasant or offensive, but both cannot surely be constituted similarly to influence so differently: separate organs may be stimulated by a scent, and others not affected by it, and hence a peculiar smell perceived; but still it must be an existent body, made up of parts, that accomplishes all this. But here, like other human investigations upon final causes, we lose ourselves in impenetrable darkness, and the mere effect alone remains. The peculiar smell arising from hay has been thought to proceed from the vernal grass growing in the crop, and it certainly at times does communicate its fragrance to the dried herbage; but yet it cannot exclusively give this well-known odour, as grasses cut down late in the year, when every blade of the *anthoxanthum* has withered away, yet possess this flavour, if well harvested. Artificial grasses, that is, crops compounded of ray-grass and clover, possess little of this smell when dry; yet we shall often observe very luxuriant plants of the vernal grass growing with the crop: the smell which distinguishes well-made hay probably proceeds from no individual plant, but is a commixture of odours arising from the various herbage of the crop. The scents proceeding from plants are confined to no one part, but may proceed from roots, leaves, flowers, fruits, &c., in various essential or gross states, or emitted from glands or secretory vessels; but when the odour proceeds from the flower, it is by no means certain what part of that organ produces it: yet, in many cases, it most probably proceeds from the petals transmitted from the plant through the claws at its base, and escaping through orifices on their surface; at others, from the nectaries, or various parts which compose the blossom. We can but conjecture the object of the smell of plants, but it seems most probably to be the signal that calls the numerous tribes of the earth to find their various wants, and supply their necessities, multiform as they may be; and probably most plants do transfuse such intimations of their presence, imperceptible to some, but manifest to those which require them for the objects of their being: — the day, the evening, and the night, each present these signals to the animated creatures of the air.

June and July, 1825. — The quantity of that sweet clammy fluid which we find upon certain leaves, and commonly call "honey-dew," was more than usually abundant during these months. In the day-time, bees, wasps, and tribes of flies, collected to feed upon it; and in the evenings moths and insects of the night frequented the fruit-trees on our walls, particularly the cherry and the plum, for the same purpose; and their presence brought the bat, so that some places were animated by the flitting about of these creatures. Aphides abounded upon all the young sprays.

June 17, 1828. — Abundance of rain has fallen during the preceding night, and in the morning of this day, about two o'clock, the sun broke out, the air becoming hot and heavy. I was soon surprised by observing multitudes of hive-bees buzzing and crawling about the foliage and young shoots of my laurel-bushes (*prunus laurocerasus*), and feeding upon some sweet matter lodged on them; the blossoms had long before fallen off: no aphides frequent this plant, nor were there any trees near them from whence any sweet matter might have fallen; we have no honey-dew upon our fruit-trees, and an aphis is scarcely to be found. Has any saccharine matter fallen, or been emitted by the plant, to entice these insects to harbour about them? It clearly appears that honey-dews arise from two causes: that a large portion of it is the discharges from insects of the genus aphis, has long since been manifested by the Abbé Sauvages, Mr. Curtis, and others; insects discharge in all days and hours during the warm months of the year. But there is another kind which we find only at particular times, and in certain states of the atmosphere, lodged on certain plants during the night in such quantities as to hang occasionally in drops from the points of the leaves. The foliage of the oak is at times lucid with this sweet liquor; and this the bees are soon acquainted with, and eagerly collect it, which they only partially do when spread upon the leaves on the wall, the evident discharge of aphides. Some of my neighbours who have hives will occasionally observe, "A heavy honeydew last night, and the bees are hard at work:" this cannot proceed from insect discharges. That some foliage may condense any matter that may fall upon it, is not improbable; or even excrete it from their pores by the impellent power of the air in certain states, is to be conceived: but all this is conjectural, and our knowledge of the causes which produce these partial honey-dews is yet to be acquired.

July 31st, 1829. A close hot night, all the young spring arising from the stocks in a neighbouring wood which was cut down last winter, was this morning covered with honey-dew, these consisted almost solely of nut, ash, privet, oak, the trees above, which were young, consisted of oak. and their foliage was perfectly free from aphis or honey-dew, the undergrowing shoots were luxuriant, nor were any insects about them from whence this clammy liquor could have proceeded, it seemed like a recent exudation from the pores of the foliage and hung in drops at their points; it disappeared in one day, and I notice it no more that summer.

In the years 1825 and 1826, the foliage of our hedges in the spring months was unusually mangled by the caterpillars of different moths; but in 1827 these creatures had increased so much, that the entire leaves of the sloe and the whitethorn were consumed by them; the hedges, when consisting of these shrubs alone, presented for miles the appearance of winter sprays, covered with a cottony web. The other hedge plants were little injured. The larvæ of several species of small creatures were concerned in this annihilation of verdure; but the little ermine moths (*phalæna evonymella*, and *ph. padella*) were the chief performers in this denuding process. In July the perfected moths swarmed about the scene of their birth in vast numbers; yet such was the retrieving power of Nature, that by the middle of August only a small portion of the injury occasioned by these creatures was to be observed, the summer-shoot bursting out, and covering the sprays with the verdure of spring. The chief singularity in all this was the appearance of the sloe-bush, all the foliage being consumed by insects, or crisped away by severe winds, leaving the sprays profusely covered with the small young fruit, perfectly uninjured, and proceeding in its growth; so that, by the time the foliage was renewed in August, it had obtained its usual size. This was the case, too, with the crab, and some of the orchard fruits, presenting the unusual sight of fruit growing alone on the boughs without leaves; so that, in fact, the offices of inspiration, transpiration, and all their consequences, usually accomplished by the leaves of plants, must have been suspended, or performed by other organs, as no deficiency of vegetative powers was apparent.

But insects alone were not the cause of all the denudation and unsightly appearance which our orchard and other trees so remarkably presented this year; for the destruction of the foliage was accomplished in part by some malignant influences, not well understood. Like the Egyptian king, we are accustomed to attribute all our evils of this nature to the "blasting of the east wind;" yet we find all aspects and places obnoxious to it: one situation may be exempted for a period of many years from such visitations, when others suffer; on a sudden, a partial or a local stream of hot, cold, salt, or what we denominate a pestilential wind, sweeps along, and it is destroyed.

Surrounded by and situate in the midst of an agricultural district, we are eager and persevering "leasers" here; and it becomes in a certain degree profitable to our poor, though they cannot hope, like the dutiful Ruth, to gather their three pecks and over in the day. It may be difficult to comprehend how the picking up a head of corn here, and another there, should be a remunerative employ; but in this case, like all other slow operations, a distant result, rather than an instant effect, must be looked for. I have found some little difficulty in obtaining intelligence sufficient to acquire a knowledge of the gain by this employ. The poor are often jealous and suspicious of the motives, when any attempts are made to procure information regarding their profits or improvements; and indeed the advantages of one year are uncertain in another. Catching, doubtful seasons, when the farmer collects in haste, and is unmindful of trifles, afford the best harvest to the gleaner. In fine, settled weather, the operation of reaping is conducted with more deliberation, and less corn is scattered about. When a woman with two or three active children lease in concert, it becomes a beneficial employ. I have heard of a family in the parish thus engaged, who have in one season obtained eight bushels of clear wheat; but this was excess. I know a single woman, also, who has gleaned in the same period four bushels and a half; but this, again, was under very favourable and partial circumstances. In general a good leaser is satisfied if she can obtain, single handed, a clear three bushels in the season, which gives her about a bushel in the week; and, if taken at seven shillings, is very reasonable, and far from being any great accession of profit — less perhaps than is generally supposed to be the emolument of the gleaner; and this may have been acquired by the active labour of eight or nine hours. Yet such is the ardour for this occupation, the enjoyment of this full association with their neighbours, the prattle, the gossip, the glee, the excitement it occasions, that I am sure the allowance of fourteen pence a day, certain and constant, would hardly be accepted by my leasing neighbours in place of it. Indeed I would not offer it, believing that this gleaning season is looked forward to with anxiety and satisfaction; and is a season, too, in which the children of the family can contribute to its support without pain or undue exertion; and viewing with much approbation and pleasure this long-established custom as a relaxation from domestic confinement, when every cottage is locked up and abandoned by its inmates, to pursue this innocent, healthful, laudable employ, where every grain that is collected is saved from waste, and converted to the benefit of a needy and laborious community. From the result of the pauper leasing, no bad criterion may be obtained of the general product of the season; for, as the collection is made from many stations, and variety of culture, these samples of all afford a reasonable average of the quality. It has been thought, but I trust and believe only in the apprehension of evil, that leasing is injurious to the morals of the poor, affording them an opportunity and initiating them in petty pilfering: but, if the disposition existed, it could be practicable but in very few instances; mutual jealousy would prevent individual success, and immediate detection would follow the filching of numbers. The commencement of many ceremonies and solemnities are lost by perversion, or in the obscurity of years; the stream of habit may trickle on from age to age, till it flows in time a steady current, yet the original source remain unknown: but this custom of gleaning the remnant of the field we know existed from the earliest periods, three thousand years and upwards for certain; for, if it were not then first instituted, it was secured and regulated by an especial ordinance of the Almighty to the Israelites in the wilderness, as a privilege to be fully enjoyed by the poor of the land, whenever their triumphant armies should enter into possession of Canaan. By this law, the leasing of three products was granted to the destitute inhabitants of the soil, — the olive, the grape vine, and corn; the olive tree was to be beaten but once; the scattered grape in the vintage was not to be gathered; and in the field where the corn grew, "clean riddance" was not to be made, the corners were to be left unreaped, and even the forgotten sheaf was not to be fetched away by the owner, but to be left for the "poor and the stranger, the fatherless and the widow." This was not simply declared once, as an act of mercy, but enjoined and confirmed by ordinances thrice repeated, and impressed with particular solemnity: "I am the Lord thy God," I have given thee all, and I command unreserved obedience to this my appointment.

Revolving in our minds, as we old-mannered people often do, the forms, rites, and usages of earlier days, we occasionally regret that fashions by gradual neglect have passed away, and can never be revived, to give that feeling of pleasure which a natural growth seemed to have inspired. Some, though probably of pagan origin, were innocent and harmless practices; the may-pole, with all its flowery wreaths, so often surrounded by the dance and the song, is now but seldom seen, where we have known it, especially in the lacemaking counties, the evening and almost sole recreation, after long hours of unhealthy occupation, for happy groups of

> Those pale maids who weave their threads with bone;

and it gave these poor villagers a transient glow of health, seen then alone; but it is gone with the rest, and we grieve to think how little remains that poverty and innocence can partake of. Others were of monkish introduction, yet seemed to keep in remembrance the revolutions of seasons and events, which, though recorded elsewhere, had become the types of written things. Yet one of them in the irritation of the moment I have at times wished, selfishly enough perhaps, consigned to oblivion with monks and monkish deeds. "Christmassing," as we call it, the decorating our churches, houses, and market meats with evergreens, is yet retained among us; and we growers of such things annually contribute more than we wish for the demand of the towns. Sprays and sprigs may be connived at, but this year I lost most of my beautiful young holly trees, the cherished nurslings of my hedgerows. The holly, though indigenous with us, is a very slow growing tree, and certainly the most ornamental of our native foresters. Its fine foliage shining in vigour and health, mingling with its brilliant coral beads, gives us the cheering aspect of a summer's verdure, when all besides is desolation and decay. It is not only grateful to the eye, but gives us pleasure, when we contemplate the food it will afford our poor hedgefaring birds, when all but its berries and those of the ivy are consumed; and we are careful to preserve these gay youths of promise, when we trim our fences: but no sooner do they become young trees, in splendid beauty, than the merciless hatchet, in some December's night, lops off their heads, leaving a naked unsightly stake to point out our loss; and we grieve and are vexed, for they never acquire again comparative beauty. These young heads, that we have been robbed of, are in especial request to form a bush, dependent from the centre of the kitchen or the servants' hall, which, in this season of license and festivity, becomes a station for extra liberty, as every female passing under it becomes subject to the salutation of her male companion. This centre bush is often the object of particular decoration, being surrounded by the translucent berries of the misletoe, and those of the ivy, dipped in blue and white starch. But at this season I have noticed one remarkable decoration among the natives of the principality. A large white turnip is stuck as full as possible of black oats, so as to hide almost the substance in which they are set, and sometimes having compartments of white oats; and being placed upon a candlestick, or some other elevation, on the mantel-tree, presents an extraordinary hedgehog-like appearance. The first adoption of this purely rural fancy, and its designation, I am perfectly unacquainted with; but, when it is well executed, it requires attentive examination to detect the device.

We hear the village fancy and we smile at the yet lingering remnants of simple times gone by, harmless and innocent never to return, the imaginations of the old and departed, for the younger scions of the hamlet, when they timidly repeat them, tell them as ancient sayings, things they have heard; foolish as so many of them were, we can in no way regret their decline and loss, and only be surprised that conceptions so void of rationality could ever have been entertained. *Daphne Laurola*, the spurge laurel, an evergreen not very uncommonly found in rocky and sequestered places, did some years past exist in a few places with us, and one might have conjectured could occasion the enmity of none but be rather in favour with our village dames by reason of the medicinal efficacy it possesses, its bright and perennial verdure; but the whole has been nearly eradicated from the idea that it was in some way prejudicial to marriage by remaining, influential in weakening the attachments of youth, and thus deserving extirpation. Our soil or situation from some cause is very favourable to the growth of the Brompton Stock, and few cottage gardens are found not able to produce superb specimens of the double flowers, which fail in becoming so with more care in better cultivated places; but my neighbours tell me that they pursue but one plan to ensure success, unmindful of the time when the day may occur they always deposit seed in the ground upon "Good Friday" – upon what principle I cannot decide – but my essays I think have not been so made, and being generally inferior, must continue perhaps heretically to admire the beautiful products of my neighbours in their dry and sheltered borders, rather than my own.

The innovations of time and example have not yet effaced a very rude but effectual mode of intimating assent, or denial, in one of our rural practices, and now noticed merely as a record of the usages of the days whose memorials are passing speedily to oblivion. With us, as in most other villages I suppose, we impound the cattle of incorrigible owners who would rather pasture them at the cost of others than themselves; when the keeper of the pound receives these trespassing beasts, he cuts from the nearest hedge a small stick three or four inches long, or a little longer if necessary, this he scores half round, with as many notches as there are creatures under his keeping, and splitting it through the scores keeps himself one half, the corresponding part he delivers to the individual who has given the cattle to his care, and should he liberate them without this given half, which is his warrant, he becomes liable to the amount of damage incurred by the strays, its delivery signifying consent to their restoration, is a note of hand that cannot be imitated; it thus becomes a kind of tally, authorizing delivery, the usage probably of a very remote and artless age, but perfectly effectual and comprehensible to the plainest capacity, when writing was the acquisition of the learned few, and reading imparted to the priest alone.

We are no votarists of fortune here, nor do we trouble ourselves concerning predestinate ordinations, or like subtilties; but when we notice passing events, we lament the ills and are pleased with the good luck of a neighbour: and a little turn happened lately to a parishioner, which in former times, when events were viewed under aspects different from those by which we now regard them, might have occasioned more wonderment and comment than it did. An industrious labouring man had been some time unemployed, and having sought an engagement at all those places most likely to have afforded it, but without success, sat himself down upon a bank in one of our potato fields, carelessly twisting a straw, and ruminating what his next resource might be; when, casting his eyes to the ground, he discovered, immediately between his feet, a guinea! a guinea perfect in all its requisites! The finding of such a coin, at such a time, was no common occurrence; but by what casualty did the money come there? The frequenters of our fields, breakers of stone, and delvers of the soil, inhabiters of the tenement and the cot, have no superfluous gold to drop unheeded in their progress, and one should have supposed that the various operations which the field had undergone in the potato culture would have brought to view any coin of that size and lustre. Upon looking at the land, however, much of our perplexity was removed by observing that the ground had been in part manured by scrapings from our turnpike road, rendering it highly probable that this golden stranger had been dropped by some traveller, not missed by him, or lost in the mire, this mortar from the road possibly so coating it about, as to secrete it for a time, some heavy rain dissolving the clod, and bringing it to view. This, I am sensible, is an incident little deserving of narration, but it has been done from two motives: we village historians meet with but few important events to detail from the annals of our district; we have no gazettes, few public records, or official documents to embellish our pages, and if we will write, must be content with such small matters as present themselves; and to point out how frequently very mysterious circumstances may be elucidated, and appear as consistent events by an unbiassed examination. We may not be able always satisfactorily to see why a tide of good fortune should flow at the desire of one, and ebb from the wishes of another; yet many of the occurrences of human life are, perhaps, not so extraordinary as they are made to appear by the suppression of facts, or our ignorance of circumstances.

The effects of atmospheric changes upon vegetation have been noticed in the rudest ages: even the simplest people have remarked their influence on the appetites of their cattle, so that to "eat like a rabbit before rain" has become proverbial, from the common observance of the fact: but the influence of the electric fluid upon the common herbage has not been, perhaps, so generally perceived. My men complain to-day that they cannot mow, that they "cannot any how make a hand of it," as the grass hangs about the blade of the scythe, and is become tough and woolly; heavy rains are falling to the southward, and thunder rolls around us; this indicates the electric state of the air, and points out the influence that atmospheric temperature and condition have upon organized and unorganized bodies, though from their nature not always manifested, all terrestrial substances being replete with electric matter. In the case here mentioned, it appears probable that the state of the air induced a temporary degree of moisture to arise from the earth, or to be given out by the air, and that this moisture conducted the electric fluid to the vegetation of the field. Experiments prove that electric matter discharged into a vegetable withers and destroys it; and it appeared to me at the time — but I am no electrician — that an inferior or natural portion of this fluid, such as was then circulating around, had influenced my grass in a lower degree, so as not to wither, but to cause it to flag and become tough, or, as they call it in some counties, to "wilt;" the farina of the grass appeared damper than is usual, by its hanging about the blades of the scythes more than it commonly does; the stone removed it, as the men whetted them, just at the edge, but they were soon clogged again. As the thunder cleared away, the impediments became less obvious, and by degrees the difficulties ceased. The observance of local facts, though unimportant in themselves, may at times elucidate perplexities, or strengthen conclusions.

That purely rural, little noticed, and indeed local occurrence, called by the country people "hummings in the air," is annually to be heard in one or two fields near my dwelling. About the middle of the day, perhaps from twelve o'clock till two, on a few calm, sultry days in July, we occasionally hear, when in particular places, the humming of apparently a large swarm of bees. It is generally in some spacious open spot, that this murmuring first arrests our attention. As we move onward the sound becomes fainter, and by degrees is no longer audible. That this sound proceeds from a collection of bees, or some such insects, high in the air, there can be no doubt; yet the musicians are invisible. At these times a solitary insect or so may be observed here and there, occupied in its usual employ, but this creature takes no part in our aerial orchestra. We investigators, who endeavour to find a reason and a cause for all things, are a little puzzled sometimes in our pursuits, like other people; and, perhaps, would have but little success in attempting an elucidation of this occurrence, which, with those circles in our pastures and on our lawns, that produce such crops of fungi (*agaricus oreades*) , and are called by the common name, for want of a better or more significant one, of "fairy rings," we will leave as we find them, an *odium physiologicum*. Yet considering the cause of these sounds in the air, an idea occurred which though only conjectural, yet as it carries a degree of probability with it may be mentioned. We hear these murmurings only during one short period of the year, and that is the time when the drones or male bees are said to quit the hive, mounting high in the air, and our lost to sight. Shortly after this migration the queen departs, and having performed various gyrations mounts aloft too, and joins their company, it appears from these circumstances, that the hummings may be the contentions occasioned by the arrival of the female of the hive in this great assemblage of the drones.

1827. — The winds of this autumn have been violent and distressing; but of all variable things, we know of none more so than our seasons and temperatures, produced probably by causes and combinations of which we have no comprehension, or power of foreseeing, "for these things come not by observation; we cannot say, Lo here! or Lo there!" What can be more extraordinary, or inexplicable by table or computation, than the sudden visitation, in the midst of storms and frosts, of such a day of brightness and warmth as we sometimes witness, cheering the aspect of all things, — a portrait of summer, brought from we know not what region, in a framework of winter. All these things assuredly have their effects upon the products of the earth, and by their means upon the creatures that are nourished by them, carrying on that imperceptible line of influences and intelligences that is maintained throughout nature. We know that vegetation and the atmosphere are in a constant state of barter and exchange, receiving and modifying; and possibly, from the unseen effects of a frosty morning, a fall of snow, or a few hours temperature of the air, a fruitful or an unproductive season may arise. We notice the effects of spring changes, because vegetation has so far advanced as to render influences manifest; but we cannot perceive the injuries or benefits accruing to a hidden circulation from particular events. Every person who has been conversant with cattle, must have remarked how uncertain their progress in improvement has been; that the abundant provision of one year did not prove equally nutritive with the scanty product of some other: this fact originates probably from the effects of atmospheric impulse, either directly upon vegetation, or upon the soil which produced the food collaterally, or upon both collectively. We may observe frequently in fields where cattle have been feeding, patches of herbage left here and there untouched: much of this originates from the manure of the animals, which has rendered the grass springing up around it rank, and it becomes in consequence rejected; but besides this, we shall see, in several places, portions of herbage apparently good, and similar in growth with that which has been closely eaten down in the field, not arising from any visible accessory matter, yet still unconsumed: and in all probability, the soil in those places is differently compounded from other parts of the pasture, communicating some flavour disagreeable to the animal, and it becomes rejected as long as grateful food is obtainable; yet probably no investigation of the soil or produce would manifest this, but that something generally distasteful does exist, is obvious by the various creatures rejecting it, consuming to a certain verge, and then proceeding forwards. My sheep again eat, with a marked preference, the dark-coloured grass springing up from those circles and semi-circles we frequently see in the dry pastures, called "fairy rings;" manifesting that it possesses an agreeable flavour, derived immediately from the soil, or communicated by atmospheric influence. The edges of pathways are favourite feeding-grounds for several creatures; but this indeed originates from the nature of the grass which vegetates in those places, trampling of the feet keeping down the coarser herbage, fining the better sorts, and especially promoting the growth of the annua poa (*poa annua*), a delicate and agreeable food to the horse and the sheep, by pressing down the stoles or runners, and levelling the worm castes, the plant thus throwing out a succession of foliage; for this poa, by reason of its abundant fibrous roots, endures drought better than many of its congeners, by which means a pathway becomes frequently a verdant line in a brown and arid pasture. In a wet season, water appears to nourish plants, or to supply their requirements principally: in a dry one, nutriment must be obtained from the soil by means of the fibre of the root, and hence particles are imbibed chemically different; a dry or a drained soil, producing short and scanty herbage, will frequently improve the condition of cattle more than an adjoining meadow having a profusion of food, though probably no chemical analysis could indicate the difference. These periodical winds again, violent and distressing as they often prove, are yet unquestionably essential in the economy of nature: our two seasons, in which these commotions of the air most usually become manifest, are about the equinoxes of autumn and spring, periods which in many respects have a similarity with each other. In the autumn of our year, the foliage of trees and plants, &c. putrefies and decays; marshes and dull waters, clogged by their own products, stagnate, and discharge large portions of hydrogen, carbonic gas, &c. injurious and even fatal to animal existence: in summer all these baneful exhalations are neutralized and rendered wholesome by the vast quantities of oxygen, or vital air, discharged from vegetable foliage: but these agents of benefit, by the autumn, are no more — consequently the discharge of oxygen is suspended, but the production of unhealthy air increased by the additional decomposition of the season. To counteract this is probably the business of the storms of wind and rain prevailing at this season, which, by agitating and dissipating the noxious airs, introduce fresh currents, and render the fluid we breathe salubrious. The same may be advanced in regard to spring: the whole decay of winter, having no neutralizing body to render it wholesome, requires some great influencing power to remove it. But all this is reasoning without actual evidence; a discursive license, from the fallibility of human judgment not often to be indulged in: yet we can so rarely perceive the purport of the movements of nature, that our conceptions, vague as they may be, are almost all that remain to us.

Having observed above, that particles are imbibed by one plant chemically different from those of another, perhaps I may point out a very remarkable instance of a plant possessing this tendency to incorporate mineral substances. Water, even in the simplest form of rain, by passing over calcareous earths will take up certain portions of them, but if impregnated with carbonic acid, as many springs are, the power of solution is greatly increased. When this gas becomes liberated, the earth subsides by mere specific gravity, filming calcareous incrustations. But when it is held suspended in the fluid without this agency, it appears to be detached by some chemical affinity, for though the water may be equally imbibed by various plants, its mineral principles are not equally retained. We have a very common plant, submerged in many pools and streams remarkable for its rank, nauseous, mud-like smell, the *Chara vulgaris*, which appears to possess this power in a notable degree. Its fine thread-like stems, and whorls are a little rough to the touch, seemingly occasioned by natural asperities, though the microscope will occasionally detect minute adherent extraneous particles. This chara, being most perfectly washed, picked, and dried, will rub down in a mortar to a pale green powder. The component parts vary, but by an analysis carefully made this day, from plants growing in a pool, not so favourably circumstanced as to possess any visible source of calcareous matter. I found 50 parts to contain 16 of mineral extract, in the proportion of 14 pure lime, $1^1/_2$ silex, $^1/_2$ Alumine.[42] So that in fact the vegetating plant

---

42 14 pure lime
 1 ½ $^1/_2$. Silex
 ½ alumine
 14 vegetable matter
 20. carbon: acid
 ———
 50.

contained 19 parts of carbonate of lime, 21 of mineral extract in 50. This is assuredly a remarkable instance of the absorption of mineral particles by a plant, and though I have found it in inferior portions generally, yet in all instances in a large quantity, sufficiently to manifest the power possessed by plants, and that many may so imbibe terrene matter undetected by us, as to become the occasion of the peculiar qualities and effects. The silica that is taken up by the bamboo, and wheat plants, constituting the varnish or coating of the rind and the straw, is immediately to the point — these mineral substances being considered to have been received through the agency of the roots. But why one kind is rejected and another received, it would be idle to investigate. From the soil in which chara vegetates, a larger portion of silica might have been expected, yet we find that the calcareous earth, which is apparently alien to the pool, is detached and preferred. Another water plant, *Potamogeton crispum* though growing in streams and pools, most unlikely from the nature of the surrounding soil to process carbonate of lime, yet will at times be so coated with it, that when dry, it will fall upon the paper in scales.

We have here so few operations of nature deserving mention, that I must not omit to notice a rather uncommon appearance in some of our claylands, which the surrounding parishes do not present. The soil of a few fields seems to cover for some depth a rock of coarse limestone, which we never burn for use. In a direction bearing nearly east and west, in a line pointing to the Severn, a number of sinkings and pits are observable, like abandoned shafts, or the commencement of mines. They are called by the country people "whirly pits." In some instances the bottoms of them are not visible, owing to the tortuous irregularity of the passages; in other cases they are only deep hollows covered with turf. These sinkings are evidently occasioned by the lowering of the surface in consequence of the removal of the support beneath. Where the under parts have been entirely displaced, the upper have fallen in, and formed a chasm; where only partially removed, deep, turfy hollows are formed. These removals have been occasioned, probably, by a stream of water running far beneath, and washing away the support; and in part by the superfluous water from the ditches and watercourses above draining into the fissures of the rock, and so gradually mining or wearing away a passage; for they are now frequently the receivers of all the running water from the land, which seems naturally to drain into them, and apparently has been so conducted for a long course of years. Some of them present dark and frightful chasms, and bushes and brambles are encouraged to grow about them, to prevent cattle from falling into the pits.

Many a fox, when hard pressed, has been known to make for these "whirly pits," as his last resource; and, secreting himself in some of the under cavities, has escaped from the pursuit of his enemies above. I once saw one of these animals dead at the bottom. Whether he perished from being unable to return up the crags after one of these retreats, or by any other means, I know not.

In particular years we are much troubled here by the luxuriant growth of a cryptogamous plant, which I believe to be the lichen fascicularis of Dillenius: it may always be found even in the driest summers, but being in those seasons shrivelled up, is in no way troublesome, nor indeed noticed, unless sought for. This lichen covers the walks of shrubberies at times in shady places, and paths in the kitchen garden, appearing like a dull olivaceous crust, most observable about October or November, and the spring months; but in the summer of 1828, the unusual moisture of that season was so favourable to its growth, that even in August we could not walk in safety in those places where it abounded, our feet sliding along upon the gelatinous, slippery foliage and tubercles. Upon the walks of our culinary gardens we sprinkle coal-ashes, and this enables us for some time to pass along with tolerable safety; but in the end it so fosters the growth of this lichen, and small mosses, which retain moisture as a sponge, that the evil we endeavour to remove is by the autumn increased: where gravel is not obtainable, paring off the crest of the walk is the only effectual remedy, and this ultimately we are necessitated to resort to. It is notable that such a very insignificant product, this hardly discernible plant, should endanger limb and life, and by circumstances become so formidable to us "lords of the creation," as to force us to devise contrivances to counteract its injurious tendencies.

There are few things in Nature or in Arts but what are deserving of attention either for instruction or amusement, the one indeed cannot be circumscribed or investigated fully, but will ever present things beyond the means of human intelligence; but the works of art will generally amuse, and however simple the fabrication may appear, yet probably upon enquiry it will be found that they could not have been accomplished and perfected, but by great assiduity and repeated trials. Some years past we had a little manufacturer of catgut for musical instruments in our neighbourhood, and a curious art it seemed to be, and though we may so often look at these twisted membranes, and listen to their various vibrations, yet have not thought that it required great attention and some adventitious circumstances to construct them, for this material so indispensable an article in stringed instruments of music, and the harmony of the orchestra, depends upon season for its perfect manufacture. For it is only in the months of July and August, the hottest periods of our year that it can be prepared; being fabricated from the intestines (gut) of sheep, not promiscuously taken, but the leanest miserable starveling possessing the least portion of internal fat, is selected, when obtained, they are long and carefully washed in the purest running water, slit into strips and accurately twisted, then dried with the utmost rapidity by exposure to the sun; in partial or insufficient heat the contraction would not be uniform, and the tone variable, hence it is said that Italian catgut is preferable to the English, the animal membrane being freer from fat, and the air drying it faster – a fine treble string consists of a single strip, a double base, will require three dozen strips, forming a mass of perhaps nearly an inch in diameter, and these to be twisted to the required size with great exactitude and force, and even when finished, the alterations of our climate so influences them, by expansion, and variable contraction, as our hygrometer makers are sensible of – that the only means of securing them in proper tone is by keeping in a bladder, with oil of almonds.

There are times when we suffer here greatly by the withering and searing up as it were of the leaves of our vegetation, which we attribute generally to an early morning's frost. That late spring frosts do occasion such injuries, and that noxious blasts, from causes which we cannot divine, occasion infinite annual mischief, if not destruction, to our wall fruit, is most manifest; yet there is great reason to suspect that a large portion of the injuries which we ascribe to blights, blasts, and frosts, are occasioned by saline sprays brought by strong western or south-western gales from King-road in the Bristol Channel, eight or ten miles distant, or from even more remote waters, and swept over the adjoining country where the wind passes. This saline wind has often been suspected by me as the evil agent that accomplishes most of our blightings here; and on November the 3rd 1825, these suspicions were corroborated — for on this and the preceding days we had strong gales from the water, in consequence of which such windows as were situate to the west and south-west were skimmed over with a light saline scurf, the brasswork of the doors was corroded and turned green, painted works of all kinds were salt to the tongue, as was everything that could condense the moisture; and the leaves of the shrubs in the hedgerows, and of trees, all turned brown, and were crisped up. A row of large elms in particular, that fronted the gale, received its full influence; the whole of the windward side, then in full foliage, became perfectly brown and seared, and the leaves shortly afterwards parted from their sprays and left them bare; while the other and sheltered side of the trees preserved its green foliage very slightly influenced by the spray that burned up the other. No period of the leafy season is exempt from these pernicious effects, more or less, if the wind be sufficiently violent and blowing from the water. Portions of the country distant from the shores often seem more influenced by these salt sprays than others more near, the wind lifting up the saline moisture, bearing it aloft to remote parts, and dropping it as it travels over the land or meets with impediments.

Our apples in some years are more inclined to become spotted than in others, from causes not quite obvious, as moist summers do not occasion it more decidedly than dry. Particular sorts are more subject to these dark markings than others. The russet, though a rough-coated fruit, seems exempt from spots; whereas some of the smooth-rinded ones, especially the pearmain, are invariably disfigured with them. These marks appear to be an æcidium, which we frequently find to be perfectly matured, the centre occupied with minute, powdery capsules, having burst through their epidermis, or covering, which hangs in fragments round the margin. This æcidium apparently derives its nutriment from the apple; for immediately round the verge of the spot the skin becomes wrinkled in consequence of the juices being drawn off by the fungus. In most cases the presence of plants of this nature is symptomatic of decay; but in this instance we find an exception to a pretty general effect, for the decay of the apple does not always commence at the spot, which does not even apparently contribute to it — for the whole fruit will shrivel up in time by the escape of its juices, without any decay by mortification. Though we are not able always to ascertain the purposes of Nature, yet this little cryptogamous plant affords a strong example of her universal tendency to produce, and every vegetable substance seems to afford a soil for her productions. We have even an agaric, with a bulbous root and downy pileus[43], that will spring from the smooth summit of another (*agaricus caseus*), which has a uniform footstalk, though not of common occurrence. Thus a plant, that itself arises from decay, is

---

43   This agaric is, I believe, unnoticed. I have called it *Agaricus surrectus*.
Pileus — convex, expanding, covered with a pile of short, white hair; centre depressed; faintly tinted with yellow; from one to three inches in diameter.
Laminæ — loose, irregular, generally four in a set, rather numerous, broad, white, changing to buff, and then pink.
Stipes — solid, tapering upwards, rather thick immediately below the pileus, three inches high, thick as a reed, white, and often downy, wrapper at the base.
Many of this species of singular plant I found in October, 1819, springing from a confluent mass of *A. caseus*. Bolton's *A. pulvinatus* is something like our plant; but he describes his under side as perfectly flat, and represents a singularity in the termination of his laminæ, which is not observable in our *A. surrectus*.

found to constitute a soil for another; and the termination of this chain of efficiency is hidden from us.

Agaricus Surrectus.

Agaricus Surrectus

But the leaves of many vegetables often become singularly spotted during some part of the summer, and such spots have not certainly been effected by the growth of cryptogamous plants, natural decay, or the punctures of insects, the usual agents in these cases. A very indifferent observer of these things, in strolling round his garden, must have remarked how uniformly and singularly the foliage of some of the varieties of the strawberry are spotted, and corroded, as it were, into little holes; whereas other kinds have seldom any of these marks visible on them. I have fancied that these spottings were occasioned by the influence of solar heat: a shower of rain falls, small drops collect and remain upon the leaf of the plant; the sun then darts out, converting all these globules of rain into so many little lenses, converging the rays, and scorching or burning a hole at the focus. This conjecture has been rather strengthened by observing, that, upon certain sorts, the hautbois, alpine, &c., the rain, when

ASH TREE

it falls uniformly, wets the leaves, yet they do not become spotted; but the smooth leaves of others, roseberry, caledonian, upon which it stands in drops, always become marked and perforated; but whatever may be the real cause of these spottings, if the foliage be touched, by way of an experiment, with the point of a heated wire, after a few days they will present an appearance very similar to what is naturally effected. There seems to be a curious analogy in their functions between the roots of plants and the moving parent of animated beings, a similar obligation being required from them both of providing for those dependent on them, and both will exert their energies in fulfilment of this ordained mandate: the roots of plants wander up and down in every direction, seeking for sustenance; and we frequently see trees, growing on rocks, extending their roots, like sensitive beings, searching for moisture; if this is not obtained sufficiently, a sickly foliage and impoverished growth point out the condition of the plant.

The notable exertions which vegetables occasionally make to obtain nutriment may be instanced by the following rude drawing of an ash, — a tree which, in consequence of the profusion of its seed, we find more often scattered in wild and singular places than any other not propagated by the agency of birds, or conveyed by the winds. This one had originally been rooted in the earth, upon the top of a wall, but nourishment being required beyond what was supplied by the precarious moisture of the scanty soil, its roots proceeded downwards, winding their way through the crevices of the

stones into the earth beneath, and remained apparently incorporated with the masonry; the materials of this wall being wanted for an adjoining work, were so pulled out, as to leave the tree with all its roots detached, much as represented, with all its vegetative powers uninjured; the root B had stretched itself along the top of the wall, but how far it had extended, when in perfection, is uncertain, being broken away when I saw it first. The wood of the ash, when burned in a green state, will emit a fragrance like that which proceeds from the violet or mezerion, and this it will diffuse, in particular states of the air, to a considerable distance— a property that, I believe, is not observable in any other British wood: it is in the country only that we can be sensible of this, and it is particularly to be perceived in passing through a village when the cottagers are lighting their fires, or by a farm-house, when this wood, fresh cloven, or newly lopped off, is burning; as the wood dries, this sweet smell is, in great measure, exhaled with the moisture, for in this state we are not sensible of any odour arising from it different from other woods.

The Year 1825.

WE ARE NATURALLY solicitous to look back upon seasons remarkable for atmospheric phenomena, and compare their results with those passing before us, though we may be fully sensible that no conclusions can safely be drawn from them, — a variety of circumstances not known, or not comprehended, combining to produce results beyond our means of calculation. There have been times when such recollections brought no pleasure with them, by displaying the injuries and sufferings that hurricanes and floods have occasioned; and thus we who were witnesses of the distress occasioned by the lamentable rains of 1793, and the several successive years, when every wheat-sheaf presented a turf of verdant vegetation, cannot recollect it without sorrow, or ever forget that famine in our land. Yet it is amusing, on some occasions, to note the extremes of weather that our island has experienced; for though in general our seasons pass away without any very considerable dissimilitude, still we have known periods of great irregularity, drought or moisture, cold or heat. The freezing of great rivers, with the roasting of animals and passage of carriages upon the ice, our calendars and diaries relate; but instances of an opposite temperature, affording less striking events, are not so fully detailed as might be wished. The winter of 1661 appears to have been remarkable for its mildness; and it is rather curious that, in the century following, the winter of 1761 should have been equally notable for the mildness of its temperature. The winter of 1795 seems to have partaken of none of the severity usual to the season; and the summer of 1765 was remarkable for its heat and dryness, and all vegetation being influenced by their effects, brought forth fruits and flowers in unusual perfection.

But, perhaps, the year 1825, taking all its circumstances, is the most extraordinary to be found in our annals. The winter of 1824-5 had been mild and wet; the ensuing spring dry, but with keen winds and frosty mornings, which greatly injured the fine blossoms that appeared on our fruit-trees; and the continued and profuse nightly fall of the honey-dew was quite unusual: the leaves of the oak, the cherry, and the plum, were constantly smeared and dropping with this clammy liquor, which, falling from the foliage on the ground, blackened it as if some dark fluid had been spilled upon it: the leaves of most of our stone fruits curled up, covered with aphides, and became deciduous; and their young shoots were destroyed by the punctures of these insects that clustered on them. This honey-dew continued to fall till about the middle of July, affording an abundant supply of food to multitudes of bees, moths, and other insects, which swarmed about the trees. We rarely begin cutting our grass before the first week in July; but in consequence of the heat of June in this year, it was so drawn up, that much hay was made and carried by the 20th of June, which commonly is not accomplished till August. Our crops, on good ground, were considered as fair, though in general the chilling season of May had occasioned a deficiency; but all our clover crops and artificial grasses were harvested in the finest order, producing good-sized ricks and mows; yet their bulk was delusive, the provender cutting out light and strawy. The heat and drought continued, with very partial and slight showers of rain, all June and July; nor had we anything like serviceable rain till the second of August. In consequence our grass lands were burned up, and our fields parched, presenting deep fissures in all parts. The heat was unusually distressing all day; and evening brought us little or no relief, as every wall radiated throughout the night the heat it had imbibed from the torrid sun of the day. Our bed-room windows were kept constantly open, all apprehension from damps and night airs, which at other times were of the first consideration, being disregarded; a cooler temperature, however obtained, was alone required; and we lingered below, unwilling to encounter the tossings and restlessness that our heated beds occasioned. Our wainscots cracked, furniture contracted and gaped with seams; a sandal-wood box, which had been in use for upwards of twenty years in dry rooms, shrunk and warped out of all form; a capsule of the sandbox tree (*hura crepitans*), which had remained in repose over a shelf above the fire-place for an unknown length of time, now first experienced an excess of dryness, and exploded in every direction; door-frames contracted, window-sashes became fixed and immovable. These are trifles to relate, but yet they mark the very unusual dryness of the atmosphere.

Monday and Tuesday, July 18th and 19th, will long be remembered as the acme of our suffering, the thermometer standing in the shade of a passage communicating immediately with the outer air, in an open situation, at 82° of Fahrenheit. A few yards nearer the air, on which the sun shone, it rose to 93°, without any influence from reflection or other causes. In towns, and more confined places, it is said, the heat was much greater. The current of air now felt like that near the mouth of an oven, heavy and oppressive, and occasioning more unpleasant sensations than such a temperature usually creates; animals became distressed, the young rooks of the season entered our gardens, and approached our doors, as in severe frosts, with open bills, panting for a cooler element; horses dropped exhausted on the roads; many of the public conveyances, which usually travelled by day, waited till night, to save the cattle from the overpowering influence of the sun. The leaves of our apple and filbert trees, in dry situations, withered up; large forest trees, especially the elm, had their leaves so scorched by the sun, that they fell from their sprays as in autumn, rustling along the ground; the larch became perfectly deciduous. In our gardens, the havoc occasioned by the heat was very manifest. The fruit of the gooseberry, burnt up before maturity, hung shrivelled upon the leafless bushes; the strawberry and raspberry quite withered away; the stalk of the early potato was perfectly destroyed, and the tubers near the surface in many places became roasted and sodden by the heat, few obtaining their natural size, and sold at this period in the Bristol market at twenty-four shillings the sack. A few choice plants were saved by watering them daily; but in general the exhalation from the foliage, by reason of the heat of the earth, was greater than the root could supply, the green parts withering as if seared by a frost.

On the 20th of July some farmers began to cut their wheat; and by the 25th reaping had generally commenced. Our bean crop presented, perhaps, an unprecedented instance of early ripeness, being usually mowed in September; but this year it was universally ripe, indeed more perfectly so than the wheat by the 1st of August. The crop, however, proved a defective one: water became scarce, and the herbage of the fields afforded so little nutriment, that the cows nearly lost their milk, eight or ten being milked into a pail that four should have filled; and one week, from July the 18th to the 24th, butter could not be made to harden, but remained a soft oleaginous mass.

This extreme heat had a favourable influence on many of our exotic plants, enabling several to perfect their seed, which do not usually in our climate; such as night-stocks, erodiums, heliotrope, groundsels, cape-asters, and such green-house plants vegetating in the open air. With me all the polyanthus tribe, especially the double varieties, suffered greatly; lovers of the cold and moisture of a northern climate, in this tropic heat, they became so parched as never properly to recover their verdure, and in the ensuing spring I missed these gay and pleasing flowers in my borders.

It was a sad destructive season for the poor butterflies, and no sooner did a specimen appear upon the wing, than the swallow and all the fly-catching tribe snapped them up, rendered eager and vigilant from the scarcity of insect food. Even that active and circumspect creature the humming-bird sphinx could not always, with every exertion of its agility, escape their pursuit.

Early in August rains fell, and continued seasonably until September; and their effect upon our scorched vegetation, from the general heat of the earth and the air, was extremely rapid. The larch, and other trees which had shed their leaves, now put forth their tender green foliage as in spring; and by the end of September the universal verdure of the country, and profusion of feed in the pastures, was so perfectly unlike what we had been accustomed to in common years, as to be astonishing. Even as low in the year as the 11th of October, there was no appearance of any change in the foliage, except a slight tinge upon the leaves of the maple; and this day was so brilliant, that the cattle were reposing in the shade, the thermometer varying from 66° to 68° F., and the general warmth to our feelings was greater than that indicated by the instrument. October the 20th, the weather changed, some sleety rain fell, and the hills were sprinkled with snow, the thermometer falling to 40°, and all our hirundines, which had been sporting about us up to this period, departed: yet still vegetation continued in all its vigour, and on the 1st of November dog-roses hung like little garlands in the hedges; the cornel bushes (*cornus sanguinea*) were in full bloom; and corn-roses (*rosa arvensis*) were decorating our hedges in a profusion equal to that of a common August. November 4th there were slight ice and partial snow, with various alternations undeserving of notice, but the weather was generally fair and mild until Christmas.

All these preceding heats and rapid changes had, I think, a manifest influence upon our constitutions. Violent catarrhs, and lingering, unremitting coughs, prevailed among all classes, both before and after Christmas, to a degree that I never remember; and children were afflicted with measles almost universally. Early in January a violent wind was succeeded by a severe frost, and in some places by a deep snow; but after about ten days' duration, a very gentle thaw removed all this, and the remainder of our winter was mild and agreeable, introducing what might be called an early spring, dry and propitious for every agricultural purpose. The trees that refoliated so vigorously in autumn seemed in no way weakened by this unusual exertion, but produced their accustomed proportion of leaves, and the sprays of every bush and tree, ripened and matured by the last summer's sun, displayed a profusion, an accumulation of blossom, that gave the fairest promise of abundance of fruit, and every product of the earth.

That the death of any creature should be required by the naturalist, to perfect his examination, or arrange it in his collection, (and without a collection the investigation of any branch of natural history can be but partially undertaken,) may be regretted; but still the epithet of "cruel employ" must not be attached to this pursuit. We do not destroy in wantonness, or unnecessarily; and that life, of which it is expedient to deprive a creature, is taken by the most speedy, and in the least painful manner known. Some of our methods, if speedy, are at the same time injurious, such as hot water, the stifling box, &c.; and some, that are not painful, such as stupefaction by spirits, ether, &c., and suffocation by carbonate of ammonia, are occasionally not effectual. But there is one process, which I believe to be neither painful nor injurious, yet decisive, and communicative with pleasure; I mean the prussic acid. This fluid may be imbibed by the insect without producing any particular effect; but, if brought to act upon the spinal cord, or what at least is analogous to that part of a vertebrate animal, whatever it may be called, and which seems to be the most vital part of the creation, instant death ensues. A crowquill must be shaped into a point, like a rather long pen, this point dipped into the prussic acid, and an incision made with it immediately beneath the head into the middle of the shoulders of the creature, so as to permit the fluid it contains to enter into the body of the insect. Immediately after this, in every instance in which I have tried it, a privation of sensation appears to take place, the corporeal action of the creature ceasing, a feeble tremulous motion of the antennae being alone perceptible; and these parts seem to be the last fortress that is abandoned by sensation, as they are the primary principle of sensibility when life is perfect: extinction of animation ensues, not a mere suspension, but an annihilation of every power, muscular and vital. As one example of the decisive effects of this fluid, I shall instance the common wasp, a creature so remarkably tenacious of sensation, or so long retaining a muscular power, that it may remain, as every one knows, for days crushed in the window, an apparently dead insect, yet upon pressing the head, the sting will be so protruded as to give a very sensible pain to the finger it should meet with; but upon the prussic acid being injected into this creature as above, when in full vigour, in the course of less than half a minute a loss of vitality ensues, the action of the muscular fibre ceases altogether, and no pressure can incite it again into action. The sudden effect of this liquor is not so generally known as from humanity and expediency might be wished. Who first devised the experiment I am ignorant; but any repetition of means whereby a necessary end can be obtained by the least painful and brief infliction, will hardly be considered as superfluous.

This subject naturally introduces the preservation of the creatures after their death, and the young entomologist is not perhaps sensible from experience of the injury some species of insects will effect in the selected specimens of others of this race, and may lament, when too late, the separation of the wings, limbs, and bodies of his collection by these tiny depredators (*ptinus fur, acarus destructor*). Mr. Waterton's recipe for preventing this evil, I have used rather extensively, and believe it to be a very effectual, and generally an innocuous preservative; but as this gentleman has not given us the exact proportions of his mixture, it may not be useless to observe, that if one part of corrosive sublimate be dissolved in eight parts of good spirit of wine, and the under side of the insect touched with a camel's-hair pencil, dipped in the liquor, so as to let it lightly pervade every part of the creature, which it readily does, it will, I apprehend, prevent any future injury from insects. A larger portion of the sublimate will leave an unsightly whiteness upon the creature when the specimen becomes dry. The under side of the board, on which the insects are fixed, should be warmed a little by the fire after the application, that the superfluous moisture may fly off, before finally closing the case. If this be omitted, the inner surface of the glass will sometimes become partially obscured by the fume arising from the mixture. The experienced entomologist needs not a notice like this; but the young collector probably will not regard it as unnecessary information, and may be spared by it from both mortification and regret. I have known insects commence their serious operations before the collections of the summer could be arranged in their permanent cases.

In noticing above, that this solution is generally harmless, it is requisite that mention should be made of the few instances in which it has been observed to be injurious. I have applied it to many specimens of foreign and British insects, and commonly observed no indication of its having been used, when the creatures had become dry. But to confine our attentions to English specimens, when the solution is made stronger than recommended, it will, after a time, injure the fine yellow of the sulphur butterfly (*papilio rhamni*), by turning parts of it brown and dirty; but even in its reduced state it has a manifest effect upon the colours of two of our moths, the Dartford emerald (*phalæna lucidata*), and what is commonly called the green housewife moth (*phalæna vernaria*), changing their plumage in several places to a red buffy hue, when at the same time the beautiful green wings of the small oak moth (*phalæna viridana*) are in no way altered by it. But notwithstanding these circumstances, it will, I apprehend, be considered as a very useful preservative, and save many specimens from destruction which other means usually fail of effecting.

There are not many of our rural practices that deserve more the disapprobation of the landed proprietor than that of pollarding trees. "It is an evil under the sun, and common among men." Here it is universal. This system of cutting off the heads of the young trees in the hedgerows is resorted to by the farmer for the purpose of forcing them, thus deprived of their leaders, to throw out collateral shoots, serving for stakes for the fences, and for firewood. These purposes are effected; but of all hopes of timber, or profit to the proprietor, there is an end. No trees suffer more in this respect than the ash. Prohibitions against mangling trees, in agreements, are usual; but, with some exceptions in regard to oak, little attention seems paid to the covenant, as is obvious on the most cursory view of the country in any direction; whereas the ash is not a less valuable tree, from its thriving more universally in all situations, and becoming saleable in a shorter period. One or two generations must pass before an oak should be felled; but the ash becomes useful wood while its more respected companion is but a sapling. These prohibitions should not simply be engrossed on the parchment, but the agent ought strictly to notice any infringement; and young ash trees should be more especially guarded, because they are the most likely to suffer, from their producing the greatest quantity of lop in the shortest time. The injury done by this practice to the present landlord and his successors is beyond estimation, as the numbers destroyed, and the vigour of their growth, must be first known: but there is not a farm of any extent, from which hundreds of ash trees might not have been felled, had their growth been permitted, making an annual return; whereas nothing can be obtained now or hereafter for the proprietor, and only a few stakes and bavins for the farmer[44]. It is by no means an uncommon thing, to observe every ash tree in a hedge reduced to stumps by successive pollardings. Many a landlord would shudder at the thought of breaking up an old productive sward, and not regard the topping of an ash; whereas this latter act is infinitely more injurious, ultimately, than the former. The land may, and will probably, recover, but the tree is lost

---

44 The ash, generally speaking, will arrive at a very serviceable age in sixty years, producing at a low rate twenty-eight feet of timber, which, at 2*s*. 3*d*. the foot, its present value, would produce a sum equivalent to 3*l*. 3*s*., a silent unheeded profit of above a shilling a year. A hundred such might have been felled annually from many farms had they not been topped, which, in consequence of this practice, have produced nothing.

for ever, as to any profitable purposes for the owner. The farmer might perhaps tell the agent when he remonstrated, that he must have firewood, and hedging stuff; but the wants of the former have decreased by the facility of obtaining other fuel, and neither is to be supplied by the landlord at such a ruinous subversion of present and future benefit. I am not so silly as to enlarge upon the beauty of what has been called "picturesque farming;" but when we cast our eyes over the country, and see such rows of dark, club-headed posts, we cannot but remark upon the unsightly character they present, and consider it neither laudable to deform our beautiful country by the connivance, nor proper attention to individual profit to allow the continuation of it. The ash, after this mutilation, in a few years becomes flattened at the summit, moisture lodges in it, and decay commences, the central parts gradually mouldering away, though for many years the sap-wood will throw out vigorous shoots for the hatchet. The goat moth now too commences its mordications, and the end is not distant. But the wood of the ash appears in every stage subject to injury: when in a dry state the weevils mine holes through it; when covered by its bark, it gives harbour to an infinite variety of insects, which are the appointed agents for the removal of the timber: the ashen bar of a stile, or a post, we may generally observe to be regularly scored by rude lines diverging from a central stem, like a trained fruit tree, by the meanderings of a little insect (*ips niger,* &c.), being the passages of the creatures feeding on the wood.

There is one race of trees, the willow, very common about us, that is so universally subject to this pollarding, for the purpose of providing stakes and hurdles for the farm, that probably few persons have ever seen a willow tree. At any rate a sight of one grown unmutilated from the root is a rare occurrence. The few that I have seen constituted trees of great beauty; but as the willow, from the nature of its wood, can never be valuable as a timber tree, perhaps by topping it we obtain its best services. In the county of Gloucester there are several remarkable trees of different species now growing, but I am not acquainted with any greater natural curiosity of this sort than an uncommonly fine willow tree in the meadows on the right of the Spa-house at Gloucester. There are two of them; the species I forget, but one tree is so healthy and finely grown, that it deserves every attention, and should be preserved as an unique specimen, an example of what magnitude this despised race may attain when suffered to proceed in its own unrestrained vigour.

Dec. 30. — A cold foggy morning, the ground covered with a white frost; about twelve o'clock the sun burst out with great brilliancy, and life and light succeeded to torpor and gloom; a steam immediately arose from our garden beds and ploughed lands, giving us a very strong example of the rapid manner in which the matter of heat (*caloric*) will at times unite with water. Half an hour before, this water was frozen and inert; but the instant that the sun's rays fell upon it, their heat was imbibed, and the icy matter converted into a body lighter than the atmosphere by which it was surrounded, and passed into it in the vapour we have just noticed. I was the more particular in observing this common event, as it afforded a forcible illustration of the invisible evaporation which is constantly going forward, the unremitting changes in operation, the action and reaction of the earth and its products with the atmosphere. During the night, and the earlier parts of the morning, water was falling on the earth in minute particles, constituting what we call fog; then out bursts the sun, and reclaimed this moisture which had fallen, and we could see it obeying the mandate, and pass away in steam. In the evening it will probably return again in fog, or in rain, when the atmosphere cools; and thus a constant visible intelligence is going on. How much insensible intercourse takes place we know not, but we can comprehend its agency by the effects and events that manifest themselves. Our country people think these "rokings" (reekings) of the earth greatly favourable to the growth of vegetation, supposing it occasioned by the internal heat of the earth producing a vapour like that from fermenting soil, thus warming the roots: but if the theory be defective, the fact may be true, by the caloric in the sun's rays promoting the decomposition of the water, or separating the component parts (oxygen and hydrogen), which, uniting with other matters contained in the earth and atmosphere (carbon and carbonic acid), become, by this means, the basis of all our fruits, our sweets, our sours, resins, &c., in the vegetable world; and hence there is a constant decomposition of water going forward by these alternations, and a constant formation of matters beneficial and necessary for the various inhabitants of the earth. When we perceive that a shower of rain has revived or promoted the increase of vegetation, we must understand that the mere wetting it has not accomplished this; but that the vegetable has, by means of its foliage, aided by light and heat, decomposed or separated the combined matters of the water, and taken from it certain portions as essential to its vigour, or been revictualled, in a manner, by the nutriment contained in the water.

Jan. 10. — The ground covered with snow, the pools with ice, trees and hedges leafless, and patched here and there with a mantle of white, present a cheerless, dreary void; no insects are animating the air, and all our songsters are silent and away; a few miserable thrushes are hopping on the ditch bank, swept bare by the wind; and the robin puffing out his feathers, and contracting his neck into his body, is peeping, with his fine bright eyes, into the windows from the cypress bough. A few evergreens are waving their sprays, and glittering in the light, yet making but poor compensation for the variety, the flutter, the verdure of our summer. Though we have little natural beauty to note or to record, we are not left without a testimony of an over-ruling Power; and, however sad and melancholy things may appear at the first view, yet a more steady observation will manifest to us a presiding Providence and Mercy. Frost and snow are but cheerless subjects for contemplation, yet I would add a reflection in my Journal of our passing event, or rather recall from memory the truth, that science has made known to us, revived by the sight of that frozen pool. There is one universal body, inherent in every known substance in nature, latent heat, which chemists have agreed to call "caloric." By artificial means bodies may be deprived of certain portions of it; and then the substance most usually contracts, and increases in weight. Water is an exception to this; for, in losing a part of its heat, the cause of its fluidity, and becoming ice, it expands, and is rendered lighter, by inclosing, during the operation, more or less of atmospheric air: consequently it swims, covering the surface. To this very simple circumstance, ice floating and not sinking, are the banks and vicinities of all the rivers, lakes, pools, or great bodies of water in northern Europe, Asia, and America, rendered habitable, and what are now the most fertile and peopled would be the most sterile and abandoned, were it not for this law of nature. Had ice been so heavy as to sink in water, the surface, on freezing, would have fallen to the bottom, and a fresh surface would be presented for congelation; this would then descend in its turn, and unite with the other; and thus, during a hard frost, successive surfaces would be presented, and fall to the bottom, as long as the frost or any fluid remained. By this means the whole body of the water would become a dense concretion of ice: its inhabitants would not only perish, but the indurated mass would resist the influence of the sun of any summer to thaw it, and continue congealed throughout the year, chilling the earth in its neighbourhood, and the winds that passed over it, preventing the growth of vegetation in the former, or blighting and destroying it by the influence of the latter.

Winter is called a dull season; and to the sensations of some, the enjoyments of others, and, perhaps, to the vision of all, it is a most cheerless period. This is so universally felt, that we always associate the idea of pleasure with the return of spring: whatsoever our occupations or employments may be, though its sleety storms and piercing winds may at times chill the very current in our veins, yet we consider it as an harbinger of pleasurable hours and grateful pursuits. We commence our undertakings, or defer them till spring. The hopes or prospects of the coming year are principally established in spring; and we trust that the delicate health of the blossoms round our hearths, which has faded in the chilling airs of winter, may be restored by the mild influence of that season. Yet winter must be considered as the time in which Nature is most busily employed; silent in her secret mansions, she is now preparing and compounding the verdure, the flowers, the nutriment of spring; and all the fruits and glorious profusion of our summer year are only the advance of what has been ordained and fabricated in these dull months. All these advances require Omnipotent wisdom and power to perfect; but perhaps a more exalted degree of wisdom and power has been requisite to call them into a state of being from nothing. The branch of that old pear tree now extended before me is denuded and bare, presenting no object of curiosity or of pleasure; but, had we the faculty to detect, and power to observe, what was going forward in its secret vessels, beneath its rugged, unsightly covering, what wonder and admiration would it create! — the materials manufacturing there for its leaf and its bark; for the petals and parts of its flowers; the tubes and machinery that concoct the juices, modify the fluids, and furnish the substance of the fruit, with multitudes of other unknown operations and contrivances, too delicate and mysterious to be seen, or even comprehended, by the blindness, the defectibility of our nature — things of which we have no information, being beyond the range of any of the works or the employments of mankind. We may gather our pear, be pleased with its form or its flavour; we may magnify its vessels, analyze its fluids, yet be no more sensible of its elaborate formation, and the multiplicity of influences and operations requisite to conduct it to our use, than a wandering native of a polar clime could be of the infinite number of processes that are necessary to furnish a loaf of bread, from ploughing the soil to drawing from the oven. This is but an isolated instance, amidst thousands of others more complicated still. How utterly inconceivable, then, are the labours, the contrivances, the combinations, that are going forward and accomplishing in this, our dull season of the year, in that host of nature's productions with which, shortly, we shall everywhere be Surrounded!

It is in a period like this, where one comfortless hue predominates over all things —

> Where all is sky and a white wilderness;
> And, here and there, a solitary pine,
> Its branches bending with a weight of snow, —

that we fully perceive the beauty, the cheerfulness, of the colours of nature, which, like so many other things in life, we do not duly appreciate, until we are deprived of them.

The splendour, the variety, of the autumnal glow is certainly magnificent; and, by reason of the diversity of tints, gives a breadth and depth to woods and glades exceeding that which an uniformity of colour would effect: but this gaiety, this lustre of the grove, however pleasing as a temporary exhibition, would probably not be agreeable as a lasting vestiture, independent of that cast of melancholy the decay of natural beings commonly conveys; but it is green, reviving green, that appears most to gratify the senses. A bank, an eminence, swept by the wind, and verdurous — the grassy streamlet, the ivy-vested trunk in the hedgerow, allure our view, as fertile islands in a desert waste; and we watch the progress of a thaw, as the subsidence of a wintry flood. Green seems to be the predominant hue of vegetation, when in health; and our faculties feel revived, reanimated by its freshness, when we view it. The wide-spread pastures, the foliage of the thicket, appear to impart their life to a wearied, languid being, and the sick man looks out on the verdant turf — his jaundiced eye imbibes the universal spirit of vitality that circulates around — he admits the hopes of restoration, as he sees the healthy existence of all things. And how is it that this colour conveys such sensations? probably not from any innate love or accordance with a particular tint; for if so, it would prevail in all things, become the influencing dye in our dress, of our fancies — which it is not. It may at times be an association of the memory — an impression on the mind of superior health and life, enjoyment, leisure, bland airs, and recreations; but, then, persons bred in cities and situations, where no early recollections of such conceptions have had birth, yet feel pleasure in the verdure of nature, an importunity for the country. Surrounded by verdure, we inhale a larger portion of oxygen, of vital air: this increases the pulsations of the heart, augmenting the flow of spirits. An analysis of a certain number of cubic inches of the atmospheres from the crowded alley and the open down would, probably, give no indication of this excess; yet we have strong presumptions that it exists. But this is an effect on the animal machine, not an operation on the mind by vision. The immediate satisfaction which the eye receives in resting upon verdure is difficult to account for: many natural colours please, but green seems to be a tint that conveys a calm exhilaration, and, as an innate disposition in many, without reasoning or deliberation, effects its influences by actions imperceptible to our understanding.

Jan. 20th. — A keen frost, and the ground covered with snow, presents a scene of apparent suffering and want to many of our poor little birds; but the preservation of the fowls of the air, which sow not, nor gather into barns, has been beautifully instanced to us, as a manifest evidence of a superintending Providence: the full force of this testimony is most strongly impressed upon us in a season like this, when winter rules with rigour, and we marvel how the life of these beings can be supported when the waters are bound up, and earth and all its products hidden by a dense covering of snow. Many of the small birds obtain subsistence by picking the refuse of our corn-stacks, by seeds scattered about our home-stalls and cattle-yards, but multitudes of others are in no way dependent upon man for shelter or support, do not even approach his dwelling, but are maintained by the universal bounty of Providence; as the wood-lark, the meadow-lark, the chats, and several others; but by what means they are maintained in a period like this is not quite manifest. The portion that they require is probably small, yet it must be insect food; and the chats, larks, and white wagtails, seem busily engaged in providing for their wants upon the furze sprays, amidst frozen grass, or upon the banks of ditches and pools; and as no insect but the winter gnat is now found in such places, it is probable that this creature, which sports in numbers in every sunny gleam, yields them in this season much of their support. Some of the insectivorous birds have at such periods no apparent difficulty in supporting their existence, finding their food in a dormant state in mosses, lichens, and crevices of trees and buildings; but for those which require animated creatures, I am sensible of none that are to be procured but this gnat, and it possibly has been endowed with its peculiar habits and dispositions for a purport like this. We have many examples in nature of similar provisions, wherein one race supports the existence and requirements of another. The molluscæ and insects of the deep continue the life of some, the feeble races of the air and waters maintain the beings of others, and the beast of the wild seeks his food amidst those which inhabit with him; but where this chain ends, human faculties will probably never be able to ascertain. The remarkable fact which our microscopes make known to us, that all infusions of natural substances in water will produce life, however extraordinary the form may be, seems to denote a continuation of beings beyond any possible comprehension, and probably subservient to the existence of each other: the minute creature that floats an hardly perceptible atom in the water of the ditch, and which subsists many of the animals which inhabit those places, feeds upon smaller than itself, and those again, possibly, upon more minute ones which the vegetable infusions of those places give existence to: here the investigation terminates, but the thread unbroken continues, probably through endless gradations, perceptible to Infinity alone.

Having applauded the operations of Nature with so much cordiality, possibly I may be called her "enthusiastic adorer," but the epithet must be disclaimed. None can respect the works of creation more, but it is not with an ecstasy that glows, fades, and expires, but with a calm deep-rooted conviction implanted in the boy, and increased by years of notice and experience. I have followed her footsteps, though far, very far distant, as an humble admirer of perfection, nor can my veneration cease whilst reason continues undisturbed.

Examining an old pane of glass from a church window, stained in patches, and corroded into holes as if eaten by insects; reflections naturally arose upon this wasting away of so apparently indestructible a substance, as we generally say, "by Time", comprising in space the action of elementary agencies which usually being slow in performance, when not in agitation, require indefinite periods to effect their objects. Hence Time is required to perform them. That all organised bodies, and fabricated things progressively decay or decrease until they are consumed, is a truth that will not possibly be controverted, but in many cases this consumption from the minuteness of the parts that fly off, the insensible nature of them, all the slowness by which combinations are effected, are not always obvious.

The air by reason of its carbonic acid, and its oxygen, becomes the principal consumer of all created things, the first seizing the alkaline, and calcareous particles, breaks up the union of the whole. The stone in the wall falls to the earth as sand or soil, the glass in the window has its principles separated, is washed into the earth, entering into new combinations and forms. The vegetable, the animal, broken up by oxygen, remodified and compounded; and this wonderful principle, this bestower of life, becomes the invested power, the organ of decay.

Nature is never at rest, when the objects of existence have been accomplished, the elements are reclaimed to commence and supply a new order of things. But there are cases in which the particles of indurated bodies become obvious to our perceptions. When a beam of light penetrates through the crevices in the shutters of a chamber along a gallery — the aisle of a church, and places when no apparent and agitation has moved the resting body, what an immense collection of floating atoms are perceived! Parts of some indurated bodies now separated, and insensibly inhaled with our breath. In process of time, most of these become by cohesion, or by gravity, heavier than the medium which suspends them, and they settle, becoming what we call dust, prepared to mingle with the soil, be imbibed by vegetable absorption, and accomplish a new combination.

Iron, which is so often the power that cements natural bodies, is yet frequently the very cause that occasions their destruction, by yielding to the influence of oxygen, deserting that which it originally combined, dividing it to become a portion to several; water, light, heat, each compounded elements, unremittingly in action. It is by daily destruction, that we daily acquire renovation of matter and form; even the very dust upon our shoes the from the lane, or floating as a cloud, disturbed by the passing animal, from whence is it? But the minute parts of siliceous grains, tenacious clay, some mighty mass of carbonate of lime, shelly of animal formation all remains as may chance, formed in the deep before the recorded notices of man, then heaved on high by some awful convulsion. Now broken down, it is ready to assume another form, enter into vegetable circulation and mingle in the animal system, or return, washed into the brook by the rains of winter, to the element from whence it proceeded. As in a moral sense, there is no life without death, so in the physical one, there is no being without decay. "Dust thou art and unto dust thou shalt return."

Thus from day-to-day I feebly pursue the tendencies of many years of my life, and find the exercise helpful and delightful still; but an old man now, the hours are numbered when those solicitudes will cease; though I view them in my passage, 'tis not with the ardour, the desire of other days as considerate things, but tempered with different reflections, use them as amusements rather than concerns. Very grateful I am, that the power yet remains enabling me to occupy my thoughts with the lowly works of his hands, the evidences of Almighty wisdom and beneficence, as I journey on seeking the pearl of great price. Yet pleasurable as they have been, I would not keep them, see them, nor let them impede my progress, but closing my eyes in the dusky eve of life, await their opening to brighter objects and contemplations, in other regions and scenes.

Nov. 9th. — A calm clear night, and the moon near her amplitude. High in the air may be heard, at intervals, the pipings of some small bird on its passage, a signal to the flights that follow: far in the distance sounds the plaintive note of the curlew, or some sea-fowl proceeding to her feed, and the brown owl calls to his fellow from some faraway oak. Obscurity and clearness is over all things, and an indefinable serenity seems to take possession of the mind; we look up and are led to wonder and contemplation, to the desire of obtaining a knowledge of those things that are hidden from us. The immortal mind that animates the body cannot be at rest; but those things which probably first incited man's inquiry, still incites him, and will be the objects of his conjecture until time shall be no more. All superior minds have been directed from the earliest ages to the consideration of the nature and object of all that splendid host which glittered in the concave of the sky above them; and yet what is the concentrated knowledge which we have acquired by all this labour of so many centuries? We may say almost nothing certain. No portion of creation which is presented to our vision creates more awful sensations than the contemplation of the heavenly bodies, — such inconceivable immensity of space, and mysterious combinations. By aid of very fine instruments we can perceive lights and shades alternating at intervals, and other changes which we consider as similar to what we experience on earth, and call diurnal or annual mutations. Here we see an assemblage of stars intermingling their rays, and forming a region of light; there, one or two faint scintillations in a dense concavity of blackness; masses of light, proceeding from we know not what, envelope others; some approach into intimate union; others glitter, unattended, in every variety of splendour, and light, glorious light, seems the very atmosphere of all! — for what a lustrous body must that be which becomes manifested to our feeble vision, from distances of which the mind can have no conception[45]! — Even this moon, our moon as we call it, so comparatively near to us, so particularly attendant upon the revolutions of our planet, is all a mystery to us! We may call certain parts alpine, or volcanic; we may map its continents, its islands, and waters, but it is little more than a vision of imagination. We lose sight of a star, and obtain a view of another, not observed before, and say a system has been removed, or a new one created — but what know we? No real knowledge of the heavenly bodies, as to their natures, laws, and objects, can be obtained but from two sources, personal visitation, or divine communication, even the former, if possible, would probably be all wonderful, but perfectly incomprehensible; if we understand so little of plain earthly things, what probability is there that stranger things should be intelligible to us? And the knowledge of them the Almighty has ever withheld from man, giving him neither science to explore, nor faculty to conceive. Those holy men of old who held communion with their Maker, if they ever received intelligence of these things, have been silent on the subject; and though it is very probable that the companions and friends of the Saviour, to whom he communicated so many heavenly truths, surrounded by celestial bodies gleaming with splendour in such a region as that of Judea, might inquire concerning them,

45 Perhaps no human mind ever yet conceived a colour different from what now exists in some known object, or which may be produced by a combination of existent ones; so probably our conception of the intensity of light is only in proportion to some visible sense of it. The planet Venus is so distant from our sphere, that the space in miles, from its immensity, we can have no conception of, as we cannot rightly entertain a sense of vast space; yet this star glitters at times with such splendour that we can scarcely look upon it: what then must be the intensity of that light, reflected light, which would proceed from this planet were we placed near it! — We have no faculties to imagine such splendour, nor power of enduring could we experience it— He only, the "Father of lights."

yet have they narrated nothing upon the subject: St. Paul, caught up to the third heaven where he heard such unspeakable things, yet thought it not lawful to utter them; all intimating that it is neither proper nor fitting that we should be acquainted with such matters. And lives the man of true faith that would wish to penetrate, to intrude his earthly vileness into the mysteries of the Holy One, the systems of other worlds, the regions of separated beings, that would tread upon the confines of the immortal ones, had he the power to do so? The Christian knows revelation to be truth, and in that perfect conviction labours only to accomplish steadily the part assigned to him. He shuts out from his mind all visionary objects, all desires but to do his makers will, and calmly wait to hear his redeemer's call to "come up hither."

The notion that the celestial bodies held intelligence with and influenced the earth and its productions was an early established belief, and scarcely are two generations gone by since the universal connexion of the stars and planets with the animal, vegetable, and mineral productions of our globe was considered so predominant, that their aspects were calculated, their favourable or baneful ascendancies inspected, before any operation upon the one ought to be performed, or collecting of the other attempted; and still this belief so lingers in the ruder minds of many of our peasantry, that some things cannot be performed unless the "sign is favourable:" if, in defiance of these admonitions, we persist, and accident ensues, it creates no surprise, warning having been given by the *vox stellarum*. That the heavenly bodies ruled the destinies and influenced all the actions of man's life, though originating, perhaps, in craft, and nourished by ignorance, was not wholly an unnatural conception, knowing the powers of the moon upon the waters of the ocean, and her probable impulses in mental aberrations; and this mysterious body may further operate collaterally upon the mind, exciting, in certain states, particular sensations. We do "not kiss our hands, or bow down before the moon," but many who should behold her as now, walking in brightness through the concave of the heavens, the stars flashing and gleaming about her with all the accompaniments of solemnity, would have their minds subdued to contemplative calmness. These awful and occult bodies have innocently been called the abode of spirits departed from the world; and though the mind cannot conceive any true notion of an after state, yet there are few, perhaps, who have meditated upon futurity but have devised some system, such a scale of happiness, humanly now judging, as would satisfy their minds to enjoy; and we muse upon the capacity of bliss that these bodies may possess, and the thoughts of future worlds and states pass before us. Where rest the souls of those loved, those dear ones, which have left us? Are they in any of those "many mansions of our Father's house?" Shall our spirit be glorified and reunited with theirs in a state of rest from earthly cares? Which of these shining orbs will receive it? What will be our vision, our condition then? And all the hopes, the fears, the trusts of man become awakened in us. Such contemplations as these, perhaps, many of us have experienced in viewing the calm, the seemingly benevolent aspect of our attendant planet, and all the mysterious lights that sparkle in the sphere above us; and if they excite this disposition to thoughtful meditation, this is surely influencing the mind, and the body through the impulses of mental action. The shining of the sun elevates and enlivens, the light of the moon calms, subdues, softens, and various minds are affected by influences imperceptible to others; but when "the light of the moon shall be as the sun, and the sun as the light of seven days," we shall perceive alike.

Sept. 8th, 1828. — A remarkable dry and exhausting day, not from any peculiar influence of the solar heat, but from the arid state of the air, which was very distressing to our feelings, and all tender vegetation became languid and suffering under its influence. I endeavoured to ascertain the power of absorption possessed by the air at the time by an experiment, rude enough, to be sure, yet it tended in some measure to indicate the rapid manner in which fluids are exhaled in particular states of the atmosphere. A linen cloth twelve inches square, which had absorbed an ounce avoirdupois of water, was suspended in the shade in a free current of air, and in the course of ten minutes it had lost 436 grains, equal to one-sixteenth of its weight. This great evaporation was principally effected by the absorbent power of the air, and manifested in some degree the exhausting influence that was passing over the earth and the vegetation exposed to the current of air; and as the roots could not derive sufficient moisture from the soil to supply what was thus drawn from the leaves, the foliage became languid and flaccid in consequence. The linen, containing the same quantity of water, was then spread upon a short turf in the sun, and in the space of ten minutes it lost 368 grains, and this was effected without any particular influencing current of air; accordingly, the evaporation from an acre of moist land covered with vegetation would exceed one hundred and twenty-two cwt. of water in an hour! As the quantity drawn from the vegetation on the soil may be equal to the shelter its foliage affords to the earth, no very accurate data can be drawn from this experiment; for different soils will give out their moisture more or less easily, and succulent vegetables be more influenced than those of a drier nature; but it served at the time to indicate the portion of moisture that was escaping from a given horizontal surface. From the invisible and insensible nature of evaporation, its influences are not always considered; but such an action on the surfaces of things as that related above, must put into operation all the inherent powers of matter susceptible of impulse, and probably would produce effects which we might suppose to be accomplished by the agency of other means.

The facility with which some plants part with their moisture, and the reluctance of others under similar treatment, may be noted, because it intimates the peculiar fitness of our grasses, such as constitute the general herbage of the field, for being dried into winter provender, whereas many of a more succulent nature in the usual temperature of our summers, could not be sufficiently dried, but would if stacked up turn mouldy, or decay. But the blades of our grasses by their thinness and separation, give out their moisture readily - an equal weight of grass, promiscuously taken from the field, and the foliage of two succulent plants were exposed for exactly an hour to a drying sun, the grass parted with 272 grains of moisture, potato leaves 225 d°, turnip, 225 d°.

Perhaps few persons are sensible how much nutritive matter is carried off from grass, when partly dried, by a single shower of rain, (which is said to do no harm) if not immediately exhaled. As long as the herbage continues green, a shower or two is of little moment. But let it approach to a dry state, and be wetted, we immediately begin to smell even at a distance that aroma so peculiar to good hay. Now this effluvia is probably the essential oil of the herbage, and if the nutritive part of vegetation in the main, consists, as it is supposed to do, of the saccharine matter, and the oleaginous particles, how large a portion of matter is now scattered away, which if retained would tend to nutrition. And after two, or three wettings, and dryings, this smell is no longer perceptible, the essential oil having been exhaled; the hay may be bright, and have the appearance of goodness, but assuredly a principal ingredient of importance has been lost.

Nov. 10. — Many effusions of the mind have been produced by the approach or existence of the seasons of our year, which seem naturally to actuate our bodily or mental feelings through the agency of the eye, or temperature of the air. The peculiar silence that prevails in autumn, like the repose of wearied nature, seems to mark the decline and termination of being in many things that animated our summer months: the singing of the bird is rare, feeble, and melancholy; the hum of the insect is not heard; the breeze passes by us like a sigh from nature, — we hear it, and it is gone for ever. But it is the vegetable tribes, which at this season most particularly influence our feeling, and excite our attention. We see the fruits of the earth stored up for our use in that dull season "in which there will be neither earing nor harvest," the termination and reward of the labours of man. But this day, November 10, presented such a scene of life and mortality, that it could not be passed by without viewing it as an admonition, a display of what had been, and is. There had occurred during the night a severe white frost; and, standing by a greenhouse filled with verdure, fragrance, arid blossom, I was surrounded in every direction by the parents of all this gaiety, in blackness, dissolution, and decay. But the very day before, they had attracted the most merited admiration and delight by the splendour of their bloom and the vigour of their growth; but now just touched by the icy finger of the night, they had become a mass of unsightly ruins and confusion. Once the gay belles of the parterre, they fluttered their hour, a generation of existent loveliness; their youthful successors, unpermitted to mingle with them, peeped from their retreats above, seeming almost to repine at their confinement; they have bloomed their day, another race succeeds, and their hour will be accomplished too. This was so perfectly in unison with the shifting scenes of life, the many changes of the hour, that it seemed inseparably connected with a train of reflection, with the precepts which all nature points out — her still small whisperings for the ears of those that can hear them.

We grieve over the wreck of nature, the confusion, the ruin of the beautiful products of our floral hours, the departure of those who have been so pleasing to us, with whom we have walked in perfect innocence of life; we look for them, and they are not, but still, friends in presence and in absence, they spread before us all summer long a volume of wisdom, splendour, and power, and in departing tell us of our change to come, and show us of things that must be hereafter.

And here is spring again! Beautiful charming spring once more! and by a merciful and good providence, I am spared from the accidents of life, the inflictions that have consumed so many around me, yet permitted to enjoy its odours, gladness, and blessings: there is a breath from the east still, that tells of wintry hours to the delicate and chambered invalid, but life is creeping on us and resurrection summons all the reposing tribes of earth: our little native birds begin again to sport amidst the sprays, or chirp upon the wall, and a movement of coming occupation is observant every where. One summer wanderer is alone in the copse, but his little eloquent tongue thrills with joy: a few butterflies, the Nettle & the Peacock, are fluttering in the sheltered lane, but their robes look aged and shabby, worn all the stormy hours of their wintry rest, yet these poor remnants animate our path, are harbingers of gayer fair ones: our flowers peep from every bank, the daisy on the turf, that brilliant star of the earth the cheerful gladsome pilewort, glistens in the nook and the brake, the primrose filling every hand, yet springing in endless profusion still, and fading in every cottage lattice, for all have their flowers, hawthorn, cowslips, blue bells. How many of us course along the roads and bustle through the highways of life and think we never notice these things, but yet we all regard them, some in real love, some with unconscious approbation, for their absence would be noticed by all; and what a dreary void, what a cheerless world would it be without our birds our insects and our flowers! Each may be a trifle in our estimation yet from habit, from association they convey insensible delight, a reminiscence possibly of something pleasing, an expectation of novelty, of pleasurable feelings! And what a lovely thing is a flower! Fresh fragrant, exquisitely formed, wonderfully diversified grateful and agreeable to the capricious fancy of the most wayward mortal, we wait for them, we love them present, and we grieve when they are no more.

The extraordinary tendency that Nature has to produce, and the vigilant perseverance she maintains to occupy all substances as a soil for her productions, when they arrive at a state fitting for her purposes, is a well-known fact, and is perfectly in consistency with the uniform habit she preserves, of letting "no fragment be lost." All things tend upwards, from some original, through an infinity of gradations, though the beginning and termination may not always be perceived, nor the links of this vast chain be found. The most obscure plants, agarics or mucor, as far as we know, perfect their seed, and give birth to other generations; but there is a fine green substance, observable upon the sprays of trees, stems of various shrubs in every hedge, upon old rails and exposed wood-work, leaving a powdery mark upon one's coat that has rubbed against such places, which I have always considered as the very lowest rudiment of vegetation. This matter, submitted to examination in the microscope, presents no foliage or plant-like form, but appears a kind of pollen, a capsule, or a perfected seed, suspended on a fine fibre; but from the extreme smallness of it I speak with hesitation, not being able to define it satisfactorily with the most powerful lens. If it be, as I have conjectured, a perfected seed, it probably is the origin of many of those minute mosses, that become rooted, we know not by what means, upon banks, stones, barks, &c. in such profusion; but here all investigation ceases: by what agency this fine seed has been so profusely scattered, or from what source it sprang, is hidden from us, and we can no more satisfactorily conjecture, than we can account for those myriads of blighting insects, which so suddenly infest our grain, our fruits, and our plants. There is an

CAPABILITIES OF MAN — CONCLUDING REFLECTIONS

inquisition, where all human knowledge terminates; the bounds of nature have never been defined.

Without considering the various sources of enjoyment and pleasure bestowed upon an intelligent creature, what a scene of glorious display might be opened to man through the agency of the eye alone! Motives we must abandon, as probably they are beyond our comprehensions; but were the powers of vision so enlarged or cleared as to bring to observation the now unknown fabrication of animate and inanimate things, what astonishment would be elicited! The seeds, the pollen of plants, the capillary vessels and channels of their several parts, with their concurrent actions, the clothing of various creatures, and all that host of unperceived wisdom around us! Yet probably the mind, constituted as it now is, would be disturbed by the constant excitement such wonders would create; but at present, though sparingly searched out by the patient investigator, and but obscurely seen, they solace and delight; "cheer, but not inebriate."

> Oh good beyond compare!
> If thus thy meaner works are fair,
> If thus thy bounties gild the span
> Of ruin'd earth and sinful man,
> How glorious must that mansion be
> Where thy redeem'd shall live with thee!

AND NOW I think I have pretty well run over my diary, the humble record of the birds, the reptiles, the plants, and inanimate things around me. They who have had the patience to read these my notes, will probably be surprised that I could take the trouble to register such accounts of such things; and I might think so too, did I not know how much occupation and healthful recreation the seeking out these trifles has afforded me, rendering, besides, all my rural rambles full of enjoyment and interest: companions and intimates were found in every hedge, on every bank, whose connexions I knew something of, and whose individual habits had become familiar by association; and thus this narrative of my contemporaries was formed. Few of us, perhaps, in reviewing our by-gone days, could the hours return again, but would wish many of them differently disposed of, and more profitably employed: but I gratefully say, that portion of my own passed in the contemplation of the works of nature is the part which I most approve

— which has been most conducive to my happiness; and, perhaps, from the sensations excited by the wisdom and benevolence perceived, not wholly unprofitable to a final state, and which might be passed again, could I but obtain a clearer comprehension of the ways of Infinite Wisdom. If in my profound ignorance I received such gratification and pleasure, what would have been my enjoyment and satisfaction, "if the secrets of the Most High had been with me, and when by His light I had walked through darkness?"

THE PROGRESS OF A NATURALIST

TO FLORA.

Tis said that poets, if they'd muse,
Must first a fancied object choose,
And, by imagination's aid,
Picture some visionary maid,
To hear the wand'rings of their brain —
Sigh to their transports, or their pain.

   Then may a fond adorer, too,
Muse on the maid he's loved so true —
The first dear idol of his praise,
And faithful still in calmer days.

   Time's finger, perhaps, has rased from view
Some fancies that he early knew;
But, Nymph, the love for thee profess'd
Has ne'er been rifled from his breast.

   The aged spaniel in his lair
Pursues in sleep the timid hare;
Too old his ferny hills to trace,
In slumbers still he leads the chace.
So, FLORA, in the vale of time
I trace the pleasures of my prime,
And find my heart beats calm and true
To the first mistress that my fancy knew.

Deep in the mist of years, I see,
Rambling alone the woodland lea,
A musing, slender, happy child,
Snatching in haste his flowers wild:
Twined in a wreath, with rushes green,
Cowslips and violets are seen;
Now both his hands he stoops to fill,
Then shouts, and smiles, and gathers still:
He seems the very king of joy —
This young, this gleeful, slender boy.

The wild bee sees, and murmurs round,
To scare him from her fav'rite ground,
In fear, lest he should gather more,
And rob her of her honied store.

In little garden see him toil,
Planting these natives of the soil:
Though tended by his daily care,
The sickly nurslings languish there.

A dusky cloud is come between —
This musing child's no longer seen.

By table now, with blossoms spread,
A pallid youth bends low his head;
Some ancient volume poring o'er,
A dark bound herbal on the floor.
The tawny page he reads with care,
Yet finds a something wanting there;
Then lifts his face with thoughtful look,
And calmly shuts his musty book.

I trace him still on Scotia's hills,
By craggy steeps, by mossy rills,
In heathy vale, o'er ferny lands,
'Long dusky shelves, on granite bands;
Each nook, each cranny, close explore,
And half the island's sandy shore;

The mead, the woodland, and the plain:
And now the wand'rer's home again.

   I see, upon a wide-spread board,
This rambler's rare and cherish'd hoard:
Mysterious grasses scatter'd o'er,
A glass, a press, and books of lore;
With grave, consid'rate care and thought,
Compares the species he has brought;
Then, ranging out each sep'rate race,
He bids their form the pencil trace.

   Ah, Vulcan! that thy hateful rage
Should moulder half his studious page.

   How wav'ring is the mind of man!
A fickle thing since time began —
Hates this to-day, to-morrow loves,
Neglecting what he most approves —

   Thy votarist can his errors see,
And, FLORA, owns his slights to thee;
E'en he could once unsteady prove,
A wand'rer from his fair one's love.

   In silent ev'ning's balmy hour,
He sought Phalæna in her shady bower;
And oft his morning walks prolong,
To hear the syren Sylvia's song;
And then he'd seek the azure main,
Where dwelt fair Doris and her train.

   The mystic nymph by Nilus' shore[46],
From thrice-fam'd Hermes named of yore,

---

46 This is rather a shadowy fair one — yet the Egyptian Minerva has ever been considered as the patroness of the liberal arts, the Hermetic, or mystic science; and then we have such veracious relations as those of Borrichius and Zosimus, as quoted by Scaliger, that a son of heaven was expelled his seat for instructing an earthly nymph in metallurgy and alchemical secrets.

With meteor flames beguiled his heart,
Seduced the man, and ruled by art.
And as her veil she coyly drew,
Combined beauties met his view;
But soon the subtle lady fled,
To beam her lights on Davie's head —
And other fancies too there are
Which turn'd him from his early fair,
Yet, not forgetful of the joy
Thy smiles in spring-time gave the boy,
He seeks — these little wand'rings o'er, —
His former love, nor quits her more.

    There, mark him now in Cambria's shades,
Panting up steeps through forest glades,
To woo thee in thy humblest seat,
Rarely disturb'd by mortal feet,
To view thee on thy mossy bed,
Where changeful agarics lift their head,
And riveted in rock, the oak
Scarcely has heard the woodman's stroke;
Whilst his grey lichen, pendent there,
Looks like some hoary peasant's hair.
The staring woodcock wakes in fright
From leafy bed on alpine height,
And flutt'ring from her foliage sere,
Steals to some silent valley near;
The squirrel peeps beside the tree,
Th' intruder on his haunts to see,
Then darts with agile leaps away
To watch him from some mossy spray.
The dark owl glares with moony eye,
As the lone wand'rer passes by,
And wonders what could bring him there,
To wake her in her beechen lair.

    But time a change to all must bring,
And wear the form of mortal thing, —
The ardour of the fire decay

As its best fuel wastes away,
And things are lost, and heeded not —
This is forsaken, that forgot.

 And where is he, that infant, fled,
Which wreath'd the wild flower round his head?

 That aged man reposing there,
On cushion soft in elbow chair,
Was he, the slender, musing boy,
Which play'd with Flora, as his toy,
And to his early passion true,
Fair nymph! still often thinks on you!
He sits in silent humble bower,
And marks the border's varied flower;
Exulting to his neighbour shows
The bright Geranium's vivid glows;
Some new Carnation from its bed
Lifts by his care her speckled head;
And other simple pretty things
As harmless recreation brings;
Tells of the seasons that are flown,
And waits in patient hope — his own. —

 Yet one spray more — one wreath would crave,
Such palmy boughs as angels wave;
In that fair pasture he would stray,
Where faith, where virtue, point the way,
With amaranthine crown to sing
Hosannas sweet to Flora's King.

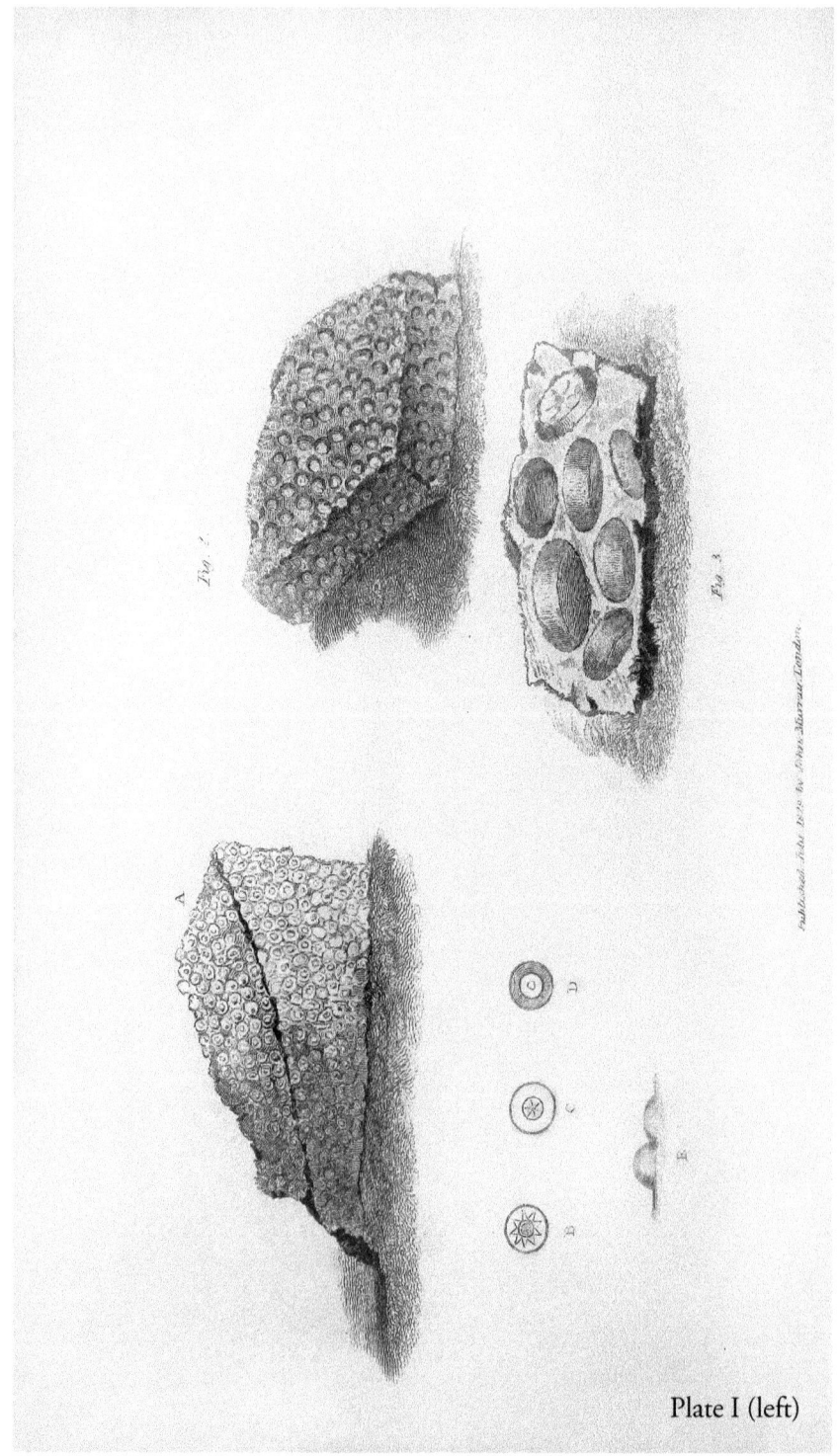

Plate I (left)

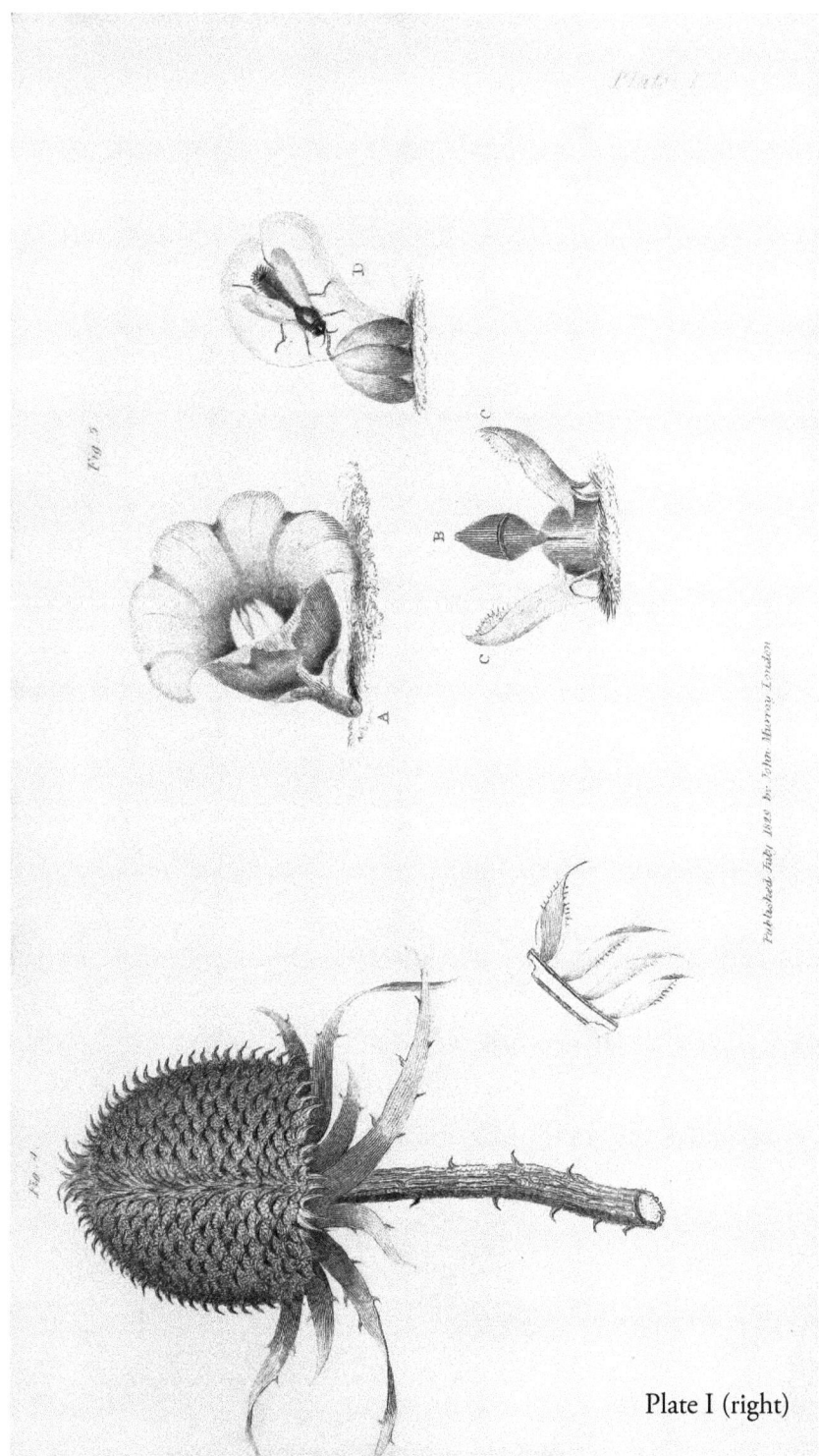

Plate I (right)

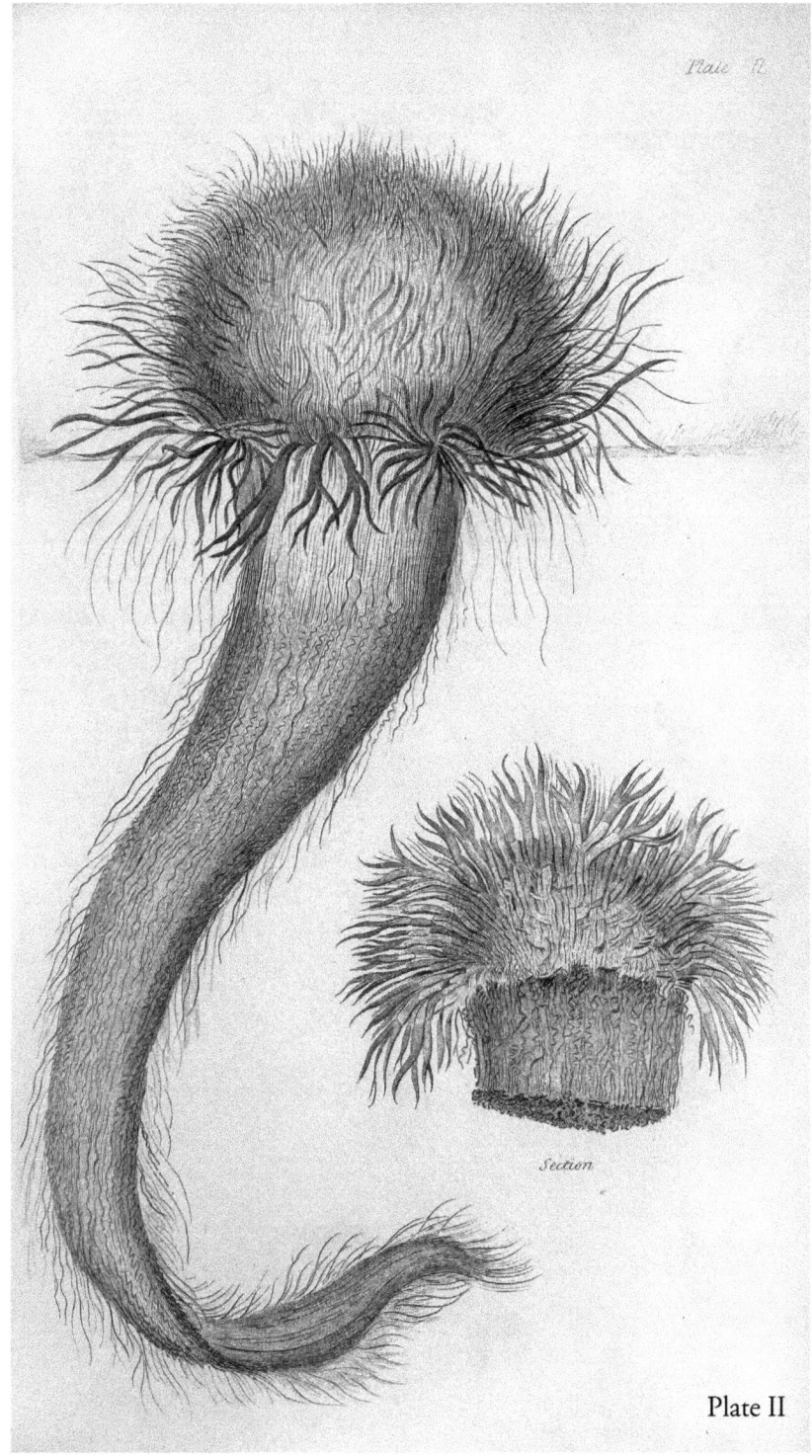

Section

Plate II

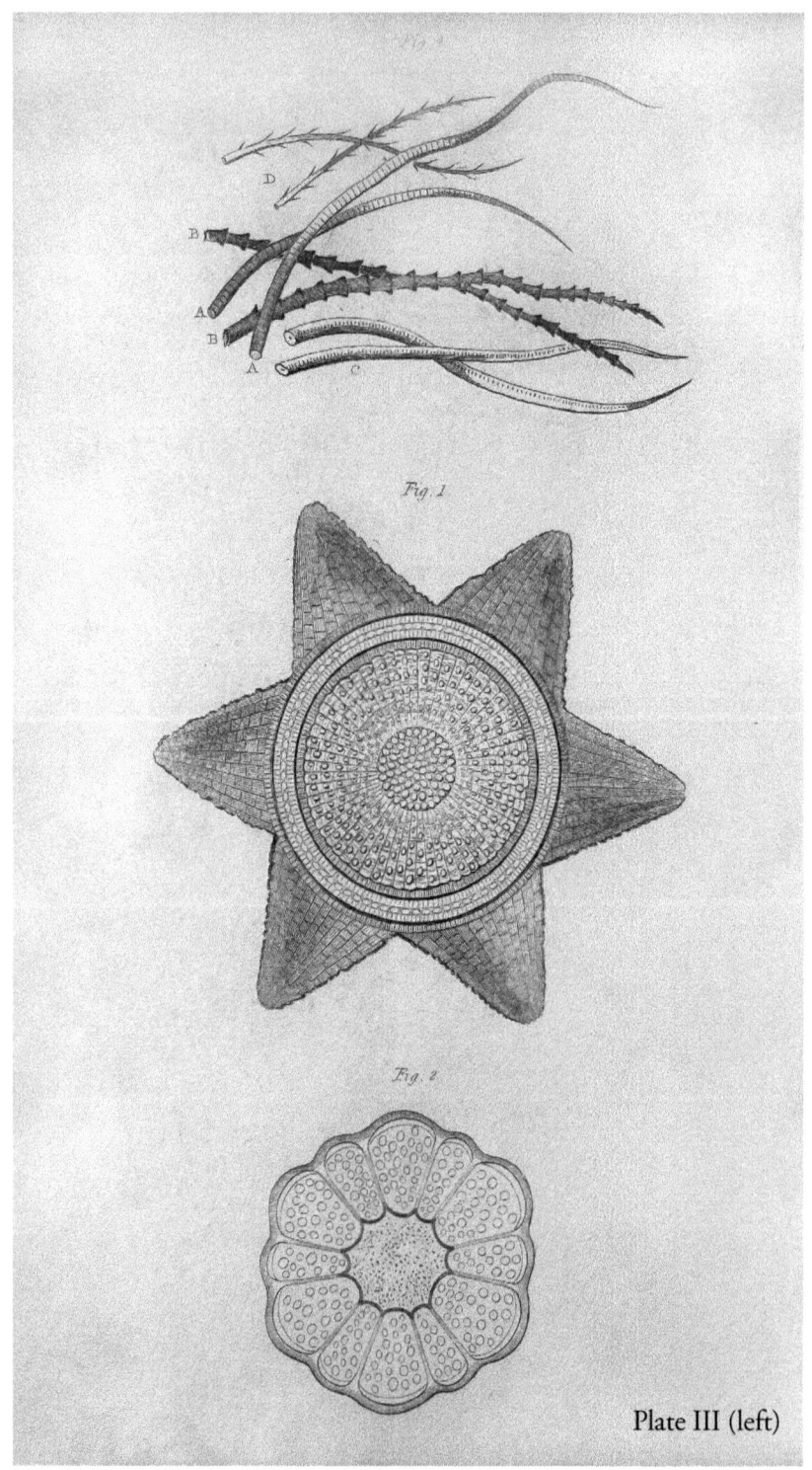

Plate III (left)

Plate III (right)

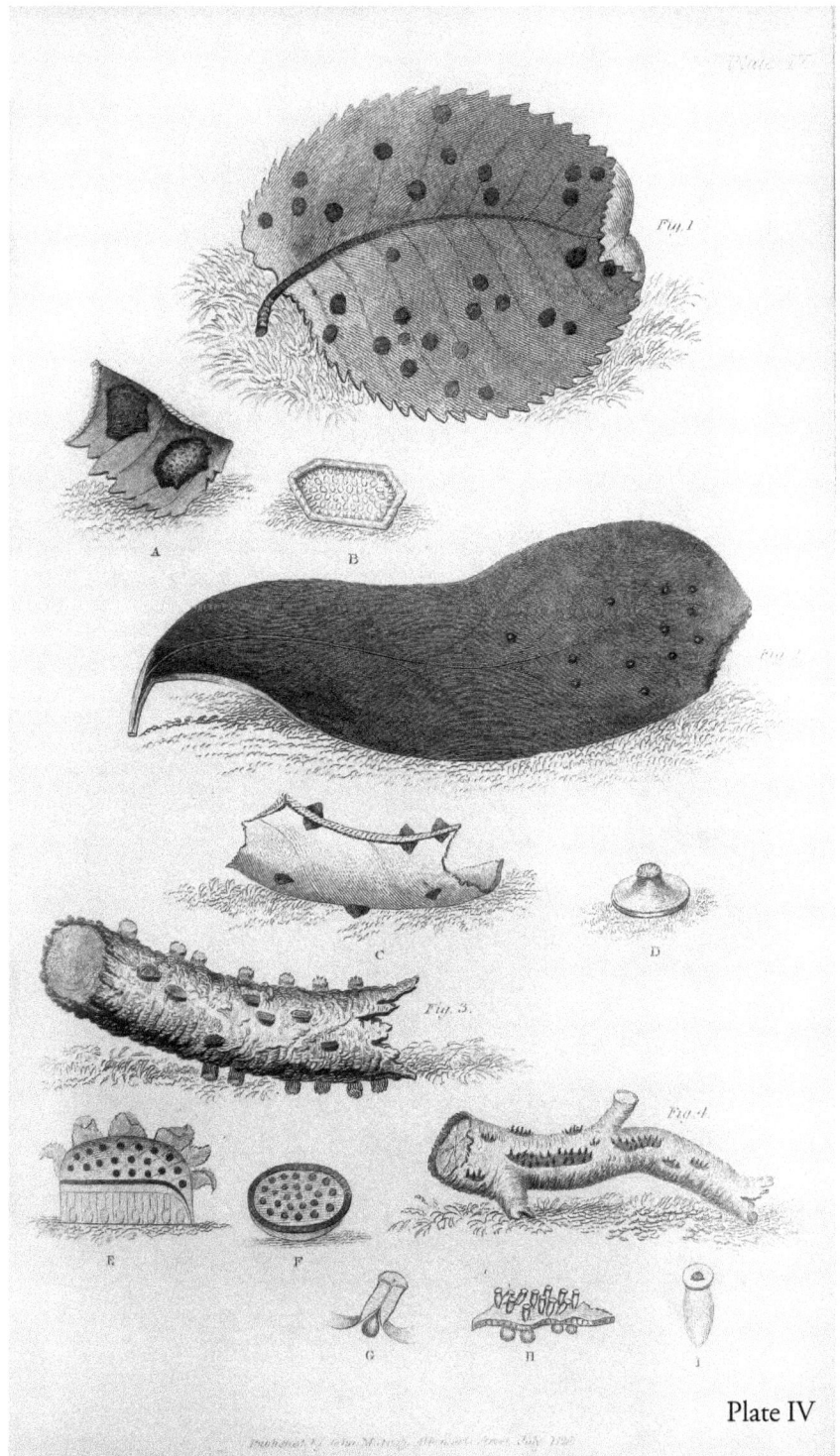

Plate IV

Plate V (left)

JOURNAL OF A NATURALIST 285

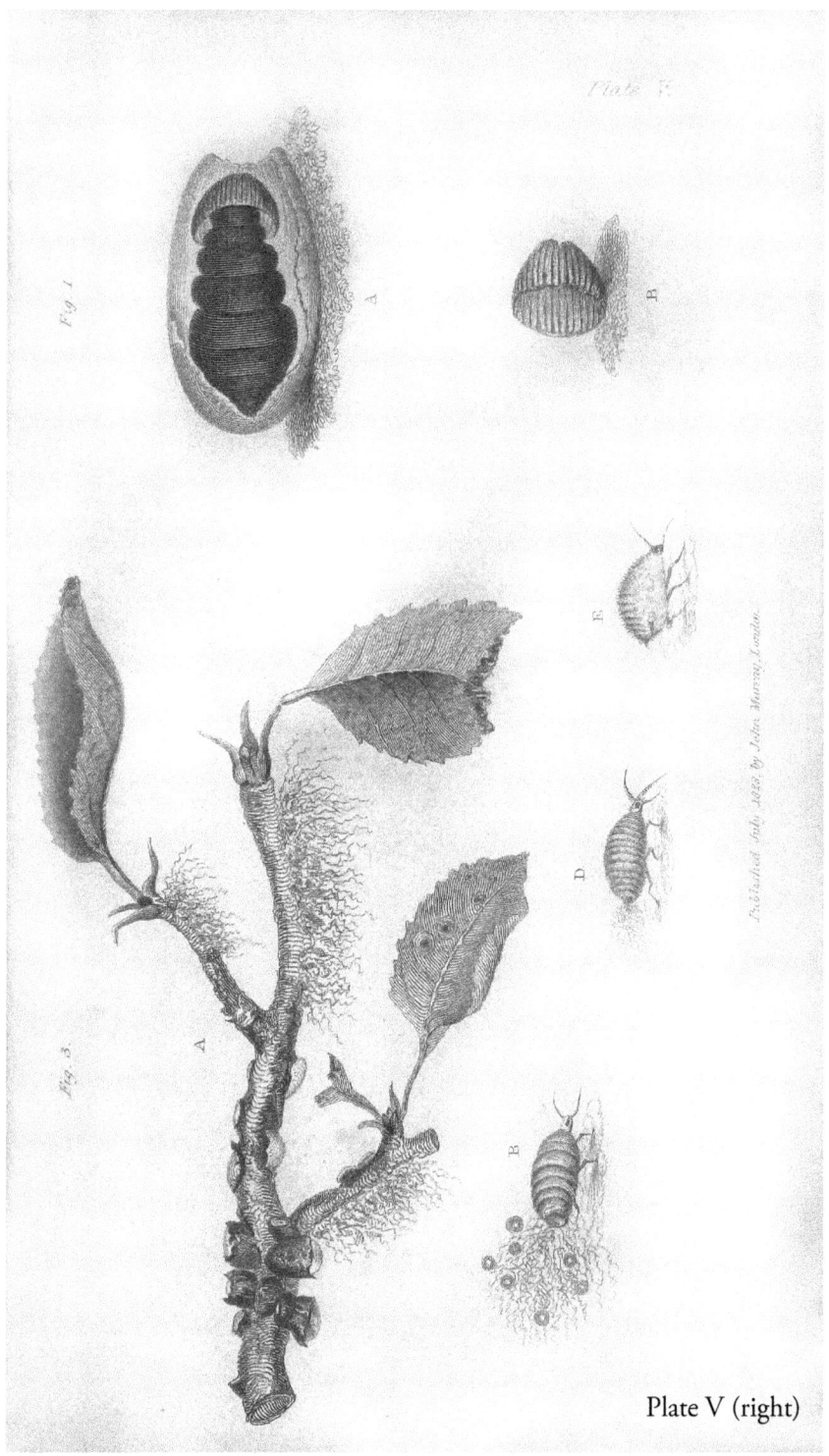

Plate V (right)

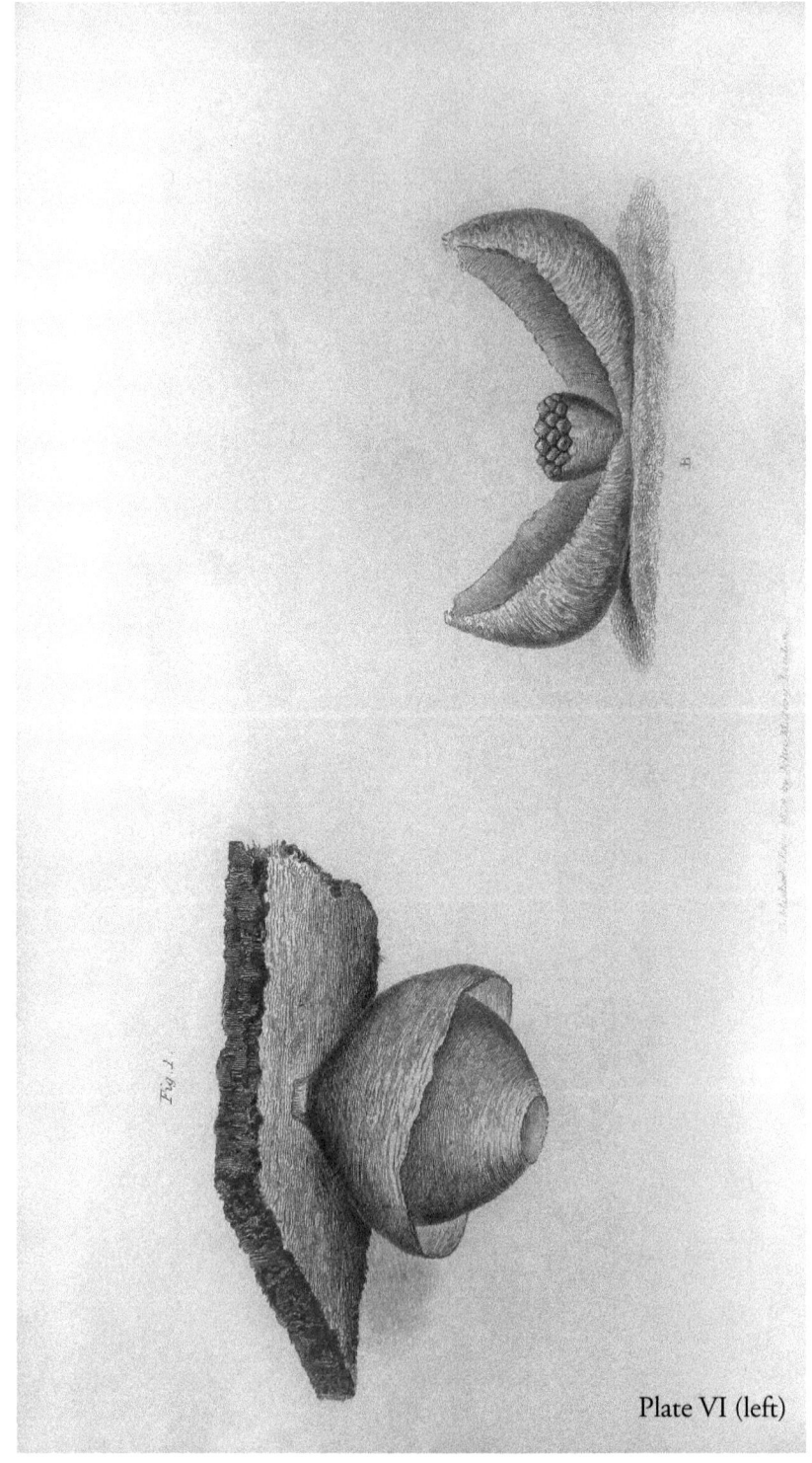

Plate VI (left)

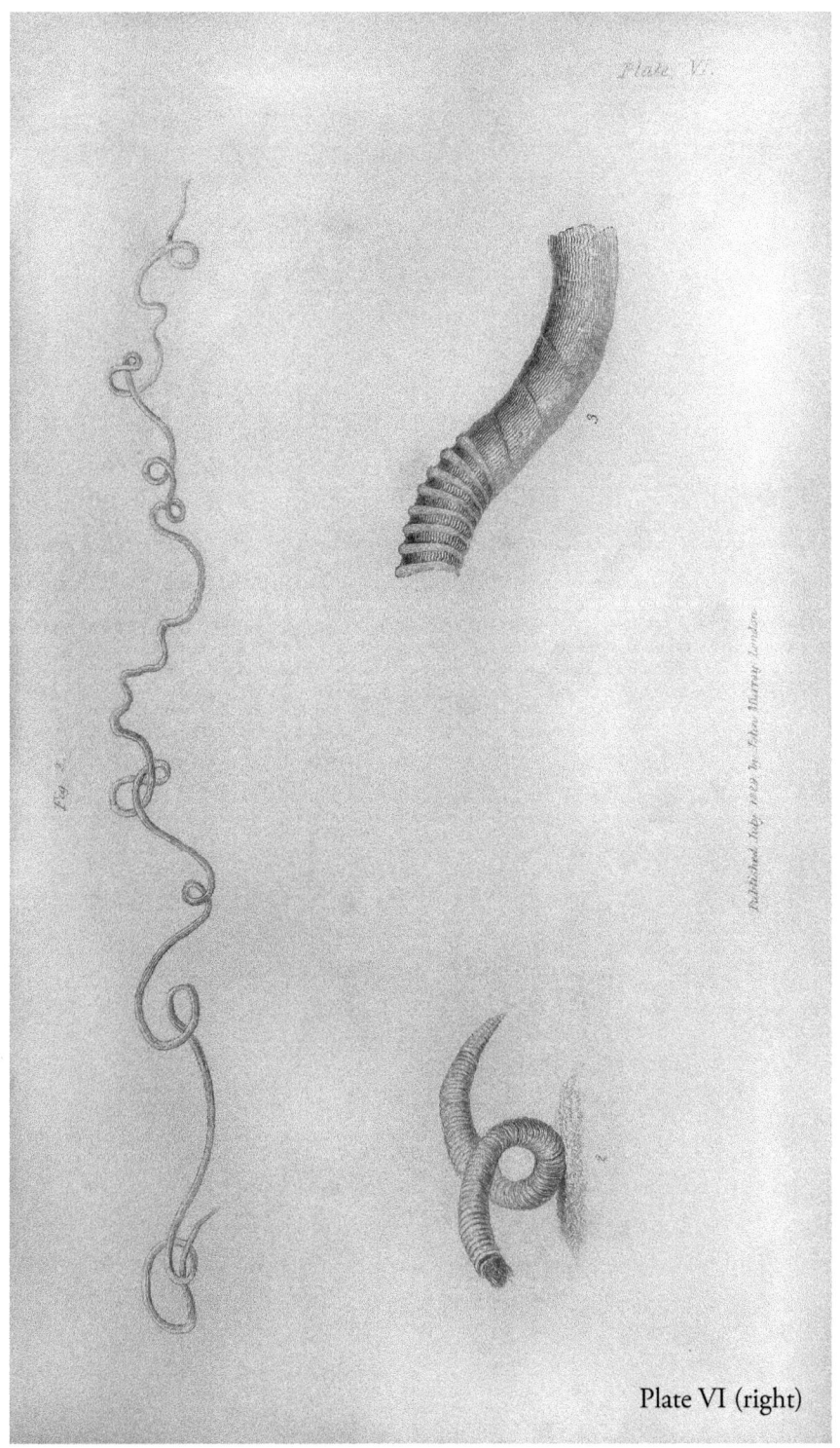

Plate VI (right)

Plate VII

BRIEF ESSAYS

These short pieces were certainly intended by John Knapp to be included in his much-hoped-for fifth edition of the Journal, but he left no instructions as to where they should be fitted into the text of the previously published fourth edition. They are therefore appended here as short essays in their own right, placed in the apparently miscellaneous order in which he wrote them in his notebooks. Brief titles to each have been added by the editor.

THE WREN.

IN THE SPRING of the year this little creature becomes greatly excited, and though his melody is brief, yet it has a shrillness and power, not expected to proceed from so small a bird, expanding his wings, raising and spreading his tail, and swelling his throat, he turns round and round upon the spray, his whole body seeming to vibrate with animation, and then soon exhausted by these efforts, he creeps away into silent concealment. Several of our small birds build, or partially finish two or three nests before they finally commence that in which they lay their eggs; in some cases this probably originates from suspicion of insecurity, or from being disturbed when at work, but generally it appears to be occasioned by a feeling of a coming season, a premature disposition to incubate, and I am the more inclined to think this the case, because we often find nests completely finished several days before they are wanted; a pied wagtail this year close to my house, finished her nest indeed so long before it was wanted that we considered the little fabric at first to have been deserted; the Wren seems particularly given to this species of experimental architecture, she will commence her labours in this hole, and then begin again in the shed, and then leave all her work for a more favoured site, and at times we find even these fragments of nests pulled out either by the jealousy of some other bird, or the caprice of the builders, and then completed and immediately made use of.

The Ivy.

Did we not know that the merciful creator of all things never abandons any creature which he has called into being, but in some way or another provides for the necessities of life and existence in all, we might be led to conclude that this year (1835) multitudes of birds would have perished for want of sustenance, but the ivy this summer produced an unusual profusion of berries; the end of February, all March, and nearly all April presented days of drought, severe winds, and frosts, consuming or rendering unattainable the worms, the slugs, the snails, and insect provender of these creatures; and here is 27th of April a day after a severe night of frost, and multitudes of starlings, thrushes, and black-birds, having consumed all these berries from every old pollard ash, barn, or ruined building, where they might be found, have now fearlessly approached my garden, near which they hesitated to come when less pressed, to feed upon the fruit of the ivy that runs about some walls in it, the red-breast too, who cannot find his worm, now makes his daily meal from the same bountiful store. The wood-pigeon, scared by our population, and persecuted by our numerous gunners, has until lately been deterred from a near approach to our dwellings, but now a pair or two, are seen snatching their food in vigilant haste, and are away again, but their rapid and wavering motions, so different from the heavy flight of autumn, tells of spare diet, and lightness of body. We have in this neighbourhood some long walling antient, and modern, covered with ivy, whose berries fall off frequently in summer unwanted by the common requirements of nature, this spring they have been so dark with fruit, as to excite the notice of indifferent persons, but not a berry this year was remaining by the middle of April, that could serve as animal food. The year, its events, mutations, the fleeting incidents of organised life, the truths that appear for the moment and perhaps come not again, and all the minutiae of a world of wonder, become the concern of the rural naturalist, pursuing "even the tenor of his way", where such things may be found: the townsman perhaps may smile at his note-book, as the register of weeds, of vermin, and of trumpery, but historians as we are, must rest the truth of our anecdotes upon contemporary evidence whilst that source exists. Memory deceives us, may exaggerate on the one side, and lose sight of leading traits of character on the other, but the transcript of the moment will be a faithful monitor, at least of the perception of our senses, the surest and best evidence we can obtain.

The Land-Rail.

The Land Rail, (*Rallus crex*) in the spring of the year gives sufficient notice of its arrival in many of our fields; and in the early dewey mornings of May, so loud and unremitting is his voice that occasionally he has awakened me from my rest: cautious in all his ways, and providing a number of eggs (10 or 18), we might suppose that this bird would be most extensively spread about, but there are counteracting causes and we find this is not the case; the abundant cover afforded by our crops of clover, and the artificial grasses leads it to prefer those places to lay its eggs in, but these being almost generally cut down before the hatch can take place, the nest is destroyed; the old birds, or a chance brood, when the corn is off the ground, seek for food and shelter in the potato crop, and are there sought for with frequent success by our sportsmen. The sexual call of the male bird can never be mistaken for that of any other when once known, and from its loudness, and seeming nearness, has occasioned the creature often to be looked for, but though I have repeatedly peeped over a fence, through the openings of a hedge, when my presence could not be known, and into herbage so scanty that a mouse might be seen running, and the bird by its creaking supposed within a few feet, yet never could observe it: I have known boys surround it when in call, and carefully closing the circle make sure of their object, but still it escaped unperceived: it probably squats upon the ground at those times just raising its head: but sounds are often as delusive as lights are at night, how long may one look for a young bird moaning upon a bush, or tree, close to us before we find it, and possibly are as much deceived as to locality by the call of the land rail, the sound varying constantly in seeming distance as the bird changes the direction of its head, and his deceptive voice, a species of ventriloquism may be appointed as a means of preservation at this period of being, when its call might otherwise invite enemies to their prey. The food of this crake, is believed entirely insects or small snails, but such food in some instances appears to be remarkably nutritive from the high condition in which animals are that usually diet upon it; three birds of very different natures may be selected as examples; the wheatear, the corn crake, and the honey buzzard. The general state of the first, especially after incubation and of the young yearlings is well known: the latter bird is too seldom obtained to form an obvious example, but in one instance before me the quantity of sebaceous matter was so copious as to be remarkable: the quiet solitary life of the land-rail may be conducive to the autumnal good condition of the bird, but the insect food must yet be the principal cause. In the preserving of a young bird, this fat is found so loosely enclosed in the membrane, that it flows as oil before the knife, and the warmth of the fingers; delicate as this bird is when brought to the table, yet the rankness of this fatty fluid the operator will long remember! It is not exceeded even by the mole, the fingers and the cloth on which they are wiped are for a length of time tainted with the odour. In an adult bird this fatty matter is much less oily, and feted.

The land-rail upon its arrival in April has occasionally fallen into my hands, and like all the rest of our visiting birds has a plumage much more fresh and delicate than that which invests them after a residence of months in our climate. It is now lean and the body flattened, the rufous brown of the scapulars and back in the autumnal bird is at this time so pale as to make the black terminations of the feathers very apparent, the throat is nearly white, and the neck approaching an ashy hue, the covert of the wings of a much brighter bay than in the fatter bird of September.

Sea swallows swept inland by gales.

Most persons must have observed that our summer visiting birds arrive in very unequal portions in particular seasons, and occasionally very few indeed reach our shores, and we are left alone to wish for the harmony of our native songsters. This year 1842 is one of those which with us is nearly destitute of these annual visitants. Not a nightingale is heard in the brake. A solitary fire-tail is flitting in the hedge; the willow wren is scarcely heard in the grove. The black-cap is rather more plentiful, but no butcher-bird has arrived. The wheatear has abandoned us from local causes, the others are rare and scattered. Supposition in these cases is, that adverse winds have overtaken the flights, diverted their course to other shores, or overpowering the weary travellers and drowned them in the ocean. This summer a remarkable instance occurred of birds being driven from their stations occasioning death and confusion to multitudes. That pretty, elegant actioned bird common in the Orkney and Shetland isles, one of the sea swallows, and very similar to the common S. Swallow (*Sterna artica*), which never appear inland, but being overtaken by the violent westerly wind of the 6th and 7th of May, were borne away by the tempest far beyond their usual stations, so that the whole estuary and waters of the Severn were covered by streams of these poor little strangers. Light in body and feeble on the wing, they were unable to resist the gale, and were driven against the rocks and perished in the waters, yet endeavouring to follow the course of the river they arrived in the float and harbour of Bristol, where hundreds were killed by sticks and stones. Far beyond their knowledge now, yet following the stream, they reached the most distant inland places – as Hereford – Tewkesbury – Devizes – and were scattered over all the immediate country. Those I saw were poor and starved, yet still they struggled on, and in numbers incalculable, at times so crowded that nine or ten were killed at a single discharge. In some following period the scarcity of these poor little birds will be noticed in places of their usual haunts, and years must pass over before the destruction occasioned by these furious winds, can be restored.

Rooks.

May 23RD – a day, in a short season of rural commotion – to repair the ravages of ill-fated years, my diminished establishment of rooks, have been allowed to bring off their broods without molestation, and they are all departing with youthful hurry and enterprise to other and distant trees, some settled upon the branches of the elm, and oak, to capture the chaffer, at rest for the hour, and are tumbling from the feeble sprays; the continual querulous calls of the young, and the harsher monitions of the parents, as they hover around their adventurous nestlings, fill the air with an incessant clamour which we hear at no other period; but this will cease ere long, as experience guides the one to depend upon individual prowess, and that indifference which visits all the wildings of nature when the necessities of carefulness ceases, at which time they heed not the offspring they have so tenderly raised, though separation does not take place, as often observable with creatures who have only one brood in the year, but dependence has ceased, and the calls of paternal aid have fully subsided. Yet the families continue in association, by the operation probably of no other cause than habit, until other objects and feelings consistent with the ordinations of nature induce separation in those species whose natural mode of life is not permanently gregarious. But this family union is an influence very different from that which leads creatures to associate for migration, or companionship in winter, and there are circumstances connected with the one, not well understood, and the object of the latter remains to be explained. In close thundery weather when the lower portion of the atmosphere is greatly heated, rooks mount into the higher regions to obtain a cooler temperature congenial to their feelings, which they seem greatly to enjoy by sailing backwards and forwards for hours, at times so high as to appear as small as swallows, and a portion even are invisible. When the great heat of the day has changed, they descend, and a sudden rushing of wings are heard above the head and the birds observed with closed wings sweeping and darting in long lines, traversing their descent with a rapidity that is amusing to observe, until they gain a resting upon the summit of some tall tree. I have noticed that these descents take place usually between three and four o'clock. Our country people who observe much the actions of animals, connecting them with some natural consequences, or fanciful events, remark that when the crows are given to playing at "break-neck" rain will ensue – a not uncommon circumstance in a climate like ours after a sultry day.

Apple trees infected with aphids.

THE INCREASE OR decrease of natural productions though so obvious as to excite attention, can seldom be satisfactorily accounted for, we may conveniently enough attribute them to season, or atmospheric mutations, but in this we do little more than guess an agency, when the mode of action, or the efficient cause, the cooperating influences remain unknown. Amidst all the witherings and honey-dews of this year (1835) and most destructive in our neighbourhood they have been, Aphis lanata has not been sparing in appearance, but has abounded with a profusion equalled perhaps, but never surpassed in our observance, and very few of our apple trees have escaped the investments of this most pernicious creature, sorts that had escaped for years received it now, though with such degrees of infection as to manifest that all are not susceptive of like injury: the Kentish codling, and kindred sorts, are almost clothed with patches and cottony wreaths, the Ribstone, and Golden Harvey, only partially so when young and vigorous, but the elderly trees of the former, received their full share of favour from the creature: the Fox-whelp (a cider fruit) resisted long but at length was sparingly visited: the Quarendon, appears quite free from the attachments of this Aphis – the Blandering, a very good culinary apple to a little beyond Christmas (we have only local names for these varieties) fully resisted contamination, or its juices were found unpalatable to the insect; one large tree in particular though all its young shoots had been seared by a morning frost and clammy with honey-dew, though all its boughs on one side were entangled with those of a Kentish codling, and on the other equally in union with those of a Court of Wick pippen, both provided with every means of supply, perfectly escaped. The bark of this apple is as rough as the others, would retain the cotton and conceal the weaver, but from some cause is avoided, in what requisite the sap is deficient or unpalatable to a creature that will imbibe that of a wilding or a crab? Our taste or analysis would probably in no way detect a variation in its circulating fluids, but the animal appetite has perceived something not congenial; and for this reason the cultivation of the Blandering might be found beneficial, when other varieties will probably perish from our orchards, though again it is obnoxious to early frost, and to pestilential winds depriving it of much of its efficiency in other respects – the crops beneath some of the badly infected trees, grass or potatoes, were scattered over with small flocks of this detached cottony matter in which the aphis is embedded, the vehicle of conveyance to other stations, yet still it waved in long loose rags from the boughs, by the constant labours of remaining myriads.

Strong blight.

June 6th. For several days we have had what the country people call a "strong blight". That is the air in the distance presents a pale smoky blueness, very thin, for it is not observable until seen in a deep mass, it has no resemblance to fog, is bluer and thinner, has no feeling of dampness, nor is it agitated by the breeze as fogs are; it is not a precipitation of moisture by the lowerment of temperature in any strata of the air, for all is dryness and harshness. That this blueness is occasioned by clouds of minute insects, or their eggs, as is the vulgar opinion, has no appearance of foundation in truth or reason, but as such a conclusion saves a great deal of puzzle and conjecture, and is a comprehensible cause and has been adopted by many. We see our roses covered in aphides, the leaves of the cherry, black current, sloe and plum shrivelled up (crokelled) with insects in the hollows beneath, whose punctures by withdrawing the vital sap have apparently occasions this, and if the leaves are carefully examined, upon the footstalks, or the midribs, the nodes or excrescences which have arisen from these woundings may be observed: but from whence came these myriads of insects which has occasioned this? From whence the cause of clamminess, scattered over foliage where insect discharges have no access? Are our suppositions, the prevalent idea of ages, based in truth, that they are more universal subsequent to these blue atmospheres, or merely fanciful? The Peach and Nectarine, have their leaves greatly withered up, and become partially deciduous, and as insects are only occasionally found beneath, this probably has been effected by the frosts of the morning, or low temperature of the evening suspending circulation which the radiation from the walls have not counteracted entirely, for had the frost been more powerful, it would have congealed the moisture in the capillary tubes of the tender shoot, expanded and ruptured them, bringing on mortification and decay, but being below this, the circulating fluid has ceased to flow, and nourishment failed from the foliage, and this is more probably the case, for when the circulation has been restored by reason of a milder temperature, though not able to restore the vigour of the injured foliage, yet it has enabled fresh leaves to spring from the termination of the spray, and life proceeds.

The blue argus.

Favourites with all, and noticed by every one as the butterfly tribe are, a race of creatures identified with the country, gentle airs, and sunny beams, a mutual association with harmony and enjoyment – few of these gay flys are better known than that little flutterer of June, the blue argus (*Polyom^s: argus*). These like the rest of them are annually produced in very uncertain numbers: selecting the short and flowery pastures, preferring the shelter of the tall hedge where they become embayed from the breeze. In 1836 their numbers far exceeded the usual proportion that appear, and it was observable that amidst hundreds of the females, how seldom the male animal was found, not that he is a scarce creature, but retiring to some adjoining pasture in seeming avoidance of the bustle and vivacity, to bask and be nourished by some humble flower. Common and scattered every where as this azure fly is, yet the larva and chrysalis are so concealed, that I am not sure of having ever seen them, they are probably hidden and perfected beneath the soil of pastures near the places frequented by the matured fly. The life of an individual butterfly is probably brief, we see a few species continually through the summer, but they are apparently the produce of successive hatches, rather than enduring ones, subject to that universal law of nature which has ordained that from the lowest grades, the invisible atom to the highest orders, one becomes the prey of another, so these creatures are the fated food of birds. We notice them continually captured, see their wings torn and shattered by their bills, and the swallow tribes will frequently cruise about us in the field snatching up immediately this little blue argus, as we put them to flight by our actions, and the main body are soon removed from the pastures.

The preservation of fruit.

AMIDST OUR RURAL concerns, and we discourse only upon rustic things, the winter preservation of our orchard and garden fruit is a subject of some consideration. All our good housewives know how liable they are to injury, keeping at time infinitely longer than at others, though the treatment and sorts are alike, these variations are probably occasioned by the influence of the summer upon the growing fruit. This year (1836) a remarkably cold and gloomy one, though some of our produce perished soon, yet generally speaking it kept remarkably well, enduring until late in the spring; and I think it will be found that the fruit obtained in ungenial seasons, though inferior in richness, will decay in less abundance and remain longer than that which is collected for store after a bright and sunny one, though higher flavoured.

For the general cause of this we must resort to the principle of fermentation, or that mysterious accomplished agency whereby nature in the vegetable world brings about the dissolution of a compounded body, disposing it to assume new combinations, which art, availing itself of, acquires the knowledge and uses of alcohol, and vinegar. The ripening of fruit is apparently entirely a chemical action, the oxygen of the sun (light) gradually converting the acids into sugar, we can readily perceive how completely this may be effected in a sunny summer, and how partially in a gloomy one. Fruit well ripened rapidly decays, the sugar soon taking a vinous fermentation, followed by the acetous, but in those cases where the action of the sun has been less influential, little or no sugar has been formed to effect a vinous fermentation, and the juices therefore long remain in an unagitated state, and the vessels not broken by internal motion. Should the heat of the sun be long continued, the juices will become thickened. When the transmission of them from the tree is cut off – the influence of oxygen, approaching the nature of concrete sugar, is little liable to fermentation as in the case of dryed fruits. Again, this fermentation may be retarded by keeping the fruit in low temperatures far below 60°. this process will not take place, but congelation by cold at the freezing point will rupture the vessels and mortification ensue. Sugar in fermentation has its elements perfectly separated, a part becomes spiritous, giving out alcohol, a compound of carbon water and hydrogen, it then after a time absorbs oxygen from the air and becomes vinegar. Now our fruits may undergo both these processes, the vinous fermentation first producing an higher temperature and bringing on the acetous. By fruit being kept in the dark from the exciting influence of oxygen, an excess of ripening is retarded, or the vinous fermentation in part suspended, still a low uniform temperature is requisite, that there may be no excitement, no disposition to fermentation. Damp must be avoided as being productive of mould, another agent of disintegration. I have kept apples very long by pitting them under the earth like potatoes, and though surrounded naturally by dampness yet from the lowness of the temperature, darkness, exclusion from air preventing the absorption of the juices, and occasioning shrivelling, the destructive processes have been suspended. Yet this is noticed merely as an experiment, for though they were kept from decay, and looked quite fresh and sound, yet their substance was tasteless and spongy, and like all such interments rapidly fell to worthlessness when removed.

There is a time, a season for all things to every purpose, and we cannot in this and many other cases, subvert the natural ordinations, the appointed disposition of events. Yet we may by comprehending causes in some measure alter them, and frequently rest more satisfied with consequences when we can contemplate the operating agencies, and instead of wondering and complaining with an ignorant mind, philosophize upon the events, the mutations of nature.

The effects of drought.

Much observation leads us to believe that we have fewer general showers of rain in our immediate district than in some other places; we look to the south west and see the advancing cloud, but in this apparent progress it becomes broken, or parted by the highlands at Knoll rising above the alluvial bank of the Severn, a portion then follows the Monmouth range of hills, and another part goes up the Avon, and Bath and Wiltshire elevations, leaving a delta unwatered, when the skirts of it receive their portion of the falling shower. This year 1835, we had a long and distressing series of dry weather, occasionally a slight shower or so, but no rain of consequence fell from about June 12th until Augt 22^d, but in consequence all our crops of clover and grass which were very abundant, were secured in a manner very seldom experienced in our dripping climate, and the harvest of corn and teasels were most generally well secured. All this dry, hot weather distressed our poor birds sadly, the vast flights of starlings, whose food appears to be exclusively insects and worms, were rambling every where for what the country could very partially supply, as all the races of winged and creeping things were unusually rare. Our rooks forsook our arid pastures and fallows, retiring probably to the meadows of the Severn which were less burned up. The harmless wagtails, entirely insectivorous birds, came in for their full share of distress. But there were circumstances relating to the animals around us, occasioned I believe by the dryness of this season, that may be related. In two of these cases, instinctive wildness and avoidance of man were subdued by the necessities of nature. The wryneck, a shy bird always, but after the breeding season when he becomes mute, lives so particularly secluded from human haunts until he leaves us about the end of September, as alone to be noticed when surprised by the naturalist in his retirement. Insects I conjecture, were scarce or hidden from the heat of the sun, however three of these birds now visited the ant-hills upon our grass plot, feeding with the common domestic birds, one of which took up his station immediately beneath our windows, collecting the emmets that frequented the place. Though looking vigilantly about, he suffered all his actions to be inspected, intruding his bill into the earth, he disturbed the dormant creatures, and seized them as they appeared. But rain came, natural shyness returned, and the visit ceased. The woodcock, is with us at all seasons a scarce bird, but one poor animal probably by reason of a winters wound unable to follow his fellows in their long migration, remained with us feeding a he might, but towards the end of July, when the earth became as iron, driven to extremity by severity of season, he sought to supply himself in a little ditch softened by the trickling of a cottage pig-sty, but so enfeebled as to permit being captured by the hand. The person into whose possession it came, not knowing how to feed him, killed it - and some difficulty to do so at this period would have been experienced, for those who have kept them say that the consumption of worms by an individual bird is beyond imagination. An insect constitutes our third example. The chalk-hill butterfly (*Polym^s: Corydon*) is I believe every where a rare creature, with us is never observed. Its station, as its name implies, is chalky, or limestone altitudes, and we have one range, Stinchcombe Hill, with Nibley & Wootton, distant perhaps eight or ten miles, from whence a few specimens have been obtained. But this year a small flight frequented a rough grassy field in our clay-lands, as unlike its favourite residence as possible. We conjecture that the burning sun had seared up every flower that could afford nourishment upon the hill, and that the creature left it to seek a supply, and thus wandered to us in a progress elsewhere, as it remained only two days; at least no other cause can we assign for the appearance of so unusual a visitant as a chalk-hill butterfly. Such are our light and passing village records, the humble concerns and accidental incidents of rural life, noticed perhaps by few; the pursuits of man are various, mine are perhaps of little use, of importance certainly to no one, yet it pleases me to note the transient events of my day, to silent glide along my rustic path, and speed my summer hour void of offence to all.

Ancient vegetation preserved beneath a Roman path.

August 8th 1835. I had this day a notable instance of the durability of vegetable substances when kept from the influence of the air, not worth recording perhaps, after the constant examples so well known, but as a local circumstance I have mentioned it. Our Roman encampment has been divided into two unequal portions by the main turnpike leading to Gloucester, and we have been lowering an elevation over which it passed near this encampment, where I had suspected we should have cut into the main communication with this station and another in the neighbourhood; the sinking in some places was four feet deep, but no vestige of any former work appeared: however this day they cut down to what had been a cross-road which had been formed outside the agger, and from its construction so unlike the massy military solidity of this nation was probably a mere path of convenience between some of the settlements connected with the establishment, a path from the suttlers huts or something of that nature. From the necessity of restoring a passage as soon as possible, we had less time for investigation than could have been desired, but the foundation appeared similar, to what it would be if now required, a quantity of brush wood, rough grass, and such vegetable productions as a rude uncultivated tract produces, the middle being covered with irregular flat stones roughly laid down, the vegetable matter projecting from the sides, and this was so perfect as decidedly to manifest the nature of it: the grass blackened, and like what neighbouring hedges produce, the unfavourable weather prevented the attempt of identifying the species. The leaves of the holly were in abundance and quite perfect, only brown sprays of the white thorn, with quantities of bramble, but though dark as charred wood, was so perfect, even the long annual shoots, as to have its bark preserved and the internal pith, with all its spongy filmy valves quite manifest; knotty pieces of holly – and hedge roots undefinable – from the great number of holly leaves, and no other foliage but grass being obvious, nor any impressions of former existence. At the time I conjectured, fancifully perhaps, that the construction had been a winters work, or a spring occupation, when the passage became miry, the leaves of the surrounding brush all cast, or not matured. If then this encampment was formed under the government of Ostorius, as conjectured, we will say 50 years after the era when our enumeration commences, here we have vegetable matter not remarkable for induration when exposed to air, preserved when excluded from its influence, though obnoxious to moisture, for above 1780 years! – one small coin of red brass, only came into my hands, of roman mintage apparently, but decayed beyond the power of deciphering.

The Village Well.

Every village has perhaps its well, the probable origin in many cases of the settlement, and often of its name: I do not mean that place of resort with its bucket, its windlass, and its chain, but the little sequestered, never freezing perennial spring, beneath the bank in the green lane, or in the "dippy-down". Many of these are walled some way below the top, from the fissures of which depend long moisture-loving mosses, and confervae, the harts-tongue drooping over the surface. Others are only margined by rude stones as preservatives against the pressure of feet, deeply worn, but enduring. Wells are often remarkable monuments of antiquity, co-eval with the oldest productions of nature, the oozings of the veins of moisture which trickle beneath the hills, the everlasting mountains upon the surface of the earth. All the fabrications of art "man and his marvels" – the castle, the abbey, have perished, but the well which supplied their inhabitants yet remains. To all indispensable, they escaped ruin in the havock of war, the reckless destruction of individual possessions. When all other roads have been lost by disuse, by inclosures of districts, or cultivation, the little grassy foot-way that leads to the well of S^t. Mary is found. The village maid which dips today her pitcher in the fount, only succeeds the employ of generations long passed away, the Roman, the Saxon, or the Norman lass. Very many of these wells yet retain some sainted appellation, having in distant times been blessed and dedicated, and its patron's blessings annually invoked by the priest, in those primitive ages when itinerant preachers from the monasteries, or the more irregular religious, passed from village to village, instructing the poor ignorant people in duties leading to salvation. Their knowledge, such as it was, their subject matter, or the morality of their conduct, doubtless often obtained the consideration of peculiar sanctity. If Father Nicholas, or Sister Bridget, blessed the water or "drank of the spring in the way", it became identified with them and hallowed by their use. Mystic ceremonies, the observances of the forms and rights of creeds often survive, being memorials of holy days, and festivals of remembrances of national institutions and laws. After the extensive miseries of long intestine feuds, the desolation of Danish, Saxon and Norman conquests, yet the feast of wells, the worship of fountains, a relick probably of Fontinalia of the Romans, or an older reverence of druidism, yet continued to be so regarded in Britain for some years after the conquest, that the great council of Westminster in 1102, found it necessary to suppress it as a pagan institution. Many of these waters were esteemed in cases of opthalmia, the coldness of the one, or the chalybeate of another, might strengthen the eye-sight, and through the purification of all, be beneficial. They are remarkable for the preservation of original properties, be they mineral in some, or thermal in others. In a neighbouring village we have a spring walled round of great antiquity, which the learned author of the "Celtic Researches" considered as certainly Roman, notable from the abundance of water welling out in all seasons. With us, our wells are chiefly artificial sinkings, but we have two of natural, or nearly natural origin, one of these is so immediately situated on the edge of a raised Roman communication between two well ascertained stations of this people, that it seems probable the way was partly guided in that direction for its convenience, or at least the spring respected for its uses; and we can easily imagine the weary legionaries of Claudius, or Vespasian, refreshing themselves from its produce. Few things mark more forcibly the mutability of worldly affairs, the shiftings and inconstancy of human inclinations, than these our rustic wells. Towns, and villages are not always permanent, some local cause, a church, a factory, a public road, draw away the inhabitants from the old hamlet, to these sites of convenience or profit, the residences decay, the spring that was their origin is abandoned – perhaps some "Sybil Gray", from a ruin in the village yet bathes her eyes in its cool waters, but the fount, though it will run, for ages run, is forgotten, long tangled grasses almost conceal it, the tall fern waves over its margin, and the well of S^t. Anne, with all its sanctity, its virtues, and its usefulness, becomes the undisturbed winter haunt of the lonely snipe, the washing-pool of the summer linnet.

Autumnal sensations.

We generally look forward to our autumnal months not only for that agreeable temperature obtained by the admixture of extremes, but for those grand and various tints that the forest acquires as the verdure of summer departs. However these gay colourings may be brought forward, be they occasioned by the drying of the leaf upon the retiring or partial flowing of the sap through the footstalk, or by a chemical action upon the foliage, the chilling air obstructing evaporation, occasioning a fermentation or generating acidity in the vessels. From whatever cause autumnal hues may proceed, we have not this year (1835) the usual glow of the reds, and yellows, that some seasons manifest. There is a brownness, a dull ochreous rustiness, as if the winds had bruised the leaf, the suns of summer parched it. Of all our seasons we have none allowed by general admission, scriptural and moral, so peculiarly fitted to awaken sober contemplations as this of autumn. There are characters, admonitions, impressions to be perceived in many of the vegetable tribes, but they are individual with separate tendencies. But in this season there is a quietness of deportment around us in every thing that naturally disposes the mind to assume the calm tones. The unison of nature, the leaf that falls in summer, drops unminded, it is only a withered part and is better away, it leaves a flourishing abundance behind it, and we hear not the sound amidst the flutter, the gaiety, the voices around us. But if the leaf that falls in this lifeless silence is looked at, we note it as inceptive of decay, as a thing that has lived in health and usefulness. Spring is an evidence of life, revival, proceeding onwards into being. Autumn closes all, "it is finished", there is a timing of action that is remarkable, a procrastination of operation. The ash, the poplar, walnut, lime and others fall in early days, healthful and green as they may be in a summer sun, they sicken and faint when the first symptoms of adverse hours approach. Perhaps in a little associated group one may feel an earlier touch, a partial influence, a hectick of disease, and noiseless cast its verdure but briefly opened to the morning ray. Another follows, becomes deciduous in its turn, the few remaining by aid of a little constitutional vigour or aliment, resisting longer, at length become denuded like their companions. A sheltered beech, the elm, or a few willows, of stronger natural qualities, remain in verdure yet for a time, then comes an icy fog and oppresses them. The breeze succeeds, scattering their foliage in showers, as a pestilential blast would sweep a city, leaving a few quivering leafets at the summit, shortly to follow those which are gone. The oaks, a family of longevity, bear vigorously their summer honours when all the green flutterers of the copse are gone, living their three score years and ten, at last becoming what others are. The larch, holly, and others which we call ever-greens, framed to endure long, are victors in the storm and defy the frosts of winter. The heat of the summer comes, their period of life has terminated, the foliage falls, giving place in succession to a race that knew them not. The little humble bushes of the wood obey the same mandate which strips the lofty ones above them, the age of verdure ceases, and winter comes alike to all. A long hibernation of months follows next, but time renews all in their season. Such occurrences, and hosts of others equally notable to an observer of nature, leads commonly to a contemplation, prevailing over admiration or every lighter feeling, taking possession of the mind inducing particular serenity – autumnal sensations.

Growths beneath oak leaves.

Upon the under surfaces of the leaves of the Oak (*Q: robur & sessiliflora*)[47] may be found in every summer, small, round, flat substances, having when young a raised margin, from this state they keep changing, the centre becoming elevated and the margin disappearing, so that by autumn they become target shaped, fringed on the edge, and the surface sprinkled with mossy specks, attached by a very short footstalk and parting from the leaf with a little snap in consequence of the increased pressure of the margin upon the foliage. In 1835, the leaves were so crowded with these substances, some solitary, and others clustered over each other, that when they parted from their stations immediately before the fall of the leaf, the ground under the trees was sprinkled over with these scales. I have always imagined that these little targets originated from the punctures of one of that numerous family of cynips which select for the cradle of their young every part of this tree from the topmost branch to the root, giving it a preference beyond others, forming the gall-nut, oak-apple &c. But my suspicions are not satisfactorily confirmed, and there are circumstances connected with this little excrescence unlike those attending the other fabrications of cynips. When insects puncture a branch or leaf with the object of preparing a nidus, they usually deposit an egg, which becomes enveloped with the matter exuding from the wound, assuming various forms, from which in process of time an animal becomes perfected. But these targets do not appear in any stage of their growth, at least for six months, to contain any ovae. There is a small central cavity above the summit of the stipes, but to my observation no rudiment of any egg. I have seen some of these shields in November more elevated than usual, and found in the little dome beneath a dark brown active larva, but with no remains of any thing indicating a living residence, which with the scarcity of the creature, for an hundred may be examined without success, led me to consider it as the casual retreat for a stranger, and not a parental formation. I may say then that I have not observed – others may have been more successful – any certain appearances to confirm the supposition that these scales are insect formation. In the autumn they fall to the ground and become concealed beneath the surface by agency of the rains of the season. Whatever may in fact produce them, the extraordinary

---

47 Q. sessiliflora – and Ulmus glabra – are so rarely found with us, that I know but a single specimen of each in the parish – the Luccombe, or fern-leaved variety, did not produce any of these scales in the few cases under my notice.

numbers that in some years are formed upon a leaf, fifty or more, and in instances every leaf has a portion, will, if they be determined to be the labours of insects, present a very remarkable instance of perseverance surpassed by no race, industrious and provident as these creatures are known to be. Batsch, has I think in his Elenchus fig:152, figured this substance under Peziza comitialis.

THE PLUMAGE OF OWLS.

THIS YEAR 1835, we were visited by a stranger bird, the short eared, or woodcock owl, as sometimes called (*Strix Brachyatus*, of Latham) – by day it concealed itself under hedges, and in the haum of the late potato crops. One, we knew of last year, but they are very uncertain visitants, probably accompanying the great flights of field-fares, redwings, and woodcocks, which arrive in November from the shores of the Baltic. Several, for I had three in one day, of these poor harmless, useful birds were destroyed, as we here destroy every thing, nothing seems allowed to live in enjoyment of the liberty given them by nature, or equally share the bounty and blessings of providence in peace. Owls are admirably beautiful birds! Not from the gay colours of their dress, but from the fabrication, disposal, and convenience of it for their destination and habits. The plumage is very thick and warm as requisite for birds which ramble about through all the cold stormy hours of night when others are sheltered on their roost, and their heads protected beneath their wings. But this creature, whose head must be exposed to every temperature, has it surrounded with a dense vesture of great variety, so that if the separate feathers and vibrissae were figured, probably no one bird would present more variety, nor in any way surpass it in disposition and convenience. The lightness of the body, undulating upon the air, and the softness of the feathers as observed by White, enables him to advance with noiseless motion, without a ruffle upon his unsuspecting victim. For silence in these birds is a means of existence, a faculty of being, and the whiteness of the under parts of the body renders him in these approaches nearly imperceptible. The prodigious orifice of the external ear gives him intimation of the faintest sounds of feet upon the fallen leaf, or the dry grass, and what marvellous security has been provided to guard from external injury these vast openings! His noble eye discovers in a moment every movement and action about him. The great wing feathers of birds in general move over each other upon smooth surfaces, the object being facility of action. But the white web of all the wing feathers of the species under consideration and the barn owl (the only ones at hand) upon the outer faces, where covered by the overlapping of the others, are of a peculiar soft cottony texture, quite manifest to the eye, occasioned by the remarkable fineness of the vanes, and delicacy of the laminae or side filaments.

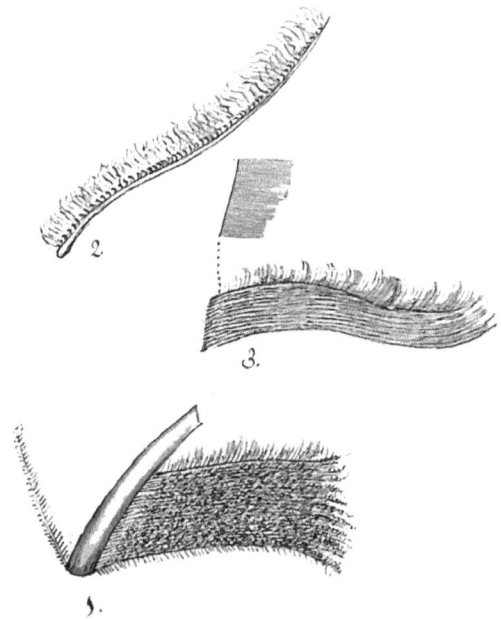

1. Magnified feather from the wing of a short-eared owl, to show the downy surface
2. A side thread representing the hooks at the base of filaments.
3. The inner surface.

These, in common, are connected nearly their whole length by minute hooks, but in the bird above noticed, they appear to be connected only close to the vane, and all the terminating parts then become long, loose, and wavy (fig 2). And this cotton uniting the vanes upon the surface by entanglements answers the same objects as the hooks; and another, for it would appear that the singular silence of flight, the noiseless action of the great wing feathers, when in motion, is obtained by their moving upon the cottony surfaces and not rustling upon unclothed hard vanes as most other birds do. There is a particular fineness and transparency in the vane, giving it almost a glassy lustre, a beautiful silky corded appearance on the inner side. But these are only common notices of the manifest provisions of a beneficent power, the exquisite ordinations of providence observable in all his creatures, their fitness for their destinations, the services of life, and we must mark them, because they delight us, give us pleasure to witness such universal wisdom and goodness; becoming certain indications of far superior intelligence and benevolence not discernible by us.[48] The wild-fowl, the snipe, and almost every other bird, with a vesture upon their wing feathers like this, would be obstructed in their progress, and the modes of life as now pursued by them, could not be maintained, but a different occupation, a dissimilar means of subsistence being appointed to these owls, it becomes eminently serviceable. Of the six or seven owls that are found with us, occasionally, or permanently, we do not know that we can distinguish more than two of them by their voices, the tawney (*S. stridula*) and the white one. Some years past when resident in a very woody part of the Kingdom where numbers of the brown, tawny or wood owl, could be seen and heard every evening, for two or three summers and autumns, we had one bird which frequently came about the premises having a voice greatly differing from that of its companions, omitting the first "Hoo" always. It uttered only the final tremulous sounds, and those in a very soft and plaintive manner, which sounded as a thing in distress, like "poo-oor ow-l", and at the time I considered it probable that an individual, of the hawk owl, had associated with the others, for I could never distinctly see it. We had here for one summer a night bird which from habits and statements seemed to be of this race, whose voice commencing at about eight o'clock in the evening was heard in a particular district, commonly through the night at intervals, until four in the morning when it ceased, sounding like the screaking of a day cart-wheel as it revolves, to be heard at a distance. It was obscurely seen and shot at, and circumstances led me to consider it as Strix scops, the little horned owl, but the bird is rarely seen any where, and I could not verify my suspicions. An intimate knowledge of these nocturnal voices, as many birds are only vocal at those times, is particularly desirable in regard to several of the marsh and water birds, which pass by us in the evening gloom, to feed or remove to distant districts.

---

48 It is this owl which possessing certain features of similarity to the ringed-tailed hawk seems to form a link of union with that race of birds, the smallness of the head, length of wings, and habit of feeding by day, are not characteristic feature of owls; the profusion of long, loose plumage of the ring tail, its markings, the ruff round the neck, give it a resemblance to the owls; the hawk however has nothing of the cottony plumage upon the surface of the feathers of the Strix, the laminae being connected as usual in feathers, and fitted for its habit of life.

The Migration of Rats.

MANY CAUSES MAY lead animals, as they do man, to change the station in which they may have been born, or resided in. This impulse or necessity has tended greatly to bring inhabitants of various kinds into many regions of the earth, and enabled those that remained behind in some instances, better fitted to obtain existence. That quadrupeds at periods did remove to other places in bodies, has frequently been suspected by the sudden increase of them without any other assignable cause. These migrations of larger species occasionally have been observed, but of the transit of smaller races, I have no remembrance of any credible testimony. I was much surprised one week late in the month of October 1835 by observing all the recent droppings in a large pasture fed by cows, trampled in and numerously impressed by the feet of animals, apparently rats. At first I conjectured that these creatures had been lately collecting a store of haws, crabs, or acorns, from the hedges, but excepting the last we never had so sparing a crop and none scarcely could be found. They were not the markings of moles, seeking for dorrs, as the droppings were not in the state as frequented by beetles, nor were they disturbed otherwise than by passing over. Upon further investigation a considerable portion of the field from different quarters was imprinted all in one direction tending to one outlet, and I concluded that a removal of these creatures for some unknown reason had taken place. That many had moved over the field was certain, but by what agency conformity was obtained, removal, and the time concerted, must remain a mystery. This appearance I never observed, or very partially before. Birds may congregate in a hedge, upon a tree, influenced by a signal call, and depart in a body, one following the example of another. But for animals like rats to collect from out-houses, and rick-yards distantly situated, where communication if obtained would only be by signal or sound, and undetected in a thickly inhabited country, for we see no destruction made by the numerous prowlers of the night, is incomprehensible! That creatures may at seasons experience a feeling, find a necessity of removal requisite, influenced by temperature, want, or other causes, will collect and move off in a body, can be understood. But how at irregular periods, this union of purpose is acquired, and time concerted, when probably no gathering in a body can take place, is not intelligible.

The Spade.

THE TOOLS OF the husbandman, those aids by which he combats the thorns the thistles of the earth, producing his sustenance by the sweat of his brow, were probably the earliest inventions of his mind and among his first requirements, retaining for ages the rude forms of original construction. Of these, there is a very plain instrument in hourly use with us country people, and though the constant companion, the friend of the cultivator of the soil, is as little considered as any we possess. The spade was perhaps the earliest tool ever used by man, and its great simplicity of form and application is a primitive character. Whatever may have been the case with others of the kind in the revolutions of time, the variations of fancy, this probably has undergone little deviation of shape. Perhaps no one instrument with exception of the axe, has wrought greater changes upon the surface of the earth. It was the first provider of food for the few (the family), as mankind increased and greater supply was required. Animal strength was called into action, and the plough contrived, but the spade was never rejected, and probably as an uncombined individual, has provided more provision, given more employment to the labouring man of the earth than any other instrument of the field. Poetry, history describe the fine gardens of Alcinous, of Cyrus, Lucullus, and the great lords of the world, but it was the workings of this tool that so mainly contributed to decorate the rugged wild, its own merits hidden in the beauties it furnished, the shades its actions produced. All the multitudinous operations of the horticulturalist, the florist are accomplished, the abundance of the market supplied, the groves of the nurseryman raised by this instrument. We may prepare the earth for our crops, by means of great strength the overturnings of the plough, but we cannot say our operations are really finished without the aid of the spade. Our trenchings, compostings, ditchings, and fifty minor aidings, beginnings and endings, are accomplished by it. Possibly ere another century is passed, if men yet increase and multiply upon the earth and ask for employment, the spade will return to the hand and be as it was at first, the chief tool of the agriculturalist, the maintainer of the cottager's household, the reliance of the husbandman, and nearly supersede the occupations of the plough. Associations have been formed for patriotic objects, increase of science and literature, good-fellowship, commemorations, but I have often been surprised that this instrument, an emblem of civilization, industry and beneficial toil, a memorial of vast antiquity, and of eminent usefulness, should remain unheeded like a common thing, instead of becoming a representative of meritorious labour, the gathering token of the village as, "Brothers of the Spade."

Instinct.

Daily and hourly examples appear of extraordinary conduct and action by creatures, enabling them to accomplish such purposes as reasoning minds are supposed only capable of contemplating. This we call "instinct", to indicate a distinction between an operation effected by reasoning upon causes, and an efficient action without previous thought. There are instances however when these differences are not perceptible. We may represent the actions of animals and ourselves as proceeding from different sources, say the one is an impulse, the other ratiocination, but call them as we may, from whence do both originate? We see one man endowed with knowledge far superior to his fellows, and in cases of need even those endowments exalted for particular purposes. Why then should we not admit that the almighty creator having denied the animal world general reasonings and the faculties of responsible beings, yet as to man, upon extremity and for particular causes, instigate them to effect purposes equivalent to what reason might dictate? To the one general conceptions, to the other temporary intelligence. The following anecdote was recently related to me by a gentleman whose judgement and veracity no question can arise. Superintending an agricultural operation, he was surprised to observe a lark alight quite near his feet, and supposing it wounded, he stooped to pick it up, but the creature avoided this and rose again, being immediately pursued by a hawk he had not observed. Watching the result, in a few minutes the lark returned to his feet, and now allowed the capture. The bird of prey thus disappointed flew away, the trembling creature being in due time restored to liberty and life. Here was an action contrary to natural habits, and without sufficient motive would have been imprudent, might have been destructive. But death without this risk was certain. The chance of safety was chosen. But natural apprehension still prevailed when danger was for a moment unseen. Peril was renewed, and the terror of man forgotten; in extremity the means of escape imparted. When we are so emphatically taught that not one sparrow can perish without permission of that Being who remains unchangeable through eternity, we must knowing this, perceive that the inferior orders of nature as well as the higher, are in every respect equally the objects of parental mercy and protection, the one influenced by temporary guidance, the other directed by permanent reason.

Danger of the Black-thorn.

BY ACCIDENT IN the spring of 1836, and how many things by casual occurrences are brought to light! A baneful property in a plant which was only before suspected, was manifested. Three men who were employed in felling and fagoting some underwood in a neighbouring copse, composed in part of the common black-thorn (*Prunus spinosa*) wounded their fingers with its spines. But disregarding such common occurrences, continued their labours. However, in a short time these pricked places began to swell and become so painful that medical advice was required. One of these persons, a hearty healthy young man, very carelessly attended to the directions, and persevered in his work, the swelling then increased ran up the arm, seized the shoulder, then the body, and shortly occasioned his death. The others after some time suffering from painful swellings and suppurations gradually recovered. They were positive that all this was in consequence of the black-thorn, having drawn the spines which had penetrated through their gloves from their hands. Most of our fences have portions of this sloe in them and every hedger has frequently pricked his hand, no doubt during the usual operations of repairing them, but no serious injury from such circumstances have ever come to my knowledge before. But this circumstance fully corroborates D^r. Witherings conjectures concerning it. In the following year another of our labourers was deprived of the use of his arm for months in consequence of a similar puncture, though the thorn was immediately pulled out.

Climate change.

SUCCESSIVE YEARS OF mild winters and the almost entire absence of snow, induced many persons to suppose that the climate of England had undergone a change in some respects, but we have every reason to conclude though occasionally a series of mild or severe seasons may occur, that the average of a certain number of years will afford the usual temperatures of any country which has been long settled. Irregular atmospheric influences may afford a temporary mutation, a warm climate may experience a diminution of heat, or a cold one have an accession of warmth, but the general condition of temperament in the progress of years not very materially. However, an impression seems to have been entertained by many persons that the winter of 1835-6 would be one of rigour, but unless it was the old favourite yet lingering, apprehension that the

Comet would unsettle the state of the elements, producing a deviation from the course of natural effects, no assignable appearance of things inducing such a supposition, was visible. The profusion of hedge-fruits has ere now raised such ideas, but in these parts, we never were more scantily supplied, excepting acorns. Consequently, very few of the migrating hedge feeding birds visited us. The mole was in constant activity and casting, manifesting that worms were near the surface and not retired beyond the influence of approaching frosts. As early as Septr. 12th the temperature of the air began to lower and aided by high winds, the thermometer declined from 80° to 62° and early in October we found thin ice upon shallow pools. Upon the 5th of December, we had a clear indication that winter had commenced in the north of Europe by the unusual number of wild geese which passed up the Severn to the flats of the shore near Slimbridge, where they rested, persons in the vicinity considering them an excess of general years, supposing the aggregate to equal 7000 birds. The frost increasing, these animals upon the 19th were hours in passing over us in detachments, for the broader and safer waters of the lower parts of the river. Upon the 4th of the same month, a rare bird was brought to me, shot inland near this place upon some flooded meadows, which proved to be the Lesser Imber of Bewick, the young of Colymbus Imber of Pennant, which is the female of the Colymbus Glacialis, the great northern diver. Few instances are recorded of this creature's leaving its usual abode in the icy seas for our confined waters, and the ungenial temperature of our streams and lakes, where its usual food is hardly to be obtained. We therefore considered it as accidentally driven from its common haunts or course, by some of the violent gales which preceded its appearance. It was in good condition but weighed under five pounds. All January and February were months of winter, intervals of open weather, but accompanied with violent winds, frost, and heavy snows, relenting rapidly, but again returning, rendering the feeling of the air to ordinary constitutions more severe than the thermometer indicated. March, piercing as its winds usually are, was more than commonly winter-like, with a train of snow, hail and tempest. One general circumstance was observable during these months, whenever the wind ceased, heavy rains commenced, when they subsided, the gale arose. All agricultural operations were suspended from the state of the soil, that important labour of cleaning the land unattended to. April was nearly a repetition of the foregoing month, an occasional few fine hours, but keen winds, commonly freezing the shallow waters, on only 15° or 20° above freezing, with flights of snow. About May 9th, a change ensued. Burning weather came on towards the middle and end of the month, and terminated a winter of unusual severity, verifying by chance, the conjecture that had been advanced. During this cheerless intemperate season, many of our summer birds arrived, and we could occasionally see them in some sheltered copse, or hedge row, but they were almost wholly mute, and those alternations, scarcity or abundance, so commonly observable in the migrating tribes, were apparent now. The last spring, we had more nightingales than usually visit us, as their evening song proved, though the cuckoo was very scarce. But this year the voice of the former is only partially heard in particular sites. Now I have even noticed more than one, but the cuckoo has abounded. The causes that occasion these variations, probably arise from accidents in their passages to the summer stations, for travelling at different periods, an adverse gale may be destructive to the one, when another might not be impeded, or even favoured. The cuckoo in the month of May, with us always frequents two or three poor pasture fields abounding with Scabiosa succisa, several in company, feeding upon the ground. At this period the larvae of the six spot Burnet sphinx[49], which uses this scabious for food, is the only creature found in these places applicable as food for a bird, and being in some years plentiful, I have always conjectured that this caterpillar formed the cuckoo's principal diet at this season.

The Water-Rail.

THAT SOBER COLOURED, but beautifully feathered bird, the water-rail, is yet found with us, lingering in its ancient abodes, but in such diminished numbers, though always thinly scattered, that we must shortly expect it wholly to disappear and no more seen with us. As man multiplies and advances, waste places are digged up, the marsh is drained, the water course is cleared, and that which secreted the wilding of the field, supplied them with food or harbour for their young, are removed as injurious to the requirements of the human race. The few which are not destroyed are driven to other retirements. Man and independent nature are ever hostile. Warfare for sport or spoliation unremitting, his power and his arts become irresistible. Flight alone remains, until they are no longer found. This timid bird frequents little sequestered rushy rivulets, the trickling water-course in the ditch, the reedy margin of some pool or stream. Perpetual vigilance and wildness, that chief protection bestowed by nature upon undomesticated creatures, and manifestly increasing with persecution, prevents us seeing them but by stealth, in turning some corner, or by wary advance, when they instantly glide into the herbage which surrounds them; and by aid of their

49 Anthrocera Filipendulae

long legs advance their slender bodies so rapidly and secrete themselves so effectually, that we seldom see them again. It has been said that the breast feathers of this bird were furnished with little bulbs at their points, but this I never observed. However, these feathers are differently constructed from the generality of birds, though several have a tendency to a similar construction, and forms a good example of the nature of these breast coverings. The vanes or side branchings are usually so contrived as to diminish gradually and the end of the feather is rounded, but in the water rail, these branchings from the main stem do not project sideways but rise uniformly with the principal stem, forming with it a level edge, or a truncated termination. The laminae or fine threads which usually connect these vanes, hardly do so, but continue upon them giving a forked appearance to the terminations, the whole having more the resemblance to hair, than downy feather, and this character in some measure predominates in the bird. I know of but two other birds in which the feathers are so exactly constructed (not having others of the species at hand) each nearly connected; the land rail and the water hen. The former, though not so aquatic in its habits has many characters in common with our bird, the same dislike to use its wings, trusting to its longer legs, and glides away with all possible speed from its pursuers. Few person have probably ever seen a land-rail but on the wing. The water hen, or gallinule, though it has more efficient wings and uses them more, yet runs and creeps away among the water herbage as fast as it can. Other water birds as far as observed, do not possess this hairy plumage, nor do they commonly run into concealment as these birds, but take wing when annoyed. Hence, as all the appointments of nature have their objects, it is possible that this smooth hair-like plumage, free from the down of the feathers, may give greater facility of escape, and the mode by which they retreat become a continued habit in these birds from experience of success.[50] These truly are only conjectural ideas upon this peculiar construction, but as a cause for this ordination exists, we are shown what a vast interminable store of hidden wisdom, beneficent contrivances remain for the future races of man to investigate, to see, to marvel at, and adore.

50 The name of 'Runner,' 'Jack Runner' – by which this bird is commonly known, indicates a character of superior swiftness, its long and slender legs are the apparent means, but the unusual looseness of the skin and slightness by which it is attached to the body, affording no kind of restraint to the agile movement of the limbs, is a remarkable instance of conformity of structure to requirement. The green-woodpecker, which walks less than any land bird, and whose destination in life requires nothing of this velocity of action, has on the contrary, its skin so closely connected with its body, as in age to be hardly removed from it.

Feathers of birds.

WHAT A SUBJECT for admiration is the feather of a bird! Common and ever before us as it may be, yet it can scarcely be considered with due attention without exciting astonishment. The mechanical contrivances of the various parts as detailed by Hook, Adams and others, manifest such consummate wisdom and power, as to call forth our awful regards. But when we consider the substance of the feather of nearly the same nature as the horn of the ox, both being apparently formed of coagulated albumen and gelatine, our astonishment is directed to that omnipotence which could construct this delicate and almost impalpable matter from such a material, and then by a refinement incomprehensible, elaborate the down of a swan and the eider! Down seems to differ from feather, chiefly in the fineness and length of the laminae that proceed from the vanes or side branches, from being unconnected by claspers as those of feathers are, and thus waving loose. Feathers seem to have a different object and are an external vesture guarding off rain or moisture. Down, proceeds from the base of the body feathers, or is spread as an independent covering over the body with the evident design of periodically guarding the animal from cold. To promote this object, it has the property of expanding with warmth, covering every part. This increase in the volume of the substance is probably occasioned by distention of the air in the innumerable cells and pores of the separate parts, which the animals appear to have the power of readily accomplishing.[51] And it is this mechanism which overwhelms us with admiration at the omnipotence of that power which could pierce a filament so excessively fine, as to be nearly imperceptible to the human eye, filling it with an elastic fluid to be condensed or rarefied as required to accomplish the comfort of his creatures! Thus gliding upon the mere surface of our gross perceptions, we apprehend something of the incomprehensible attributes of infinity, and learn this lesson of humility, that the poor wandering bird upon northern wave, is alike with him who was created in the image of his maker, an object of parental mercy and wisdom.

---

51 That the animal has the power of distending this down at will, I think we may infer from the habit of many birds swelling out their feathers in severe weather, which though seeming to admit a cold atmosphere, is in fact an expansion of the down upon the body for warmth, which necessarily would cause the feathers to diverge. When the bird is wet, we observe the same puffing out, in order to keep the body warm, and dry the external covering.

The Great Bat.

WE KNOW HERE only the ordinary species of bats, the common, the long eared, and the great one. This last (*V. noctule*) visits us partially, apparently more in progress seeking the haunts of his prey than from choice, frequenting elevations rather superior to his ordinary habits. Though his flight is always high in the air, he will descend at times from his altitudes, but it seems only in pursuit of his food, immediately regaining his station. The swift and other hirundines feed at times in the higher regions of the air, but not exclusively. But this bat always pursues his occupations high aloft. It would be curious to be acquainted with the peculiar race of insects which he diets upon, that take their pastime in the altitudes. The winged ants, bees and other insects will periodically mount on high, but again return to the earth. Yet the creatures he feeds upon must permanently or very usually reside there, as he alone haunts in those places. It is plain from the size of his body and condition, that there is no lack, and from what motive is it that this species seeks for these high flyers when the others of his race are content with humbler dwellers? I once killed this bat in its flight of full size, its wings when extended measuring fourteen inches. Very intimately acquainted as I have been with this creature in former days, I have always associated the idea of fine summery weather and the enjoyments of that brief period of our year, viewing him in my mind pursuing his calling with animation and speed in the fine warm saffron rays of twilight, and uttering his shrill squeak of apparent delight, when from altitude and obscurity the animal was barely visible.

The Abbey Camp.

THIS SPRING (1837) we enclosed and divided a large piece of land called the Abbey, in which considerable earth works of some encampment remained tolerably perfect. From its form it had been originally British, but there were many reasons for conjecturing that the Romans long made use of it, and from its magnitude and situation was probably an important post. Often as I have trodden in various pursuits these turfy elevations, reflections and images have arisen, as by some influence they so commonly do when we rest upon the tokens of a past generation, possibly the awakening of historical recollections. But now the operations of the various scattered parties formed a busy and most interesting scene. Their levellings, trenchings, and overturnings, became a subject of unavoidable contemplation, presenting through the long perspective of ages the Roman soldier heaping the lofty agger, the labours of this mighty race forming the rampart of war, the helmed and cuirassed chief directing or approving the array of the warrior, the pavilion, the spear. All has now passed away, history alone telling that they have been, the sword giving place to the mattock, and the spade, the jest and merriment of a peaceful peasantry, where once was heard the legionary tumult. What dynasties and nations have passed away since these bulwarks were raised! But to everything there is a season and a time for every purpose under heaven. The very food now planting in the soil manifests the ever changing course of time and manners: the potato, a root unknown to these antient lords of the earth and from a region where the war cry of the Latin had never been heard, the name of Caesar unknown. The small depth of earth removed for these first crops afforded little expectation of any remains of the ancient possessors being found, but yet several coins fell into my hands, copper and brass of small size, in general greatly decayed, but some less so, but all of little interest or value. Of the few that could be identified as F. Jovian, V. Constantinus, Valentinian?, ranged about 300 to 400 years after the Christian era, thus proving that the encampment had been used by the Romans, and that apparently until they finally quitted the island in 426. The almost invariable circumstance of finding pecuniary mementos of this wonderful people, whenever their defensive works have been removed, may not appear so extraordinary when we consider the trifling intrinsic value of much of this scattered coin. Not less than 40 workmen are now hacking and delving the site, not one of whom will leave a vestige of value behind them. Our smallest coin, the farthing is little in circulation with the agricultural peasantry. Our other copper money is large and if dropped easily recovered, but the legionary workmen had in circulation their uncia, semi-uncia, their sextula, the fourth, eighth and twenty-fourth part of our smallest coin and when lost would be hardly missed or regarded. We found likewise many fragments of the coarse and fine pottery of the period, the broken stones of the little military mills, the quern of the northern nations, alluded to in Saint Luke's gospel, where two women were grinding together. But our movements were only superficial, and unproductive of any important discovery. The whole of this turfy ground was turned over by the spade upon an average of four inches in depth, and it was curious to note how slowly even this shallow surface had accumulated. I had twenty or more coins brought me, and yet from the natural decay of the vegetable matter for fourteen centuries, the pressure of heavy cattle, and all the accession of substance from the pasturage of creatures, from immemorial time, yet these coins were covered with a soil barely four inches in depth! The metal from which some of these ancient coins were fabricated, must have been very free from alloy, or it is difficult to account for the remarkable fine preservation of some of them. The copper and brass have commonly perished more or less, but the silver ones preserve the original impressions, with considerable perfection. I have a Valens, from this abbey ground, of silver as perfect as if immediately from the mint, every feature and letter clear and defined. Unless we can suppose that by accident this piece was concealed from the atmospheric influence beneath some flat stone, it can probably be alone the purity of the metal that enabled it to retain this perfect impression after the destructive action of elementary corrosion for possibly 14,60 years.

Changes in plumage.

THE YELLOW-HAMMER (*EMBERIZA citronella*) one of our best known, and most generally diffused birds, affords us good example of that habit with which some birds are endowed, changing their plumage towards the winter season, which we conclude to be in some instances a provision of nature to conceal them from observance of their enemies. But such an object can hardly be the only purport of this provision, or it would be universally afforded to all those which are obnoxious to the same contingencies, rather than be a partial appointment, for many birds preserve the same coloured plumage through the year, only varying in freshness after the moult. The chaffinch, the linnet, and few others lose the brilliancy of their tints as the cold weather advances. The latter indeed is deprived of its gay colours by moulting entirely for a time. The yellow-hammer preserves but little of the finery of its summer dress, for out of a flock of some scores of these birds hanging about the barley mow, or hopping among the chaff in every farm yard in January, the sexes though so strongly marked in summer, can hardly be distinguished now. In March he begins to put on his nuptial suit, and by the end of the month is generally arrayed in that fine golden dress which distinguishes him from every British bird. But his gentle mate remains in the same sober unornamented vesture she has worn through the season. They quit the barn-door and disperse through the hedge rows. We can hardly consider natural causes alone to accomplish this change, for if so, it would be operative in the male sex of every species. The combe, the wattles, the caruncles of several of our birds manifest a change in spring, yet the feathers in most instances exhibit no brilliancy superior to the ordinary character. To view this change as administering to sexual attraction or preference, is perfectly inconsistent with the order of things observed through all the kingdom of irrational nature. Possibly no instance of exalted estimation arising from such a motive is manifest, and was it so partially, we should be much perplexed for such influences in one case and indifference to it in another. It seems rather a fact to be observed than reasoned upon or accounted for. The same season too brings forward the voice of gratulation, which ceases before or with the moult, and may in some cases be connected with, or influenced by similarity of agency with the alteration of plumage. Both however exist, but the object seems not known to us.

Larvae of the lepidoptera.

The extraordinary appearance of the larvae of some of the lepidoptera (puss moth, lobster moth, pale and dark tussock &c), must always excite attention, but any conclusion that such will produce as remarkably constructed flys, is as little to be depended upon, as that the colours upon the eggs of birds will indicate the hues of the creature to be hatched from them. We seem to be equally ignorant of the object of nature in the colouring of birds' eggs, as we are of the unusual formation of the caterpillars of moths. I think I may add to those remarkable larvae well known, another perhaps less so, and the fly produced from it, a plain, and sober creature.

Acronycta alni

A year or two past, the larvae of some lepidopterous creature was found upon a post near Bristol, very unlike any we were previously acquainted with, but as the fly was not perfected, we could ascertain nothing. However upon the 13th of August 1836, a similar creature was brought me, and a drawing of it immediately made. It fed upon the leaf of a plum for a few days and when the fly appeared, the species I apprehend to be *acronycta alni*. The larva must be of very uncommon occurrence, or otherwise would have been noted in some of the authors I have referred to. Larva, with alternate bands of yellow and dark brown, free from hairs. But from several of the lower yellow bands proceeded two spatula formed, fine, smooth, horny substances. From the upper band beneath the head, were two dark filaments, and two of the spatula appendages. From a few of the yellow bands these appendages were wanting, as if injured, or broken away by accident. Beneath the body, dark.

Vulnerability of the robin and wren.

Two of our domestic birds, so familiar about us that we hourly observe them, though prolific in their natures, do not annually increase their numbers: the robin and the kitty wren. The first lays six or eight eggs, breeding more than once in the season. The latter provides not fewer eggs, generally bringing off two broods in the year. Yet in the spring we see them only sparingly scattered about our hedges. The fearless and the sociable nature of the robin, occasions often premature death, and even the food it obtains about our dwellings is occasionally pernicious, especially its crumbs of bread. But it is in its roost that the titty wren chiefly suffers, assembling together as the nights get cold, we see them at the close of the evening popping in to some little hole in the hay-mow or under the thatch of the rick, successively following each other. Apparently, the whole family cuddling at the extremity for mutual warmth, is here detected by vermin, whose keen sense of smelling leads them to their prey. They all perish, and the domestic cat in her progress may be observed to stop, scenting up at this asylum from the ground and meditating means to obtain access to it.

High winds provide welcome food for birds.

High winds when trees are in full foliage, provide a plentiful treat for such birds as capture their insect prey on the wing, the swallow tribes and the grey fly catcher most particularly. By disturbing the night roaming creatures which have sought an asylum beneath their leaves, these are now shaken off by the buffeting of the boughs, drowsy and obscurely seeing in the day-light, they become a ready resource to their enemies. The birds, coursing up and down in the shelter of the hedge or avenue, manifest by their evolutions and the constant snapping of their bills what success they have. This is the more welcome now, as by reason of the turbulence of the air, those insects which sport in day-light and constitute their usual food, are unable to combat with the wind and remain in concealment and rest.

The destruction of birds' eggs.

Every hedge-hunting boy, and those who have been familiar with birds, the little interesting flutterers about them, observing and preserving their nests in their gardens and shrubberies, must have been aware how often these rudiments of future families are rendered abortive, by their eggs being broken, pecked in holes whilst in the nest, and the contents extracted. In the poultry-yard this sometimes is so frequently observed with individuals, as to preclude them ever producing young ones. Our good house-wives say the bird is impatient, that the calcareous matter of the shell is required to correct crudities in the system or indurate the osseous parts. This is an idea probably inclining rather to fancy than to fact, but it is manifest that birds will occasionally for some reason destroy their own eggs. This circumstance was so frequent in one year with us, connected with the nests of those which usually resort to our premises, that neither thrush, blackbird, finch, hedge-sparrow, or robin, produced their general numbers. A single young one was here or there observed, but the incubation of the year was almost wanting. From the security and concealment of many, we considered it not the action of any hostile bill, but the work of the owners themselves. Noted but not accounted for, the injury has been attributed to probably an harmless creature, and the cuckoo been called a "suck-egg", which no observation has justified. If the birds are the destroyers of their own produce, the violation of an imperative tendency in nature must be occasioned by some remarkable inducement, not defined by us. Seasons vary so little at this period of the year as to present no apparent motive for the action, and I have seen birds, thrushes in particular, with eggs in their bills, and scared them to ascertain the truth. Desertion from a nest from insecurity may be comprehended, the want of food at the time may be implied but hardly conclusive of such a departure from the objects of nature. Summer came, and it did not require a very observant eye to be sensible of the result of this action of the parent birds. The numerous little fruit eaters so assiduous in our gardens in common cases, came not. The flocks of finches and linnets that assemble upon our grass-swaths, and twitter to the ring of the whetting scythe, could not now be seen, and a sensible feature of innocent enjoyment was missed.

The capture of woodpeckers.

REDUCED IN NUMBERS as the green woodpecker (*P. viridis*) is with us, like all his congeners, by the advance, and spread of man, yet we see him or hear him commonly. But he is a very wary bird and only occasionally obtained by means of the gun. Having had them however, repeatedly brought to me alive, about the month of May, I was very inquisitive to know by what means they were obtained, not apprehending that any trap could be successfully used, and as very similar accounts were given me by all, we probably have the truth. It seems that boys watch the bird as she fixes on a tree and wait beneath it until she commences boring her hole. When they see that her head is concealed in the cavity, they ascend by the aid of the branches on the opposite side, the creature being so busied with her employ as to disregard the approaching danger, until the hand is suddenly clapped upon her back. We have very few instances of such an undivided attention to the objects of their duty in birds, as to permit capture by hand, though all the offices of incubation, occasion so often the death of these poor creatures. One would almost hope that our woodpeckers who have regained their liberty, might acquire more vigilant habits in the future.

Large gatherings of creatures.

WE SEE ALL the orders of sensitive nature pursue systematic congregation, the bird, the beast, the reptile and the fish, with man as a responsible being. These associations, if for change of condition, political purposes, innovation, or alteration, are accomplished by hopes, by fears, by representations artful or just; and are so far comprehensible. But the object of these unions in many cases, where there is no mind to direct, or apparent power to prescribe, become inexplicable. Search for food, or change of temperature, may influence these assemblages in many cases assuredly, but not the cause universally of stimulating these associations. These great unions, are eminently serviceable to man, rendering often his food of easy attainment, but we cannot indulge for a moment that full the purposes of this being, this favoured being, actions so adverse to the habits of many creatures, and in the smaller objects utterly useless, should be solely ordained. Man's wisdom may incite him to avail himself of these occasions, but the vast combinations we may be assured, are devised by infinite Providence for purpose is not obvious to us. The writer, during the earlier period of the French revolution, when famine afflicted Europe, and England most severely, was up on the coast of York, when the remarkable and unusual supply of fish congregated along the shore, saving the lives of famished thousands. This was so manifestly the bounty of him who pityeth us in distress, as to excite the gratitude of many, whose thoughts in general were directed to none of these things.

Cuckoo's eggs.

THE DIRECTING POWER that influences the operations of creatures and the unwearied perseverance, which is manifested by them to effect their objects, are daily brought to our observation. Yet I may possibly without offence relate a very curious instance of vigilance and success, which was this year (1830) observed in the garden of a friend in this county. Some beds of sea kale, having been cleared of the produce, the earthen pots which had been used in bleaching the foliage were placed on one side merely with a view to removal. One of these happening to be raised up on another, a robin descended into the hole of the upper one and formed her nest upon the basement of that beneath it. A cuckoo, upon the watch observed this, and contrived to deposit her egg, in the robin's nest, and in this singular situation was the young cuckoo hatched and tended to maturity by the poor robin. Cuckoos generally select the nests of such birds as build open nests in the hedge, or on the ground, and we had this year a knowledge of eight examples. But in this instance the hole in the pot was too small to permit the cuckoos descending into it easily and we cannot be readily satisfied that the egg was dropped into the robin's nest, as from the remarkable thinness of the shell of that egg, it is probable it would have broken. Naturalists appear to be by no means certain that the cuckoo deposits her eggs in the usual manner of birds.

Fossil remains of plants and animals.

WE CONSTANTLY OBSERVE the remains of animals that once existed upon the earth, which having accomplished the objects of their creation, or become no longer expedient, were not probably preserved in the ark at the general deluge. Fossil fragments too of a vegetation in magnitude proportionate to the mighty creatures which were sustained by them, have been brought to light. These not according with any existing species, render it manifest that some of this race are not in being at this day; but the vegetable would by the endurance of their seeds in most instances, appear exempted from destruction.

The Gansel Bergamot pear.

AMIDST THE MANY sketches of rural things and occurrences which I have noticed, perhaps the horticulturalist may allow me to impart a hint, as a means of increasing the fertility of an useful and excellent fruit. The idea does not originate with me, but as the experiment has fully answered, perhaps another may benefit by it. Gansels Bergamot pear, is an esteemed fruit, but classed amidst the "shy bearers" in whatever aspect it may be placed. Perhaps a northern one suits it best, but then the fruit is greatly diminished in size. It blooms well, but this promise is rarely succeeded by a proportionate produce of pears, seeming to intimate some deficiency in the fertilizing pollen of the blossom. In accordance with this supposition, a branch was cut away from the middle of the tree, and a Colmar pear, which usually is a productive sort, and blooms at the same period as the Gansel, was grafted in its place. This scion produced two or three tufts of blossom the second year, and nearly the whole side of the tree which could be influenced by the Colmar pollen was furnished with fruit, the other parts deficient as usual. Pears, as much as any fruit, are influenced by the agency of the seasons, but whenever a favourable period has arrived, that part of the original tree which can receive any of the pollen from the graft, has never failed to produce a crop. At times it becomes so covered as to require thinning out. A single graft is sufficient for a large tree, if a greater stimulus is given, the fine size of the pear is diminished by the quantity produced.

Index

AGARICS, 31, 69-72, 75-7, 239, 245, 246, 363, 268, 274 — the pale grey species of, 70 — the verdigris, 71— not easily investigated, ibid. — the odorous agaric, 74 — the scented, ibid. — the " stainer," 75 — the surrectus springing from another species, 246.

Agriculture, practice of, at a village in Gloucestershire, 11 — bad custom of the farmers there, 26.

Alder moth, the, 320, 321.

Alms-house at Bristol endowed by Alderman Stevens, 108.

Amusements, heretofore holyday ones, wearing out, 234.

Animals, increase and decrease of, 85 — strength of 83 — longevity of, 107 — what dependent on man, 134 — what independent of man, 135 — usefulness to man, 134 — affection of, to their young, 155 — vesture of, 175, 176, 177, — mercy to, a scriptural command, 156 — instinct of, 311.

Ant, the black, 195 — the red, 196 — the yellow, 196 — common hill, possibly heliotropes, 197.

Aphides, 62, 221, 222, 225, 231, 249, 296.

Apples, 295 — Crofton pippin, 221, Russet, 245 — injured by aphides, 221, 295 — spottings on, how occasioned, 245.

Apples, crab, 64, 79, 94, 114, 224, 232, 296, 309.

Artichoke, Jerusalem, 20.

Ash trees, 33, 187, 209, 232, 247, 248, 254, 291, 304.

Aspen, 42.

Atmospheric influences on vegetation, 237 — observations, 248, 255, 265.

Auger-worms, 187.

Aust-ferry, the " Trajectus" of the Romans, 1.

Autumn, pleasure of a morning's walk in, 68 — sensations in, 303.

Backwell, Somerset, crossbills at, (note) 109.

Badger, 64, 224.

Barking of the oak tree, 31.

Bat, the great, 317.

Bee, the carpenter, 37, 211, 212 — its nest, 211.

Beech, 29, 33, 69, 72, 74, 304.

Beetle, the rose, 37 — the dorr, 191, 192, 201, 202 — the great water species, 206, 218.

Birds, partiality of the author to, 91, 102 — migration of, 94, 126 — injurious to plants, 99 — various foods of, 93, 161 — song and voices of, 163, 164, 167— nests of, 103, 104 — great destroyers of insects, 109 — species of, diminishing in number, 119 — labour of, to feed their young, 92, 121 — friendship of, 129 — roosts of, 132 — eggs of, 137, 138, 139, 140, 155, 322, 323 — dislike of man, 142, 143 — stratagem of one, 145 — early rising of, 145 — boldness of, 155 — solitary and congregating ones, 156 — language of, unvarying, 168 — pleasure afforded by, to man, 170 — feathers of, 316 — changes in plumage of, 319, 320 — winds provide food for, 321, 322 — For the different kinds, see their respective English names.

Blackbird, 103, 112, 130, 152, 165, 322 — song of the, 98, 146, 169.

Blackcap, the, 104, 128, 142, 144.

Blight, American, 221 — some trees not affected by, ibid. — manner in which this insect propagates itself, ibid. — whence derived, uncertain, 223 — remedy for, ibid.

Blighting, saline, 244.

Blind-worm, 79, 194.

Blossoms of plants, their utility not obvious, 37.

Blue Argus, the, 297.

Bombylius, the, 194, 195.

Bones of horses and human beings dug up, conjecture respecting, 2.

Book-worm, the, 47 — its delight in sweets, ibid.

Bouquets, no longer popular, 40.

Bramble, the common, almost an evergreen, 63, 64, 68, 301 — lines on its leaves, ibid. — its uses, 64.

Briony, black 11.

Buff-tip moth, the, 33, 174.

Bull-finch, the, 96, 97, 98, 112, 152 — destructive to gardens, ibid., 97 — its nest, 104, death of a, 125.

Bunting, 95, 152, — cirl, 95 — peculiar practice of the, 153, 154.

Burnet, conjecture as to its lasting verdure, 48, 314.

Butcher-bird, 115, its use of a shamble-hook, 116.

Buttercup, 39, 59.

Butterfly, the peacock, 49 — the sulphur, 59 — the blue argus, 176 - 178 — common white, 62, 100, 181 — the phlæas, 177, 178 — the azure, ibid. — the painted lady, 179 — the marble, 180 — the meadow brown, 181 — van. Atalanta, 182 — attracted to favourite flowers, 183 — admirable or admiral, 49, 181 — Melitæa artemis, 184.

Buzzard, 78, 141, 193 — honey, the, 292.

Caloric, effects of, on bodies, 256.

Catgut, made locally for musical instruments, 243, 244.

Cerealia, 21.

Chaffinch, the, 92, 93, 98, 104, 165, 167, 319 — fondness for pea sprouts, 99.

Changes in Nature, 203.

Cheese, cheap kind of, 11 — for toasting, 11— 'heaving of the curd', 11, 12.

Cherlock, 14, 59, 152.

Cherry, 41, 96-97, 114, 228, 231, 249, 296.

Christmassing, 235.

Cinquefoil, 69.

Cherry, 41, 96-7, 114, 228, 231, 249, 296.

Chrysalides of insects, 63, 100, 171, 182, 184, 205, 297 — a singular one, 172.

Cleanliness of animals, 203.

Clematis, the wild, 67 — sticks of, used by boys for smoking, ibid.

Climate change, 312, 313.

Clocks, name given to the great dorr beetle, 318.

Clover, 13, 14, 181, 184, 230, 249, 292, 299.

Cock, crowing of the, 272.

Cockchafer, 110, 201, 204, 205.

Coins, 100 — dug out of the earth, 2, 302, 318, 319 — found by chance, 237.

Colemouse, 101.

Colour, probably reflected light, 44 — effects of the rays of the sun upon, 45.

Coral polypi, 5.

Corn flower, 14, 19.

Cornel, 251.

Corn rose, 251.

Cowslip, 39, 59, 268, 272.

Crocus, 59.

Crossbill, the, 108, (note) 109, 152.

Crow, 78, 110, 113, 151, 155, 202, 204, 295 — Royston, the, 120.

Crowfeet, 227.

Cruelty, 193, a vice of the ignorant, 79, 80 — in a herb, 46.

Cuckoo, 146, 152, 163, 167, 169, 314, 322 — eggs of, 324.

Daffodil, 55.

Dandelion, the, 59, 60, 61, 110, 227.

Daisy, or day's eye, the, 55, 60, 268 — ox-eye, 13.

Death's-head moth, 174, 205.

Dewberry, 69.

Dew-worm, the, 79, 90, 109.

Dibble, the, 16.

Digestion, power of, in birds, 128.

Dock, wayside, 64.

Dog, 79, 81, 90, 164, 204, — usefulness to man, 134, 135 — excrement of, 220.

Dogsbane cruelly destructive to insects, 46 — the flower of, ibid.

Dog-rose, 97, 251.

Dogstail, 13, 227.

Drought, effects of, 299, 300.

Dry-rot, 72.

Dyers, 31, 43, capricious in their art, 44.

Dyers' broom, 43— gathering of, ibid. — uses of, 44 — dyers' weed, 58.

Elm tree, the wych, a singularly beautiful one, 31 — value of, 32 — uses of, ibid. — soon decays, 33 — leaves of the elm marked with plague spots, 73.

Empiricism, 219.

Entomology, a difficult study, 170.

Ermine moth, 232.

Evaporation from the earth, effect of, 49.

Fairfax, General, supposed skeletons of some of his foragers, 2.

Fairy rings, 239, 240.

Fern, 14, 31, 62, 69, 226, 303.

Fescue, spines of the hard, bearing no flowers, 63.

Fieldfare, the, 120, 126, 152, 155, 161, 162.

Fire-tails, 123, 138, 293.

Fish, 76, 81, 93, 95, 176, 177, 203, 217, 323 — shell-fish, 200 — miraculous appearance of, 324.

Flax and silk, difference in the fibre of, 44.

Flowers of plants, 37 — what most abundant, ibid —pleasures afforded by, ibid. — use and application of, 38, 39, 40 — natural love of, 38 — the natural playthings of children, 39.

Fly, 46, 172, 196 — great black, 49 — the house, 134 190, 198 — the biting, ibid. — the four-spotted dragon, 172 — ichneumon, 212.

Flycatcher, the grey, 142, 150, — nest of, 104 — shading her young, 127, 128.

Foliage of trees and plants, its great utility, 35, the destruction of by insects, 232.

Fossil remains, 324, 325.

Foxglove, 52, 58.

Friendship between birds, instance of, 129.

Frost, early, effect of, on flowering plants, 267.

Fruit, preservation of, 298, 299.

Fungi, beauties of, 70 — varieties of, 71 — uncertain appearance of, ibid. — mutations of, ibid. — agents of decay, 72 — propagation of, 76.

Fur of animals, 89.

Gallinaceous birds, 122, 139, 152.

Gamma moth, the, 183.

Gatherings of creatures, 323, 324.

Ghost moth, 172.

Gipseys, 81.

Glauber, obtained gold from common sand, 10.

Glaucous birthwort, 46.

Gleaning, profits of, to the poor, 233— antiquity of the custom, 234.

Globeflower, 59.

Glowworm, the, 190 — curious contrivance about its eyes, ibid.— its light, ibid.

Gnat, the winter, 170, 198.

Goat moth, 186, 255.

Goldfinch, the, 104, 115, 147, 148, 152, 165.

Gooseberry, 52, 96, 104, 130, 144, 145, 250.

Grass crops, nature of, in the Author's village, 12, 13, 14 227, 228 — vernal, 11 — agrostis, 13, 228 — ray grass, 13 — holcus lanatus, 13 — cocksfoot, 13, 227 — bent, 13 — foxtail, 13, 14 — meadow-fescue, 13 — meadow poa, 13, 227 — phleum, 13 — quaking, 13 — matt, 14 — carnation, 14 — festuca sylvatica, 14 — festuca uniglumis, 14 — coarse, 85 — turfy hair, 110 —certain grasses attached to certain soils, 13 — grass balls, 62, — affected by the weather, 56, 249 — essential oils in, 265, 266.

Grasshopper, 95.

Green colour for illustrating foliage, how to obtain, 44.

Green finch, 104, 112, 152, 154.

Guinea, anecdote of finding one, 237.

Hair of animals, 89.

Hairworm, the clay, 208.

Hawk, the sparrow, 125 — the kestrel, ibid. — hawk owl, 308, — outwitted by a lark, 311.

Hawkmoth, 178.

Haws and Heps, 161.

Hay, crops of, 12, 228, 230, 249, 266, — method of saving, in bad seasons, 14 — sneezing occasioned by, 230.

Hay-chat, 102.

Hazel tree, 64 — liable to decay, 71, 73.

Hedgehog, the, 79, 80, 81 — sold alive to bakehouses, 81.

Hellebore, the fetid, 36— its medicinal uses, ibid. — fostered in cottage gardens, ibid.

Helvella, the mitred, 71.

Heron, 87.

Holly trees, 235.

Honey dew, 231, 232, 249, 295.

Hornet, the, 182, 194, 213.

Horse, afflictions of, 80, — instance of the longevity of one, 108, — overcome by heat, 250 — skeletons of, 2.

Horse-chestnut, introduction of, 41.

Horsetail, 85.

Horticulture, the first occupation of man, 37 — greatest amusement, 40, 41 — effects of cultivation, 61.

Human life, length of, 108.

Hummingbird hawkmoth, the, 178.

Hummings in the air, 238, 239.

Hydnum fungus, the beautiful floriform, 71.

Ice, cause of its swimming instead of sinking, 256, 257.

Industry, profitable fruits of, to an agricultural labourer, 9.

Insects, inhabiting the deep, 4 — feeding on plants, 36, 37 — entrapped by the snapdragon, 45 — destroyed by the sundew, 46 — destroyed by the dogsbane with great suffering, ibid. — paths of, through leaves, 63 — their manner of puncturing, 66 — agents of decomposition, 72, 73 — consumption by hedgehogs, 81 — consumption by birds, 93, 100, 102, — destruction of grass by, 111 — profusion of, near water, 124 — but little attended to or studied, 171 — mechanical function of, 172 — chrysalis of, ibid.— profusion of, 198 — possible purpose of, 198, 199 — speedy methods of killing for display, 252 — best mode of preserving specimens, 253.

Insensibility to pain, striking instance of, 7.

Instinct, 311.

Ips niger, 255.

Iris, fetid, 37.

Ivy, 49, 51, 52 — adornment of ruins, 50 — unusual profusion of, 291.

Jack snipe, 156.

Jack Runner, alternative name for the water-rail, (note). 315.

Jay, the, 69, 114, 115, 152, 228.

Jerusalem artichoke, 20.

Kings Weston Park, 132.

Kirby, Rev. Mr, 63.

Kite, its numbers greatly on the decline, 141 — extraordinary capture of a number, ibid.

Knowledge, slowly attained, 170.

Labour of the peasantry in the author's village, profits of, 8.

Lady-bird, 181 — note of a song-thrush, 168, 169.

Land-rail, the, 292.

Language of birds, 163.

Larch, famous one brought to Rome in the reign of Tiberius, 30 — deciduous in extreme heat, 250, 251.

Lark, 115, 152, 260 — tit-lark, 166 — pure white, 158 — woodlark, 165, 166, 259, meadow-lark, 259.

Laurel tree, the, 72, — visited by multitudes of bees, 231 — spurge laurel believed prejudicial to marriage, 236.

Leasing, 233, 234.

Leveret, 78, 79.

Lichen, 31, 62, 69, 99, 101, 104, 149, 153, 226, 243, 260, 274 — farinaceous, 102 — fascicularis, 243 — prunastri, 102.

Life, duration of, 107, — vital principle of seeds retained, 55.

Lily, blossoms of, supposedly indicative of the price of wheat, 154.

Lime, nature and uses of, 4 — its abundance, 5 — formation and origin of, ibid. — analysis of, 6 — residences upon lime supposed to be healthy, 8.

Limekiln, frightful consequences of a traveller's sleeping on one, 7, — rejected rubbish of, 58.

Lincoln, Great Swan Pool at, 119.

Linnet, the, 96, 156, — remarkable milk-white example, 159, — voice of, 165 — washing in village well, 303 — change of plumage colour in cold weather, 319 — eggs destroyed 323.

Loggerheads, 13.

Longevity, 107.

Lousewort, 58.

Magpie, the, 113 — its behaviour, ibid. — a cause of terror, 122, — its eggs, 138, — preys on new hatches of poultry, (note) 151.

Manure, picking from grass lands, a bad practice, 26.

Maple tree, 65 — the under sides of the leaves of, a beautiful microscopic object, ibid.

Marten cat, the, 78.

Marten or martin, house, failure of nests, (note) 105, — example of industry and affection, 121, found near rivers, 124, — late to arrive and eager to depart, 163.

May, 214, 232.

May-poles, now seldom seen, 234.

Migration of birds, 94, 95, 96, 112, 124, 126, 162.

Migration of rats, 309.

Mineuses, 63.

Mistletoe, 57.

Moles, wants, mouldwarpers or mouldturners, 87 — sense of smelling, 88 — concealed ears, (note) ibid., — fur, 89, — rankness of their flesh, 90, 293, — diet of worms, 217 — white coloured, 159, constant activity, 313.

Montague, Mr, 95.

Morell, the stinking, 75 — the esculent, 77.

Moth, probable cause why woollen vestments are a prey to, (note) 48 — Phalæna potamogeta, 86 — the ghost, 172, 174 — the hawk, 178 — the yellow under-wing, 174, 181 — found frozen in ice, 197 — the gamma, 183 — the tiger, 185— the goat, 186, 187, 188, 255— small oak, 254 — the death's head, 174, 205, 206 — the ermine 232 — Dartford emerald, 253 — Green housewife, 253.

Mother Cary's chickens, 117, 118.

Mouse, 77 — hamster, xvi, 90. harvest, 82 — short-tailed meadow, 85 — the common, 133 — the meadow, and long-tailed, ibid.

Mucor or mould, species of, 66.

Mustard, wild, 14, 54, 184.

Natural affection, 113, 115, 121, 129, 156, 186.

Natural history, a delightful study, 27, 28 — contemplation of, 28, 29.

Naturalist, pleasing occupations of the, 68, — The Progress of a, 271.

Nature, designs of, 188 — ordained for wise purposes, 166 — changes in, 203 — delight in, 226, 227 — never at rest 260, 261, 262.

Nests of different birds, 50, 78, 91, 96, 103.

Nest of the carpenter-bee, 37.

Newt, the common, 199, 200 — a small shellfish often attached to its toes, 200 — spawn of, 200.

Nidularia, the bell-shaped, 77.

Nightingale, the, 62, 128, 153, 293, 314 — less common than heretofore, 119— its song, 168, croaking of, 169.

Nightjar, 202.

Nosegays, 40.

Nuthatch, 69.

Oak of Mamre, under which Abraham sat, 31.

Oak tree, 274, — description of one, 29 — several of extraordinary magnitude, 30 — suffers from the "shake", 34 — the oak less fruitful than formerly, 30— its value, from its various uses, 31 — copied for ornamentation, 40 — insect wounds to, 66, — foliage devoured by cockchafers, 204, — longevity of, 304 — growths under the leaves of, 305.

Oat grass, 56.

Onion, liquor of, abounds with sugar, 47.

Orchises, 39.

Owls, 151, 164, 202, 274 —brown, 90, 166 — fern, 173 — plumage of, 306, 307, 308.

Pain, instance of insensibility to, 7.

Passerine birds, 152.

Peacock butterfly, 49, (note) 90, — survive the winter, 182, — harbingers of spring, 267.

Pear, Gansel Bergamot, the, 325.

Peewit, the, 158.

Pennant, Mr. 86, 149, 157.

Phallus, the stinking, 75, 76.

Pick-a-bud, name given to the bull-finch, 97.

Pigeon, Rock, 111 — Wood, 50, 52, 104, 291.

Pilewort, 59.

Pimpernel, the, a prognosticator of fine weather, 154.

Pismire, black, 45.

Plantain, 110, 140.

Plants, blossoms of, 38, 39 — sugar in, 47 — names given to them anciently, from their supposed qualities, 53 — pores of, 67, 231, 232 — decomposition of, 72 — dispersion of the seeds of, 76, 84 — emission of scents from, 126, 183, 230, 267. — chemical composition of, 241, 242.

Plants, wild, 14, 36, — foliage of, 60— flowers of, 60, 272.

Plants, tendency of, to revert to their original creation, 22 — preference given by, to particular soils, 36.

Plum, Bullace, 55.

Polecat, 193.

Pollarding trees, 49, 187, — 'an evil under the sun' 254.

Polypi of the coral, 5.

Pony, a favourite, 108 — death of, (note) 108.

Poor, employment of, in the Author's

village, 8.

Poplar tree, 42, 304.

Poppy, 19.

Population, 15, 62, 77, 228, 291 — excess of, 16, 119.

Potato, culture of the, 15 — different sorts, 16 — profits, 17 — effects of, on soils, 18 — history of this extraordinary root, 19 — value of, as an article of food, 21.

Primrose, 39, 55, 59, 125, 268.

Prognostications of wind and weather, 161, 164.

Providence, inattention to the works of, 220.

Puff, the gray, 71 — the turreted, 76— the starry, 76.

Rabbit, 78, 79.

Rapacious birds, 151, 155.

Rats, migration of, 85, 309 — other particulars of, 134.

Raven, the, 106.

Red archangel, 99.

Redshanks, 58.

Redstart, 122, 128.

Redwing, 126, 155, 161, 166 (note), 306.

Reeking or roking of the earth, 256.

Revelation, truth of, 263, 264.

Robin, the, 98, 116, 118, 146, 159, 163, 256, 321, 322, 324.

Romano British encampment, 1, 2, 58, 317, 318, 319 — vegetation beneath a Roman path, 301, 302.

Roman well, 302, 303.

Rook, the, 91, 110, 112, 132, 145, 157, 158, 250, 294, 299 — its affection, 109 — appears to be decreasing in numbers, 119 — its nests and habits, 159.

Rose, 20, 42, 62, 65, 211, 296 — damask, 37, 55 — white moss, (note) 223 — wild, 66, 161.

Rose beetle, 37.

Royal forest, indications of one in Gloucestershire, 3.

Rural sounds, 165.

Sallow, 50.

Sand piper, green legged, 157.

Scabious, common field, 13, 184.

Seasons, variableness of, 125 —effect of, 227.

Seeds, preserved in ancient lime kiln, 55 — propagation of, 84 — dissemination of, 224.

Sex, increase of, in 1825, (note) 86.

Shellard's lane oak, 29.

Shrew, the water, 86— the common, 87 — new species of, ibid. — white, 159.

Shrike, or Butcher-bird, 115, 116.

Shuffle-wing 132.

Silk and flax, difference in the fibre of, 44.

Silkworm, 139.

Sinking of the earth, 242.

Skylark, the, 115, 165, 166.

Slowworm, 194.

Slug, 47, 76, 103, 113, 152, 158, 203, 215-216, 218, 291.

Smelling, question of the sense of, 78, 81, 88, 111, 216, 321.

Smoke wood, sticks of the wild clematis so called, 67.

Snail, the common, 214, 216, — high mortality rate of, 214, 215 — the banded, 216 — supposed super-natural arrival of, 217 — hortensis, 214 — nemoralis, 214 — halotideus, 223 — wreathed, 227 — remarkably nutritive for birds, 292.

Snakes, eggs of, 192 — harmlessness of, 193 — general aversion to, ibid.

Snapdragon or 'bull-dog', peculiarities of, 45 — a perfect insect trap, ibid. — ability to survive drought, 48.

Snipe, 106, 124, 303, 308 — no longer on Glastonbury marshes, 119, — the jack, its habits, 156 — supposed the male of the larger snipe, 157.

Snowdrop, the, 54 — a melancholy flower, 55.

Soil, of the parish in which the Author resides, 3, 7, 8, 32, 55, 56 — various sorts of, 9, 10 — analysis of, considered fallacious, ibid. — goodness of, 11 — genius of, 12 — suitability of 14 — for cultivation of the potato, 16, 18, 20 — for cultivation of the teazle, 25 — reinvigorated by autumn leaves 36 — trenching up, 62.

Song of birds, 98, 145, 163, 165, 168.

Spade, the, 310.

Sparrow, the hedge, 91, 146, — the common, 130 — house, 103, 132 — eggs of, 138, 322.

Spider, nest of unknown, 208.

Spottings, on hazel-nut leaves 64 — on elm leaves, 73 — on apples, 245 — on strawberry leaves, 246.

Spring, joy of, 267.

Squirrel, 228, 229.

Starling, the common, 120, 121, 132, 133, 299.

Steaming of the earth, 255, 256.

Stinking phallus, the, 75.

Stoat, 78.

Stormy petrel, the, 117, 118.

Strawberry, sometimes called stray-berry, (note) 229.

Strong blight, 296.

Strontian, 7.

Sugar, 20, 148, 199 — in plants, 47, 298 — the basis of animal nutriment, 48 — contains above half its weight of oxalic acid, ibid.

Sulphur butterfly, 59, 253.

Sun, effects of the rays upon colours, 44, 45.

Sundew, destructive to insects, 46.

Superstition, 107, 140, 205, 216, 235.

Swallows, 124, 126, 153, 167, 321 — killed in wanton sport, 137 — sea swallows swept inland, 293.

Swift, 123, 124, 128, 153, 317.

Sycamore tree, 72 — singularity of its leaves, 73.

Tally stick, used in village pound, 236.

Taste, 44, the most variable of our senses, 47.

Teazle, 24 — its cultivation, 23— its profits, 25— its uses, 26.

Thistle, 69, 83, 104, 147, 152, 183, 310.

Thorn, 232, the white, uniform in its blossoming, 126 — use in floral decoration, 39 — used as shamble-hook, 116 — found beneath a Roman path, 301 — black-thorn, danger of, 312.

Thrush, 103, 130, 214, the common, 129, 339 — the missel, 154 — song of, 168.

Tiger moth, 185.

Timidity of animals, 102, 142, 155, 179.

Toad, 79.

Tobacco, 21.

Tokens of wind and weather, 154.

Tomtit, or titmouse, the little blue, 101, 128, — parish rewards for the destruction of, 99, 100 — perishes in severe winters, 100 — the long-tailed, 102, 222 — instance of its intelligence in the care of its young, 105.

Totnes, house swallows overwintering at, 124, 125.

Traveller's joy, name given to the wild clematis, 67, 68.

Trees, 250 — uncertainty of the age of. 29 — prodigious size of several, 30 — attractors of humidity, 33 — air under, 34 — condense fogs, ibid. — cause of the verdure beneath, 34 — 'shake', a malady in some, ibid. — mischief of pollarding them 254.

Tree creeper, the, 149.

Turnip, 99 — singularly-decorated one, as a holyday amusement, 235 — parting with moisture, 266.

Uredo, the two-fronted, a substance attached to the leaves of the laurel, 73.

Vegetation, atmospheric effects on, 237 — preserved beneath a Roman path, 301.

Vermin, parish reward for the destruction of, 99, 100.

Vervain, 56 — respect paid of old to this plant, 57. — its supposed powers and qualities, ibid.

Vetch or cinquefoil, 69

Village clubs, 40.

Violet, 42, 125, 248, 272 — Spring, 38, — Autumn, 39.

Voices of birds, 163.

Wagtail, 150, 290, 299 — yellow, the, 128, 149, 153, 181 — grey, the, 128, 153 — white, the 260.

Walnut, 109, 304.

Want, the, 87.

Wasp, the common, 194 — securing wings in winter, 180 — the solitary, 209 — its nest, 210 — prey of hornets, 213 — tenacious of sensation, 252.

Water, stagnant and putrescent, favourable for the residence of insects, 70 — supreme value of, 220.

Water hemlock, 58.

Water hen, 140.

Water flea, 200, 201.

Water parsnip, 58.

Water-rail, the, 314, 315.

Water rat, 85.

Water shrew, 86.

Weasel, 78-79, 90-91, 167.

Weather, 248, 249, 250, 251, 252, 299, 312, 313, 314.

Weevil, 47, 225, 255.

Weld or wold, the dyers' weed so called, 58.

Well, the village, 302, 303.

Wheat, crops of, method of saving, 14.

Wheat-ear, the, 136, 137, 166, 153, 292, 293.

Whirly pits, what, 242.

White thorn, 126, 214, 232 — use in floral decoration, 39 — used as shamble-hook, 116 — found beneath a Roman path, 301.

White, Rev. Gilbert, Selborne, 128.

White throat, 101, 102, 128 — pure white, 159, 159.

Wild plants, 14.

Willow tree, 50, 209, 255, 304.

Winds and Weather, 49, 239 — old tokens of, 154, 161 — saline winds, 244.

Winter, 8, 49, 51, 78, 100, 121, 124, 129, 141, 170, 227, 248, 249, 257, 258, 312, 313 — the season of, depicted, 257.

Winter subsistence of birds, 259, 260.

Withy-wind, name given to clematis, 68.

Woodlark, the, 165, 166.

Woodlouse, 196.

Woodpecker, 164, capture of, 323.

Woody night shade, 12.

Worm, 188 — the hair, 208 — the common, 35, 36, 216, 217, 218, 219, 220.

Wren, the willow, 93, 94, 95, 126, 144, 293 — the golden-crested, 95, 100, 104, 222 — the common, 100, 167, 290, 321 — instance of its stratagem to preserve its nest, 145.

Wryneck, the, 118, 128, 152, 195, 300.

Wych elm, 31, 32, 33, 106.

Yarrell, Mr. 176.

Year 1825, singular increase of sex in the, 86, (note), 132 — other peculiarities of, 17, 48, 68, 71, 109, 162, 214, 223, 227, 231, 232, 244, 248.

Yellow hammer, 116, 125, 138, 319.

Yellow under-wing moth, 174, 181.

Yellow weed, name given to dyers' weed, 58— yellow, the prevailing colour of the flowers of plants in spring, 59 — and in autumn, 60.

Yew, 52.

Young, provision of creatures for their, 185.

THE END

-product-compliance